Desalination

Scrivener Publishing
100 Cummings Center, Suite 541J
Beverly, MA 01915-6106

Publishers at Scrivener
Martin Scrivener (martin@scrivenerpublishing.com)
Phillip Carmical (pcarmical@scrivenerpublishing.com)

Desalination

Water from Water

Edited by

Jane Kucera

Scrivener Publishing

WILEY

Co-published by John Wiley & Sons, Inc. Hoboken, New Jersey, and Scrivener Publishing LLC, Salem, Massachusetts.
Published simultaneously in Canada.

For general information on our other products and services or for technical support, please contact our Customer Care Department within the United States at (800) 762-2974, outside the United States at (317) 572-3993 or fax (317) 572-4002.

Wiley also publishes its books in a variety of electronic formats. Some content that appears in print may not be available in electronic formats. For more information about Wiley products, visit our web site at www.wiley.com.

For more information about Scrivener products please visit www.scrivenerpublishing.com.

Cover design by Kris Hackerott

Library of Congress Cataloging-in-Publication Data:

ISBN 978-1-118-20852-6

Printed in the United States of America

10 9 8 7 6 5 4

In fond memory of Julius "Bud" Glater, my thesis advisor and mentor, and a pioneer in the development of water desalination technologies

Contents

Section IV Non-Traditional Desalination Processes 425

Preface

The world-wide demand for "fresh" water is growing exponentially, while the supply of readily-available fresh water is dwindling nearly as quickly. Several diverse techniques have been implemented to try to meet the growing demand for fresh water, with variable degrees of success. One technique that is growing in application is desalination. Desalination encompasses a host of technologies such that clean water may be generated regardless of location, make-up source, and/or energy source.

This book explores numerous desalination technologies. Some of the technologies that are covered here are highly commercialized and are in extensive use today, while others are under development and may be commercially-viable tomorrow. This book also touches on renewable energy sources as drivers for desalination.

World-renounced experts have contributed to this book. The authors' experience, which ranges from about 10 to over 41 years in their respective fields, covers the gamut from academia to real-world application. I thank the authors for contributing their time and sharing their expertise to help us explore the possibilities within the realm of desalination.

Contributing Authors

Chapter 1

Jane Kucera (volume editor)

Jane Kucera is a chemical engineer with over 31 years' experience in the area of membrane technology. Jane began her work with membranes in 1982 while working in the Seawater Laboratory at UCLA, where she received her M.S. Degree in Chemical Engineering in 1984. She then went to work for Bend Research, Inc., as a research engineer, where she spent 7 years developing novel membrane materials and process designs, including serving as lead researcher for several NASA research contracts to develop membrane-based water-recovery and reuse systems for the International Space Station. While with USFilter Corporation, she designed the 2 MGD Honouliuli Water Reclamation Facility for Oahu, Hawaii, which involves treating secondary municipal effluent with microfiltration and reverse osmosis desalination for reuse as boiler feed for refineries and power generation facilities. Jane is currently a Senior Technical Consultant in the Engineering and Project Development Group at Nalco/an Ecolab Company where she is responsible for developing industrial desalination treatment equipment and system designs. She has authored the book, Reverse Osmosis: Design, Applications and Processes for Engineers, and has contributed chapters to several other volumes covering membranes, reverse osmosis, and desalination. Jane also received a BA in Chemistry, Summa Cum Laude, from Linfield College in McMinnville, OR.

Chapter 2

Joachim Gebel

Dr. Joachim Gebel received his Doctorate in Mechanical Engineering, Summa Cum Laude, from RWTH Aachen University in 1990. For more than 20 years, Joachim has worked as a consultant in different functions; since 2003, he is co-owner and acting partner of S.T.E.P. Consulting GmbH, Aachen, Germany. Joachim has focused on topics such as seawater and brackish water desalination, waste water treatment, dual-purpose plants and cogeneration throughout his career. With their book "An Engineer's Guide to Seawater Desalination," published in 2008, Joachim and his co-author Dr. Süleyman Yüce reported on important topics of main desalination technologies, from the basics up to material problems, as well as the design and the operation of facilities. Joachim is currently a Professor in the Department of Technology and Bionics at the Hochschule Rhein-Waal in Kleve, Germany, where he presents lectures on thermodynamics, process engineering and plant design.

Chapter 4

Mark Wilf

Dr Mark Wilf has been involved in process development, system design, project execution, plant operation and maintenance of large, commercial desalination plants in US, Europe and Middle East since 1977. Mark is a regular contributor to professional journals, wrote chapters on membrane technology processes and applications to a number of books. He edited and wrote with other coauthors The Guidebook to Membrane Desalination Technology that has been published in 2006. The second book on membrane technology he edited and contributed to: The Guidebook to Membrane Technology for Wastewater Reclamation has been published in 2010. Mark regularly presents and teaches desalination and membrane technology subjects to engineers and water professionals. His teaching activity includes teaching course on membrane technology and desalination for the European Desalination Society at the Genoa University, Italy and in San Diego.

Chapter 5

Christopher Bellona

Dr. Christopher Bellona is an Assistant Professor in the Department of Civil and Environmental Engineering at Clarkson University.

Dr. Bellona's past research has largely been focused on membrane science and the potable reuse of wastewater effluent. His research has spanned both fundamental and applied aspects of water reuse including the fate and removal of constituents of concern (e.g., organic contaminants, nutrients), membrane fouling, modeling, and pilot-scale testing of membrane technologies. His current research is focused on membrane systems for anaerobic wastewater treatment, extraction of energy and valuable materials from waste streams, and development of novel and hybrid membrane systems. He received his M.S. and Ph.D. degrees in Environmental Science and Engineering from the Colorado School of Mines, and his B.S. in Environmental Science from Western Washington University.

Chapter 5

Jeff McCutcheon

Nhu-Ngol Bui

Chapter 6

Hong-Joo Lee

Prof. Hong-Joo Lee has been working on the membrane process development for the application of bioenergy in the Dept. Bioenergy Sci. & Technol. at Chonnam National University. Major areas of his research are related to membrane applications on the bioethanol separation, the removal of inhibitors, and the recovery of acid catalysts. Also, his interesting research area expands to membrane fouling, and development of integrated membrane process. Furthermore, he also concerns climate change, the pretreatment for the use of lignocellulosic biomass, the development of membrane bioreactors, and biorefinery. He gives lectures on organic chemistry, physical chemistry, analytical chemistry, fermentation engineering, biochemical engineering, and bioprocess engineering.

Seung-Hyeon Moon

Prof. Seung-Hyeon Moon has been working on synthesis, characterization, and process application of electromembranes in School of Environmental Science and Engineering at Gwangju Institute of

Science and Technology. He gives lectures on transport phenomena, electrochemistry, and water chemistry. In the area of environmental applications, the membranes studies are focused on electrodialysis, electrodeionization, and capacitive deionization. Further his recent interests are expanded to charged membranes for energy conversion systems including fuel cells, rechargeable battery and redox flow battery based on polymer physics and nanoionics to innovate the electromembrane processes.

Chapter 7

Jonathan Wood

Jonathan Wood is Manager of Process Development for Siemens Water Technologies in Lowell, MA. He has a B.S. in Chemical Engineering from Worcester Polytechnic Institute in Worcester, MA and an M.S. in Environmental Engineering from Northeastern University in Boston, MA. Jon has over 30 years experience in water purification, encompassing research, product development, product management and technical service. He started his career with Barnstead Company, working with distillation, pure steam generation, and deionization equipment, and then joined Millipore Corporation, where he was part of the team that commercialized electrodeionization. He has continued to work in this field with USFilter, Veolia, and Siemens.

Joseph Gifford

Joseph Gifford is Director of R&D - Industrial for Siemens Water Technologies in Lowell, MA He has a B.S. in Chemical Engineering from Worcester Polytechnic Institute in Worcester, MA and an M.S. in Chemical Engineering from the University of Massachusetts in Lowell, MA where he was awarded the Dean's Gold Medal and the Outstanding Graduate Student award. He is currently responsible for all Research and Development activities for the Industrial Division of Siemens Water Technologies.

Chapter 8

Xing Yang

Dr. Xing Yang is currently a membrane researcher in Singapore Membrane Technology Centre (SMTC). With a background on chemical engineering, she has been involved in membrane technology, wastewater treatment and desalination industry for close to eight years. Xing received her PhD In 2012 in civil and environmental engineering from Nanyang Technological University (NTU), Singapore. Major areas of her research are related to membrane applications on membrane contactors for liquid/gas and liquid/liquid separation, membrane distillation (MD) and membrane distillation crystallization (MDC). During her PhD study, she has authored more than 10 publications in Tier-1 journals, specifically on membrane characterization, membrane surface modification, process design and optimization, membrane module design, flow dynamics and mass/heat transfer analyses using computational fluid dynamics (CFD) modeling, and process simulation and energy analysis of membrane separation for water systems using Aspen Plus. Also, her recent work has covered matrix modular design and economic analysis for RO brine processing and interdisciplinary research for non-invasive flow studies in hollow fiber modules with the aid of nuclear magnetic resonance imaging (MRI) instruments.

Anthony G. Fane

Tony Fane is a Chemical Engineer from Imperial College, London, who has worked on membranes since 1973. His current interests are in membranes applied to environmental applications and the water cycle, with a focus on sustainability aspects of membrane technology, including desalination and reuse. He is a former Director of the UNESCO Centre for Membrane Science and Technology at UNSW. Since 2002 he has directed membrane research in Singapore as Temasek Professor (2002-2006) at Nanyang Technological University, then as Foundation Director of the Singapore Membrane Technology Centre (SMTC) at NTU. He is now Director-Mentor at SMTC. The SMTC has a group of over 85 researchers dedicated to fundamental and applied research into membranes for the water cycle. He is on the Advisory Board of the Journal of Membrane Science (former editor from 1992 to 2005) and Editorial Board of Desalination. He is the Patron of the Membrane Society of Australasia.

Rong Wang

Dr. Rong Wang is currently an Associate Professor in the School of Civil and Environmental Engineering, Nanyang Technological

University, Singapore. She also holds the position of Director, Singapore Membrane Technology Centre (SMTC).

Dr. Wang has 25 years of experience in Chemical Engineering, Water, Environment and Energy related R&D. She specializes in novel membrane development for water and wastewater treatment, liquid purification and gas separation, and in development of novel hybrid membrane systems and process simulation. Her research interests cover (1) novel forward osmosis (FO)/pressure retarded osmosis (PRO) hollow fiber membranes, low pressure nanofiltration (NF) hollow fiber membranes, bio-mimetic membranes, hydrophobic hollow membranes and mixed matrix membranes for seawater desalination, wastewater treatment, membrane contactor for CO2 capture, etc; (2) simulation and optimization of various membrane processes such as membrane contactor, membrane distillation and separation of mixed gases in membrane modules, etc.

Dr. Wang has over 130 SCI-tracked journal publications (cited times: >2600, h-index: 32 @Scopus) and over 100 conference contributions. She is the inventor of 20 patents/ technical disclosures for novel membrane fabrication. She has co-authored 6 book chapters. She is the editorial board member for two journals of Journal of Membrane Science, and Desalination. She was a Guest Editor for two special issues of Desalination in 2011 and 2013. As PI and Co-PI of projects, she has been successfully awarded research grants of over S$15 millions by the governmental agencies and industry in the past five years.

Chapter 9

Prakash Narayan Govindan

Dr. Prakash Narayan Govindan is a recent PhD from Massachusetts Institute of Technology, and he was also a MIT Legatum fellow for the year 2010-2011. Prakash received the de Florez prize at MIT for graduate design and a best paper award at the IDA World Congress on Desalination and Water Reuse. His research experiences include pool boiling heat transfer studies of nanoparticle suspensions, thermodynamic design of combined cycle (gas and steam turbine) power plants, and thermal design of seawater desalination systems. He has co-authored 13 international journal publications and has filed 14 patent disclosures at the US PTO.

John H. Lienhard V

Dr. John Lienhard is the Samuel C. Collins Professor of Mechanical Engineering at MIT. During more than 25 years on the MIT faculty, Lienhard's research and educational efforts have focused on heat transfer, desalination, thermodynamics, fluid mechanics, and instrumentation. He has also filled a number of administrative roles at MIT. Lienhard received his bachelors and masters degrees in thermal engineering at UCLA from the Chemical, Nuclear, and Thermal Engineering Department, where he worked on thermal instabilities in solar collectors and evaporating meniscus measurements for desalination systems. He joined MIT immediately after completing his PhD in the Applied Mechanics and Engineering Science Department at UC San Diego, where he did experimental work on thermally stratified turbulent flows. Since coming to MIT, Lienhard has worked on liquid jet impingement, buoyant instabilities, high heat flux engineering, electronics thermal management, glass fiber formation, and thermally-driven desalination processes. He is a recipient of the 1988 National Science Foundation Presidential Young Investigator Award, the 1992 SAE Teetor Award, a 1997 R&D 100 Award, and the 2012 ASME Technical Communities Globalization Medal. He has been the Director of the Rohsenow Kendall Heat Transfer Laboratory since 1997, and he is a Fellow of the American Society of Mechanical Engineers. He serves on the editorial boards of several international journals, including the *International Journal of Thermal Sciences, Desalination and Water Treatment, Desalination, Experimental Heat Transfer,* and *Frontiers in Heat and Mass Transfer.* Lienhard is the co-author of textbooks on heat transfer and on measurement and instrumentation. At MIT, Lienhard has served as Associate Head of the ME department, Undergraduate Officer, Head of the Fluids, Energy & Transport Division of ME, and as chair or member of innumerable committees. Professor Lienhard is currently the Director of the Center for Clean Water and Clean Energy at MIT and KFUPM, and he visits the Arabian Gulf often.

Chapter 10

M. Shafiur Rahman

Dr. Mohammad Shafiur Rahman, Professor at the Sultan Qaboos University, Sultanate of Oman, is the author or co-author of over

250 technical articles including 100 journal papers and 7 books. He is the author of the internationally acclaimed and award-wining Food Properties Handbook, and editor of the popular Handbook of Food Preservation published by CRC Press, Florida. First editions received the bestseller recognition and the second edition is now released. He is one of the co-authors of the Handbook of Food Process Design, which is recently published by Wiley-Blackwell, UK. He has initiated the International Journal of Food Properties and serving as the founding Editor-in-Chief for more than 10 years, and serving in the editorial boards of 10 international journals and book series. In 1998 he has been invited and continued to serve as a Food Science Adviser for the International Foundation for Science, Sweden. He received the HortResearch Chairman's Award, BRAP Award, CAMS Outstanding Researcher Award 2003, and SQU Distinction in Research Award 2008. In 2008, Professor Rahman has ranked among the top five Leading Scientists and Engineers of 57 OIC Member Countries in the Agroscience Discipline. He received the B. Sc. Eng. (Chemical) in 1983 and M. Sc. Eng. (Chemical) in 1984 from Bangladesh University of Engineering and Technology, Dhaka, M. Sc. in food engineering in 1985 from Leeds University, England, and Ph. D. in food engineering in 1992 from the University of New South Wales, Sydney, Australia.

Mohamed Al-Khusaibi

Dr. Mohamed Al-Khusaibi, Assistant Professor at the Sultan Qaboos University, Sultanate of Oman. He has completed his B. Sc. (Food Science) in 2000 and M. Sc. (Food Science) in 2005 from the Sultan Qaboos University. He has completed his Ph. D. degree in 2011 from the University of Reading, UK.

Chapter 11

William E. Bornak

Bill Bornak has specialized in water treatment for over 40 years, developing particular expertise in resin- and membrane-based demineralization for high pressure boilers and microelectronics applications. In 2001, he formed a new corporation, Recirculation Technologies Inc. (RTI) with former Betz vice president, Bob Finley, to purchase the mobile resin cleaning operation from the company.

Bill is a member of several technical societies and has lectured extensively in North and South America, Europe, Asia and the Middle East. He has published numerous papers on specialized topics in water chemistry and ion exchange. Trained as an analytical chemist, he graduated with a B.A. in chemistry from LaSalle University in Philadelphia and did graduate work in biochemistry at the University of Pennsylvania. He has worked in ion exchange for both (now) Dow/Rohm and Haas and (then) Betz Laboratories, where he was water plant consultant and training director.

Chapter 12

Pei Xu

Dr. Pei Xu is an Assistant Professor in the Department of Civil Engineering at New Mexico State University. Her research has focused on water and wastewater treatment engineering; treatment and beneficial use of produced water; desalination; membrane processes; potable and non-potable water reuse; water quality; and photocatalysis. She has authored and co-authored over 100 papers, book chapters, technical reports and conference presentations. She received her B.S in Environmental Engineering from Xi'an University of Architecture & Technology, M.S. in Water and Wastewater Engineering from Lanzhou Jiaotong University, China, and Ph.D. in Hydrosciences from Ecole Nationale du Génie Rural, des Eaux et des Forêts (ENGREF), France. Dr. Xu is a proposal reviewer for RPSEA/Department of Energy, Bureau of Reclamation, Civilian Research and Development Foundation (CRDF), Water Research Foundation (Formerly AwwaRF), NASA EPSCoR panel, and Qatar National Research Fund (QNRF). She is also a regular reviewer for scientific journals such as Environmental Science and Technology, Water Research, Journal of Membrane Science, Desalination, and Separation and Purification Technology, etc.

Chapter 13

D. Yogi Goswami

Dr. Yogi Goswami is a University Distinguished Professor, the John and Naida Ramil Professor in the Chemical Engineering

Department and Co-Director of the Clean Energy Research Center at the University of South Florida. He conducts fundamental and applied research on solar thermal energy, thermodynamics, heat transfer, HVAC, photovoltaics, hydrogen, and fuel cells. Professor Goswami is the Editor-in-Chief of the *Solar Energy* journal. Within the field of RE he has published as author/editor 16 books, 16 book chapters, 6 conference proceedings and more than 300 refereed technical papers. He has delivered 51 keynote and plenary lectures at major international conferences. He holds 18 patents. A recognized leader in professional scientific and technical societies, Prof. Goswami has served as a Governor of ASME-International (2003-2006), President of the International Solar Energy Society (2004-2005), Senior Vice President of ASME (2000-2003), Vice President of ISES and President of the International Association for Solar Energy Education (IASEE, 2000-2002).

Elias (Lee) Stefankos

Dr. Lee Stefanakos is presently Professor of Electrical Engineering and Director of the Clean Energy Research Center (CERC) at the University of South Florida (USF) located in Tampa, Florida, USA. Up to August 2003 and for 13 years he was Chairman of the Department of Electrical Engineering at USF. He is an Associate Editor of the *Journal of Solar Energy*; Co-Editor-USA of the *Journal of Asian Electric Vehicles*; and Vice President for Academic Affairs of the American Foundation for Greek Language and Culture. He has published more than 140 research papers in refereed journals and international conferences in the areas of materials, renewable energy sources and systems, hydrogen and fuel cells, and electric and hybrid vehicles. He has received more than $14 million in contracts and grants from agencies such as the National Science Foundation (NSF), US Department of Energy (USDOE), National Aeronautics and Space Administration (NASA), Defense Advanced Research Projects Agency (DARPA), and others. He has been a consultant to a number of companies and international organizations. CERC is an interdisciplinary center whose mission is the development of clean energy sources and systems with emphasis on technology development and technology transfer.

Mohammad Abutayeh

Dr. Mohammad Abutayeh holds a PhD in Chemical Engineering from the University of South Florida and is a Florida-registered professional engineer. He wrote a thesis on predicting the citrate soluble–loss of the dihydrate process as well as a dissertation on simulating a novel passive vacuum solar flash desalination process. He worked in several engineering areas designing process control systems, optimizing unit operations, customizing process equipment, employing renewable energy, administering US patent laws, and others. He is currently a mechanical engineering professor at Khalifa University of Science, Technology, and Research in Abu Dhabi, United Arab Emirates where he continues to conduct research in employing renewable energy sources to develop more efficient sustainable processes such as water desalination and power generation.

Chennan Li

Dr. Chennan Li holds a PhD degree in Chemical Engineering mentored by Dr. Goswami and Dr. Stefankos. He is a Registered Professional Engineer and worked in water industry with experience on desalination plant design and pilot desalination plant construction/operation. His PhD research focused on developing innovative desalination system using renewable energy; and his master study was polymer/composite membranes related. Currently he is a lead simulation and control engineer in GE Power and Water.

Chapter 14

Patrick V. Brady

Dr. Pat Brady is a Senior Scientist at Sandia National Laboratories and has authored or co-authored several dozen peer-reviewed journal articles, books, book chapters, and nine patents, in the fields of water treatment, contaminant chemistry, and climate change. He joined Sandia in 1993 and also serves as adjunct Assistant Professor of Civil and Environmental Engineering at New Mexico Institute of Mining and Technology, Socorro, New Mexico. Education: Bachelors - Geology, University of California -Berkeley (1984); Ph.D. - Geochemistry, Northwestern (1990); Post-doc - ETH-EAWAG, Dübendorf, Switzerland.

M. Michael Hightower

Mike Hightower is a Distinguished Member of Technical Staff at Sandia National Laboratories. His Bachelor's and M.S. degrees are both from New Mexico State University in Las Cruces, New Mexico.

Notice from the Publisher

This book is intended for informational purposes. To the best of the Publisher's knowledge, the information contained in this book is accurate.

The Publisher assumes no responsibility or liability for errors or any consequences arising from the use of the information contained herein. Final determination as to the suitability of any information, procedure, or product for use contemplated by any user, and the manner of that use, is the sole responsibility of the user. Expert advice should be obtained at all times when implementation is being considered.

Mention of trade names or commercial products does not constitute endorsement or recommendation for use by the Publisher.

SECTION I

INTRODUCTION

1

Introduction to Desalination

Jane Kucera

Nalco/an Ecolab Company

Abstract
The availability of fresh water for a growing, industrialized planet is quickly becoming scarce. Methods to attain more fresh water to meet the increasing demand include techniques such as conservation and reuse, desalination, and moving water from water-rich regions to water-poor regions. Of these techniques, the potential for desalination is most promising in that it can be applied virtually anywhere in the world. And, desalination has the potential for generating relatively large volumes of "fresh" water from a host of feed stocks. This chapter discusses the need for desalination and provides the framework for the detailed discussions about desalination techniques that are presented in this book.

Keywords: Desalination, Water Scarcity, Conservation, Reuse, Thermal Desalination, Reverse Osmosis

1.1 Introduction

Desalination: from the root word *desalt* meaning to "remove salt from" [1]. By convention, the term *desalination* is defined as the "process of removing dissolved solids, such as salts and minerals, from water" [2]. Other terms that are sometimes used interchangeably with desalination are *desalting* and *desalinization*, although these terms have alternate meanings; *desalting* is conventionally used to mean removing salt from other more valuable products such as food, pharmaceuticals, and oil, while *desalinization* is used to mean removing salt from soil, such as by leaching [2].

The first practical use of desalination goes back to the sixteenth and seventeenth centuries, when sailors such as Sir Richard Hawkins

Jane Kucera (ed.) Desalination , (3–38) 2014 © Scrivener Publishing LLC

reported that their men generated fresh water from seawater using shipboard distillation during their voyages [3]. The early twentieth century saw the first desalination facilities developed on the Island of Curaçao and in the Arabian Peninsula [3]. The research into and application and use of desalination gained momentum in the mid-twentieth century, and over the last 30 years has witnessed exponential growth in the construction of desalination facilities.

One could ask the question, "Why desalination?" Desalination has become necessary for several reasons, the most compelling of which may be: 1) the increased demand for fresh water by population growth in arid climates and other geographies with limited access to high-quality, low-salinity water, and 2) the per capital increase in demand for fresh water due to industrialization and urbanization that out paces availability of high-quality water. Research and development over the last 50 years into desalination has resulted in advanced techniques that have made desalination more efficient and cost-effective. Desalination is, and will be in the future, a viable and even necessary technique for generating fresh water from water of relatively low quality. In this chapter, and in this entire book, we make the case for desalination as one of the major tools for meeting the fresh water needs of a growing planet. Thus, the title of this book, Desalination: Water from Water.

1.2 How Much Water is There?

The allocation of the world's water is shown in Figure 1.1. More than 97%, or about 1338 million km^3, of the world's water is seawater [3, 4]. Eighty percent of the remaining water is bound up as snow in permanent glaciers or as permafrost [4]. Hence, only 0.5% of the world's water is readily available as low-salinity groundwater or in lakes or rivers for "direct" use by humans.

1.2.1 Global Water Availability

Some regions of the world are blessed with an abundance of fresh water. This includes areas with relatively low populations and easy access to surface waters, such as northern Russia, Scandinavia, central and southern coastal regions of South America, and northern North America (Canada, Alaska) [2, 5]. More populated areas and areas with repaid industrialization are experiencing more water

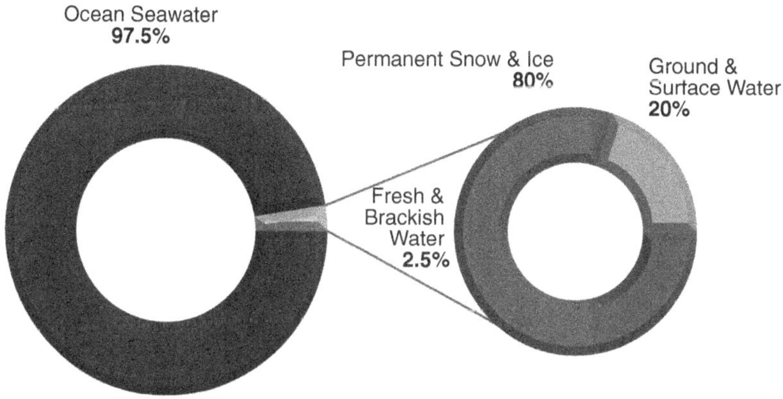

Figure 1.1 Allocation of the world's water resources.

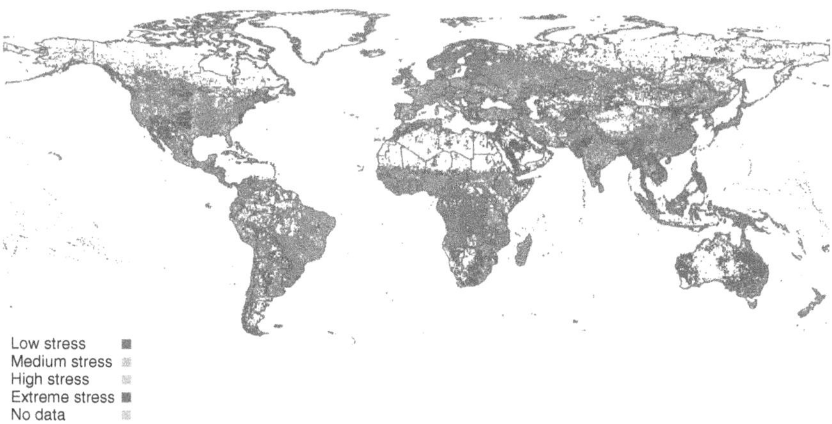

Figure 1.2 Water availability estimates for 2011. *Courtesy of the Water Stress Index 2011 by Maplecroft.*

stress, particularly when located in arid regions. (Water stress is typically measured by comparing the amount of water used to that which is readily available.)

There are numerous water maps available that measure current and predict future water stress. Figure 1.2 shows the Water Stress Index 2011 by Maplecroft [5] that estimates current water stress by comparing water use to the available, renewable supply for regions around the world:

"The Water Stress Index evaluates the ratio of total water use (sum of domestic, industrial, and agricultural demand) to renewable water supply, which is the available local runoff (precipitation

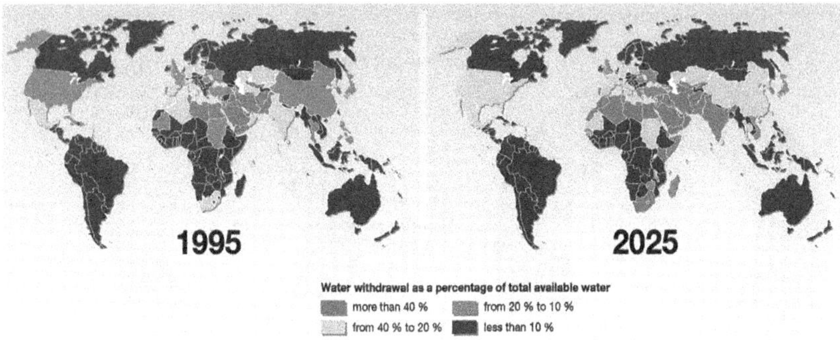

Figure 1.3 Global water stress in 1995 and predicted for 2025. *Courtesy of Philippe Rekacewicz (Le Monde diplomatique), February 2006.*

less evaporation) as delivered through streams, rivers, and shallow groundwater. It does not include access to deep subterranean aquifers of water accumulated over centuries and millennia.

The application of the index is to provide a strategic overview of the current situation of physical water stress at global, continental, regional, and national levels. It does not take account [any] future projection, [or] water management policies, such as desalination, or the extent of water re-use" [5].

The areas of the world that are not rich in water resources and that also experience un-stable and rapid population growth and industrialization will see water stress significantly increase in the future. Figure 1.3 compares the global water stress in 1995 with that predicted for 2025 [6]. As many as 2.8 billion people will face water stress or scarcity issues by 2025; by 2050, that number could reach 4 billion people [6]. Water stressed areas will include the south central United States, Eastern Europe, and Asia, while water scarcity will be experienced in the Southwestern United States; Northern, Southern, and Eastern Africa; the Middle East; and most of Asia [2].

1.2.2 Water Demand

In addition to population growth, another pressure being exerted on water supply is fact that the per capita water demand is increasing faster than the rate of population growth [7]. According to *Global Water Intelligence* [8], the per capital water demand has outpaced population growth by a factor of 2.

The demand for water in developed nations is relatively high. Demand in the United States is about 400 liters (105 gallons) per person per day [4]. Some Western countries that have been successful

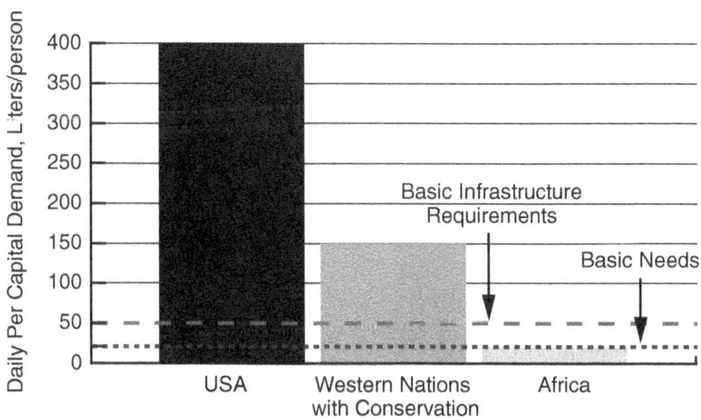

Figure 1.4 Global demand for water and World Health Organization basic water requirements (2010).[4, 9].

in implementing conservation and reuse measures have seen their demand for water drop to about 150 l (40 gallons) per person per day [4, 9]. However, the limited availability and access to water in some parts of the world, results in much lower consumption in such regions. For example, per capita freshwater consumption in Africa is only about 20 l (5.3 gallons) per day due to the shortage of suitable water [9]. The World Heath Organization (WHO) deems 15 to 20 l (4 – 5.3 gallons) per person per day is necessary for survival, while 50 l (13 gallons) per person per day is estimated to be needed for operation of basic infrastructure such as hospitals and schools (see Figure 1.4) [4]. The WHO estimates that by 2025, the world-wide demand for fresh water will exceed supply by 56% [9].

1.2.3 Additional Water Stress Due to Climate Change

While population growth and per capita increase in demand are two major water stressors, the impact of climate change on global water stress cannot be ignored. The effects of climate change actually work synergistically with population growth and increasing demand to strain water supply as population and industrialization grow, climate change accelerates, leading to more drastic climate events such as drought. A study by the National Center for Atmospheric Research (NCAR) indicates that severe drought is a real possibility for many populous countries [10]. Regions that are projected to experience considerable drought include most of Latin America, the Mediterranean regions, Southeast and Southwest Asia, Africa, the

Southwest United States, and Australia [7]. Coincidentally, many of these regions will also experience increases in population and industrialization and urbanization, and the corresponding increase in per capita water demand. The United Nations forecasts that the world will have 27 cities with populations greater than 10 million by the year 2020, and all but 3, New York City, Moscow, and Paris, will be in regions under the threat of significant drought [7].

1.3 Finding More Fresh Water

For much of the world's urbanized population, fresh water is an afterthought, a commodity that has been easy to find and always there at the tap. However, water in some parts of the world is increasingly considered a "product" that must to be found and developed to meet growing demand. Depending on the specific circumstances in a particular geography, one or more methods may need to be implemented to find and develop water sources to meet future water needs. Some of these methods are summarized below.

1.3.1 Move Water from Water-Rich to Water-Poor Areas

Moving water from water-rich areas to water-scarce regions, while sounding extreme, is not a new idea. Witness the diversion of water to the desert southwest United States for drinking, power, and irrigation uses. Los Angeles currently imports 85% of its water demand from outside sources: the Sierra Nevada Mountains, the Delta in Northern California, the Los Angeles Aqueduct, and the Colorado River Aqueduct [11].

However, moving water is not always palatable. Public outcry against moving water from a water-rich region can be a formidable obstacle. Consider the Columbia River in the Pacific Northwest United States. "Water is Oregon's oil," declared Oregon State Senator David Nelson in his 2007 white paper, "Columbia River Diversion as a Public Revenue Source." Diversion of the Columbia River to other western states has been a topic of discussion in the State of Oregon for over 35 years. Not much has come of this discussion to date however, as water-poor areas in the region have found other sources for water, and, more to the point, Oregonians have routinely declined to give up their supply of inexpensive fresh water that also serves as their source for relatively inexpensive hydroelectric power.

Politics can also play a role in how water supplies are dispersed. In Spain, different political parties are having a tug-of-war over how to supply the south-eastern area of Spain with water. The conservative party in Spain is advocating moving water from the Ebro River (an eastern river whose delta into the Mediterranean Sea is about half way between Barcelona and Valencia) to the Community of Valencia, which lies approximately 200km from the delta. The Socialist Party in power has commissioned the Torrevieja Seawater Reverse Osmosis (SWRO) facility, the 6th largest SWRO facility in the world, which is located in Alicante, Municipality of Torrevieja, about 75 km from Valencia. Backers of the Ebro river project have denied a permit for concentrate discharge from the SWRO facility, thereby preventing the construction of the seawater intake and outfall pipelines [11]. The Valencia region has 1.5+ million people with 4 more SWRO projects under way that could encounter the same political stalemate.

While moving fresh water makes sense in some cases, public and political pressures, as well as technical issues, such as moving water long distances, particularly when elevation changes are involved, will not make moving water supplies feasible or even possible to meet the requirements of all regions in need of fresh water.

1.3.2 Conservation and Reuse

Conservation is a term that has been used for decades to mean more efficient usage and savings of a resource, in this case, water. The twenty-first century equivalent terms for conservation are *sustainability* and *green*. Regardless of which term is used, the need to conserve through more efficient usage, recycling, and reuse has become popular in today's culture. While these techniques are oft times the first choice of populations located in arid areas or far from an ocean as a means of finding more fresh water, all populations can benefit from these techniques.

For example, consider the City of Los Angeles, California, an arid, coastal city that receives only about 40cm (15.7 in) of rain a year. Los Angeles is one large metropolitan area that has selected conservation to supply a portion of its future water needs. The Los Angeles area has a current population of about 4 million people and is expected to grow to at least 10 million inhabitants by 2020; water demand is expect to rise by 123 million m^3 ***per year*** [7, 12]. The Los Angeles Department of Water and Power (LADWP) describes the future of the city's water philosophy: "Conservation will continue to be a foundation of LADWP water resource management policy, and will be implemented to the fullest extent concurrent with further consideration of alternative water supplies" [13].

Additionally, the LADWP has developed a new Recycled Water Master Plan which relies heavily on recycling highly-treated wastewater as a cost-effective solution to meet some of the future demands of the city [14]. The Edward C. Little Water Recycling Facility (ELWRF) located in the City of El Segundo, Los Angeles County, CA (commonly referred to as "West Basin"), is a model for water conservation, recycling, and reuse. The facility, funded in 1992 following the severe drought in California in the late 1980's and early 1990's, today produces about 114,000 m^3/d of recycled water (41.5 m^3 per year) at a current investment of $500 million [15]. Five grades of water, known as "designer" water, are produced by the facility to match the needs of local industry (water type listed roughly from lowest to highest quality):

1. Tertiary wastewater (known as Title 22 Water) for general industrial and irrigation uses, such as irrigating golf courses,
2. Nitrified water for use in industrial cooling towers,
3. Softened reverse osmosis water for ground water recharge,
4. Reverse osmosis water for low-pressure boiler feed water at local refineries, and
5. Ultra-pure reverse osmosis water for high-pressure boiler feed water at local refineries.

The goal of the LADWP Recycled Water Master Plan is to recycle a total of about 62 million m^3 of water per year by 2019 at an estimated cost of $715 million to $1 billion [11, 13]. The ELWRF (West Basin) has already achieved 2/3 of that water recycling goal.

Conservation and recycling wastewater, using West Basin as the example in Southern California, will require treatment, such as desalination, to produce water that is suitable for reuse. Conservation and recycling has the potential to slow the rate at which new, future supplies of fresh water may need to be developed, but will not, by itself, meet the total need for fresh water.

1.3.3 Develop New Sources of Fresh water

Developing sources of fresh water other than traditional sources, such as lakes, rivers, or relatively shallow wells, is another method for meeting the demand for more fresh water. The most logical new sources for developing new fresh water supplies are seawater and deep wells or saline aquifers.

Seawater is the traditional source water when of thinks of desalination. Seawater represents the feed water source for the majority of desalination facilities in the world (58.85%), and an explosion in growth of large seawater facilities is responsible for the steep increase in desalination capacity since 2003 [16]. The majority of these facilities were developed in the Arabian Gulf region, Algeria, Australia, and Spain; for sea-bounded and generally arid areas such as the Gulf Coast and Australia, turning to the sea for water is only natural.

Seawater supply is only suitable as a source for coastal areas; inland areas would need to rely on sources such as saline aquifers for new water supply. Figure 1.5 shows a United States Geologic Survey map of US saline aquifers; the map was generated in the early 1960's, and has not been updated since its initial publication. Note that most current activity involving saline aquifers centers on using them as storage for greenhouse gases, primarily carbon dioxide, rather than as sources for fresh water [17]. This is presumably due to the need to treat the water to generate fresh water from the

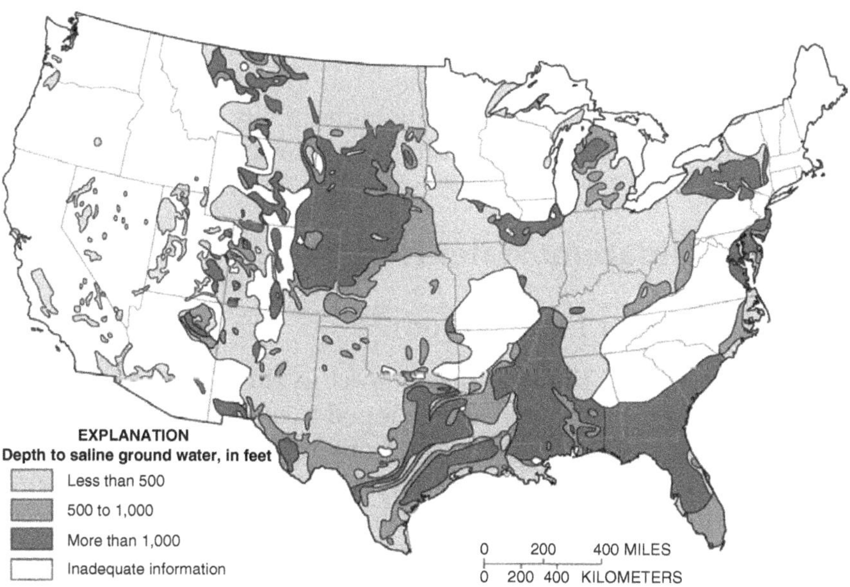

Figure 1.5 United States Geologic Service map of aquifers in the United States, cir. 1965.

Table 1.1 Classification of sources waters as a function of total dissolved solids (TDS).

Source Water	Total Dissolved Solids (ppm)	Classification
Drinking Water*	< 500	Fresh
Fresh	< 1,000	Fresh
Brackish	1,000 – 5,000	Mildly Brackish
	5,000 – 15,000	Moderately Brackish
	15,000 – 35,000	Heavily Brackish / Seawater
Seawater	35,000	Standard Average Seawater
	35,000 – 45,000	Seawater

* World Health Organization [8].

saline brines as opposed to the relative ease of injecting greenhouse gases, a process that does not require treatment, into the aquifers.

Seawater and other saline sources present an opportunity to meet the growing water needs of the world. Table 1.1 lists generally-held classifications of water as a function of salinity (note that saline aquifers are generally considered to be at least moderately brackish). Of these classifications, even the higher-salinity "fresh" water would require treatment for potable or industrial use to reduce the concentration of dissolved minerals. Thus, desalination techniques are once again necessary to generate high-quality water from water that is, without treatment, not suitable for direct use.

1.4 Desalination: Water from Water

1.4.1 Drivers for Desalination

One can conclude from the discussions in this chapter that new sources of fresh water must be developed to meet the growth in the demand. Apart from moving water from location to location, reuse of wastewater and use of alternate sources of water will require treatment to yield water that is suitable for potable or industrial use. And, since wastewaters and alternate source waters are generally high is dissolved solids, desalination technologies will most certainly be required as part of the treatment scheme. Thus, the driver for desalination is clear: future demand for high quality water will require desalination of water sources that are lower in quality (higher in dissolved solids) than are commonly utilized today (and which may not be available tomorrow).

Desalination of various water sources to provide a supply of fresh, usable water has been growing almost exponentially since 1965, when global commissioned desalination capacity was less than 2 million m^3/d [18]. By 2011, the global commissioned desalination capacity was over 71 million m^3/d [16]. New, on-line capacity for the year 2010 was about 6.2 m^3/d, and new, on-line capacity has increased year over year since 1995, with steep increases in new capacity since 2003, as demonstrated in Figure 1.6a; Figure 1.6b shows the cumulative on-line capacity since 1980 [16].

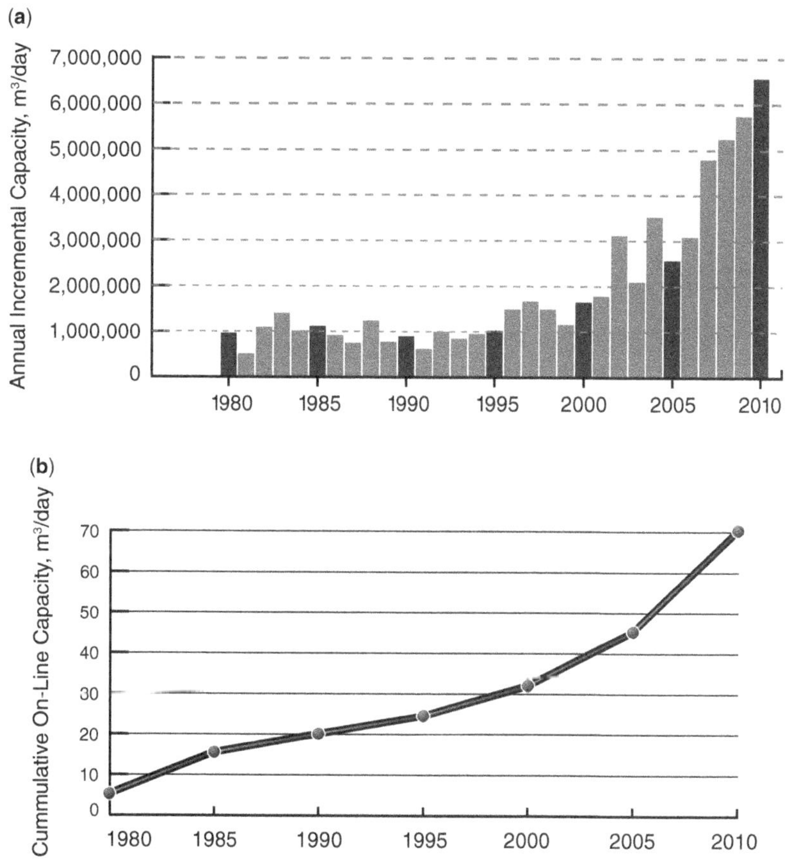

Figure 1.6 Growth of new, on-line desalination capacity. *Courtesy of Global Water Intelligence.*

1.4.2 Feed Water Sources for Desalination

Feed water sources for desalination are varied. As previously discussed, feed sources can range from seawater and saline aquifers to wastewater for recycle and reuse. While seawater represents the feed water source for the majority of desalination facilities, the use of other feed water sources, such as brackish water, saline aquifers, and wastewater, has been growing steadily since 2000 [16]. Figure 1.7 shows the growth in annual new contracted capacity by feed water type, and Figure 1.8 shows total worldwide installed capacity by feed water type through 2010 [16, 18].

Although the alternative sources appear to be limited in number, composition of specific examples of the various make-up source classifications can differ greatly depending on their hydrologic origin. Table 1.2 demonstrates some of the variability in well and surface waters, with a standard seawater and a sample grey water source included for comparison (the well, river, and grey waters shown in Table 1.2 either are currently being used as feed water sources for desalination facilities or have been considered for use as source water for such facilities) [3].

Despite variations in quality among the various feed water sources, they all share the characteristic of being relatively high in salinity of total dissolved solids (TDS). High salinity (and, in some

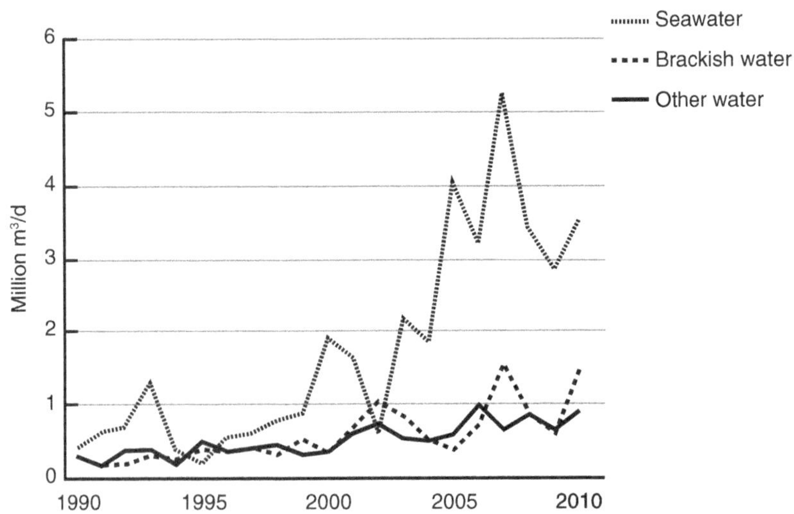

Figure 1.7 Annual new contracted desalination capacity by feed water type [16]. *Courtesy of Global Water Intelligence.*

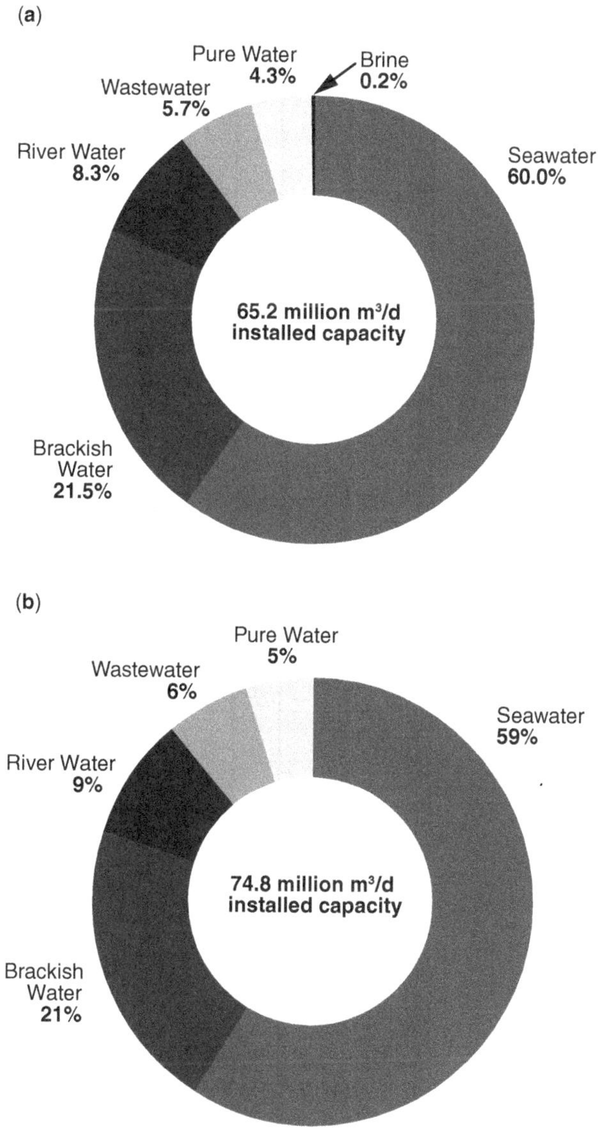

Figure 1.8 Total, global installed capacity by feed water source as of 2010 [16]. *Courtesy of Global Water Intelligence.*

cases, other constituents) makes the water unsuitable for potable and industrial use. Therefore, demineralization or "desalination" treatment to reduce the concentration of TDS must be part of the treatment system employed if these sources are to be used to supplement or replace existing fresh water supplies.

Table 1.2 Sample water composition of seawater, well water, surface water, and grey water sources.

Species (ppm)	Standard Seawater[a]	Well Water[b]	Well Water[c]	Colorado River Water[d]	Lake Michigan Water[e]	Grey Water[f]
Total Dissolved Solids	35,000	3,170	350	1,021	210	650
Chloride	19,345	1,370	3.9	181	16	67
Sodium	10,752	745	38	185	8.3	66
Sulfate	2,701	301	110	342	27	200
Magnesium	1,295	38.4	16	38.3	12	34
Calcium	416	176	50	104	39	88
Potassium	390	15.9	2.8	5.7	1.7	21
Bicarbonate	145	75	220	160	134	244
Bromide	66	0.05				
Strontium	8		1.6	1.4	0.13	1.2

Silica	6.4	29.4	29	14.2	2.6	42
Boron	4.5					
Nitrate	3.0	0.11	1.7	2.6	1.7	4.4
Fluoride	1.3	0.61		0.5	1.1	0.56
Iron	0.0034		0.01			1.3
Arsenic	0.003			0.0035		
Uranium	0.003			0.0038		
Manganese	0.0004					0.03
Selenium	0.00009			0 0023		

[a]References 3, 4, and 9
[b]El Paso Water Utilitites, El Paso Airport Wells [3]
[c]Water from Benson, Arizona, Kevin O'Leary, Nalco Company, personal communication, April 7, 2008
[d]Colorado River Water near Andrade, CO [3]
[e]Water from Chicago, Illinois, Anne Arza, Nalco Company, personal communication, May 16, 2011
[f]Hotel hygiene wastewater, Las Vegas, Nevada, August 21, 2007

1.4.3 Current Users of Desalinated Water

The primary user for desalinated water is the municipal sector; nearly two-thirds of desalinated water is used for potable application. Industrial and power users together accept another third of the worldwide desalination capacity [18]. Although potable applications account for nearly twice the total volume of desalinated water used than industrial applications, the number of industrial facilities (including power) out number municipal facilities by almost 2 to 1 (8,715 to 4,415 respectively), indicating that the size of industrial desalination facilities are considerably smaller than municipal facilities [18]. The remaining 6% of desalinated water is used for irrigation, tourism, military, and other applications [18A].

Figure 1.9 provides a breakdown of users of desalinated water for 2010 and 2011 [18, 18A]. These data shows little change over the year, but do show the trend toward more industrial users relative to municipal users.

1.4.4 Overview of Desalination Technologies

The world-wide installed desalination capacity in 2011 was about 74.8 million m^3/d [18A]. Figure 1.10 shows the relative breakdown of installed capacity of various desalination technologies for 2010 and 2011 [18, 19].

Membranes are currently outpacing traditional thermal technologies in total installed desalination capacity. Prior to 1980, membrane technologies made up less than a third of the global desalination capacity, while today, membranes account for just under 2/3 of the total installed desalination capacity [14]. As shown in Figure 1.10, installed capacity of reverse osmosis (RO) grew from 60% of total installed capacity in 2010 to 63% in 2011, an increase of 5% over the year, while installed capacity of multi-stage flash (MSF) evaporation decreased by over 14% for the year; installed capacity of multi-effect distillation (MED) remained steady at 8%. Membrane technologies such as RO offer the advantage of smaller infrastructure, and RO total treated water cost is becoming competitive with traditional thermal processes [19].

Membrane-based systems are popular in rising markets such as Algeria, Spain, and Australia, while thermal processes are still strong in traditional markets such as the Middle East, where energy costs are lower. Saudi Arabia accounts for 34.8% of the total

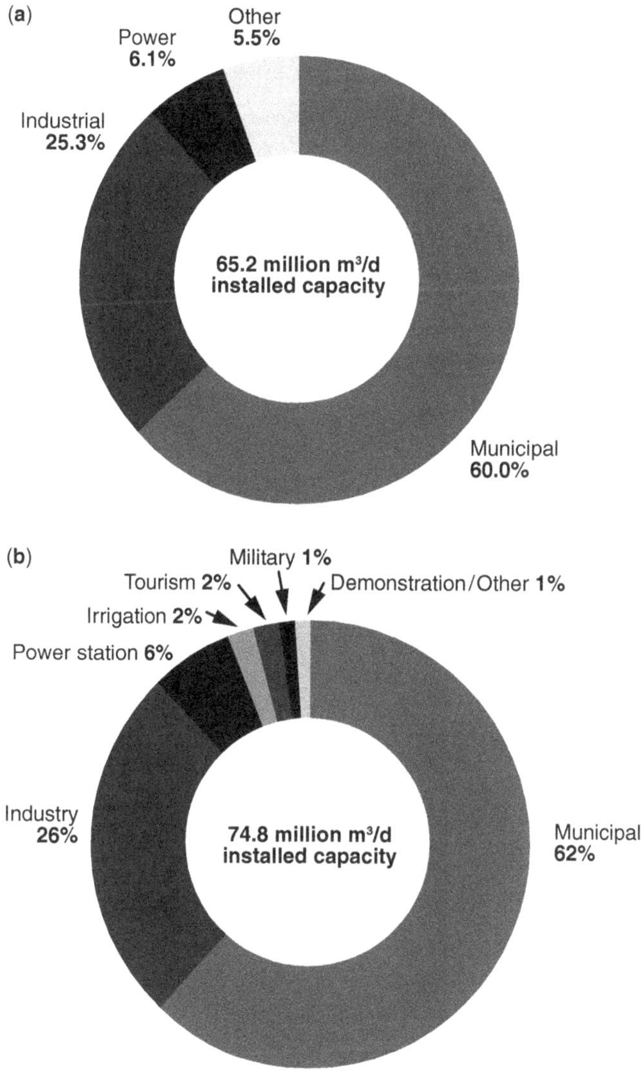

Figure 1.9 Total, 2010 and 2011 global installed capacity by type of user [18, 19]. *Courtesy of Global Water Intelligence.*

installed thermal capacity in the world at 5.9 m^3/d, followed closely by UAE at 5.8 m^3/d, and then by Kuwait, Qatar, Libya, Bahrain and Oman (0.5 to 2.1 m^3/d) [18]. By contrast, the United States is ranked 9th in total installed thermal capacity (~0.35 m^3/d), but 1st in total installed membrane capacity at 7.5 m^3/d (second is Saudi Arabia

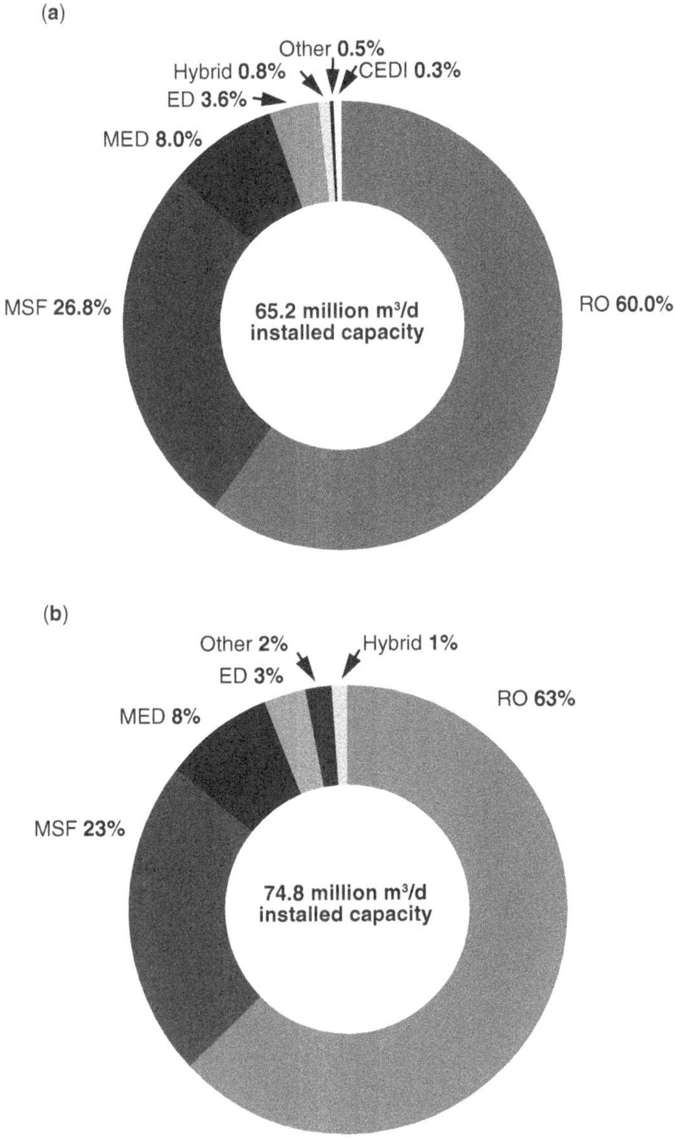

Figure 1.10 Global installed desalination capacity by technology for 2010 and 2011 [18, 19]. *Courtesy of Global Water Intelligence.*

at just under 5 m^3/d) [18]. Figure 1.11 shows the fraction of total global desalination capacity by region since 2003 [16]. Since 2003, growth in desalination technologies is still led by the Middle East (Saudi Arabia and UAE). However, 3 – 7 in rankings of total growth

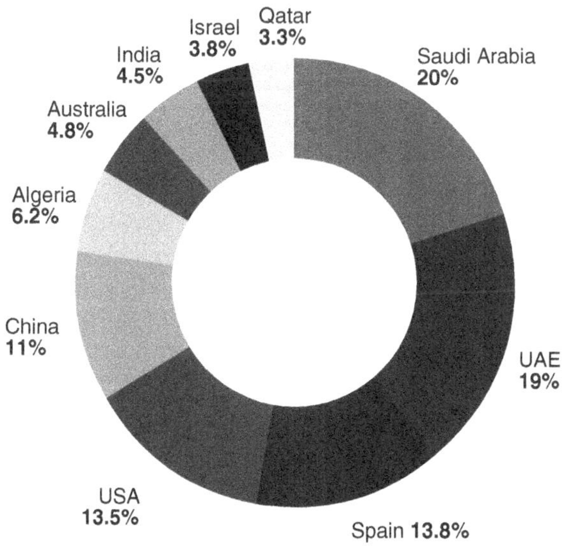

Figure 1.11 Fraction of total global desalination capacity by region since 2003 [16]. *Courtesy of Global Water Intelligence.*

in desalination capacity since 2003 are Spain, United States, China, Algeria, Australia, Israel, and India, respectively [18].

1.4.5 History of Desalination Technologies

Desalination has grown substantially since the mid 1960s. In 1952, there were only about 225 desalination facilities world-wide with a total capacity of about 100,000 m^3/d. **(either** [2] **or Evans 1969 (roadmap))** Today, there are over 15,000 desalination facilities globally with a total global capacity of over 65 million m^3/d [18].

While there are many desalination technologies in use or being developed today, desalination began using thermal processes. Membrane-based processes, such as RO, helped to further promote desalination over the last 30 years. The history of these pioneering technologies is outlined below.

1.4.5.1 History of Thermal Desalination

Thermal desalination techniques were recognized as early as 320 B.C. when Aristotle wrote, 'saltwater, when it turns into vapor, becomes sweet and the vapor does not form saltwater again when it condenses.' Shipboard distillation beginning in the sixteenth century is the first practical use of distillation to generate fresh water from seawater [20]. In 1843, Rillieux successfully patented, built, and sold multi-effect evaporators [20].

The number of thermal desalination installations has grown rapidly over the last 100 years. However, while reverse osmosis and other membrane technologies were revolutionary in development, the development of thermal desalination technologies over the last 40 years has been more evolutionary than revolutionary [21].

Multi-effect distillation (MED) was the first thermal desalination technology employed [20, 21]. The first units were designed with submerged tube evaporators the exhibited low heat transfer rates and high scaling rates. Vertical- and horizontal-tube evaporators (also known as falling-film evaporators) used in modern MED facilities provide higher heat transfer coefficients and lower specific heat transfer surface area requirements than their older counterparts. Drawbacks of the current MED technology are the complexity and production capacities [21]. Also, the relatively low, maximum-brine temperature of MED (~65°C) due to scale-forming issues, is another limitation of MED. However, the use of membrane pretreatment such as nanofiltration (NF) prior to MED to remove the calcium that contributes to calcium-sulfate scale in MED units has been considered as a way of allowing higher temperature operation of MED and, thereby enhance the use of MED for desalination [21].

Due to the early issues with MED (e.g., scaling and low heat transfer rates), multi-stage flash (MSF) distillation technology was developed in the late 1950s and early 1960s as an alternative. Flashing distillation was first commercially employed by Westinghouse in Kuwait in 1957 [20]. That same year, the MSF distillation patent was issued and in 1959/60, the first commercial MSF facilities were installed in Kuwait (19 stages, 4550 m^3/d) and the Channel Islands (40 stages, 2775 m^3/d) [22, 23]. In 1973, the "standard" MSF units, that produce 27,277 m^3/d and consist of 24 stages, were developed [20].

Recent developments in the thermal desalination technology have focused on scale and corrosion control techniques and on the increase in distiller production capacity [21]. Early, pre 1980, MSF units were primarily constructed using carbon steel for the shell and the internals [24]. Corrosion of the metal due to seawater resulted in the use of thicker materials of construction, which made the units larger and heavier. Units built after 1980 use stainless and duplex stainless steel to reduce corrosion, allowing for lighter and smaller MSF units. Future advances in the technology will most likely focus on improvements in thermodynamics and material selection [24].

1.4.5.2 *History of Membrane Desalination*

While the earliest recognition of thermal desalination was a few hundred years B.C., the earliest recorded documentation of semi-permeable membranes was in 1748, when the phenomenon of osmosis was observed by Abbe Nollet.[25] Osmotic phenomenon was also studied in the 1850's, and then in the 1940's, when Dr. Gerald Hassler began investigation of the osmotic properties of cellophane [26]. "Modern" RO technology truly began in the late 1950's when C.E. Reid and E.J. Breton at the University of Florida and Sidney Loeb and Srinivasa Sourirajan at the University of California at Los Angeles (UCLA) independently demonstrated RO using polymeric membranes.

The United States was the early leader in desalination research in the 1960s and 70s due in most part to government funding. The Saline Water Conversion Act of 1952 established the Office of Saline Water (OSW) in 1955, which later became the Office of Water Research and Technology (OWRT) in 1974. It was under such governmental programs that Loeb and Sourirajan developed the first commercially-viable reverse osmosis membrane while at the UCLA [26]. In 1965, the tubular, cellulose acetate membrane developed at UCLA became the membrane used in the first commercial reverse osmosis facility located in Coalinga, California [27].

Government funding also lead to the development in 1965 of the solution/diffusion membrane transport model by Harry Lonsdale, U. Merten, and Robert Riley at the General Atomic Division of General Dynamics, Corp [28]. This model has become the basis of research and development of new membrane materials since that time.

It was under similarly-funded governmental programs that John Cadotte, while at North Star Research, prepared the first interfacial polyamide membrane that soon after became the basis of the FilmTec FT30 membrane (now part of Dow Water and Process Solutions) [29]. The original FT30 membrane chemistry is the basis of the majority of reverse osmosis membranes in use today [22].

The OWRT was abolished in 1982 and government funding of desalination research in the United States dropped considerably. By that time, however, much of the foundation for reverse osmosis,

Table 1.3 Advances in brackish water membrane performance [23].

Year	Pressure (psi)	Relative Flux	Rejection (%)	Membrane Material
1970's	435	1	97	Cellulose acetate
1980's	290	1.9	99.0	Cross-linked polyamide composite
1987	220	3.0	99.7	Cross-linked aromatic polyamide composite
1988	145	4.2	99.7	Cross-linked aromatic polyamide composite
1996	110	5.6	99.7	Cross-linked aromatic polyamide composite
1999	75	8.0	99.7	Cross-linked aromatic polyamide composite

membrane-based desalination had been laid. Since then, incremental membrane improvements have been made in the areas of flux, rejection, and operating pressure requirements, as shown in Table 1.3 [23]. However, no major "breakthroughs" in terms of higher membrane selectivity with high water flux and chlorine tolerance have occurred in 30+ years since the revolutionary early developments. Research is continuing, however, and the development of nanotechnology and "nanocomposite" membranes circ. 2005 has raised hopes that reverse osmosis membranes with higher selectivity, water flux, and chlorine tolerance may be on the horizon, all of which will reduce costs and improve efficiency associated with membrane desalination [22].

1.4.5.3 Developments in Desalination Since 1980

Since 1980, the world-wide development of desalination techniques has been driven out of necessity due to water scarcity and population growth. The private sector has led the investment in research and development as they began to see water not as a commodity, but as a product to be sold at a profit [3]. This development by the private-sector has lead to a significant drop in cost of water generated through desalination techniques. An example of such is the 80% reduction in price of reverse osmosis membrane elements

Table 1.4 Decline in membrane cost relative to 1980 [31].

Year	Relative Cost
1980	1.00
1985	0.65
1990	0.34
1995	0.19
2000	0.14

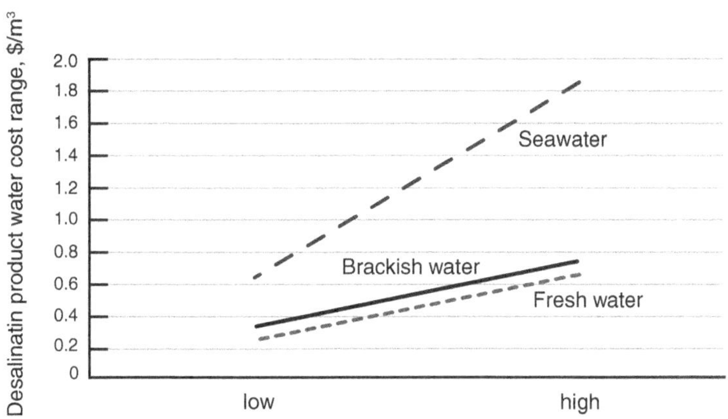

Figure 1.12 Cost range of desalinated water [19]. *Courtesy of Global Water Intelligence.*

over the last 15 – 20 years, while incremental improvements in flux, selectivity, and operating pressure were realized (Tables 1.3 and 1.4 [21, 29]. In 1991, the cost to produce water at the SWRO Santa Barbara desalination facility was about $8/kgal; in 2007, the estimated cost had dropped to about $3.40/kgal [2, 24, 30]. Figure 1.12 shows the general cost range of desalinated water (variability in cost is due to factors such as size of plant and degree of pretreatment employed) [19].

There have been other desalination techniques developed over the years, some more commercially successful than others, and none more commercially successful than traditional thermal and reverse osmosis desalination processes. Table 1.5 lists a selection of desalination technologies and their current status.

Table 1.5 Sample desalination technologies. technologies covered in this volume are noted in italics.

Technique	Technology	Driving Force	Status	Comment
Membrane	*Continuous Electrodeionization (CEDI)*	*Electric Field*	*Mature*	*Used primarily for polishing RO permeate; some process (non-water purification) applications.*
	Dialysis	Osmotic pressure	Commercial application limited to kidney dialysis	Osmotic process is inherently not selective and transport by diffusion is slow such that application to desalination Is not economically feasible. {32}
	Electrodialysis / Electrodialysis Reversal (ED / EDR)	Electric Field	Mature	Best suited to charged, low molecular weight species; less efficient with low salinity feed streams or when low-conductivity product streams are required.
	Forward Osmosis (FO)	Osmotic pressure	Gaining commercial application for desalination, particularly since 2010 [33]	Requires additional treatment such as low-pressure RO to yield purified water suitable for use. Differs from Pressure-Retarded Osmosis (PRO) in that PRO Is used primarily to generate power.
	Membrane Distillation (MD)	Temperature (vapor pressure)	On-going study of fundamentals; little industrial-scale study to date [35]	Capital cost of capillary membranes used for MD is higher than RO membranes, thereby limiting commercial application [32, 36]

	Nanofiltration(NF)	Pressure	Mature	Used for "softening" or partial demineralization of salt water and as pretreatment to seawater desalination using RO.
	Piezodialysis	Pressure	Little commercial interest today(32)	Researched in the 1970s [34]. Salt separation (enrichment) decreases with increasing feed salt concentration; uses charge mosaic (ion exchange) membranes.
	Reverse Osmosis (RO)	***Pressure***	***Mature***	***Research into fouling resistance and chlorine tolerance and well as reduction in energy while maintaining or improving rejection is in progress.***
Thermal	***Carrier-gas***	***Temperature***	***Under Development***	***Also known as humidification dehumidification desalination (HDH); process has high specific energy requirements, usually related to the dehumidification step*** [38]
	Evaporation	Temperature	Mature	Surface scaling, improved heat transfer, and corrosion- resistant materials primary focus of current development [24]

(Continued)

Table 1.5 (*Cont.*)

Technique	Technology	Driving Force	Status	Comment
	Freezing-Melting (FM)	*Temperature*	*Under Development, currently limited to food processing where high-temperature treatment could adversely affect product quality*	*Relatively high capital cost and process complexity; potential for desalination using freezing-melting technique in concert with other desalination technologies* [37]
Other	*Capacitive Deionization (CDT)*	*Electric field*	*Under Development since the 1970s*	*Older carbon aerogel electrodes (made from resorcinol) were expensive and their ion storage capacity was relatively low (39); newer hierarchical carbon aerogel monoliths show improved performance* [40]
	Ion Exchange	Ionic Charge	Mature	Limited to relatively low-salinity, brackish water desalination applications.

1.4.6 The Future of Desalination

Desalination today is still a capital and energy-intensive proposition. Methods to reduce costs are necessary to make the desalinated water more affordable. To that end, developments to increase the efficiency and thereby reduce costs of desalination are needed. Some areas of current development include:

- Energy: renewable energy to drive desalination projects, such as wind and solar, are being considered to reduce the energy footprint and costs of desalination (today's contribution of renewable energy sources to desalination processes is only about 0.05%) [41]. Table 1.6 shows some of the renewable energy sources (RES) currently considered to power desalination processes.
- Materials: materials of construction for thermal processes that resist corrosion and reduce the size and weight (and costs) of units need to be developed. New membrane materials that resist attack by chlorine and show high selectivity and high flux are needed.
- Chemicals: antiscalants for both thermal and membrane –based desalination processes to increase the degree of water recovery per given unit size.

Current desalination techniques can also have a significant impact on the environment. Some of these issues are described below [41]. Before desalination can become sustainable, these environmental issues must be addressed. The two most pressing issues are concentrate disposal and airborne emissions:

- Concentrate disposal—very high salinity wastewater is generated through thermal and reverse osmosis membrane desalination. Total dissolved solids can be as high 100,000 ppm in the wastewater from the desalination of seawater. Furthermore, concentrate can be highly turbid and be at elevated temperatures (thermal desalination plants), and can contain chemical additives such as polymers/coagulants, acid, biocides, corrosion inhibitors, and cleaners. The issue becomes how to dispose of the wastewater in an environmentally-safe manner

Table 1.6 Renewable energy sources (RES) used to replace traditional fossil-fuel based energy sources to drive traditional desalination processes. These RES play a relatively small role in desalination today [41], but given the concerns regarding fossil fuels (e.g., availability, carbon footprint (see Table 1.7)), these sources show potential for future application in powering desalination. Solar energy is covered in this volume.

Renewable Energy Source	Status	Comments
Solar Energy	*One of the first desalination power sources used to generate fresh water (via evaporation of salt water) [42]*	*Can be used to power thermal desalination process directly and to generate power for RO desalination processes.*
Wave Energy	Under Development since 1980s; relatively few desalination installations today [43]	Used to generate electricity (using techniques known as "Delbuoy" RO concept, "McCabe" wave pump or "water hammer") which can then be used to power reverse osmosis desalination. [43]
Wind Energy	Under Development for desalination since 1982 [43]	Limited to relatively small, less than 100 m^3/h installations; most significantly smaller. (43)
Nuclear Energy	Proven for power- generation; growing application for cogeneration of electricity and desalinated water	Can reduce CO_2 emissions by thermal desalination plants compared to those using fossil fuels by about 90% [44]

that is also cost effective. Seawater desalination facilities currently discharge to the ocean, which some argue is damaging to local flora and fauna via increases in seawater temperature, salinity, and turbidity, which may be harmful to marine life and cause them to migrate away and at the same time enhancing the populations of algae and nematodes.[45, 46] Inland brackish water facilities use other disposal methods, including discharge to surface water (45%), discharge to sewer (27%), deep well injection (16%), land application (8%), and discharge to evaporation ponds (4%) [46]. Each of these disposal methods has its own environmental concerns that are directly related to the high salinity and other components of the water for discharge.

- Carbon footprint and airborne emissions—Table 1.7 lists airborne emissions for fossil fuel-powered desalination technologies. The carbon dioxide emissions from thermal desalination processes are a full order of magnitude higher than that for RO, when powered by fossil fuels. And, energy to power the desalination facilities has the largest impact on the carbon footprint of the process [47]. The emissions for desalination technologies can be reduced significantly when these processes are powered by waste heat or renewable energy sources such as solar radiation rather than by fossil fuels [41].

Table 1.7 Current airborne emissions per cubic meter of water generated by various desalination technologies when powered by fossil fuels and/or waste heat [41].

	MSF		MED		RO
Energy Source	Fossil Fuel	Waste Heat	Fossil Fuel	Waste Heat	Fossil Fuel
CO_2 (kg)	24	2.0	18	1.1	1.8
Dust (g)	2.0	2.0	1.0	1.0	2.0
NOx (g)	28	4.1	21	2.4	3.9
NMVOC (g)	7.9	1.2	5.9	0.6	1.1
SOx (g)	28	15	26	16	11

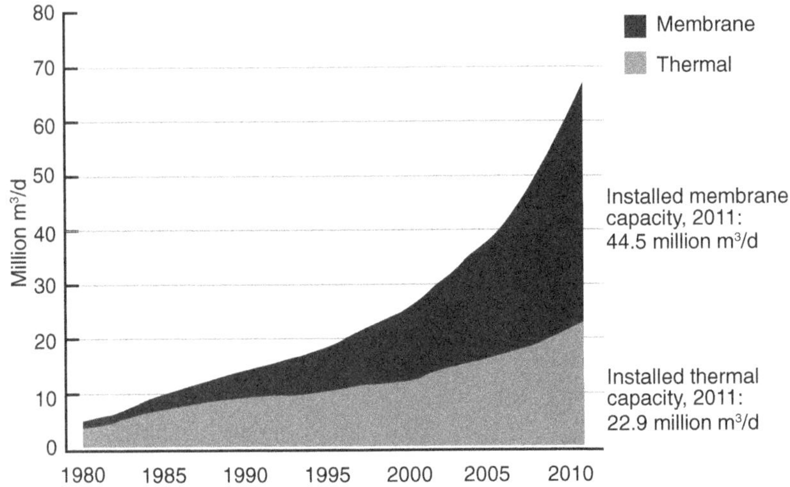

Figure 1.13 Growth of installed membrane and thermal desalination capacity. *Courtesy of Global Water Intelligence.*

Future demand for fresh water will likely result in an estimated \$40-\$50 billion expenditure on global desalination projects over the next ten years [41]. It is apparent that RO will be the primary mode of desalination for the foreseeable future, [48], as Figure 1.13 [16] demonstrates the relative growth of RO to traditional thermal processes. Thermal processes, while in decline, still have a foothold, primarily in the Gulf Region.(16)

For desalination to be a sustainable method to develop sources of fresh water in the future, desalination technologies need to be efficient, cost effective, and have low environmental impact. Indeed, concerted efforts on three fronts have resulted in significant advances in desalination [49], each of which is developed in this book:

- development of innovative new technologies,
- improvements in performance and design of conventional technologies,
- the marriage of desalination technologies with renewable energy sources (RES).

1.5 Desalination: Water from Water Outline

The objective of this volume is to present the case for desalination, describe conventional and innovative new desalination technologies, touch on RES for desalination, and conclude with a discussion of future directions.

Section I of this book discusses the need for finding new sources of fresh water to meet the ever growing demand. The case for desalination as a method of preparing fresh water from salty or impaired water, that has proven to be successful, is developed.

In addition to conventional desalination technologies (thermal and RO), there are many new technologies under development, as listed in Table 1.5. Current and several of the more promising desalination technologies under development are discussed in Sections II through IV of this book:

- **Section II**: this section covers traditional thermal desalination technologies, including MED, and MSF. These technologies are very mature, but do have limitations that may be overcome with through future development of new materials to improve corrosion resistance and heat transfer, and through the development of antiscalants.
- **Section III**: this section describes several membrane-based technologies, including RO, continuous electrodeionization, and membrane distillation, as well as some membrane-based desalination technologies that have only recently been commercialized, namely forward osmosis. Significant research is on-going to surmount the limitations of relatively mature membrane technologies such as RO, while also developing less commercialized technologies such as membrane distillation.
- **Section IV**: this section details non-traditional desalination technologies. Technologies covered in this section include freezing-melting desalination processes, capacitive deionization, and ion exchange. These technologies may be limited for desalination applications, and some require additional development to become viable for commercial or industrial desalination.

The future need for renewable energy sources (RES) to power desalination may be as great as the need for desalination itself, if the issues associated with fossil fuels become even more acute. A common RES, solar energy, is described in **Section V** as an alternative to fossil fuels to power desalination. The feasibility of solar energy is relatively high as an RES to power desalination today, due to the availability, and shows great future potential, particularly in arid areas where sunshine is plentiful, but fresh water is not.

The book concludes with a discussion about the future prospects for desalination in **Section VI**. This final section discusses future water sources for desalination, including traditional seawater sources, and other, more impaired sources, such as industrial wastewater. Future water demand for desalination water, including traditional municipal users and emerging water users, such as oil field hydrofracking, is profiled. Finally, research needs to develop additional desalination technologies that are efficient and cost effective are presented along with some of the more promising desalination techniques to come out of research and development.

Abbreviations

CDT	capacitive deionization technology
CEDI	continuous electrodeionization
ED	electrodialysis
EDR	electrodialysis reversal
ELWRF	Edward C. Little Water Recycling Facility
FM	freezing-melting
FO	forward osmosis
HDH	humidification dehumidification
LADWP	Los Angeles Department of Water and Power
MD	membrane distillation
MED	multi-effect distillation
MSF	multi-stage flash
NCAR	National Center for Atmospheric Research
NF	nanofiltration
OSW	Office of Saline Water
OWRT	Office of Water Search and Technology
PRO	pressure-retarded osmosis
RES	renewable energy sources
RO	reverse osmosis
SWRO	seawater reverse osmosis
TDS	total dissolved solids
UCLA	University of California at Los Angeles
WHO	World Health Organization

References

1. http://www.Merriam-Webster.com. 2001.
2. National Research Council, Committee to Review the Desalination and Water Purification Technology Roadmap, Review of the Desalination and Water Purification Technology Roadmap, The National Academies Press, Washington, D.C., www.nap.edu/catalog/10912.html, 2004.
3. National Research Council, Committee on Advancing Desalination Technology, Desalination, A National Perspective, The National Academies Press, Washington, D.C., 2008.
4. G. Micale, A. Cipollina, and L. Rizzui, *Seawater Deslination for Freshwater Production, Seawater Desalination, Conventional and Renewable Energy Processes*, G. Micale, A. Cipollina, and L. Rizzui (Ed.), Springer, Heidelberg, Germany, (2009).
5. J. McGeown, *Personal Communication*, March 5, (2012).
6. Rekacewicz, Philippe, http://www.grida.no/grphicslib/detail/increased-global-water-stress_5694, accessed October 15, 2012.
7. Kerschner, M. Edward, and M W. Peterson, *Peak Water: The Preeminent 21st Century Commodity Story*, Morgan Stanley Smith Barney Global Investment Committee, November, (2011).
8. *Global Water Intelligence*, March, (2010).
9. The World Health Organization, *Water Facts & Water Stories from Across the Globe*, http://www.theworldwater.org/water_thefacts.php (2010).
10. Dai, Aiguo, *Drought Under Global warming: A Review, Wiley Interdisciplinary Reviews: Climate Change 2*, October, (2010).
11. SWRO Falls Victim to Politics, *World Desalination Report*, vol. 47, number 19, 16 May, (2011).
12. LADWP, *Securing L.A.'s Water Supply*,"City of Los Angeles Water Supply Action Plan, http://www.ladwp.com/ladwp/cms/ladwp010587.pdf , May, (2008).
13. http://www.ladwp.com/ladwp/cms/ladwp001620.jsp, (2011).
14. Mabry, Kat, City Seeks Public Input on New Recycled Water Program, Encino Patch, http://www.encino.patch.com/articles/city-seeks-public-input-on-new-recycled-water-program, May 19, (2011).
15. West Basin Municipal Water District http://www.westbasin.org/water-reliability-2020/recycled-water/water-recycling-facility, accessed October 15, (2012).
16. Pankratz, Tom, *IDA Desalination Yearbook 2011-2012, for Global Water Intelligence*, Media Analytics, Ltd. Publishers, Oxford, United Kingdom, (2011).
17. *Mission 2013: Carbon Sequestration MIT*, http://igutek.scripts.mit.edu/terrascope/index.php?page=index, accessed October 15, (2012).

18. Pankratz, Tom, *IDA Desalination Yearbook 2010-2011, for Global Water Intelligence*, Media Analytics, Ltd. Publishers, Oxford, United Kingdom, (2010).
19. Pankratz, Tom and T. John , deslination.com, an environmental primer, 2nd ed., Lone Oak Publishing, Houston, Texas, (2004).
20. Al-Mutaz, S. Ibrehim , *Water Desalination*, http://faculty.ksu.edu.sa/Almutaz/Documents/ChE-413/Desalination_Introduction.pdf.
21. Hamed, A. Osman, *Evolutionary Developments of Thermal Desalination Plants in the Arab Gulf Region*, http://www.swcc.gov.sa/files/assets/Research/Technical%20Papers/Thermal/EVOLUTIONARY%20%20DEVELOPMENTS%20%20OF%20%20%20THERMAL%20DESALINATION%20%20PLANT.pdf, Beruit Conference, (2004).
22. Kucera, Jane, *Reverse Osmosis, Industrial Applications and Processes*, Scrivener Publishing, Salem, Mass., (2010).
23. Norman Li, F. Anthony , W.S. Winston Ho, and M. Takeshi , *Advanced Membrane Technology and Applications,* Eds. John Wiley & Sons, Inc., Hoboken, NJ, (2008).
24. Sommariva, Corrado, and R. Vekates, MSF Desalination: A Review of its Development, *Watermark*, issue 18, December, (2002).
25. Cheryan, Munir, *Ultrafiltration and Microfiltration Handbook*, 2nd ed. 1998
26. Koenigsberg, Diana, Water Warriors, *UCLA Magazine*, http://www.magazine.ucla.edu/features/waterwarriors, July 1, (2006).
27. Glater, Julius, Professor Emeritus, UCLA, personal communication, February 24, (2009).
28. H.K. Lonsdale, U. Merten, and R.L. Riley, Transport properties of cellulose acetate osmotic membranes. *J. Appl. Polym. Sci* **9**, (1965).
29. Cadotte, John, R.S. King, R.J. Majerle, and R.J. Peterson, Interfacial synthesis in preparation of reverse osmosis membranes. *J. Macromol. Sci. Chem* **15**, (1981).
30. Desalination, County of Santa Barbara Public Works Online. http://www.countyofsb.org/pwd/water/desalination.htm, January 27, (2007).
31. Baker, and Richard, *Membrane Technology and Applications*, 2nd ed., John Wiley & Sons, Ltd., Chichester, West Sussex, England. (2004).
32. Baker, and Richard, *Membrane Technology and Applications*, 3nd ed., John Wiley & Sons, Ltd., Chichester, West Sussex, England. (2012).
33. Thompson, A. Neil. and P.G. Nicoll, Forward Osmosis Desalination: A Commercial Reality, IDAWX/PER11-198, presented at the 2011 IDA World Congress, Perth, Australia, September. (2011).
34. F.B. Leitz and W.A. McRae, An engineering analysis of piezodialysis, *Desalination* **10**, issue 3, (1972).
35. Alkhudhiri, Abdullah, D. Haif, and H. Nidal, Membrane Distillation: A Comprehensive Review, *Desalination* **287**, (2012).
36. Gullinkala, Tilak, D. Brett, G. Colleen, H. Richard and C.E. Isabel, Desalination: Reverse Osmosis and Membrane Distillation, in

Sustainable Water for the Future, E. Isabel and S. Andrea (Ed.), Elsevier, Amsterdam, The Netherlands, (2010).

37. Rahman, M. Shafiur, A. Mushtaque and X. Dong Chen, Freezing-melting process and desalination: Review of present status and future prospects, *Int. J. Nucl. Desalination* **2**,Number 3 (2007).
38. Narayan, G. Prakash, H. R H. McGovern and H.J. Lienhard, Helium as a Carrier Gas in Humidification Dehumidification Desalination Systems," ASME Conference Proceedings, Paper number IMECE2011-62875, http://dx.doi.org/10.1115/IMECE2011-62875, accessed November 28 (2012).
39. Dietz and Steven, Improved Electrodes for Capacitive Deionization, SBIR contract DMI-0216299, presented at Proceedings of the 2004 NSF Design, Service and Manufacturing Grantees and Research Conference, Birmingham, AL, January (2004).
40. http://www.rwlwater.com/capacitive-deionization-promising-desalination-technology-resdesigned/, accessed November 28 (2012).
41. Al-Ansari, S. Mohammed and Nader Al-Masri, Future Sustainable Desalination Technologies for the GCC, Presented at the Third SQU-JCCP Joint Symposium, Muscat, December. (2010).
42. M.T. Chaibi and Ali M. El-Nashar, Solar thermal Processes, A review of solar thermal energy technologies for water desalination, in *Seawater Desalination*, A. Cipollina, G. Micale and L. Rizzuti (Ed.), Springer-Verlag, Berlin, GE. (2009).
43. E. Tzen, Wind and wave energy for reverse osmosis, in *Seawater Desalination*, A. Cipollina, G. Micale and L. Rizzuti (Ed.), Springer-Verlag, Berlin, GE. (2009).
44. B.M. Misra and I. Khamis, Nuclear desalination, A review of desalination plants coupled to nuclear power stations, in *Seawater Desalination*, A. Cipollina, G. Micale, and L. Rizzuti (Ed.), Springer-Verlag, Berlin, GE, (2009).
45. Birkett and Jim, Desalination at a Glance, *IDA White Paper*, http://www.idadesal.org. (2011).
46. Mickley and Mike, Concentrate Management, in *The Guidebook to Membrane Desalination Technology*, Mark Wilf (Ed.), Balaban Desalination Publications, L'Aquila, Italy (2007).
47. Emerging Trends in Desalination: A Review, UNESCO Centre for Membrane Science and Technology, University of New South Wales, Waterlines Report Series No. 9, October. (2008).
48. Matsuura, Takeshi, and F. Tony, Editorial: New Directions in Desalination, *Desalination* **308** (1). (2013).
49. http://www.desalination.biz/news/news_story.asp?id=6318&title=Patents+indicate+'significant+advances'+in+desalination%2C+says+report, posted January 31, 2012, accessed December 11, (2012).

SECTION II

TRADITIONAL THERMAL PROCESS

2

Thermal Desalination Processes

Joachim Gebel

Abstract

This chapter starts with an extract of fundamentals of engineering science such as thermodynamics and heat transfer. Building on this, mass- and energy balances for single-effect and multiple-effect distillation processes are introduced. A complete set of design equations for MED, MSF and mechanically as well as thermally driven vapor compression plants is presented. In order to be able to compare the different processes in terms of energy demand, the so-called Gained Output Ratio as a performance indicator is introduced and discussed. Based on a vivid description of history of thermal seawater desalination, future prospects and challenges for thermal desalination technologies are discussed at the end of the chapter.

Keywords: Mass- and Energy Balances, Single-Stage Evaporation, Multiple-Effect Distillation (MED), Multi-Stage-Flash - Evaporation (MSF), Multiple-Effect Distillation with Thermally Driven Vapour Compression (TVC), Single-Stage Evaporation with Mechanically Driven Vapour Compression (MVC), Gained Output Ratio (GOR), Performance Ratio, Primary Energy Consumption, Historical Review

2.1 Thermodynamic Fundamentals

2.1.1 First and Second Rule of Thermodynamics

It is mandatory to know the first and second rule of thermodynamics for the energy-related balancing and design of seawater desalination plants, in particular of thermal plants. The first rule is concerned with the conservation of energy, whereas the second rule makes a statement on the direction in which the process runs.

Jane Kucera (ed.) Desalination , (41–154) 2014 © Scrivener Publishing LLC

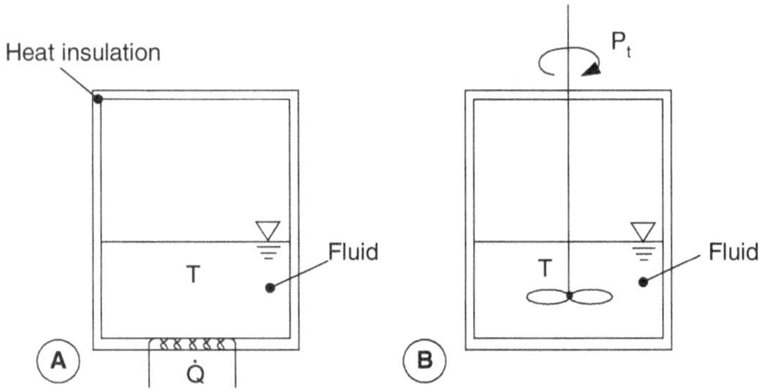

Figure 2.1 First rule of thermodynamics (Closed system).

The first rule may be explained with the help of a simple system. Let us examine system A as represented in Figure 2.1 This is a closed system, in other words there are no mass flows passing across the borders of the system. Furthermore the system is insulated, i.e. there are no heat flows passing across the system borders. Within the system there is a homogeneous fluid, for example water. At a particular time t, heat is introduced from outside into the system. If the temperature of the fluid is measured, a steady increase in temperature will be observed until the heat flow is switched off. The temperature increase is equal to an increase in the internal energy of the system.

Now system B should be examined. This system is also closed and insulated. However, instead of the introduction of heat a stirrer is placed in the fluid. At a particular time t, the stirrer is switched on. What happens? How does the system react to this introduction of energy? The mechanical energy which is introduced into the system via the stirrer leads to a temperature increase comparable to the introduction of heat, in other words, to an increase in the internal energy of the system. The energy transfer takes place through friction at the blades of the stirrer. This process is described as "dissipation". In daily life everyone has surely observed that brake pads heat up when braking. The brake pads are then finally cooled down by the surrounding air, i.e. the whole of the mechanical energy which is contained in the braking process ends up in the surrounding area. This point will be more closely examined in the discussion of the second rule.

So the introduction of both heat and mechanical energy has a completely identical effect on the system: there is an increase in the temperature. The system cannot differentiate between the different forms of energy introduced.

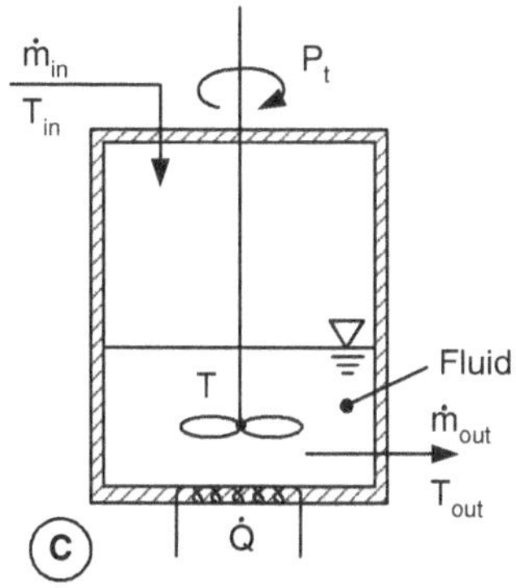

Figure 2.2 First rule of thermodynamics (Open system).

We now allow mass flows to pass across the system borders: the system is thus "open" (see Figure 2.2). In addition to having a mass flow that enters and a mass flow that leaves it, system C is characterised by a heat flow and the additional energy input through the stirrer. Steady state conditions mean that there are no changes over time for the system, in other words the temperature and the mass flows are constant. This means that the mass flow leaving the system captures the energy which is added to the system: thus the mass flow leaving the system has a higher temperature than the flow which enters ($T_{out} > T_{in}$), or in other words, the mass flow cools the system.

The energy of a fluid flow is known as enthalpy. The enthalpy is the product of mass flow, specific heat capacity and absolute temperature:

$$H = \dot{m} \cdot h = \dot{m} \cdot c_p \cdot T \tag{2.1}$$

The energy which the mass flow absorbs on its path through system C is the difference between ingoing and outgoing enthalpy:

$$\Delta H = H_{out} - H_{in} = \dot{m}_{out} \cdot c_P \cdot T_{out} - \dot{m}_{in} \cdot c_P \cdot T_{in} \tag{2.2}$$

Assuming that the mass flows entering and leaving are equal and that the dependency of the specific heat capacity on the temperature is disregarded, the following is derived:

$$\Delta H = \dot{m} \cdot c_P \cdot (T_{out} - T_{in}) = \dot{m} \cdot c_P \cdot \Delta T \tag{2.3}$$

This difference in enthalpy is also equal to the total of the energy added to the system, i.e.:

$$\dot{Q} + P_t = \Delta H \tag{2.4}$$

This relationship is known in thermodynamics as the first rule or the rule of energy conservation. In words it reads:

The sum of heat and technical work is equal to changes in the enthalpies of the mass flows entering and leaving.

The unit of the individual terms is [Watt] or [Joule/s].

This is valid under the following conditions:

1. Steady state conditions
2. Neglecting change of external energies

Neglecting the external energies is allowed in thermal seawater desalination plants, but there is one exception: When designing a thermal vapour compressor, the kinetic energy of the mass flows plays a significant role. The first rule of thermodynamics for a steady state process with external energies reads as follows:

$$\dot{Q} + P_t = \dot{m}_{out} \cdot h_{out} - \dot{m}_{in} \cdot h_{in} + \dot{m}_{out} \cdot e_{ex,out} - \dot{m}_{in} \cdot e_{ex,in} \tag{2.5}$$

The external energy is the sum of kinetic and potential energy (in this case the symbol "H" means "Height", not enthalpy):

$$\dot{E}_{ex} = \dot{m} \cdot \frac{v^2}{2} + \dot{m} \cdot g \cdot H \tag{2.6}$$

When taking into account specific external energies:

$$e_{ex} = \frac{\dot{E}_{ex}}{\dot{m}} = \frac{v^2}{2} + g \cdot H$$

the first rule reads as follows:

$$\dot{Q} + P_t = \dot{m} \cdot (h_{out} - h_{in}) + \dot{m} \cdot \left[(\frac{v^2}{2} + g \cdot H)_{out} - (\frac{v^2}{2} + g \cdot H)_{in} \right] \tag{2.7}$$

If the difference $H_{out} - H_{in}$ of the potential energy between inlet and outlet is set to zero, which is always the case for a thermal compressor, we can derive the following equation:

$$\dot{Q} + P_t = \dot{m} \cdot c_p \cdot (T_{out} - T_{in}) + \dot{m} \cdot (\frac{v_{out}^2}{2} - \frac{v_{in}^2}{2}) \tag{2.8}$$

The substitution of an enthalpy difference to a temperature difference as per Eq. 2 is only permissible if there is no phase change, that means water does not vaporize or that vapour does not condense. The enthalpy can only be expressed as a product of specific heat capacity and temperature for a single phase. The enthalpy change as a function of the temperature is shown in Figure 2.3. At ambient pressure the enthalpy of water will increase linearly (on condition that the specific heat capacity is not dependent on the temperature) between the ambient temperature and the boiling temperature of 100°C. The gradient is the specific heat capacity of liquid water. When the water has reached a temperature of 100°C, the boiling process begins. The energy which is introduced into the system is employed for the phase change liquid to vapour. During this process the temperature remains constant. The vapour thus created has a greater enthalpy than the boiling water. This difference in enthalpy is known as "heat of evaporation". If the vapour is further heated the increase in enthalpy follows a straight line once again, whereby the gradient is the specific heat capacity of vapour.

The total enthalpy change between T and $T_{ambient}$ is thus calculated as follows:

$$\Delta H = \dot{m} \cdot c_{P,water} \cdot (T_V - T_{ambient}) + \dot{m} \cdot \Delta h_V + \dot{m} \cdot c_{P,vapour} \cdot (T - T_V) \quad (2.9)$$

In an evaporation plant the vapour is not superheated but is condensed and is discharged out of the process as the product, or distillate. The minimal energy which must be utilised for the evaporation

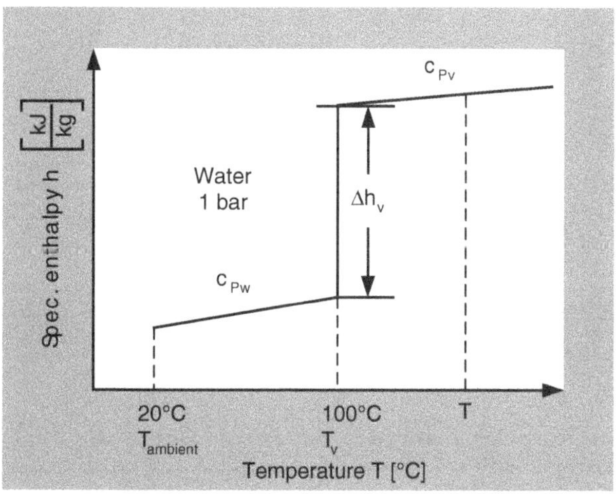

Figure 2.3 Enthalpy – temperature diagram for water.

of a certain amount of distillate can be determined from the above equation as follows:

$$\Delta H = \dot{m}_D \cdot c_{P,water} \cdot (T_V - T_{ambient}) + \dot{m}_D \cdot \Delta h_V \qquad (2.10)$$

The following prerequisites must be taken into consideration for the calculation of exact values:

1. The specific heat capacity depends on temperature and salt content.
2. The heat of evaporation is a function of the evaporation temperature.

The first rule (Eq. 2.4) for an evaporation process as per Figure 2.4 can now be formulated as follows, assuming that no technical work is introduced to the system:

$$\dot{Q} = \dot{m}_D \cdot c_{P,water} \cdot (T_V - T_{ambient}) + \dot{m}_D \cdot \Delta h_V \qquad (2.11)$$

The heat transferred into the process from outside is the sum of the preheating of the distillate mass flow to evaporation temperature and the energy for the evaporation itself. This equation is the starting point for the design of the seawater desalination plants in Chapter. 2.2.

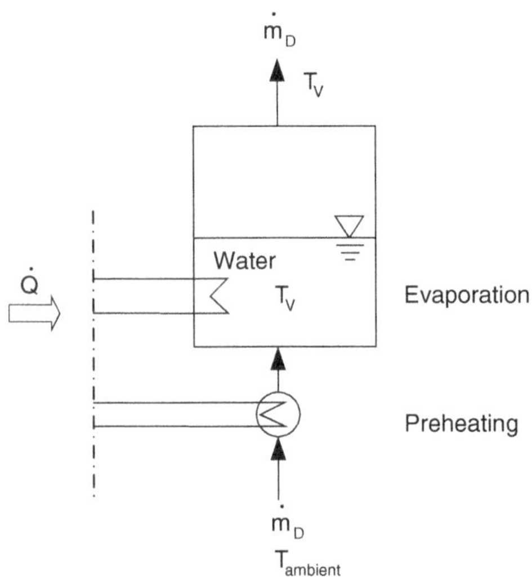

Figure 2.4 Flow scheme of an evaporation process with preheating.

Disregarding the technical work is of course only permissible in certain cases, for example in the overall energy balance of a thermal seawater desalination plant. However if we only set up the balance, for instance, for the seawater intake, which mainly comprises pumps, or the thermal vapour compression, it is the technical work which is the decisive part in the first rule of thermodynamics and the heat which is introduced or leaves the system can be disregarded.

Example: Seawater intake

Assumption: $\dot{Q} \approx 0$

first rule: $$P_t = \Delta H = \dot{m}_{out} \cdot h_{out} - \dot{m}_{in} \cdot h_{in} \quad (2.12)$$

At this point the second rule of thermodynamics should be explained using the following example (see Figure 2.5):

Let us assume that there is an insulated box with a homogeneous fluid separated in two chambers at different temperatures. When the separating plate is removed the fluids will spontaneously mix. At the end of the mixing process it is possible to measure an average temperature in the whole box. This process is irreversible, that is to say noone has yet observed that the fluids separate again and that different temperatures can be measured on the different sides of the box.

Most processes in nature run in a certain direction when not influenced from outside and are therefore irreversible. So, for example, is the braking process mentioned above, in which mechanical energy is transformed into thermal energy which then dissipates into the environment. This energy can no longer be used: a process in the order "brake pads heat up → brake works mechanically against the pressure of the pedal" will never take place.

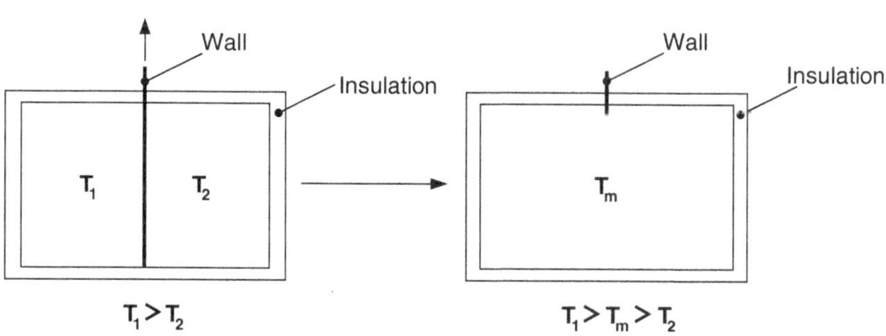

Figure 2.5 Irreversible mixing as an example for the second rule of thermodynamics.

The process of heat transfer is also subject to the second rule of thermodynamics:

Heat can only flow from a hot to a cold region.
The driving force for this process is a temperature difference.

As will be shown at various points in this book this "trivial-at-first-glance" sentence is of decisive importance for the design and costs of multiple-effect seawater desalination plants.

2.1.2 Boiling and Boiling Point Elevation

Everyone knows from daily life experience that water boils at 100°C. In doing this it is easily overlooked that the boiling point temperature is dependent on pressure. If the pressure is increased or lowered, then the boiling temperature also changes accordingly. Water boils on Mount Everest at a height of about 8,800 m at about 60°C because of the low atmospheric pressure. If the boiling points are entered into a pressure – temperature diagram a characteristic curve is obtained, the so-called vapour pressure curve or boiling point curve (Figure 2.6). The non-linear progression of the boiling

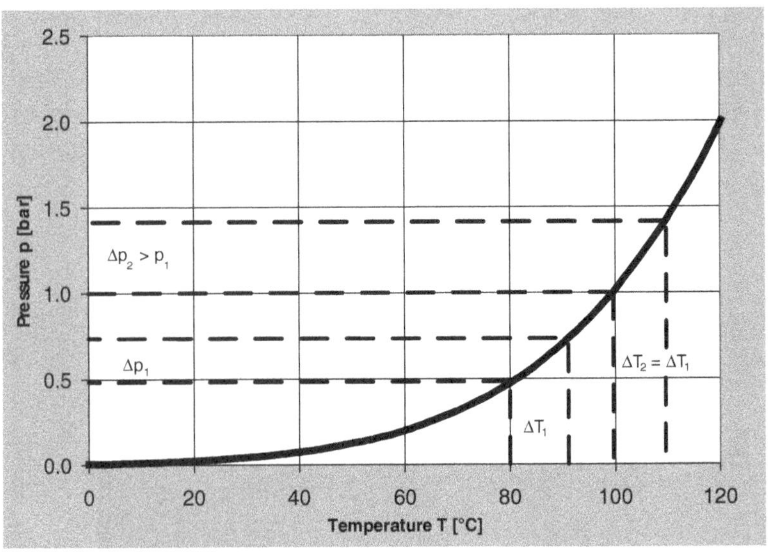

Figure 2.6 Vapour pressure curve or boiling point curve for pure water.

point curve has wide-reaching consequences for the design of multi-effect evaporation plants:

1. As the temperature increases, the pressure does so also, but over-proportionately. Hence, evaporation plants which work at high temperatures must be designed to withstand high pressures, which requires a greater amount of material for walls and reinforcements and thus has an effect on the investment costs; (a temperature increase from 100°C to 120°C means a twofold increase in pressure, a temperature increase from 100°C to 180°C means a tenfold increase in pressure!)
2. Similar differences in temperature bring various differences in pressure, which have an effect on the mass flows between and the levels in the individual evaporation stages.

The progression of the vapour pressure curve follows a logarithmic pattern. Eq. 2.13, named after its originators Clausius-Clapeyron Equation, provides a connection between pressure, temperature and evaporation enthalpy:

$$\Delta h_V = T \cdot (v'' - v') \cdot \frac{d\,p}{d\,T} \tag{2.13}$$

The parameters v'' *and* v' are the specific volumes at saturated vapour and boiling liquid conditions respectively. With the help of the Ideal Gas Equation and the assumption that v' is small compared to v'', this results in:

Ideal gas: $$p \cdot v'' = \frac{p}{\rho''} = R \cdot T$$

$$\Rightarrow \quad \Delta h_V = \frac{R \cdot T^2}{d\,T} \cdot \frac{d\,p}{d\,p} = -R \cdot \frac{d\ln p}{d\left(\frac{1}{T}\right)} \tag{2.14}$$

As a consequence the vapour pressure curve is a straight line with a negative slope in a double-logarithmic graph. The gradient is given by the evaporation heat and the gas constant. Figure 2.7 shows the vapour pressure curves of various substances, among which is also water. Since, for instance, the vapour pressure curves

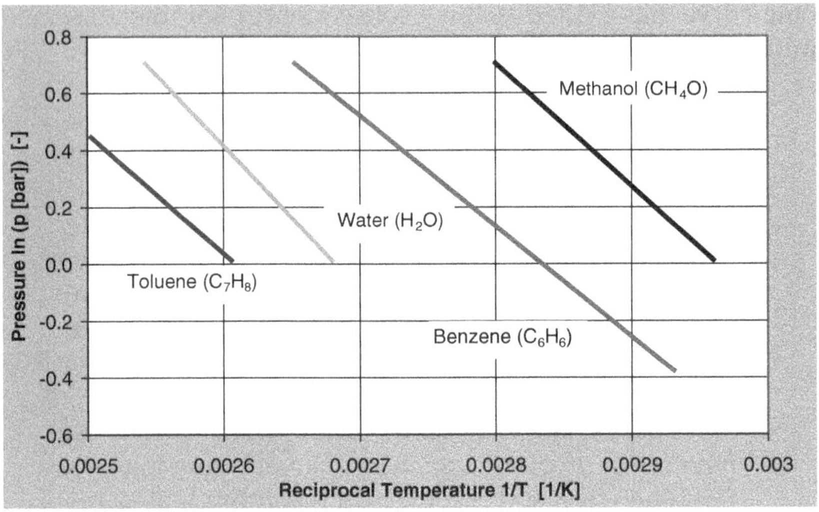

Figure 2.7 Vapour pressure curve for various substances (ln p vs. 1/T diagram).

of toluene and water are in the same region, the use of several stages is mandatory for the separation of these two substances. The corresponding process is called "distillation" or "rectification" and takes place in columns.

The vapour pressure of salt, i.e. sodium chloride is too small as to be shown in this diagram. This means for the evaporation process that even in a single-stage process the required separation of water from salt is possible. The multi-stage nature of evaporation plants therefore solely has the aim of reducing the energy demand; the multi-stage principle is of no importance for the purity of the distillate.

The water–vapour equilibrium is one of the most comprehensively investigated systems as a result of its overwhelming importance for process and energy technology. In Appendix B there is a clearly arranged table with the essential water–vapour properties sufficient for the calculation of seawater desalination plants. Regression equations for the water–vapour properties are also summarised in this appendix. Should more precise specifications be required, reference can be made to the extensive appropriate literature (and computer-supported data bases) on this subject.

Does it make a difference if either water or seawater is evaporated? This question may be answered with a definite "Yes". The salt dissolved in seawater has an influence on the evaporation process: it reduces the vapour pressure of the water and increases the boiling point. The difference in the boiling points is known as Boiling Point Elevation ΔT_{BPE}. Figure 2.8 shows the Boiling Point Elevation for water as a function of temperature and salt concentrations.

The diagram can be read as follows: Water with a salt content of 40 g/kg boils at an ambient pressure of p ≈ 1 bar not at 100°C, but at 100.6°C.

With this the fundamentals necessary in order to be able to understand the thermodynamics of a thermal seawater desalination plant have been explained.

2.1.3 Heat Transfer

In principle there are three separate forms of heat transfer:

1. Thermal conduction
2. Convection
3. Radiation

Although the sun is the impetus of the natural water cycle, we will not be dealing with radiation at this point, as this form of heat

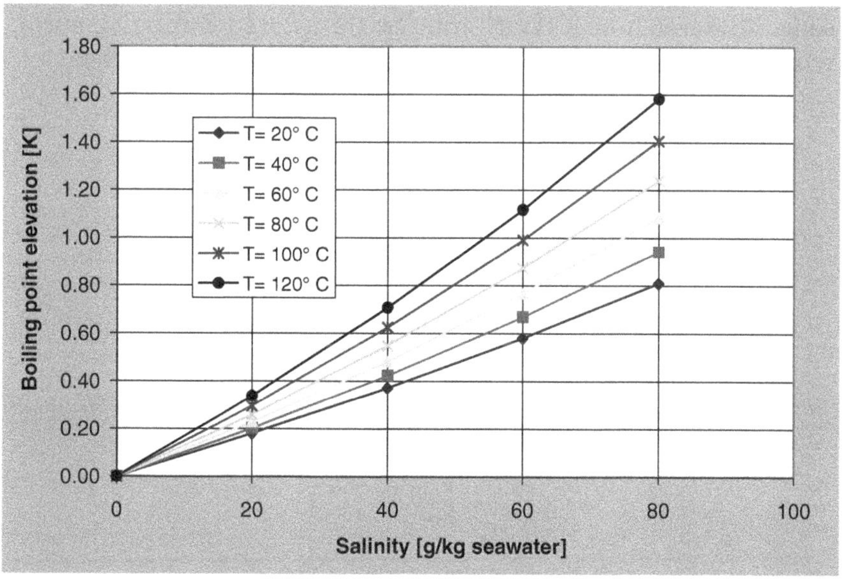

Figure 2.8 Boiling point elevation as a function of salt content and temperature[1].

transfer does not play any great part in technical seawater desalination plants. This does not mean that radiation has no significance for seawater desalination: in the future increasing attention have to be paid to solar energy in order to be able to provide seawater desalination plants both with thermal and electrical energy; however, as soon as heat is introduced into the desalination plant, the processes are run by thermal conduction and convection.

Figure 2.9 shows a typical cross-section of a heat exchanger. A fluid which may be assumed to be homogenous flows on both sides of a wall (mostly in counter current), and the heat flows from the hot to the cold fluid according to the second rule of thermodynamics. The impermeable wall physically separates the two fluids. Thus this process is known as an indirect heat exchange.

The heat transfer from the fluid to the wall and from the wall to the fluid takes place using convection, through the wall by thermal conduction. The specific heat flow q'' relative to the surface A is directly proportional to the temperature gradient. The heat conductivity λ is introduced as a proportionality factor by definition:

$$\dot{q}'' = \frac{\dot{Q}}{A} = -\lambda \cdot \frac{d\vartheta}{dy} \tag{2.15}$$

The heat conductivity λ is a property of the material and represents how much heat (Watt) may be transferred through a surface with an area of 1 m^2 at 1 K temperature difference per meter.

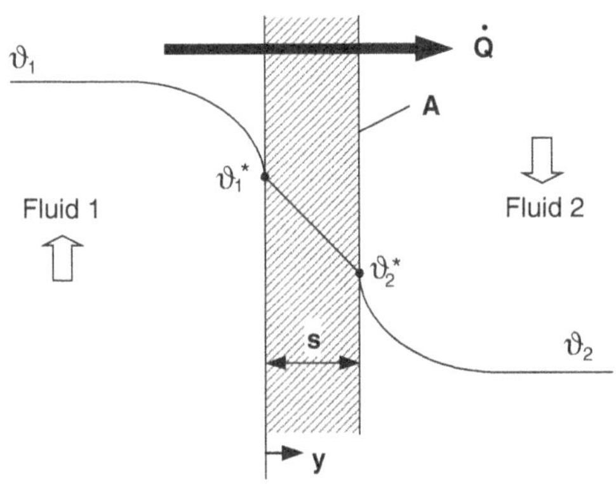

Figure 2.9 Heat transfer by convection and by conduction through a wall.

Table 2.1 Heat conductivity of various materials

Material	Heat conductivity λ [W/m K]
Copper	**444**
Iron	**66.9**
Concrete	**2.1**
Glas	**0.8**
Water	**0.5**
Wood	**0.13**
Insulator	**0.06**
Air	**0.0012**
Rubber	**0.16**
PVC	**0.16**
Scaling (gypsum)	**0.6-2.3**
Scaling (silica)	**0.08-0.18**

Material	Heat conductivity λ [W/m K]
AlMn	**232**
CuNi 10 Fe	**50**
CuNi 30 Fe	**29**
CuSn 6	**62.7**
CuZn 30	**111.5**
G-CuSn 10	**66**
GG-20	**46**
GGG-NiCr 20 2	**12.8**
NiCu 30 Fe	**26.7**
NiCr 21 Mo	**11.1**
NiMo 16 Cr 16 Ti	**10**
St 35	**54.6**
Ti 99.6	**16.3**
X5 CrNi 18 9	**14**
X 10 Cr 13	**26.7**

Table 2.1 lists the heat conductivity of various materials. For values below 0.1 $W/m \cdot K$ the materials are referred to as insulators, for values above 10 $W/m \cdot K$ the range of conductive materials begins. Therefore, depending on the task involved, a specific material may be selected.

The heat conductivity is constant for a homogenous wall and the temperature profile in the wall is linear. Therefore the integral between the temperatures on each of the surfaces can be solved:

$$\dot{q}'' \cdot \int_0^s d\,y = -\lambda \cdot \int_{\vartheta_1^*}^{\vartheta_2^*} d\,\vartheta$$

$$\dot{q}'' = \frac{\lambda_w}{s} \cdot (\vartheta_1^* - \vartheta_2^*) \tag{2.16}$$

According to Eq. 2.15 the gradient on the surface must be known in order to be able to determine the specific heat flow from fluid 1 into the wall and from the wall to fluid 2:

$$\dot{q}'' = -\lambda_1 \cdot \left.\frac{d\,\vartheta}{d\,y}\right|_{y=0} = -\lambda_2 \cdot \left.\frac{d\,\vartheta}{d\,y}\right|_{y=s} \tag{2.17}$$

However, as the gradients on the wall are only measurable with great effort but the temperature in the bulk of the flowing fluids are easily measurable, a coefficient α is defined which makes the problem manageable for the engineer. The coefficient α is known as the heat transfer coefficient and is defined as:

$$\alpha = \frac{-\lambda \cdot \frac{d\vartheta}{dy}}{\Delta\vartheta}$$

Thus the following is derived from Eq. 2.17:

$$\dot{q}'' = \alpha_1 \cdot \Delta\vartheta_1 = \alpha_2 \cdot \Delta\vartheta_2 \tag{2.18}$$

Combination of Eqs. 2.18 and 2.16 with the relevant temperatures yields:

$$\dot{q}'' = \alpha_1 \cdot (\vartheta_1 - \vartheta_1^*) = \frac{\lambda_w}{s} \cdot (\vartheta_1^* - \vartheta_2^*) = \alpha_2 \cdot (\vartheta_2^* - \vartheta_2) \tag{2.19}$$

In this equation it is only the bulk temperatures which are measurable and thus known. Now a proportionality factor k can be introduced which couples the heat flow density with these temperatures:

$$\dot{q}'' = k \cdot (\vartheta_1 - \vartheta_2) \tag{2.20}$$

Through skilful transformation of the bracket term and with the assistance of Eq. 2.19 the following is finally obtained:

$$\dot{q}'' = k \cdot \left(\frac{\dot{q}''}{\alpha_1} + \frac{\dot{q}''}{\lambda_w / s} + \frac{\dot{q}''}{\alpha_2}\right)$$

and

$$\frac{1}{k} = \frac{1}{\alpha_1} + \frac{s}{\lambda_w} + \frac{1}{\alpha_2} \tag{2.21}$$

k is known as the overall heat transfer coefficient. Its reciprocal is the sum of the reciprocals of the individual heat transfer coefficients. The heat flow from fluid 1 to fluid 2 can thus be expressed as:

$$\dot{Q} = k \cdot A \cdot (\vartheta_1 - \vartheta_2) \tag{2.22}$$

Thus if the overall heat transfer coefficient is known, then this equation can be solved according to the quantity sought (heat flow, heat transfer surface, temperatures or temperature difference). However, the knowledge of the overall heat transfer coefficient assumes knowledge of the heat transfer coefficients and thus of the flow condition on both sides of the wall – and here things become complicated!

There are a large number of different cases, starting with simple tubes up to flows around aircraft and space vehicles, for example the Space Shuttle. However, as a result of many years of working with this problem there are a number of experiences and standard works available, to which the engineer can refer in the planning of a seawater desalination plant. Reference should be made to the appropriate extensive literature on the subject of heat transfer[2,3].

2.1.4 Compression of Gases

For the calculation of the power consumption of a mechanical vapour compressor installed in a seawater desalination plant reference must be made to the fundamentals for the compression of gases.

The specific work in the compression of a gas from a state 1 to a state 2 may be expressed in terms of thermodynamics as an integral over the product specific volume and pressure change:

$$w_{comp,12} = -\int_1^2 v \cdot dp \tag{2.23}$$

In order to be able to solve the integral the change in state must be more closely observed. In reality a compression process involves friction and heat exchange with the surroundings. This may be easily recognised by the fact that the housing of a compressor in operation becomes hot and emits heat. In thermodynamics such a process is described as polytropic. If the so-called polytropic exponent n is introduced, this gives:

$$p_1 \cdot v_1^n = p_2 \cdot v_2^n \tag{2.24}$$

Polytropic change in state: $dq + dw_{friction} \neq 0$

$$\left. \frac{v}{p} \cdot \frac{dp}{dv} \right|_{polytropic} = -n \tag{2.25}$$

Using these formulas the integral above can be solved as follows:

1. $p \cdot dv = -\frac{1}{n} \cdot v \cdot dp$

2. $d(p \cdot v) = v \cdot dp + p \cdot dv$

$$\Rightarrow d(p \cdot v) = v \cdot dp \cdot \frac{1-n}{n}$$

$$\Rightarrow w_{comp,12} = -\int_1^2 \frac{n}{1-n} d(p \cdot v) = \int_1^2 \frac{n}{n-1} d(p \cdot v)$$

$$\Rightarrow w_{comp,12} = \frac{n}{1-n} \cdot (p_2 \cdot v_2 - p_1 \cdot v_1)$$

Using the ideal gas equation this term may be further transformed:

Ideal gas equation: $p \cdot v = R \cdot T$

$$\Rightarrow w_{comp,12} = \frac{n}{1-n} \cdot R \cdot T_1 \cdot \left[\left(\frac{p_2}{p_1}\right)^{\frac{n}{n-1}} - 1 \right] \quad (2.26)$$

This equation enables the calculation of the power consumption of a seawater desalination plant with mechanically driven vapour compression (see Chap. 2.2.5)

2.2 Mass- and Energy Balances

2.2.1 Single-Stage Evaporation

As a result of the large difference in the vapour pressures, a single-stage process is sufficient to separate water and salt. Theoretically the salt concentration in the distillate produced by evaporation is zero, however in practice it is always higher than zero as a result of the seawater droplets entrained. As the salt content in the distillate in comparison with the feed concentration in the seawater can be kept low using a well-operating demister, in the following mass balance the salt concentration has been set to zero. It must be emphasised at this point that this is only valid for thermal processes, in reverse osmosis we may always assume that there will be some flow of salt through the membrane, so that it is not possible to set the permeate concentration to zero.

Figure 2.10 shows a schematic of a single-stage evaporation process. There are three mass flows entering and leaving the systems which are subscripted as feed F, brine B, and distillate D. If it is assumed that the relevant pumps are outside of the system there are only two energy fluxes to be found: heat supply for the production of the vapour (subscript "H") and heat extraction in the condensation of the vapour (subscript "C"). In a first step we would like to make a mass- , a salt - and an energy balance for the system using the following assumptions:

- Pure water is evaporated.
- The temperature of the feed flow is the same as the temperature in the evaporation chamber.
- There is no cooling of the distillate or of the brine.

With these assumptions we obtain the following equations:

1. Mass balance

$$\dot{m}_F = \dot{m}_B + \dot{m}_D \tag{2.27}$$

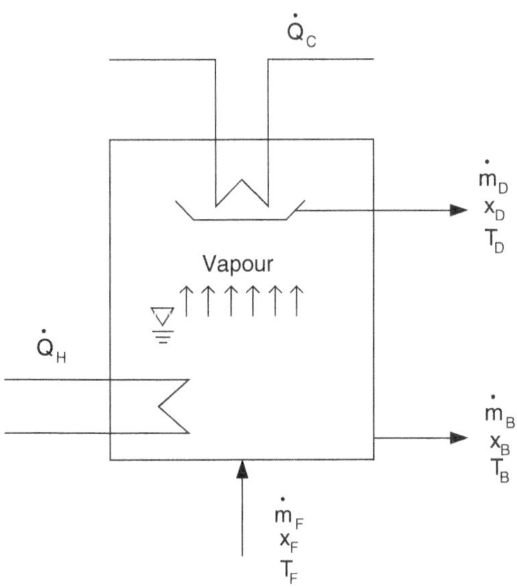

Figure 2.10 Flow scheme of a single- stage evaporation process.

2. Salt balance

$$\dot{m}_F \cdot x_F = \dot{m}_B \cdot x_B \tag{2.28}$$

3. Energy balance (1st rule of thermodynamics)

$$\dot{Q}_H = \dot{Q}_C \tag{2.29}$$

The maximum salt content in the brine is determined by the solubility of the salts. In order to avoid salt precipitation leading to encrustations and blockages (key word: scaling), the concentration in the brine must be limited. The concentration factor CF being defined as the ratio of brine concentration to feed concentration is usually used as the parameter which gives the limitation of the evaporation, i.e. concentration process:

$$CF = \frac{x_B}{x_F} = \frac{\dot{m}_F}{\dot{m}_B} \tag{2.30}$$

It may immediately be derived from the salt balance (Eq. 2.28) that the concentration factor can also be interpreted as the ratio of the feed mass flow to the brine mass flow. As the capacity of a seawater desalination plant is defined by the distillate production, it is helpful for our analysis to introduce the yield or recovery rate φ. Starting with the definition of the concentration factor CF (Eq. 2.30) we obtain:

$$CF = \frac{\dot{m}_F}{\dot{m}_B} = \frac{\dot{m}_F}{\dot{m}_F - \dot{m}_D} = \frac{1}{1 - \frac{\dot{m}_D}{\dot{m}_F}} = \frac{1}{1 - \varphi} \tag{2.31}$$

$$\varphi = \frac{CF - 1}{CF} \tag{2.32}$$

The significance of the concentration factor and the yield may be illustrated by the following example. Assumed figures are:

$x_F = 35{,}000\ mg/l$ (Mediterranean Sea)

$x_B = 60{,}000\ mg/l$ (max. concentration in brine due to scaling)

$$CF = \frac{x_B}{x_F} = \frac{60,000}{35,000} = 1.7$$

$$\varphi = \frac{CF - 1}{CF} = \frac{1.7 - 1}{1.7} = 0.4$$

This means: 40 % of the seawater, which is pumped into the plant, pre-treated and heated up to evaporation temperature, will flow out of the plant as a product, in other words as distillate. 60 % is rejected as brine. For a plant in the Arabian Gulf the situation is different:

$x_F = 42,000\ mg/l$ (Arab Gulf)

$x_B = 60,000\ mg/l$ (max. concentration due to scaling)

$$CF = \frac{x_B}{x_F} = \frac{60,000}{42,000} = 1.43$$

$$\varphi = \frac{CF - 1}{CF} = \frac{1.43 - 1}{1.43} = 0.3$$

Here only 30 % of the seawater can be recovered as distillate, 70 % is rejected as brine. This means extra cost for each m^3 of product due to additional pumping energy and additional chemicals for the pre-treatment.

In conclusion it may be said that: The concentration factor and the yield should be chosen to be as high as possible. A limiting factor here is the salts which tend to produce precipitates, i.e. scaling. A significant increase in both of these parameters can only be achieved in the end by avoiding scaling, for example by the use of chemicals or through the selective removal of the salts which form the salt-crusts (for example using Nanofiltration).

The energy balance for the system represented above gives evidence that the heat which is introduced will also be released in the condenser. This is in line with the statement in the 1st rule of thermodynamics which postulates the conservation of energy. However, for the production of fresh water from seawater the amount of energy supplied is of most particular interest. If only the sub-system "vapour production" is observed, the following balance can be made:

$$\dot{Q}_H = \dot{m}_D \cdot \Delta h_{V,T_V} \tag{2.33}$$

In this case, is Δh_V is the evaporation heat for pure water under evaporation pressure or at evaporation temperature. If evaporation takes place under ambient pressure, then the specific heat demand of single-stage evaporation may be derived as:

$$\frac{\dot{Q}_H}{\dot{m}_D} = \Delta h_{V,100°C} = 2,257\frac{kJ}{kg} = 627\frac{kWh}{t}$$

Now, in order to gradually approach the reality of seawater desalination it is necessary to distance ourselves from the following assumption:

- It is not pure water which will be evaporated but rather salty seawater.
- The feed flow has ambient temperature and has to be pre-heated to evaporation temperature.

The schematic of this extended plant is represented in Figure 2.11. While nothing changes in the mass and salt balance using these assumptions, the heat flow for the pre-heating of the feed to evaporation temperature now appears in the energy balance. In taking this into account it must be noted that the evaporation temperature in the seawater is increased as a result of the saline nature in comparison with pure water. In terms of figures the pre-heating of the feed flow may be divided into the pre-heating to boiling point of the pure water (given by the evaporation pressure in the evaporation chamber) and the additional heat input as a result of the elevation of the boiling point.

$$\dot{Q}_H = \dot{Q}_{Evap} + \dot{Q}_{PH} + \dot{Q}_{BPE} \tag{2.34}$$

$$\dot{Q}_H = \dot{m}_D \cdot \Delta h_{V,T_V} + \dot{m}_F \cdot c_P \cdot (T_V - T_{SW}) + \dot{m}_F \cdot c_P \cdot \Delta T_{BPE} \tag{2.35}$$

$$\dot{Q}_H = \dot{m}_D \cdot \Delta h_{V,T_V} + \dot{m}_F \cdot c_P \cdot (\Delta T_{PH} + \Delta T_{BPE}) \tag{2.36}$$

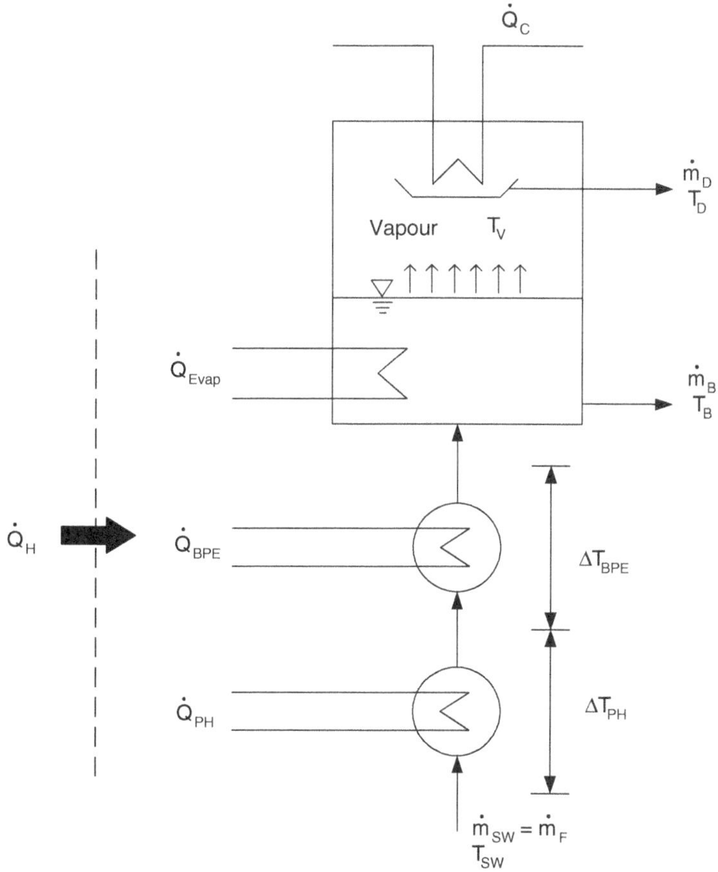

Figure 2.11 Flow scheme of a single-stage evaporation process with seawater pre-heating.

If the concentration factor CF, as defined above, is introduced, this leads to the specific heat demand of a single-stage seawater desalination plant with feed pre-heating of:

$$\frac{\dot{Q}_H}{\dot{m}_D} = \Delta h_{V,T_V} + \frac{CF}{CF-1} \cdot c_P \cdot (\Delta T_{PH} + \Delta T_{BPE}) \qquad (2.37)$$

According to the law of energy conservation the heat which is introduced must also be emitted. This happens on the one hand through the brine having an increased temperature equivalent to the elevation in boiling point. On the other hand the vapour is overheated by the elevation in boiling point. The heat being emitted is therefore composed of evaporation heat and the superheating. In a multiple-effect plant the superheating is reduced by losses on the path to the next stage, i.e. the vapour condenses in the evaporator of the next stage as saturated vapour at the relevant evaporation temperature. Hence the overall heat balance exactly reads as follows:

$$\dot{Q}_H = \dot{Q}_C + \dot{Q}_{Losses}$$

How high the thermal energy demand of such a plant is, may be illustrated in the following example. Given is:

Seawater temperature: $T_{SW} = 20°C$

Evaporation temperature: $T_V = 100°C$

Heat of evaporation: $\Delta h_V = 2257 \frac{kJ}{kg}$

Boiling point elevation: $\Delta T_{BPE} = 1.0\ K$

Concentration factor: $CF = 1.4$

Specific heat capacity: $c_P = 4.0 \frac{kJ}{kgK}$ *(seawater)*

This specification leads to:

$$\frac{\dot{Q}_H}{\dot{m}_D} = 2,257 \frac{kJ}{kg} + \frac{1.4}{1.4-1} \cdot 4.0 \frac{kJ}{kgK} \cdot ((100-20)+1.0)\,K$$

$$\frac{\dot{Q}_H}{\dot{m}_D} = 2,257 \frac{kJ}{kg} + 1,134 \frac{kJ}{kg} = 3,391 \frac{kJ}{kg} = 942 \frac{kWh}{to}$$

For a plant with a capacity of for instance 15,000 to/d, a thermal energy demand of almost 600 MW may thus be computed!

$$\dot{Q}_H = \dot{m}_D \cdot \frac{\dot{Q}_H}{\dot{m}_D} = 15,000 \frac{to}{d} \cdot \frac{1\ d}{24\ h} \cdot 942 \frac{kWh}{to} = 588,750\ kW$$

Without going into the details of the costs of the fuel required to provide this amount of heat, it may be said that the production of fresh water in such a seawater desalination plant would be extremely expensive and thus uneconomic. Energy saving measures must therefore be undertaken. Eq. 2.37 gives a hint of these measures:

- Reduction in the energy demand for pre-heating by recovery of condensation heat
- Reduction in the energy demand by lowering of the evaporation temperature (evaporation under vacuum)
- Reduction in the energy demand for the evaporation using multiple-effects

To 1: Recovery of condensation heat

In Figure 2.12 the flow scheme of a single-stage plant with energy recovery is shown. The seawater flows through the condenser and

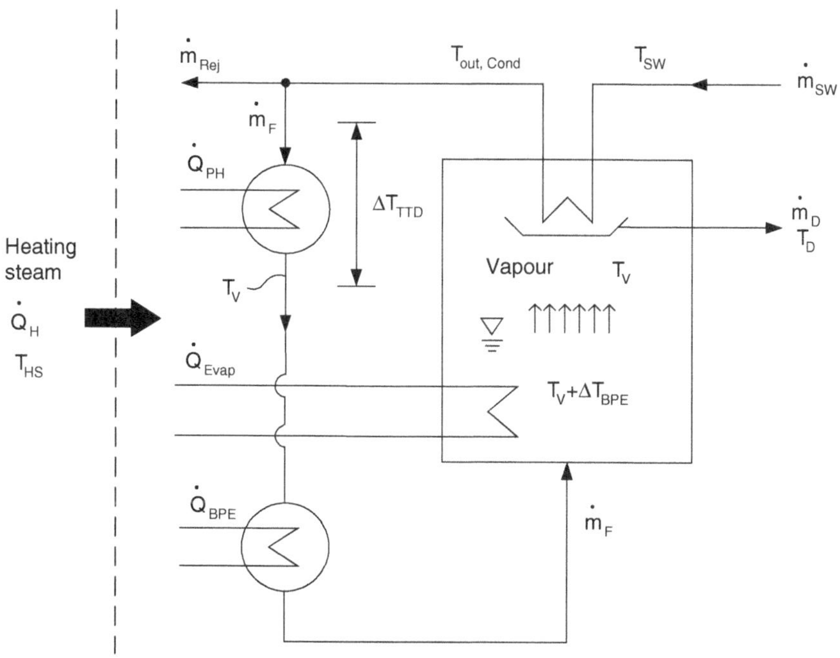

Figure 2.12 Flow scheme of a single-stage evaporation process with energy recovery.

absorbs the latent heat of the vapour. A part of this water leaves the system and flows back into the sea as "reject" ($\dot{m}_{Rej}$), the rest is introduced into the plant as feed (or so-called "make-up"). The heat balance for the condenser is as follows:

$$\dot{m}_D \cdot \Delta h_{V,T_V} = \dot{m}_{SW} \cdot c_P \cdot (T_{out,Cond} - T_{SW}) = (\dot{m}_{Rej} + \dot{m}_F) \cdot c_P \cdot \Delta T_{PH,Cond} \quad (2.38)$$

As the amount of distillate, i.e. the product of the plant should normally be maintained at a constant figure, the reject flow may be adjusted according to the inlet temperature of the seawater (summer or winter operation). The feed flow is coupled to the capacity of the plant through the concentration factor CF and is thus given. The reject flow related to the distillate volume is determined as:

$$\frac{\dot{m}_{Rej}}{\dot{m}_D} = \frac{\Delta h_{V,T_V}}{c_P \cdot \Delta T_{PH,Cond}} - \frac{CF}{CF-1} \quad (2.39)$$

The temperature at the condenser outlet is below the condensation temperature of the vapour as a result of the temperature difference required for the heat transfer (2nd rule of thermodynamics). Figure 2.13 illustrates this process by use of a temperature vs. heat transfer area diagram. The temperature difference at the so-called hot end of the heat exchanger is called Terminal Temperature Difference (ΔT_{TTD}). ΔT_{TTD} is a figure which should be stated in the

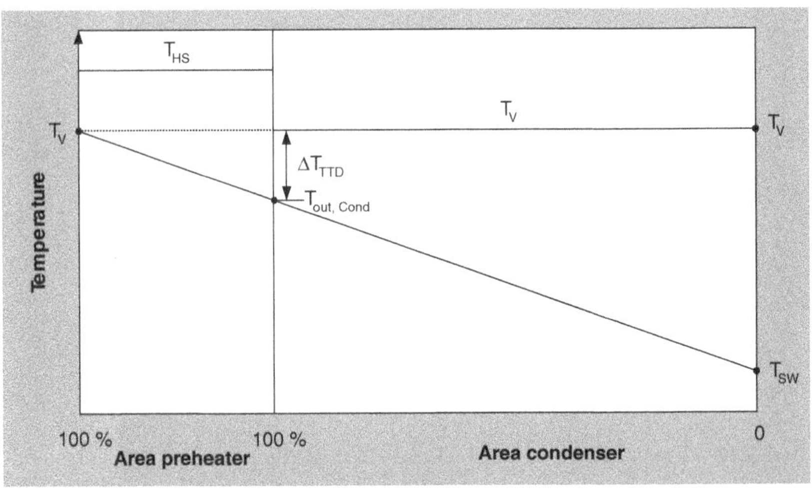

Figure 2.13 Temperature vs. area chart for condenser and preheater.

design phase from both an economic and technical point of view. As may be seen from the chart, the condenser area increases with a decreasing difference in temperature, ΔT_{TTD}. Values for ΔT_{TTD} between 2 and 6 K are appropriate for modern seawater desalination plants.

In order to pre-heat the feed to evaporation temperature an additional external heat flux is needed to close the gap between the temperature at the condenser outlet and the evaporation temperature in the evaporation chamber (see Figure 2.13). Eq. 2.36 and Eq. 2.37 may thus be modified as follows:

$$\dot{Q}_H = \dot{m}_D \cdot \Delta h_{V,T_V} + \dot{m}_F \cdot c_P \cdot (\Delta T_{TTD} + \Delta T_{BPE}) \tag{2.40}$$

and

$$\frac{\dot{Q}_H}{\dot{m}_D} = \Delta h_{V,T_V} + \frac{CF}{CF-1} \cdot c_P \cdot (\Delta T_{TTD} + \Delta T_{BPE}) \tag{2.41}$$

If a terminal temperature difference of 3 K is assumed, the heat demand in a single-stage evaporation plant thus reduces using the values listed above by more than 30 % according to the following calculation:

$$\frac{\dot{Q}_H}{\dot{m}_D} = 2257 \frac{kJ}{kg} + \frac{1.4}{1.4-1} \cdot 4.0 \frac{kJ}{kgK} \cdot (3+1.0)K$$

$$\frac{\dot{Q}_H}{\dot{m}_D} = 2257 \frac{kJ}{kg} + 56 \frac{kJ}{kg} = 2313 \frac{kJ}{kg} = 642,5 \frac{kWh}{to}$$

Despite this reduction the amount of heat is too high at 650 kWh per tonne of fresh water in order to be able to utilise the process commercially. A remedy for this is the use of multiple-effect distillation.

2.2.2 Multiple-Effect Distillation (MED)

Before going into the design of multiple-effect distillation, it should be clearly stated once again that the multiple-effect nature of a seawater desalination plant should not be confused with the multiple-stage nature of a distillation column. In the latter case the multiple-stage nature is necessary in order to be able to generate a pure product, as the vapour pressures of the elements to be separated are close to each other (e.g. water – alcohol). In seawater desalination a pure product, i.e. salt-free water, can be produced in one stage – the multiple-effect nature is solely present in order to minimise the energy requirement!

Figure 2.14 shows the simplified scheme of a multiple-effect evaporation plant. In order to explain the principle of the multiple-effect nature more clearly no feed pre-heating has been shown, i.e. it is assumed that the feed flow, which enters the first stage, has already reached evaporation temperature. It is similarly assumed that the distillate produced in the individual stages will be collected and drained off without any heat recovery. The concentrate flows to the subsequent stage without taking into regard the flashing which will be explained later.

In multiple-effect distillation the succeeding stage serves as condenser for the steam generated in the previous stage. Thereby the first stage is supplied with energy from external sources, normally with heating steam from a boiler or waste steam from a steam turbine. The amount of distillate produced in a plant with N stages is the sum of the distillates of the individual stages, i.e.

$$\dot{m}_D = \dot{m}_{D,1} + \dot{m}_{D,2} + \ldots + \dot{m}_{D,N-1} + \dot{m}_{D,N} \tag{2.42}$$

Disregarding flashing, superheating and losses, the amount of heat transferred in each condenser/evaporator unit simply is the product of distillate mass flow and heat of evaporation:

$$\begin{aligned} \dot{Q}_1 &= \dot{m}_{D,1} \cdot \Delta h_{V,T_{V,1}} \\ \dot{Q}_2 &= \dot{m}_{D,2} \cdot \Delta h_{V,T_{V,2}} \\ &\ldots\ldots \\ \dot{Q}_{N-1} &= \dot{m}_{D,N-1} \cdot \Delta h_{V,T_{V,N-1}} \\ \dot{Q}_N &= \dot{m}_{D,N} \cdot \Delta h_{V,T_{V,N}} \end{aligned} \tag{2.43}$$

According to the law of energy conservation all the heat fluxes are equal, so that the following equation is valid:

$$\dot{Q}_1 = \dot{Q}_2 = \dot{Q}_{N-1} = \dot{Q}_N = \dot{Q}_{Evap} \tag{2.44}$$

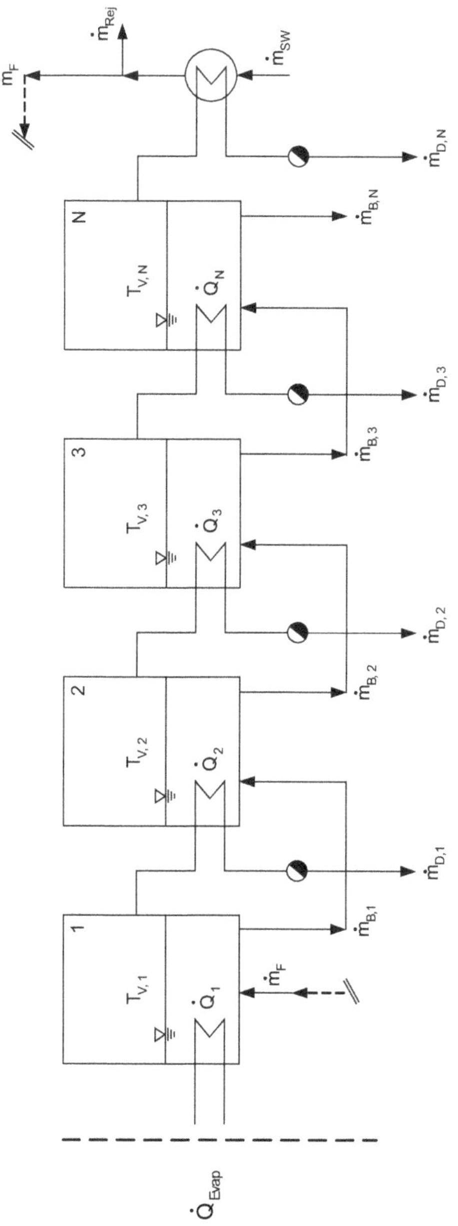

Figure 2.14 Simplified flow sheet of a multiple- effect distillation plant (MED) (feed already preheated, without flashing).

The specific heat demand of a multiple-effect plant may thus be expressed with the help of the Eqs. 2.42 and 2.43 as follows:

$$\frac{\dot{Q}_{Evap}}{\dot{m}_D} = \frac{\dot{Q}_{Evap}}{\dot{m}_{D,1} + \dot{m}_{D,2} + \ldots + \dot{m}_{D,N-1} + \dot{m}_{D,N}}$$

$$\Rightarrow \quad \frac{\dot{Q}_{Evap}}{\dot{m}_D} = \frac{\dot{Q}_{Evap}}{\dfrac{\dot{Q}_1}{\Delta h_{V,T_{V,1}}} + \dfrac{\dot{Q}_2}{\Delta h_{V,T_{V,2}}} + \ldots + \dfrac{\dot{Q}_{N-1}}{\Delta h_{V,T_{V,N-1}}} + \dfrac{\dot{Q}_N}{\Delta h_{V,T_{V,N}}}}$$

With Eq. 2.44 it follows that:

$$\frac{\dot{Q}_{Evap}}{\dot{m}_D} = \frac{1}{\dfrac{1}{\Delta h_{V,T_{V,1}}} + \dfrac{1}{\Delta h_{V,T_{V,2}}} + \ldots + \dfrac{1}{\Delta h_{V,T_{V,N-1}}} + \dfrac{1}{\Delta h_{V,T_{V,N}}}}$$

This equation may be further simplified if an average value for the evaporation heat is introduced. The subsequent result for the specific energy demand of a multiple-effect distillation plant may then be seen as:

$$\frac{\dot{Q}_{Evap}}{\dot{m}_D} = \frac{\Delta h_{V,T_{V,m}}}{N} \text{ with } T_{V,m} = \frac{\sum T_{V,i}}{N} \tag{2.45}$$

Assuming that the average temperature in a multiple-effect distillation plant is 40°C (the temperature profile of the plant will be examined at a later point), and that the number of stages N equals 10, a specific energy demand may be calculated as:

$$\frac{\dot{Q}_{Evap}}{\dot{m}_D} = \frac{\Delta h_{V,T_{V,m}}}{N} = \frac{2407 \ \dfrac{kJ}{kg}}{10} = 2{,}407 \ \frac{kJ}{kg} = 66.86 \ \frac{kWh}{to}$$

The specific thermal energy demand in a multiple-effect distillation plant, as shown in Figure 2.15, is a hyperbolic function of the number of stages, whereby the greatest percentage savings may be made in the step from one to two stages.

At first glance it appears worthwhile, when examining the process from the energy saving point of view, to select the highest possible

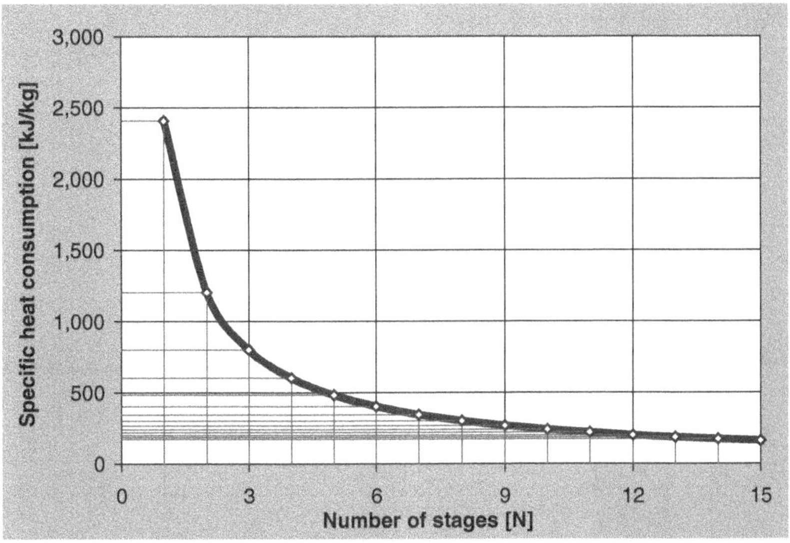

Figure 2.15 Specific heat consumption of a multiple- effect distillation plant.

number of stages. The constraint is given similarly to comparable energy-saving measures, through the investment cost which are needed. (The insulation of a tube should not be so thick that the cost for the insulation material and its fitting consume the cost savings on the energy side.) As every extra stage in a multiple-effect distillation plant increases the investment costs approximately linearly, this results in the optimisation task represented in Figure 2.16 which says: At which number of stages is the sum of investment costs and energy costs at a minimum?

In order to calculate the investment cost of a multiple-effect distillation plant it is necessary to know the size of the heat transfer areas of the evaporator / condenser unit.

Here we would like to move on discussing the following questions:

1. How is the pre-heating of the feed flow to be incorporated into the multiple-effect plant?
2. How can the heat, which is contained in the distillate and brine of the individual stages, be recovered?

It is no coincidence that both questions are linked, because the following paragraph will answer both questions at the same time.

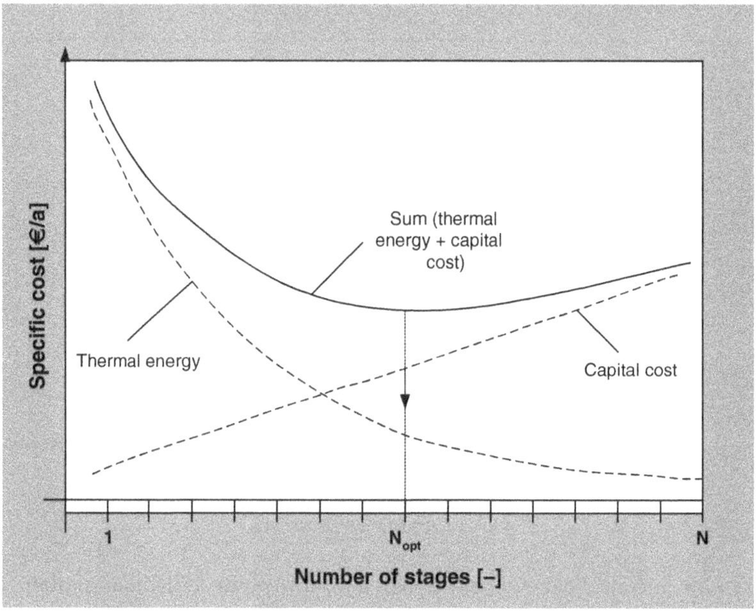

Figure 2.16 Specific energy and investment cost per year vs. number of stages.

The starting point is the schematic of a single-stage evaporation plant with energy recovery (see Figure 2.12). Figure 2.17 shows the logical extension of this single-stage nature into a multiple-effect plant. This arrangement is known in the expert jargon as a "counter current process", as the feed flows counter current to the distillate and brine.

Characteristics of this arrangement are one pre-heater and one evaporator in each stage. Additionally one external pre-heater and one external final condenser are required. The "external" because they are not incorporated into the stages, heated or cooled by an external energy source and therefore usually to be designed separately.

As in the design for a single-stage plant, the external pre-heater exists in order to heat up the feed flow to evaporation temperature. Hence the terminal temperature difference and the boiling point elevation must be bridged.

According to Eq. 2.40 this means:

$$\dot{Q}_{H,PH,1} = \dot{m}_F \cdot c_P \cdot (\Delta T_{TTD} + \Delta T_{BPE}) \tag{2.46}$$

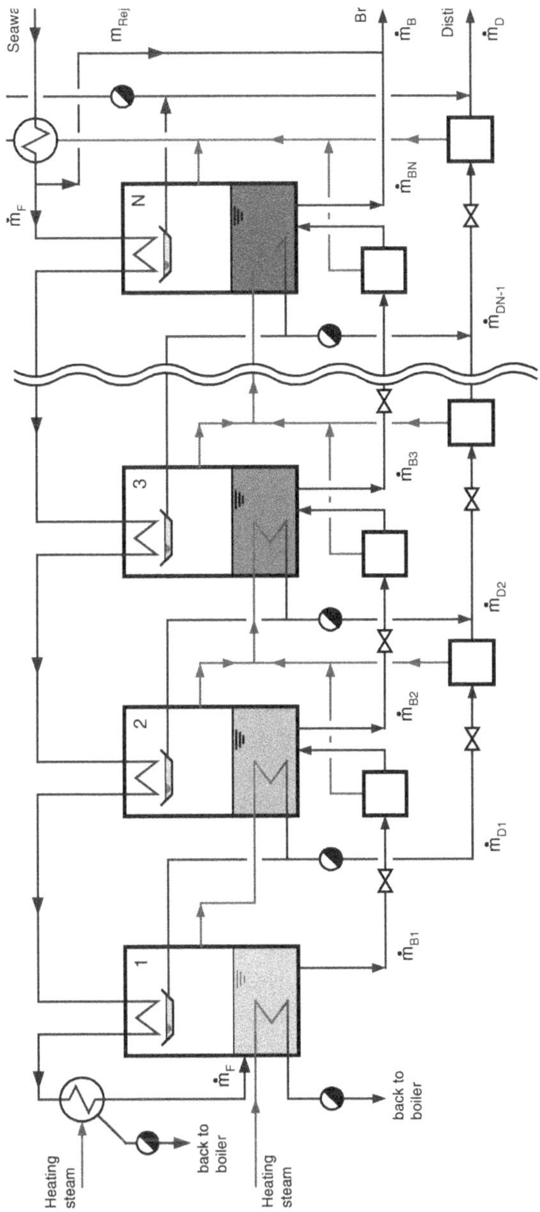

Figure 2.17 Flow sheet of a multiple- effect distillation plant (MED) (Counter current process design).

The heat necessary for the evaporation in the 1st stage is given by Eq. 2.43 as follows:

$$\dot{Q}_{H,E} = \dot{m}_{D,1} \cdot \Delta h_{V,T_{V,1}} \tag{2.47}$$

The vapour mass flow $\dot{m}_{D,1}$ condenses on the surface of the 1st stage pre-heater and also in the 2nd stage condenser / evaporator unit and in turn generates vapour whose volume, however, is reduced by the amount of the flow which condenses in the pre-heater: the multiple-effect principle explained above, by which the distillate flow is equal to the product of the heating steam flow multiplied by the number of stages, is thus incorrect. The solution to this dilemma is that the energy stored in the brine and in the distillate is utilised.

In order to demonstrate this we have to follow the path of both of these flows through the plant. In a multiple-effect plant, the brine and the distillate are fed from stage to stage. Both brine and distillate are leaving the stage as "boiling water". The boiling temperature and the boiling pressure are determined by the conditions in each stage. To transfer the heat in the condenser/evaporator unit a driving temperature difference must be present. This means that both brine and distillate enter a chamber in which a lower pressure and a lower boiling point temperature exist. The only possibility to release this surplus energy present under these conditions is through spontaneous vapour production. This process is known as "flash evaporation" or "flashing".

The process is illustrated systematically in Figure 2.18 in an enthalpy-entropy (h-s) diagram. The flashing process is a process at constant enthalpy between the pressure levels ϕ and $\kappa \rightarrow$ pressure levels 1 and 2. Since state 1 lies in the two-phase region (liquid /vapour) saturated steam 3 and boiling liquid 4 are produced.

Figure 2.19 shows two stages of an evaporation plant with the relevant temperatures and mass flows. If boiling point elevation is neglected the mass and energy balance leads to two equations for the mass flows produced by brine and distillate flashing:

$$\dot{m}_{B,Flash,i} \cdot \Delta h_{V,T_{V,i}} = \dot{m}_{B,i-1} \cdot c_{P,B} \cdot (T_{V,i-1} - T_{V,i})$$

$$\Rightarrow \quad \frac{\dot{m}_{B,Flash,i}}{\dot{m}_{B,i-1}} = \frac{c_{P,B} \cdot (T_{V,i-1} - T_{V,i})}{\Delta h_{V,T_{V,i}}} = \frac{c_{P,B} \cdot \Delta T_{stage,i}}{\Delta h_{V,T_{V,i}}} \tag{2.48}$$

$$\dot{m}_{D,Flash,i} \cdot \Delta h_{V,T_{V,i}} = \dot{m}_{D,i-1} \cdot c_{P,D} \cdot (T_{V,i-1} - T_{V,i})$$

$$\Rightarrow \quad \frac{\dot{m}_{D,Flash,i}}{\dot{m}_{D,i-1}} = \frac{c_{P,D} \cdot (T_{V,i-1} - T_{V,i})}{\Delta h_{V,T_{V,i}}} = \frac{c_{P,D} \cdot \Delta T_{stage,i}}{\Delta h_{V,T_{V,i}}} \tag{2.49}$$

The difference in the boiling temperatures between the stages is known as the stage decrement ΔT_{stage} and is derived from the

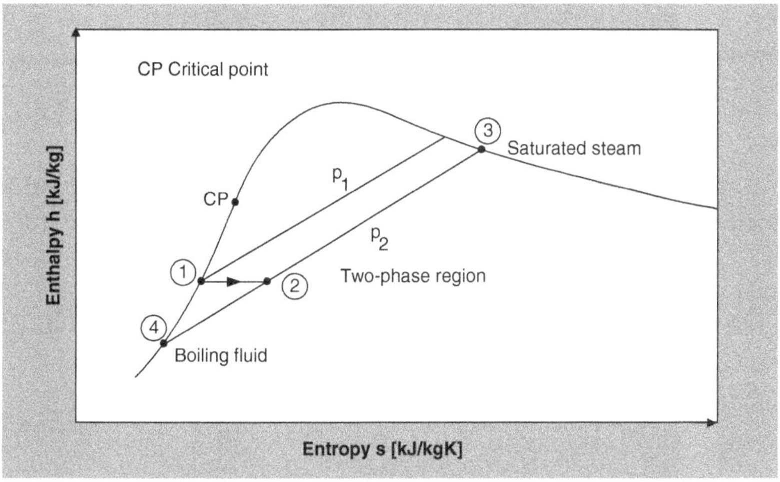

Figure 2.18 Enthalpy-entropy (h-s) diagram for flash evaporation process.

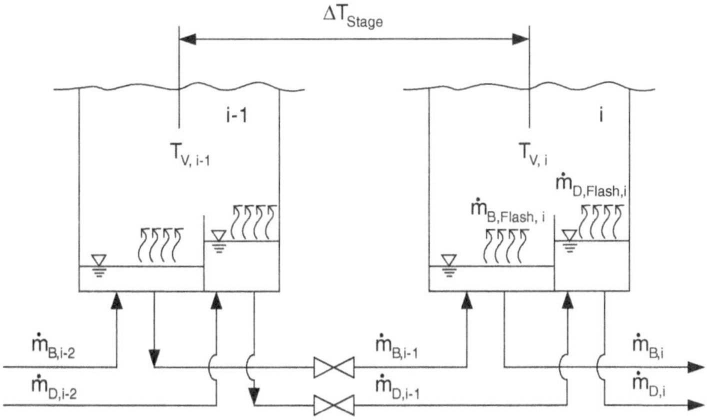

Figure 2.19 Illustration of the flashing process in a thermal desalination plant.

overall temperature difference over the whole plant ΔT_0 and the number of stages:

$$\Delta T_{stage} = \frac{\Delta T_0}{N} \tag{2.49a}$$

The temperature profile of an MED plant is represented in Figure 2.20 with the relevant temperature differences.

The vapour produced by flashing is added to the vapour produced by evaporation, so that the total distillate produced in a stage comprises 3 parts:

$$\dot{m}_{D,i} = \dot{m}_{D,E,i} + \dot{m}_{D,Flash,i} + \dot{m}_{B,Flash,i} \tag{2.50}$$

Assuming that the stage decrement is for instance 3.5 K and taking into consideration a stage with an evaporation temperature of 40°C, the ratio of vapour produced by flashing to the brine flow into stage i may be calculated from Eq. 2.48 as follows:

$$\frac{\dot{m}_{B,Flash,i}}{\dot{m}_{B,i-1}} = \frac{c_{P,B} \cdot \Delta T_{stage,i}}{\Delta h_{V,T_{V,i}}} = \frac{4.0 \frac{kJ}{kg \cdot K} \cdot 3.5\,K}{2{,}407 \frac{kJ}{kg}} = 0.0058 \approx 0.6\,\%$$

This means that only about 0.6 % of the brine evaporates during the flash process. This amount is relatively small in regard to the total mass balance, that is to say the mass of distillate produced, however the vapour produced by flashing is important for the energy balance. While the vapour generated in the evaporation unit $\dot{m}_{D,E,i}$ condenses in the condenser/evaporator unit of the succeeding stage, the flash steam $(\dot{m}_{D,Flash,i} + \dot{m}_{B,Flash,i})$ condenses on the surface of the stage pre-heater.

So that steam is also available for the pre-heating of the feed in the 1st stage, the feed flow must be pre-heated by the stage decrement to be able to flash at the inlet into the 1st stage. This can take place in the external pre-heater. The share of the heat needed for this process is given by:

$$\dot{Q}_{H,PH,2} = \dot{m}_F \cdot c_P \cdot \Delta T_{Stage} \tag{2.51}$$

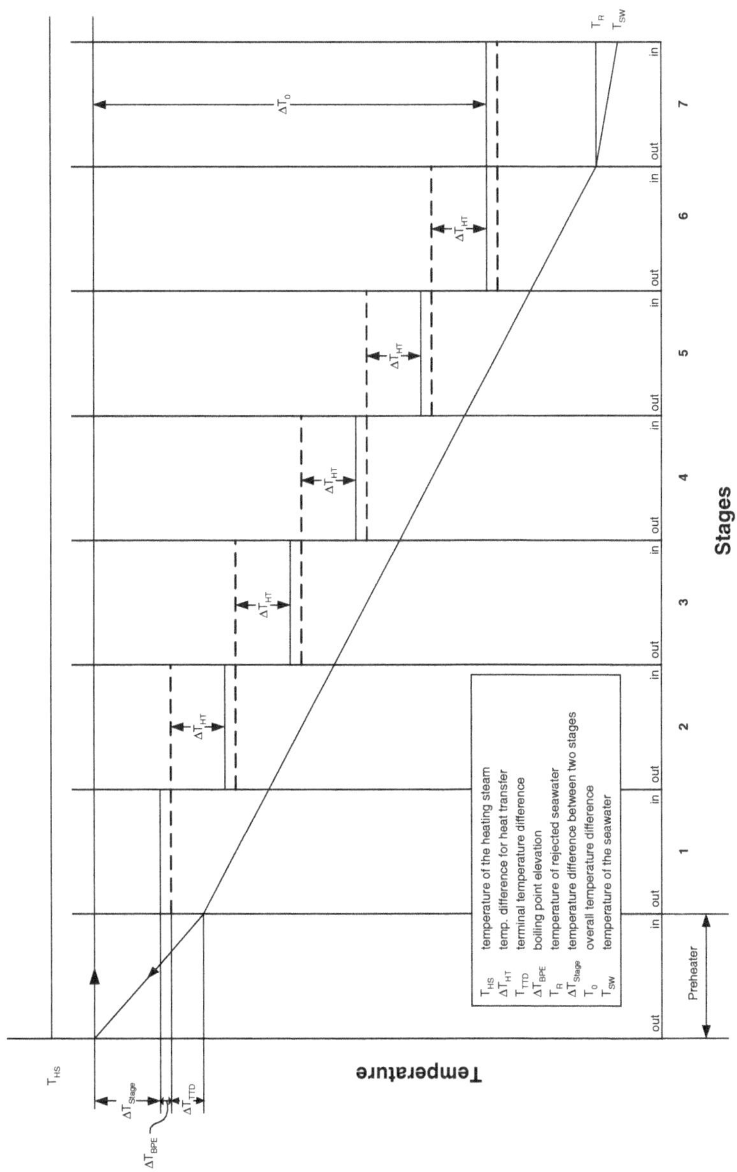

Figure 2.20 Idealized temperature profile of an MED plant (counter current flow).

If all the heat flows which are necessary for the process are added together this gives:

$$\dot{Q}_H = \dot{Q}_{H,E} + \dot{Q}_{H,PH,1} + \dot{Q}_{H,PH,2} \tag{2.52}$$

According to the multiple-effect principle the distillate mass flow generated by evaporation in the individual stages may be calculated approximately from the amount of total distillate and the number of stages. As the volume of distillate produced by flashing is relatively small (see the sample calculation above), this is permissible.

Thus it is possible to arrive to an approximation for the specific heat demand of a multiple-effect distillation plant with counter current flow arrangement:

$$\frac{\dot{Q}_H}{\dot{m}_D} = \frac{\Delta h_{V,T_{V,m}}}{N} + \frac{CF}{CF-1} \cdot c_P \cdot (\Delta T_{Stage} + \Delta T_{TTD} + \Delta T_{BPE}) \tag{2.53}$$

The share which is due to the pre-heating may be calculated in an example as follows:

Stage decrement: $\Delta T_{Stage} = 3.5\ K$

Terminal temperature difference: $\Delta T_{TTD} = 2.0\ K$

Boiling point elevation: $\Delta T_{BPE} = 0.8\ K$

Concentration factor: $CF = 1.4$

$$\Rightarrow \quad \frac{\dot{Q}_{H,PH}}{\dot{m}_D} = \frac{1.4}{1.4-1} \cdot 4.0 \frac{kJ}{kgK} \cdot (3.5 + 2.0 + 0.8)\ K = 88.2 \frac{kJ}{kg}$$

The share which may be ascribed to the evaporation was previously calculated above as:

$$\frac{\dot{Q}_{H,E}}{\dot{m}_D} = \frac{\Delta h_{V,T_{V,m}}}{N} = \frac{2{,}407 \frac{kJ}{kg}}{10} = 240.7 \frac{kJ}{kg}$$

Thus the total heat demand is:

$$\frac{\dot{Q}_H}{\dot{m}_D} = 240.7 \frac{kJ}{kg} + 88.2 \frac{kJ}{kg} = 328{,}9 \frac{kJ}{kg} = 91.36 \frac{kWh}{to}$$

The proportions for the evaporation and for the pre-heating in relation to the total energy demand are 73 % and 27 % respectively.

Occasionally in the literature a rule of thumb is used which does not explicitly formulate the pre-heating but rather takes it into consideration as an exponent for the number of stages, i.e.:

$$\frac{\dot{Q}_{H,E}}{\dot{m}_D} = \frac{\Delta h_{V,T_{V,m}}}{N^{0.85}} \tag{2.54}$$

If we insert the values given in the example above, we arrive at:

$$\frac{\dot{Q}_{H,E}}{\dot{m}_D} = \frac{2,407}{10^{0.85}} \frac{kJ}{kg} = 334 \frac{kJ}{kg} = 92.78 \frac{kWh}{to}$$

The error is smaller than 5 %, so that the rule of thumb is sufficient for a quick calculation, particularly as the energy demand is estimated conservatively, i.e. larger than in the exact calculation.

However on no account should any calculation be made using a formula which uses the number of stages without an exponent, because, as may easily be seen, the result would be 240.7 kJ/kg and thus more than 26 % too low compared to the exact calculation which includes the pre-heating.

At this point, before the mass and energy balance for an MSF plant is discussed in Chap. 2.2.3, the determination of the required heat exchanger area should be examined. We thus return to the optimisation task formulated above regarding the optimum number of stages for an MED plant. It is necessary to calculate the capital costs first. These in turn are directly proportional to the costs for the heat exchanger areas, so that if the number of square metres required is known, this can be projected onto the capital costs[4].

The total heat exchanger area of an MED plant according to Figure 2.16 is divided into 3 parts: the pre-heaters, the evaporators and the final condenser:

$$A_{ME} = A_{PH} + A_E + A_C \tag{2.55}$$

The calculation procedure can start with the following well-known general equation for the heat transfer:

$$\dot{Q} = k \cdot A \cdot \Delta T_{HT}, \tag{2.56}$$

whereby the overall heat transfer coefficient k and the driving temperature difference ΔT_{HT} must be adjusted for each individual case.

The steam coming from the previous stage condenses in the evaporator of the succeeding stage and produces the same amount of steam, i.e. the heat transferred in the evaporator unit of stage i may be expressed as:

$$\dot{Q}_{E,i} = \dot{m}_{D,i} \cdot \Delta h_{V,i} \tag{2.57}$$

If Eq. 2.56 is used to eliminate the heat flow $\dot{Q}$ in Eq. 2.57, then the following term is given for the evaporator area of a stage i :

$$A_{E,i} = \frac{\dot{m}_{D,i} \cdot \Delta h_{V,i}}{k_{E,i} \cdot \Delta T_{HT,i}} \tag{2.58}$$

If it is assumed that the evaporator areas of the individual stages are equal in size, the total area can be calculated by multiplication of the above equation by the number of stages N:

$$A_E = N \cdot A_{E,i} = N \cdot \frac{\dot{m}_{D,i} \cdot \Delta h_{V,i}}{k_{E,i} \cdot \Delta T_{HT,i}} \tag{2.59}$$

Using the following steps this equation may be further simplified:

1. A mean value over the number of stages for the specific evaporation heat and the heat transfer coefficient is introduced:

$$\Delta h_{V,i} = \Delta h_V ; \qquad k_{E,i} = k_E$$

2. The product of the number of stages and the amount of vapour per stage is equal to the total distillate volume (i.e. disregarding flashing):

$$N \cdot \dot{m}_{D,i} = \dot{m}_D$$

3. The driving temperature difference for the heat transfer is derived from the stage temperature difference minus boiling point elevation and is the same for every stage (Eq. 2.49a):

$$\Delta T_{HT} = \Delta T_{HT,i} = \Delta T_{Stage,i} - \Delta T_{BPE,i} = \frac{\Delta T_0}{N} - \Delta T_{BPE}$$

Thus a simple formula is derived for the specific evaporator area:

$$\frac{A_E}{\dot{m}_D} = \frac{\Delta h_V}{k_E \cdot \Delta T_{HT}} = \frac{N \cdot \Delta h_V}{k_E \cdot (\Delta T_0 - N \cdot \Delta T_{BPE})} \tag{2.60}$$

The temperature profile in Figure 2.20 and a temperature vs. heat transfer area diagram such as Figure 2.21 can be used for the determination of the area of a pre-heater. The feed flow is heated up by the stage temperature difference in one pre-heater:

$$Q_{PH,i} = m_F \cdot c_{P,i} \cdot \Delta T_{stage,i} \tag{2.61}$$

With the help of Eq. 2.56 the heat transferred may be explicitly formulated in relation to the area. In this case the mean logarithmic temperature difference must be used as the driving temperature gradient.

$$\dot{Q}_{PH,i} = k_{PH,i} \cdot A_{PH,i} \cdot \Delta T_{HT,i} = k_{PH,i} \cdot A_{PH,i} \cdot \Delta T_{\ln,i} \tag{2.62}$$

The mean logarithmic temperature difference for a stage i is determined according to Figure 2.20 as:

$$\Delta T_{\ln,i} = \frac{\Delta T_{Stage,i}}{\ln \dfrac{\Delta T_{TTD,i} + \Delta T_{BPE,i} + \Delta T_{HT,i}}{\Delta T_{TTD,i}}} \tag{2.63}$$

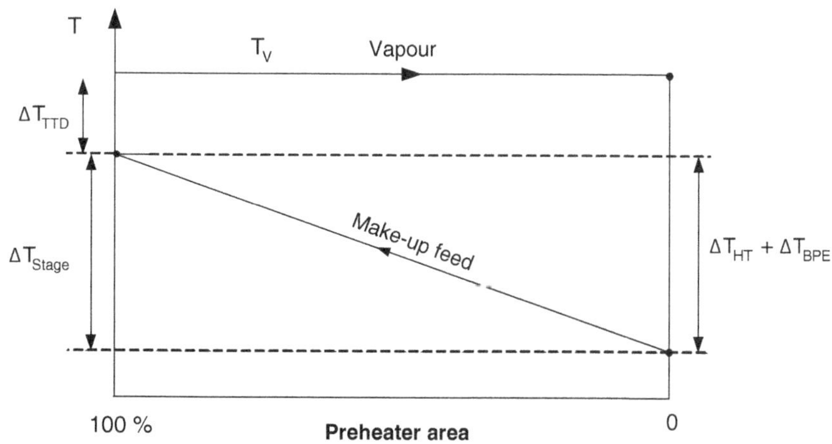

Figure 2.21 Temperature vs. heat transfer area for a pre-heater in an MED plant (counter current flow).

The area of a pre-heater can now be determined from Eqs. 2.61 and 2.62:

$$A_{PH,i} = \frac{\dot{m}_F \cdot c_{P,i} \cdot \Delta T_{stage,i}}{k_{PH,i} \cdot \Delta T_{\ln,i}} \tag{2.64}$$

If this equation is multiplied by the number of stages, then the total pre-heater area of an MED plant (counter current flow) is derived as:

$$A_{PH} = N \cdot A_{PH,i} = \frac{N \cdot \dot{m}_F \cdot c_{P,i} \cdot \Delta T_{stage,i}}{k_{PH,i} \cdot \Delta T_{\ln,i}} \tag{2.65}$$

The equation can now be simplified similarly to the procedure for the evaporator area. In addition the relationship between the feed flow, the distillate flow and the concentration factor as per Eq. 2.31 can be used.

$$\Rightarrow \quad A_{PH} = \dot{m}_D \cdot \frac{CF}{CF-1} \cdot N \cdot \frac{c_P}{k_{PH}} \cdot \frac{\Delta T_{Stage,i}}{\Delta T_{\ln,i}} \tag{2.66}$$

If the mean logarithmic temperature difference as per Eq. 2.63 is now introduced, the specific pre-heater area is computed as:

$$\frac{A_{PH}}{\dot{m}_D} = \frac{CF}{CF-1} \cdot N \cdot \frac{c_P}{k_{PH}} \cdot \ln\left(1+\frac{\Delta T_{BPE} + \Delta T_{HT}}{\Delta T_{TTD}}\right) \tag{2.67}$$

or

$$\frac{A_{PH}}{\dot{m}_D} = \frac{CF}{CF-1} \cdot N \cdot \frac{c_P}{k_{PH}} \cdot \ln\left(1+\frac{\Delta T_0}{N \cdot \Delta T_{TTD}}\right) \tag{2.68}$$

It should be noted that the external pre-heater, which is supplied by heating steam from the boiler, is also included in this equation through the definition of the overall temperature difference, but not the final condenser.

When designing the final condenser the varying temperature of the seawater in the course of the year must be taken into account. Normally allowance is made for the change in the seawater temperature through a change in the mass flow which flows through the final condenser as a coolant. Figure 2.22 shows a typical flow chart. The layout can be made on the basis of the following calculation scheme:

$$\dot{Q}_C = \dot{m}_{SW} \cdot c_{P,SW} \cdot (T_{SW,out} - T_{SW,in}) = k_C \cdot A_C \cdot \Delta T_{ln}$$

$$\Rightarrow A_C = \frac{\dot{m}_{sw} \cdot c_{P,SW} \cdot \left(T_{SW,out} - T_{SW,in}\right)}{K_c \cdot \Delta T_{ln}}$$

with

$$\Delta T_{ln} = \frac{T_{SW,out} - T_{SW,in}}{\ln\left(\dfrac{T_v - T_{SW,in}}{T_v - T_{SW,out}}\right)}$$

Thus the equations for the determination of the heat exchanger areas needed in an MED plant (counter current mode) are given. For an estimation of the capital costs in the preliminary design phase it is of assistance if the set of formulae is further simplified. This takes place in a way similar to that for the specific energy consumption, through taking into account the pre-heating by a correction factor. If the areas are calculated which are required for the evaporation, then this results in more than 90% of the total area

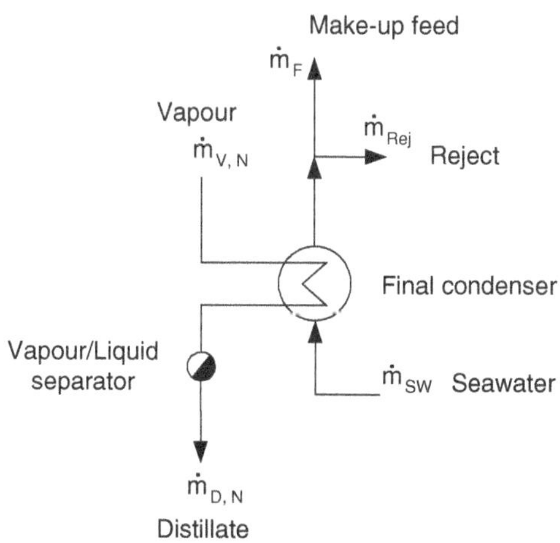

Figure 2.22 Simplified flow sheet of the final condenser.

being required for the evaporation and the pre-heating (including the final condenser).

The formula can therefore be simplified as:

$$A_{ME} = A_E + A_{PH} + A_C$$

$$A_{ME} = 1.1 \cdot A_E$$

Thus the following equations may be used for a quick and simple determination of the specific energy consumption and the specific area of an MED plant (Eqs. 2.54 and 2.60):

$$\frac{\dot{Q}_{H,E}}{\dot{m}_D} = \frac{\Delta h_{V,T_{V,m}}}{N^{0.85}} \tag{2.54a}$$

$$\frac{A_E}{\dot{m}_D} = 1.1 \cdot \frac{\Delta h_V}{k_E \cdot \Delta T_{HT}} = 1.1 \cdot \frac{N \cdot \Delta h_V}{k_E \cdot (\Delta T_0 - N \cdot \Delta T_{BPE})} \tag{2.60a}$$

Figure 2.23 shows the specific heat consumption and the specific area as a function of the number of stages for the following sample values:

Overall temperature difference $\Delta T_0 = 30\ K$

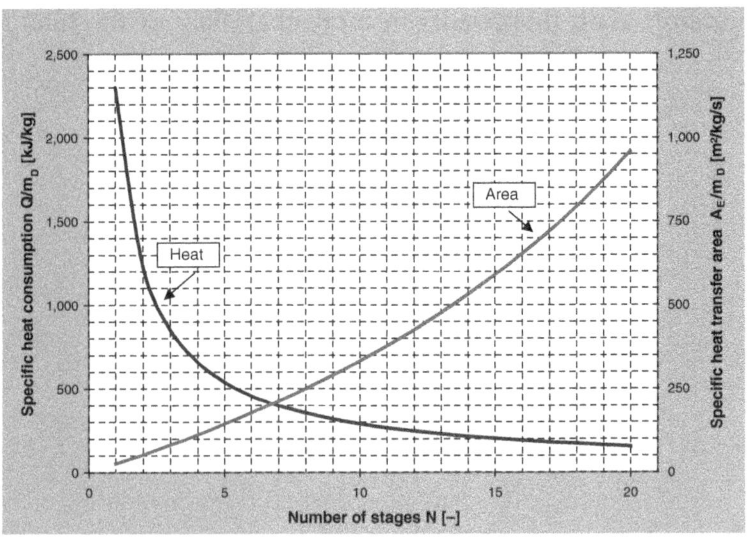

Figure 2.23 Specific heat consumption and specific area of an MED plant as a function of the number of stages (Eq. 2.54a, Eq. 2.60a).

Heat of evaporation $\Delta h_V = 2,300 \frac{kJ}{kg}$

Overall heat transfer coefficient $k_E = 3.0 \frac{kW}{m^2 \cdot K}$

Boiling Point Elevation $\Delta T_{BPE} = 0.7\ K$

The different shape of the two curves (with an increased number of stages the heat consumption decreases and the area increases) indicates that there must be an optimal number of stages when the sum of the energy costs (directly related to the heat consumption) and the capital costs (directly related to the area required) is calculated[4].

2.2.3 Multi-Stage-Flash – Evaporation (MSF)

The flow sheet of a Multi-Stage-Flash (MSF) evaporation plant in the so-called once-through mode is illustrated in Figure 2.24. In this mode the feed flows through the pre-heaters and enters the first stage where the flash process starts. The flashing progresses from stage to stage. The vapour generated condenses on each of the pre-heaters; the distillate is collected, led into the next stage and removed from the last stage as a product. The concentrated brine flow is rejected from the last stage into the sea.

An MSF plant may also be operated in the so-called brine-recycle mode. This will be discussed in more detail later. The kind of operation mode is not significant for the calculation of the heat which has to be supplied. From the flow sheet it may be recognised that only one external pre-heater is present in an MSF plant. This pre-heater is usually designated as a brine heater. An externally heated evaporator or evaporators in the individual stages are not present – in contrast to the multiple-effect distillation plant.

Figure 2.25 shows the temperature profile of an MSF plant across the stages. The seawater enters the final condenser, absorbs the heat of the condensing vapour from the last stage. A part of the incoming seawater is rejected back into the sea. The remainder, the so-called make-up feed or simply make-up flows through the pre-heaters of the stages to the top of the plant. In principle this is a long tube in

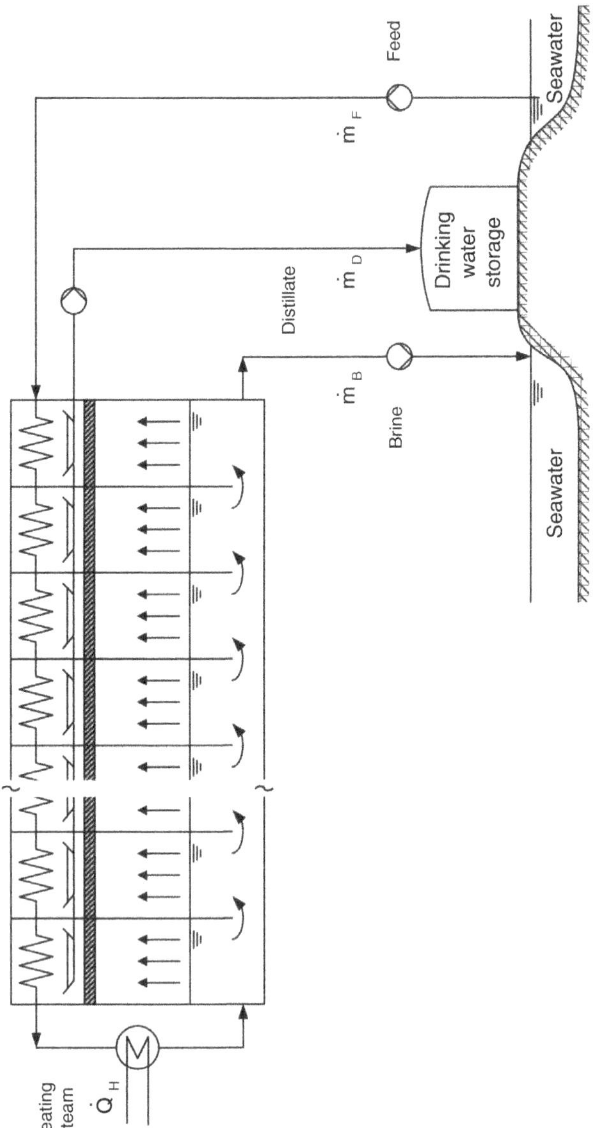

Figure 2.24 Flow sheet of a Multi-Stage-Flash (MSF) evaporation plant (Once through mode) .

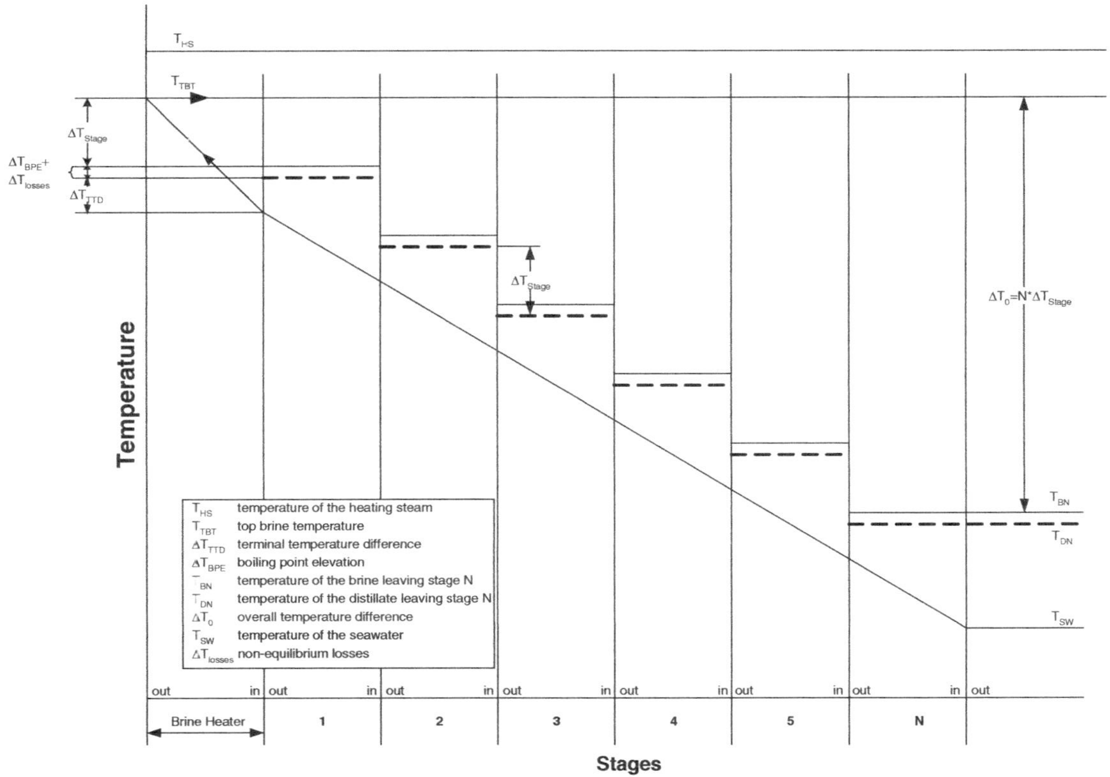

Figure 2.25 Idealized temperature profile of an Multi-Stage- Flash evaporation plant (once through mode).

which the feed is successively heated up through the condensing vapour produced by brine flashing in the chambers. In the brine heater the feed flow is then raised to the temperature level required.

An energy balance for the brine heater gives the following result for the heat flow which needs to be supplied:

$$\dot{Q}_H = \dot{m}_F \cdot c_P \cdot (\Delta T_{Stage} + \Delta T_{TTD} + \Delta T_{BPE} + \Delta T_{Losses}) \tag{2.69}$$

As with the MED plant several temperature differences must be considered:

ΔT_{Stage} Stage temperature difference, or stage decrement in order to have a driving force for the flash evaporation in the 1st stage.

ΔT_{TTD} Terminal Temperature Difference, because a temperature gradient must exist at the pre-heater outlet in order for the heat to be transferred

ΔT_{BPE} Boiling Point Elevation, because of the salt content of the seawater

ΔT_{Losses} Losses, because as a result of the finite length of the chamber the flash process cannot continue until equilibrium is reached (non-equilibrium losses)

The temperature is reduced from stage to stage by flashing, by boiling point elevation and by non-equilibrium losses. The brine leaves the plant at the temperature of the last stage.

The difference in the highest temperature at the inlet of the first stage, the so-called Top Brine Temperature and the brine temperature of the final stage is designated as the Overall Temperature Difference:

$$\Delta T_O = T_{TOP} - T_{B,N} \tag{2.70}$$

If we divide the Overall Temperature Difference by the number of stages we obtain the stage temperature difference or stage decrement:

$$\Delta T_{Stage} = \frac{\Delta T_O}{N} \tag{2.71}$$

As derived in Eq. 2.48 the mass of vapour produced by flashing, that means the distillate produced, can be calculated from the mass flow of the boiling liquid, the temperature difference from stage to stage and the heat of evaporation.

For the whole plant we obtain:

$$\frac{\dot{m}_D}{\dot{m}_F} = \frac{c_P \cdot \Delta T_O}{\Delta h_{V,Tm}} \tag{2.72}$$

If this equation is inserted into Eq. 2.69, the specific heat demand becomes:

$$\frac{\dot{Q}_H}{\dot{m}_D} = \frac{\Delta h_{V,T_m}}{c_P \cdot \Delta T_O} \cdot c_P \cdot (\Delta T_{Stage} + \Delta T_{TTD} + \Delta T_{BPE} + \Delta T_{Losses})$$

and also with Eq. 2.71:

$$\frac{\dot{Q}_{H,MSF}}{\dot{m}_D} = \frac{\Delta h_{V,T_m}}{N} \cdot (1 + N \cdot \frac{\Delta T_{TTD} + \Delta T_{BPE} + \Delta T_{Losses}}{\Delta T_O}) \tag{2.73}$$

For comparison here once more the equation for the specific heat demands of an MED plant as per Eq. 2.53:

$$\frac{\dot{Q}_{H,ME}}{\dot{m}_D} = \frac{\Delta h_{V,T_{V,m}}}{N} + \frac{CF}{CF-1} \cdot c_P \cdot \frac{\Delta T_O}{N} (1 + N \cdot \frac{\Delta T_{TTD} + \Delta T_{BPE}}{\Delta T_O}) \tag{2.74}$$

As the flashing in an MED plant is only a "by-product" and the size of the chamber is determined by the size of the evaporator, the losses through non-equilibrium are not important. Therefore, such as term as " ΔT_{Losses} " does not occur. Furthermore the stage temperature difference was replaced as in an MSF plant by the Overall Temperature Difference divided by the number of stages (see also Figure 2.21).

The equations for MED and MSF are similar only at first sight. An example can help to be able to recognise the difference. Figure 2.26 shows a comparison for the following given data:

Overall temperature difference	$\Delta T_O = (70 - 35)\ K = 35\ K$
Terminal temperature difference	$\Delta T_{TTD} = 2.0\ K$
Boiling point elevation	$\Delta T_{BPE} = 0.8\ K$
Non-equilibrium losses	$\Delta T_{Losses} = 0.5\ K$
Concentration factor	$CF = 1.4$

Specific heat capacity $c_p = 4.0 \frac{kJ}{kgK}$

Heat of evaporation $\Delta h_{V,T_m} = 2,376 \frac{kJ}{kg}$

We can see that with the number of stages the MED plant becomes more and more efficient concerning the specific heat consumption in comparison with the MSF plant. In any case this does not mean that it is always the best solution to build an MED plant with many stages, because for finding the best solution a calculation of the economic efficiency (energy cost + capital cost) has to be carried out (see Figure 2.16). For example, we have to take into account that an effect of an MED plant is much more expensive than an MSF stage due to the evaporators in each stage.

Figure 2.27 shows the flow sheet of an MSF plant in brine-recycle mode. This option differentiates itself from the once-through mode as a brine flow is circulated. One of the main advantages of brine-recycle mode is that the costs for the pre-treatment can be reduced. Pre-treatment[4] typically involves degassing and conditioning with

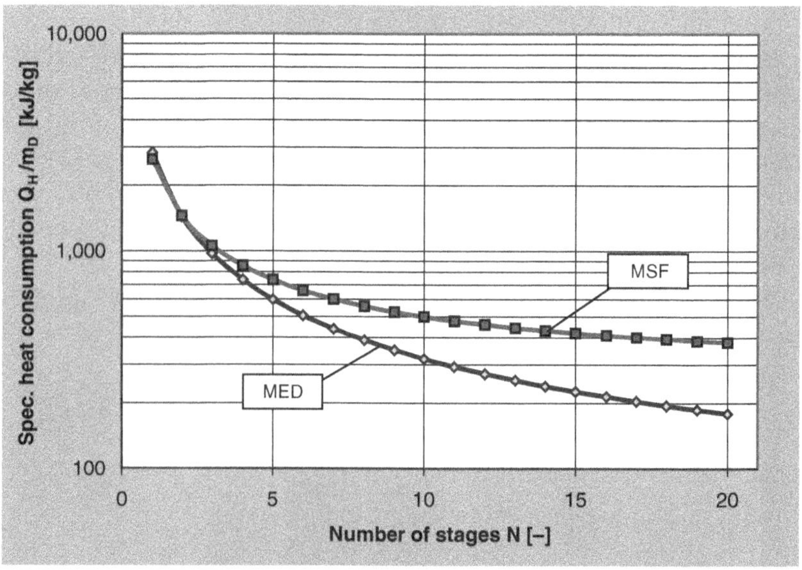

Figure 2.26 Comparison of the specific heat consumption of an MSF with an MED plant as a function of number of stages.

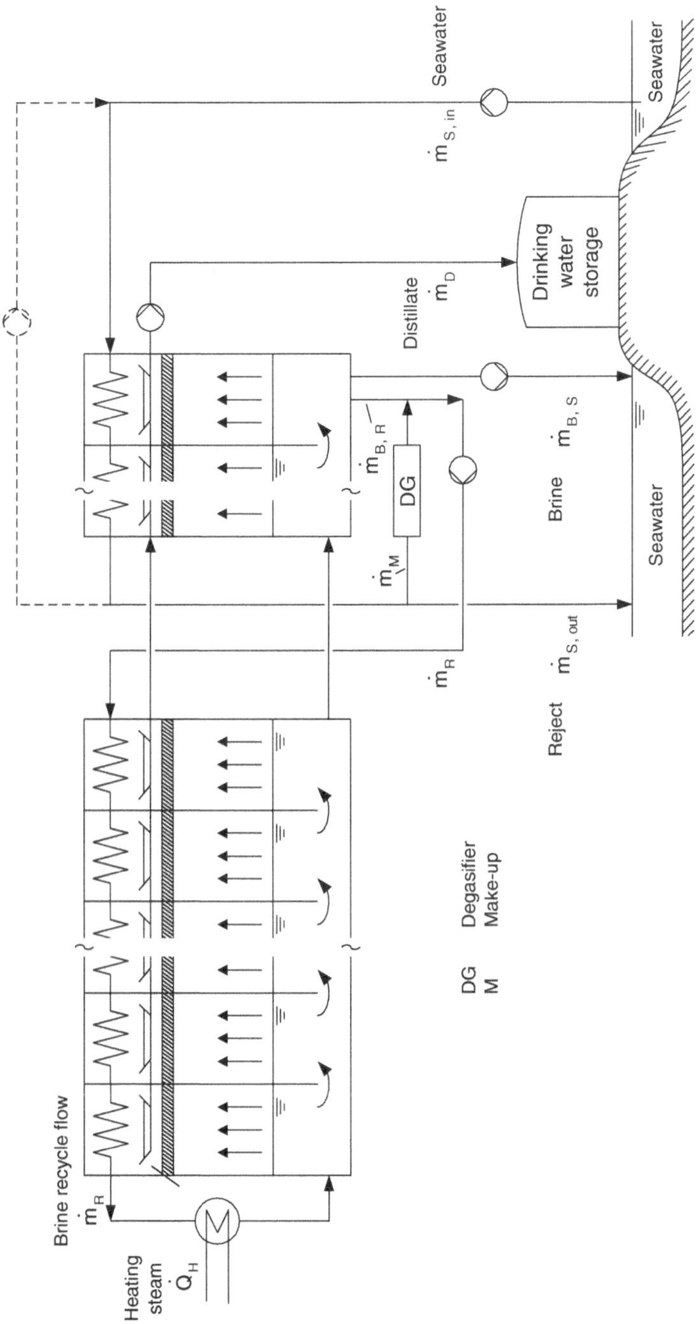

Figure 2.27 Multi-Stage-Flash evaporation plant in brine-recycle mode.

anti-scaling and anti-foaming chemicals. While in the once-through process the whole of the feed flow must be pre-treated, in the brine-recycle mode only the so-called make-up flow is treated. Significant cost savings can be achieved. A further advantage is the flexible operation of the plant through the additional degree of freedom given in the recycling.

There is no meeting of seawater desalination experts at which the advantages and disadvantages of both of the processes are not controversially discussed. Factually the brine-recycle mode has the following disadvantages:

- a higher energy demand for the large brine-recycle pump
- a high number of pumps and pipes with the disadvantage to be susceptible to faults as well as the requirement for maintenance and repair
- a complex lay out with all the associated difficulties

The advantages and disadvantages of the once-through process are:

- high costs for the pre-treatment
- a low level of flexibility in operation
- reduced thermal efficiency by rejection of hot brine (40 °C)
- a lower amount of pumps and pipes and thus lower maintenance and repair costs
- a simple and reliable operation
- overall a higher level of availability

A conclusive evaluation cannot be given at this point. However, when selecting the optimum process in a specific project both alternatives should always be examined along with the prevailing conditions under the heading "MSF".

The specific heat demand of a brine-recycle plant does not differentiate itself from that of a once-through plant. However, the following should be observed:

1. the mass flow passing through the brine heater is the brine recycle flow,
2. based on the temperature profile for the brine-recycle plant shown in Figure 2.28 the recovery and the rejection part of the plant have to be used as an effective driving overall temperature difference for the plant.

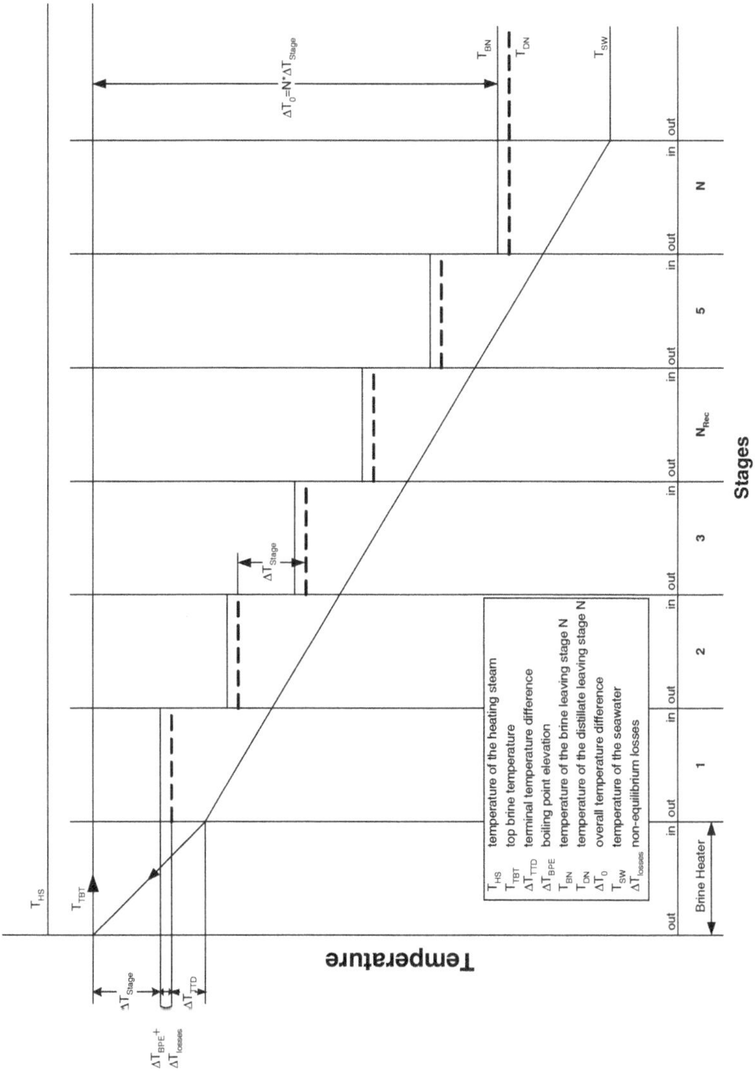

Figure 2.28 Idealized temperature profile of an Multi-Stage-Flash evaporation plant (brine-recycle mode).

Thus the following is arrived at with the help of Eqs. 2.71, 2.72 and 2.73:

Brine-recycle mode:

$$\frac{\dot{m}_D}{\dot{m}_R} = \frac{c_P \cdot \Delta T_O}{\Delta h_{V,Tm}} \tag{2.75}$$

$$\Delta T_{Stage} = \frac{\Delta T_O}{N} \tag{2.76}$$

$$\frac{\dot{Q}_{H,MSF}}{\dot{m}_D} = \frac{\Delta h_{V,T_m}}{N} \cdot (1 + N \cdot \frac{\Delta T_{TTD} + \Delta T_{BPE} + \Delta T_{Losses}}{\Delta T_O}) \tag{2.77}$$

As briefly mentioned above, the brine-recycle plant has an additional degree of freedom through the recycling in comparison with the once-through plant. Through this the balances for mass, salt and energy are more complicated. Based on the relationships and representations in the flow sheet (Figure 2.27) the following set of equation is derived:

Mass balance

Overall $\quad \dot{m}_{S,in} - \dot{m}_{S,out} - \dot{m}_{B,S} - \dot{m}_D = 0 \quad$ (1)

Mixing Point (1) $\quad \dot{m}_{S,in} - \dot{m}_{S,out} - \dot{m}_M = 0 \quad$ (2)

Mixing Point (2) $\quad \dot{m}_{B,R} + \dot{m}_M - \dot{m}_R = 0 \quad$ (3)

Mixing Point (3) $\quad \dot{m}_{B,N} - \dot{m}_{B,S} - \dot{m}_{B,R} = 0 \quad$ (4)

(2) in (1) $\quad \dot{m}_D = \dot{m}_M - \dot{m}_{B,S} \quad$ (5)

(3) and (6) $\rightarrow \quad \dot{m}_D = \dot{m}_R - \dot{m}_{B,N} \quad$ (6)

Material balance (salt balance)

from (5) $\rightarrow \quad 0 = \dot{m}_M \cdot x_M - \dot{m}_{B,S} \cdot x_{B,N}$

from (6) $\rightarrow \quad 0 = \dot{m}_R \cdot x_R - \dot{m}_{B,N} \cdot x_{B,N}$

Two different concentration factors may be defined from these equations:

$$CF_M = \frac{x_{B,N}}{x_M} = \frac{\dot{m}_M}{\dot{m}_{B,S}} \tag{7}$$

$$CF_R = \frac{x_{B,N}}{x_R} = \frac{\dot{m}_R}{\dot{m}_{B,N}} \tag{8}$$

from (7) $\rightarrow \quad \dot{m}_M = CF_M \cdot \dot{m}_{B,S}$

from (6) $\rightarrow \quad \dot{m}_M = CF_M \cdot (\dot{m}_M - \dot{m}_D)$

$$\dot{m}_M = \frac{CF_M}{CF_M - 1} \cdot \dot{m}_D \tag{9}$$

The relationship of the make-up mass flow to the distillate flow is designated as the "make-up factor", i.e.:

$$\beta = \frac{\dot{m}_M}{\dot{m}_D} = \frac{CF_M}{CF_M - 1} \tag{10}$$

Energy balance

Brine heater $\quad \dot{Q}_H = \dot{m}_R \cdot c_P \cdot \Delta T_{BH}$ (11)

Flashing $\quad \dot{m}_D \cdot \Delta h_{V,T_m} = \dot{m}_R \cdot c_P \cdot \Delta T_0$ (12)

Thus a set of equations is now available, with the help of which a complete balance for a MSF plant in brine-recycle mode is possible. The procedure is as follows:

Given:

$\dot{m}_D$	Distillate mass flow = product
$x_M = x_{SW}$	Salt concentration of the make-up flow = salt concentration of the seawater
$x_{B,N}$	Maximum concentration in the final stage
	(Given by the solubility of the salts)
ΔT_O	Overall Temperature Difference
T_{TBT}	Top Brine Temperature

Solution algorithm:

1. Eq. 12 $\dot{m}_R = \dfrac{\dot{m}_D \cdot \Delta h_{V,T_m}}{c_P \cdot \Delta T_O}$

2. Eq. 7 $CF_M = \dfrac{x_{B,N}}{x_M}$

3. Eq. 9 $\dot{m}_M = \dfrac{CF_M}{CF_M - 1} \cdot \dot{m}_D$

4. Eq. 10 $\beta = \dfrac{\dot{m}_M}{\dot{m}_D}$

5. Eq. 5 $\dot{m}_{B,S} = \dot{m}_M - \dot{m}_D$

6. Eq. 6 $\dot{m}_{B,N} = \dot{m}_R - \dot{m}_D$

7. Eq. 4 $\dot{m}_{B,R} = \dot{m}_{B,N} - \dot{m}_{B,S}$

8. Eq. 8 $x_R = x_{B,N} \cdot \dfrac{\dot{m}_{B,N}}{\dot{m}_R}$

In the following section the required condenser area for an MSF plant will now be calculated. Because, analogous to the MED plant an optimal number of stages, must also be derived for MSF plant. However this can only be done by analysing the costs and for this the required area must be known. The equations for the heat transfer in a pre-heater/condenser in an MSF plant are the same for those in an MED plant. This results in the following procedure in line with Figure 2.21:

$$\dot{Q}_{MSF,i} = \dot{m}_{D,i} \cdot \Delta h_{V,i} \tag{2.78}$$

$$\dot{Q}_{MSF,i} = k_{MSF,i} \cdot A_{MSF,i} \cdot \Delta T_{\ln,i} \tag{2.79}$$

From this the condenser area for a stage i is given as:

$$\Rightarrow \quad A_{MSF,i} = \frac{\dot{m}_{D,i} \cdot \Delta h_{V,i}}{k_{MSF,i} \cdot \Delta T_{\ln,i}} \tag{2.80}$$

The condenser area of the whole plant (without the brine heater) can be calculated by multiplying Eq. 2.80 by the total number of stages:

$$A_{MSF} = N \cdot A_{MSF,i} = \frac{N \cdot \dot{m}_{D,i} \cdot \Delta h_{V,i}}{k_{MSF,i} \cdot \Delta T_{\ln,i}} \tag{2.81}$$

This equation may be further modified in the following steps:

1. A mean value for the heat of evaporation and the heat transfer coefficient over the stages is introduced:

$$\Delta h_{V,i} = \Delta h_V ; \qquad k_{MSF,i} = k_{MSF}$$

2. The product of the number of stages and the amount of distillate per stage is equal to the total mass of distillate:

$$N \cdot \dot{m}_{D,i} = \dot{m}_D$$

3. The mean logarithm temperature difference to be used for the heat transfer is the same for every stage i (Eq. 2.63):

$$\Delta T_{\ln} = \frac{\Delta T_0}{N \cdot \ln(1 + \frac{\Delta T_0}{N \cdot \Delta T_{TTD}})} \tag{2.82}$$

With these assumptions the total specific condenser area of an MSF plant is:

$$\frac{A_{MSF}}{\dot{m}_D} = \frac{\Delta h_V}{k_{MSF}} \cdot N \cdot \frac{\ln(1 + \frac{\Delta T_0}{N \cdot \Delta T_{TTD}})}{\Delta T_0} \tag{2.83}$$

The area required in an MSF plant can now be put in relation to the energy demand. For this reference is made to Eq. 2.77. This reads, slightly modified:

$$\frac{\dot{Q}_{H,MSF}}{\dot{m}_D} = \Delta h_{V,T_m} \cdot (\frac{1}{N} + \frac{\Delta T_{TTD} + \Delta T_{BPE} + \Delta T_{Losses}}{\Delta T_0}) \tag{2.84}$$

The so-called Gained Output Ratio (GOR) is defined as a sort of efficiency measure of a thermal seawater desalination plant. It is assumed that the energy supply for the plant uses steam from a boiler or a turbine. The GOR is defined as the ratio of the distillate produced to heating steam provided:

$$GOR = \frac{\dot{m}_D}{\dot{m}_{HS}}$$

If an average heat of evaporation is assumed then the heat introduced into the MSF plant may be expressed as:

$$\dot{Q}_{H,MSF} = \dot{m}_{HS} \cdot \Delta h_{V,T_m} \tag{2.85}$$

If this equation is entered into Eq. 2.84, the following term for the GOR is computed:

$$GOR = \left[\frac{1}{N} + \frac{\Delta T_{TTD} + \Delta T_{BPE} + \Delta T_{Losses}}{\Delta T_O} \right]^{-1} \tag{2.86}$$

As well as the number of stages N the equations for the specific area (Eq. 2.83) and for the GOR (Eq. 2.86) contain the terminal temperature difference ΔT_{TTD} as a free parameter: the system consists therefore of two equations with two unknowns and can be solved using a substitution procedure, resulting in GOR as a function of the specific area and the number of stages:

$$GOR = f\,(N, \frac{A_{MSF}}{\dot{m}_D})$$

The run of this function is shown in Figure 2.29 for the following sample values.

Overall temperature difference: $\Delta T_O = 70\ K$

Heat of evaporation: $\Delta h_V = 2{,}340\ \frac{kJ}{kg}$

Overall heat transfer coefficient: $k_{MSF} = 3.0\ \frac{kW}{m^2 \cdot K}$

Boiling point elevation: $\Delta T_{BPE} = 0.48\ K$

Non-equilibrium losses: $\Delta T_{Losses} = 0.2\ K$

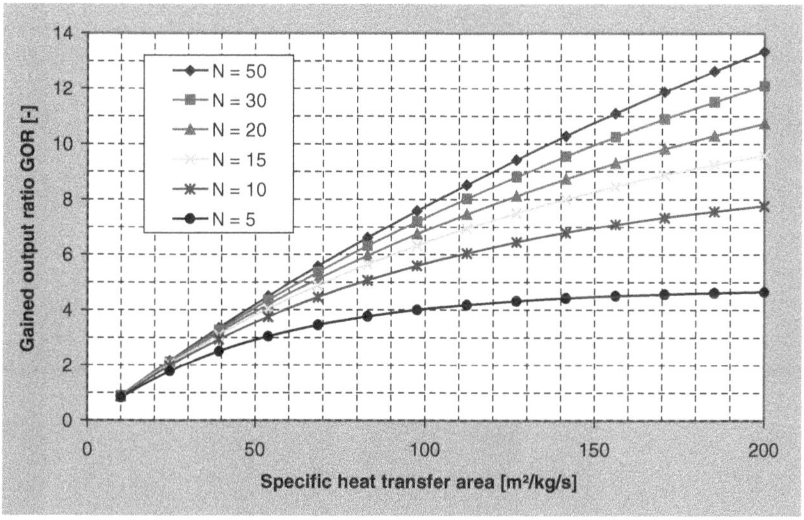

Figure 2.29 Gained output ratio of an MSF plant vs. specific condenser area (parameter: number of stages N).

The resultant set of curves has far-reaching consequences for the design of an MSF plant. When a GOR is specified, then a correct solution can be obtained from the diagram at various numbers of stages and various condenser areas. We only can obtain the actual number of stages and the condenser area by a cost calculation.

2.2.4 Multiple-Effect Distillation with Thermally Driven Vapour Compression

The various possibilities to provide the necessary heat for the evaporation process with each of their advantages and disadvantages will be examined more closely in Chap. 2.3. If steam at a relatively high temperature and pressure level is available then thermal vapour compression (TVC) can be utilised to improve the efficiency of the seawater desalination plant.

Figure 2.30 shows the flow sheet of a multiple-effect distillation plant with thermal vapour compression. The compressor takes in part of the vapour from the last stage, compresses it with the help of the motive steam and adds the mixture to the 1st stage as heating steam. Normally the motive steam will be generated in a steam boiler, which means that the mass flow of motive steam is passed back to the boiler as condensate.

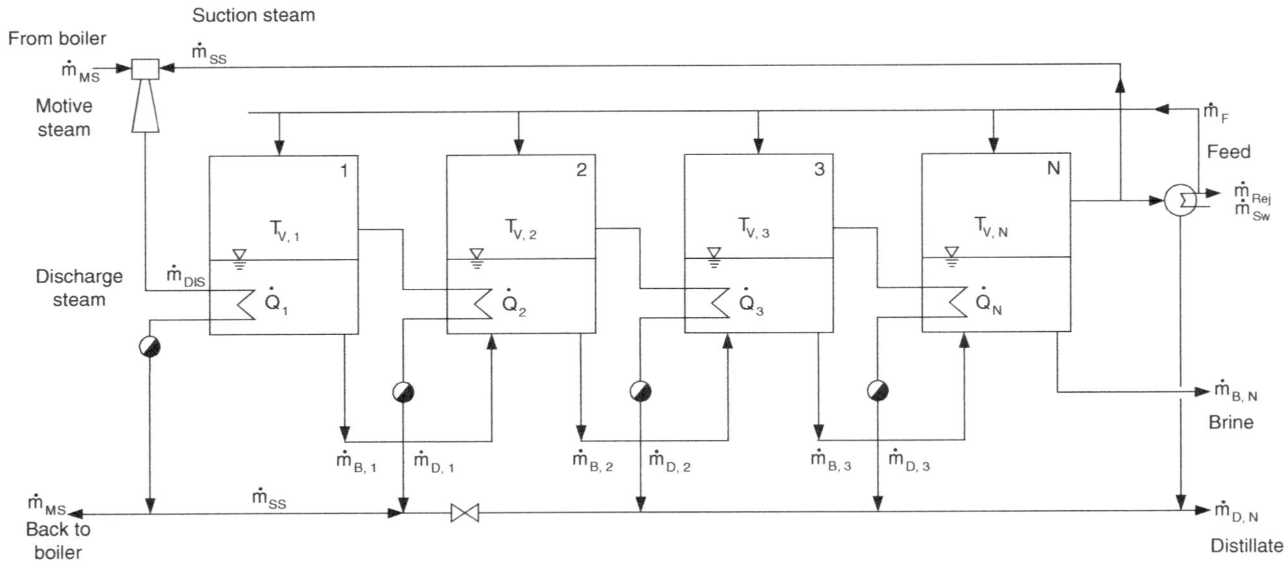

Figure 2.30 Flow sheet of an MED evaporation plant with thermal vapour compression (gross balance).

The following balances may be derived from the flow sheet:

Mass of distillate (total) $\dot{m}_D = (N-1)\cdot \dot{m}_{D,i} + \dot{m}_{D,N} + \dot{m}_{SS}$ (2.87)

Energy balance (1st stage) $\dot{m}_{Dis} \cdot \Delta h_{V,T_{Dis}} = \dot{m}_{D,1} \cdot \Delta h_{V,T_1}$ (2.88)

$$\Delta h_{V,T_{Dis}} \approx \Delta h_{V,T_1}$$

$$\dot{m}_{D,1} = \dot{m}_{Dis} \quad (2.89)$$

Mass of distillate 1st stage) $\dot{m}_{D,1} = \dot{m}_{Dis} = \dot{m}_{SS} + \dot{m}_{MS}$ (2.90)

Mass of distillate (stage N) $\dot{m}_{D,N} = \dot{m}_{Dis} - \dot{m}_{SS} = \dot{m}_{MS}$ (2.91)

From these equations follows that:

$$\dot{m}_D = (N-1)\cdot \dot{m}_{Dis} + \dot{m}_{MS} + \dot{m}_{SS}$$

$$\dot{m}_D = (N-1)\cdot (\dot{m}_{MS} + \dot{m}_{SS}) + \dot{m}_{MS} + \dot{m}_{SS}$$

$$\dot{m}_D = N \cdot (\dot{m}_{MS} + \dot{m}_{SS}) \quad (2.92)$$

The ratio of the motive steam to the suction steam is usually called the entrainment factor w. The entrainment factor is a characteristic feature of the compressor itself depending upon the various pressure levels and the internal efficiencies of the device. The entrainment factor can be obtained from a performance curve of the compressor usually published by the manufacturers (as an example see Figure 2.36) or calculated by thermodynamic means as being done later in this chapter.

Assuming that we do know the entrainment factor, Eq. 2.92 may be transformed as follows:

Entrainment factor $w = \dfrac{\dot{m}_{SS}}{\dot{m}_{MS}}$

Total mass of distillate $\dot{m}_D = N \cdot \dot{m}_{MS} \cdot (1+w)$ (2.93)

Specific mass of motive steam $\dfrac{\dot{m}_{MS}}{\dot{m}_D} = \dfrac{1}{N\cdot(1+w)}$ (2.94)

If the entrainment factor and the number of stages of the MED plant are known the amount of motive steam necessary for the compression can be calculated from this equation. As the mass of motive steam is provided by the steam boiler this figure may be equated to the thermal energy demand of the MED – TVC process.

Thermal vapour compression is an open process: the motive steam which comes from the boiler mixes with the suction steam; the mixture is fed into the 1st stage evaporator and appears after the condensation as liquid distillate. A part of this distillate returns to the boiler as boiler feed water, while the rest is added to the distillate of the remaining stages, in other words it becomes "product". If this product is utilised as drinking water there is the danger that the chemicals which were added to the boiler feed water, and which in this way can end up in the product, pollute the drinking water and make it unfit for human use. Should this be the case and if no other chemicals are able to be used which are harmless to man, then the part of the condensate from the 1st stage which is not returned to the boiler must be rejected or otherwise used. For this special case the following balances may be derived from the flow sheet in Figure 2.31:

Total mass of distillate

$$\dot{m}_D = (N-1) \cdot \dot{m}_{D,i} + \dot{m}_{D,N}$$

$$\dot{m}_D = (N-1) \cdot \dot{m}_{Dis} + \dot{m}_{MS}$$

$$\dot{m}_D = (N-1) \cdot (\dot{m}_{MS} + \dot{m}_{SS}) + \dot{m}_{MS}$$

$$\dot{m}_D = \dot{m}_{MS} \cdot \left[(N-1) \cdot (1 + \frac{\dot{m}_{MS}}{\dot{m}_{SS}}) + 1 \right]$$

$$\dot{m}_D = \dot{m}_{MS} \cdot [(N-1) \cdot (1+w) + 1]$$

$$\dot{m}_D = \dot{m}_{MS} \cdot N \cdot \left[(1+w) - \frac{w}{N} \right]$$

Specific mass of motive steam

$$\frac{\dot{m}_{MS}}{\dot{m}_D} = \frac{1}{N \cdot (1 + w - \frac{w}{N})} \tag{2.95}$$

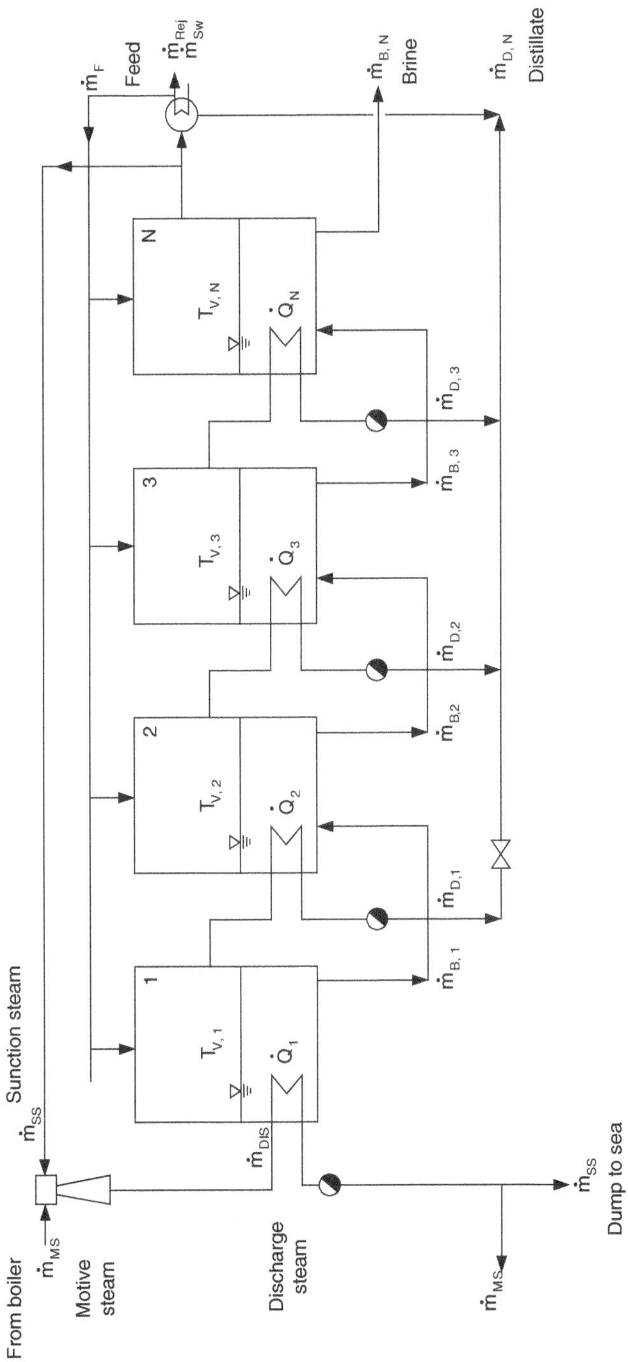

Figure 2.31 Flow sheet of an MED evaporation plant with thermal vapour compression (net balance).

In comparison with the "gross balance" (Eq. 6.68) the term $(-\frac{w}{N})$ appears here, which increases the motive steam but whose influence becomes smaller as the number of stages is increased.

In the above discussed case vapour is sucked up from the final stage N. This choice is already the result of an optimisation process taking into consideration the thermodynamic efficiency of the compressor chosen. Assuming that the compressor sucks up vapour from any stage N_{TVC} the balances presented above will change as follows:

Gross balance

Total mass of distillate $$\dot{m}_D = N \cdot \dot{m}_{MS} + N_{TVC} \cdot \dot{m}_{SS}$$

$$\frac{\dot{m}_D}{\dot{m}_{MS}} = N + N_{TVC} \cdot w$$

Specific mass of motive steam $$\frac{\dot{m}_D}{\dot{m}_{MS}} = \frac{1}{N + N_{TVC} \cdot w} \tag{2.96}$$

Net balance

Total mass of distillate $$\dot{m}_D = N \cdot \dot{m}_{MS} + (N_{TVC} - 1) \cdot \dot{m}_{SS}$$

$$\frac{\dot{m}_D}{\dot{m}_{MS}} = N + (N_{TVC} - 1) \cdot w$$

Specific mass of motive steam $$\frac{\dot{m}_D}{\dot{m}_{MS}} = \frac{1}{N + (N_{TVC} - 1) \cdot w} \tag{2.97}$$

In order to determine the mass flow of motive steam as a measure for the energy consumption of an MED / TVC - plant the entrainment factor w must be known. For this it is necessary to deal with thermal vapour compression in more detail. Figure 2.32 shows the schematic construction of a thermal vapour compressor. In order to explain the thermo- and hydro-dynamical processes three zones may be differentiated:

1. nozzle
2. mixing chamber
3. diffuser

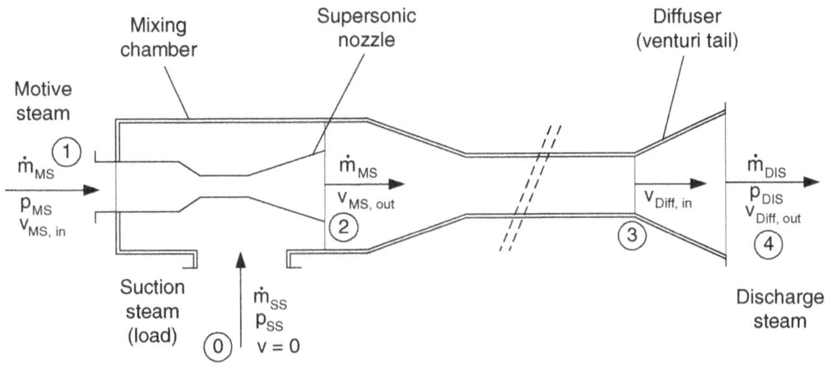

Figure 2.32 Scheme of a thermal vapour compressor.

We can understand the basic principle of the compressor with the assistance of the Bernoulli equation which states that for incompressible fluids the sum of pressure energy, kinetic energy and potential energy on a streamline remains constant:

$$p + \frac{\rho}{2} v^2 + \rho \cdot g \cdot z = const. \tag{2.98}$$

At first we should discuss the diffuser shown schematically in Figure 2.33. Assuming that the potential energy between state 1 and 2 remains constant (horizontal arrangement), Eq. 2.98 may be transformed to:

$$p_1 + \frac{\rho}{2} v_1^{\,2} = p_2 + \frac{\rho}{2} v_2^{\,2} \tag{2.99}$$

With the help of the equation of continuity for the mass flow entering and leaving the diffuser, the pressure at the diffuser outlet may be derived as follows:

$$\dot{m}_{in} = \dot{m}_{out} = \rho_1 \cdot v_1 \cdot A_1 = \rho_2 \cdot v_2 \cdot A_2 \tag{2.100}$$

With $\rho = \rho_1 = \rho_2$ we obtain

$$\Rightarrow \quad p_2 = p_1 + \frac{\rho}{2} v_1^{\,2} \cdot \left[(1 - (\frac{A_1}{A_2})^2 \right] \tag{2.101}$$

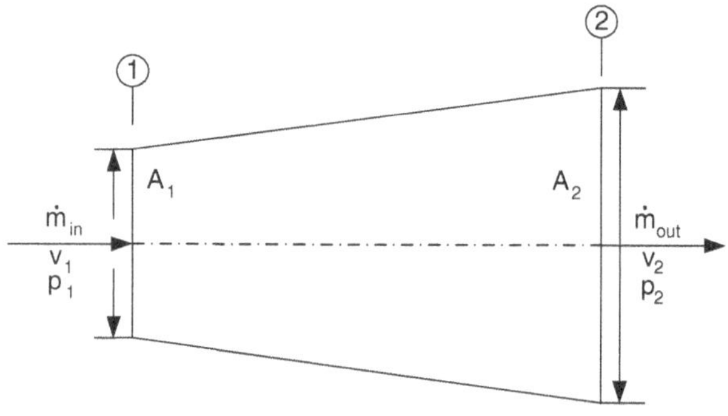

Figure 2.33 Schematic representation of a diffuser

This equation means in words:

1. If the cross-section A_2 is larger than A_1, then the pressure p_2 is greater than p_1!
2. If the cross-section A_2 is smaller than A_1, then the pressure p_2 is lower than p_1!

or:

3. If the velocity decreases in the diffuser, then the pressure increases!
4. If the velocity increases in the diffuser, then the pressure decreases!

With the help of these statements the processes in the compressor may be described as follows:

- in the nozzle the motive steam which has a high pressure and a relatively low velocity accelerates; at the nozzle outlet the motive steam has a very high velocity, whereby the prevailing pressure in the immediately surrounding region is very low;
- as a result of this low pressure vapour from the evaporation plant is sucked up into this part of the compressor (usually this vapour is called "suction steam" or "load");
- in the mixing chamber motive steam and suction steam mix together;
- In the diffuser the velocity of the steam mixture is gradually lowered, whereby the pressure increases again.

Figure 2.34 shows a characteristic velocity and pressure profile in a thermal compressor[5]. Two phenomena are of particular importance:

1. the velocity at the nozzle outlet is supersonic (!)
2. in the diffuser the velocity is reduced by a shock wave ("sonic boom")

The Bernoulli equation explains very well the functioning of the TVC process, but the law is only valid for incompressible fluids. Since steam is compressible the use of the Bernoulli equation is not quite correct for an energy balance around the nozzle, the

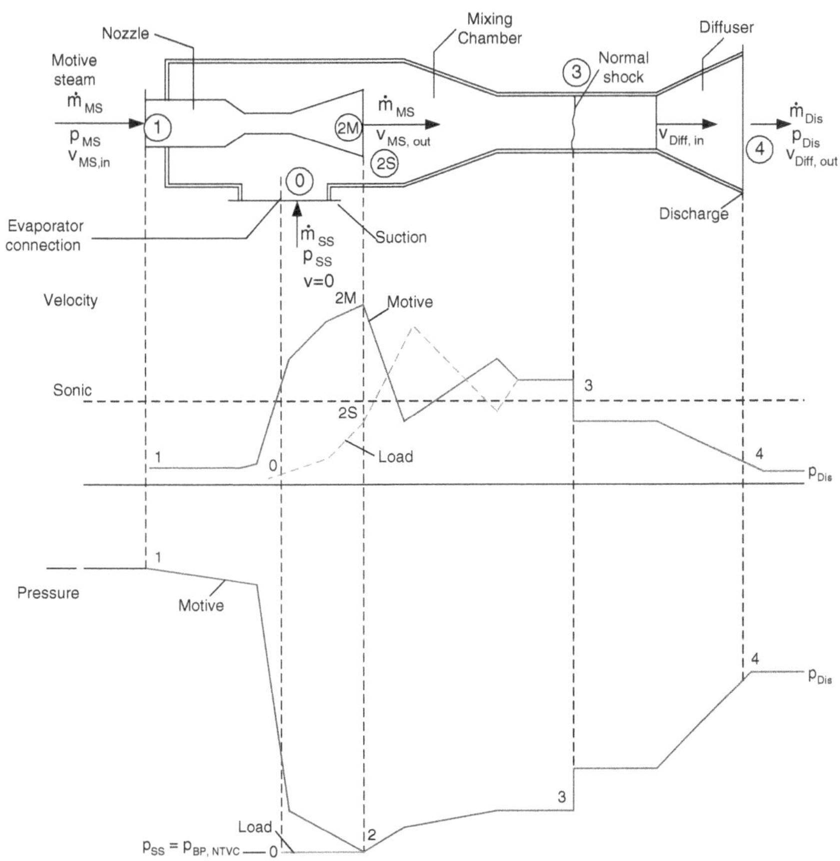

Figure 2.34 Pressure and velocity profile in a thermal vapour compressor.

mixing chamber and the diffuser. The 1st rule of thermodynamics and the principle of momentum transport must be drawn upon. Furthermore steam behaviour is not described accurately by the ideal gas law for the operation conditions in such ejectors. Calculations are best done by using an enthalpy vs. entropy (h-s) - diagram (see Figure 2.35).

The different states are mainly defined by three pressures:

p_{MS}	Motive pressure = steam produced by the steam boiler
$p_{SS} = p_{BP,N_{TVC}}$	Suction pressure or load pressure = steam pressure in the suction stage of the MED/TVC plant
p_{Dis}	Discharge pressure = condensation pressure in the 1st stage evaporator

If the temperature profile for an MED plant is given as in Figure 2.20, then the various pressures may be derived from the

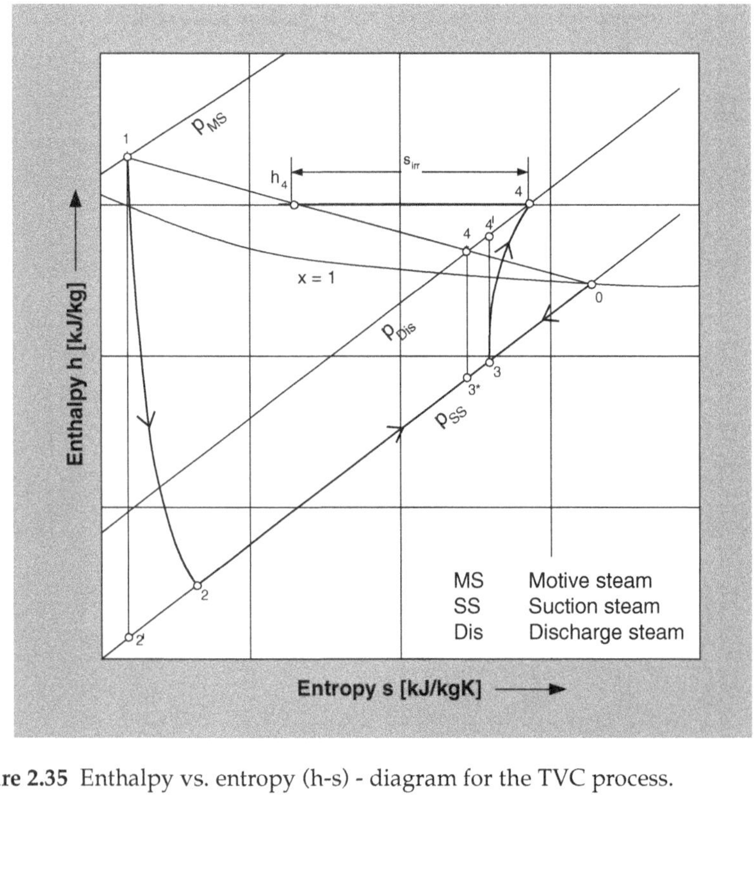

Figure 2.35 Enthalpy vs. entropy (h-s) - diagram for the TVC process.

boiling pressure curve. Since the motive steam pressure and the motive steam temperature is usually given by the performance of the steam boiler, the starting point 1 in the h – s – diagram is also defined. In the nozzle (1 → 2) the velocity increases and the pressure decreases down to suction pressure p_{ss}. This change of state is polytropic. The mixing point 3 is located on the line of constant pressure p_{ss}. To obtain the exact place of point 3 on this line, point 1 and point 0 can be connected by a straight line as to motive steam and suction steam are mixed up. The mixing point 4* is located at the crossing of the pressure line p_{Dis} and the mixing line. 4* leads to 3* but not to 3 or 3´! Due to the irreversibilities in the nozzle and the diffuser the changes of state are not isentropic (s = const.).

A solution is given by the introduction of efficiencies for the nozzle and the diffuser. Solving the energy balance for the different parts of the compressor leads to the following equation:

$$\frac{\dot{m}_{SS}}{\dot{m}_{MS}} = \sqrt{\eta_{irr,noz} \cdot \eta_{irr,dif} \cdot \frac{h_1 - h_{2'}}{h_{4'} - h_3}} - 1$$

Since the isobars are nearly parallel (see h-s – diagram), it can be assumed that

$$h_{4'} - h_3 = h_{4*} - h_{3*}$$

and this leads to the entrainment factor w:

$$w = \frac{\dot{m}_{SS}}{\dot{m}_{MS}} = \sqrt{\eta_{irr,noz} \cdot \eta_{irr,dif} \cdot \frac{h_1 - h_{2'}}{h_{4*} - h_{3*}}} - 1$$

To get familiar with this procedure we would like to calculate the following example:

Given:

Motive pressure	$p_{MS} = p_1 = 10\ bar$	Saturated steam
Suction pressure	$p_{SS} = p_0 = 0.074\ bar$	$T_V = 40°C$
Discharge pressure	$p_{Dis} = p_4 = 0.2\ bar$	$T_V = 60°C$
Efficiency of the nozzle	$\eta_{irr,noz} = 0.9$	
Efficiency of the diffuser	$\eta_{irr,dif} = 0.7$	

Solution: (by using the h-s diagram in Figure 2.35)

$$h_1 = 2{,}776.1\frac{kJ}{kg} \quad h_{2'} = 2{,}060.0\frac{kJ}{kg} \quad h_{4*} = 2{,}610.0\frac{kJ}{kg} \quad h_{3*} = 2{,}470.0\frac{kJ}{kg}$$

$$\Rightarrow \quad w = \sqrt{0.9\cdot 0.7\cdot\frac{2{,}776.1-2{,}060}{2{,}610-2{,}470}} - 1 = 0.795$$

With this figure the specific steam consumption of a 4 - staged TVC – plant can be calculated as follows (gross balance):

$$\frac{\dot{m}_{MS}}{\dot{m}_D} = \frac{1}{N\cdot(1+w)} = \frac{1}{4\cdot(1+0.795)} = \frac{1}{7.8} = 0.139$$

The Gained Output Ratio GOR of such a plant is given by:

$$\Rightarrow \quad GOR = \frac{\dot{m}_D}{\dot{m}_{MS}} = N\cdot(1+w) = 7.8$$

Data sheets may be found in the literature or produced by manufacturers with the help of which it is possible to determine the entrainment factor w at given pressures. Figure 2.36 shows such a diagram for a thermal compressor which is in principle similar to a performance curve for centrifugal pumps (Δh *vs.* $\dot{V}$)[5].

For the above discussed example we obtain:

$$\frac{p_{DIS}}{p_{SS}} = 2.7$$

$$\frac{p_{MS}}{p_{SS}} = 135$$

From diagram (Figure 2.36) we get

$$\Rightarrow \quad w = \frac{1}{R_S} \approx \frac{1}{1.25} = 0.8$$

And according to the formula used for calculating the GOR we finally obtain:

$$\Rightarrow \quad GOR = \frac{\dot{m}_D}{\dot{m}_{MS}} = N\cdot(1+w) = 7.2$$

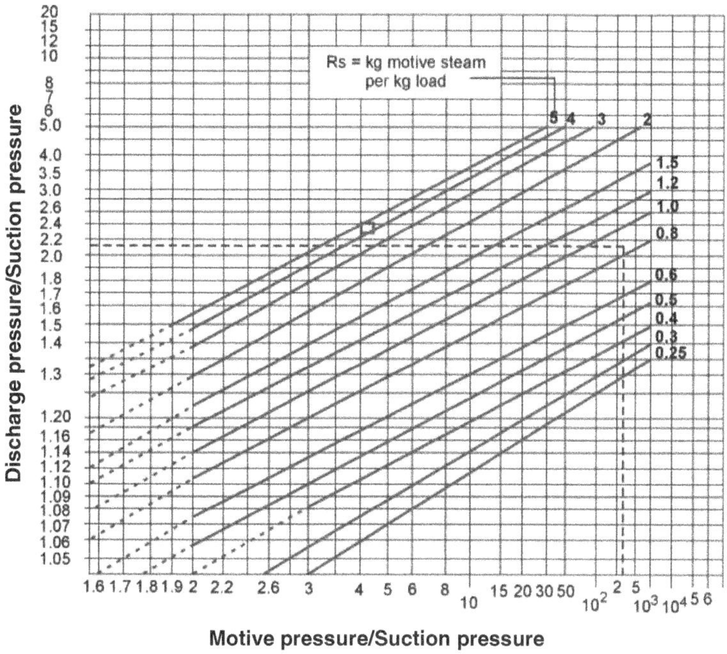

Figure 2.36 Performance diagram for a thermal vapour compressor.

The difference between this value and the more exactly calculated value using the h-s-diagram is less than 10% (7.2 to 7.8) Therefore it is appropriate using a performance diagram for rough calculation of the GOR.

2.2.5 Single-Stage Evaporation with Mechanically Driven Vapour Compression

In this desalination process the thermo-compressor described in the previous paragraph is replaced by a mechanically driven vapour compressor. With regard to the energy supply it is a process which requires solely mechanical energy and thus is comparable with reverse osmosis. However, the actual desalination process is of a thermal nature (separation of salt and water using evaporation), whereas in reverse osmosis the separation of salt and water takes place as a result of a trans-membrane pressure difference and is therefore purely mechanical.

Figure 2.37 shows the principle of a single-stage evaporation plant with mechanical vapour compression (MVC) and a picture of a typical plant. The process is essentially composed of the following elements:

- Evaporation unit
- Vapour compressor
- Feed – distillate – preheater (1)
- Feed – brine – preheater (2)
- Circulating pump
- Externally heated preheater (3) for the circulation flow

In order to be able to better understand the thermodynamics of mechanical vapour compression, a representation in the h-s diagram is helpful (see Figure 2.38). Part of the water which is usually sprayed into the evaporation chamber is continuously evaporated ($1 \rightarrow 2$). This vapour, which has a certain temperature according to the pressure in the evaporator, is sucked up by the compressor and brought up to a higher pressure ($2 \rightarrow 3$). In the enthalpy – entropy diagram the state 3 after the compression is in the superheated region.

Through the injection of a small amount of distillate saturated steam according to the pressure present at the compressor outlet ($3 \rightarrow 4$) is produced. As this pressure, the saturation temperature is higher than that on the outside of the evaporator tube bundle, and thus this vapour can be used as heating steam. The condensation ($4 \rightarrow 5$) takes usually place on the inside of a horizontal tube bundle.

It follows from an energy balance around the evaporation chamber that the condensation of 1 kg of heating steam on the inside of the tubes produces 1 kg of steam outside the tubes. However, energy is still needed in order to preheat the seawater to the evaporation temperature. Heat exchangers are used for this, through which the seawater flows and which absorbs the sensible heat of the distillate and the concentrate. As a driving temperature difference (i.e. a terminal temperature difference ΔT_{TTD}) is needed for the heat transfer, the pre-heating by the distillate and the brine is insufficient for bringing the feed up to evaporation temperature. Usually the recycle flow is heated up by electric heating. Additionally the external heating also covers the insulation losses which amount to about 1% of the evaporation heat.

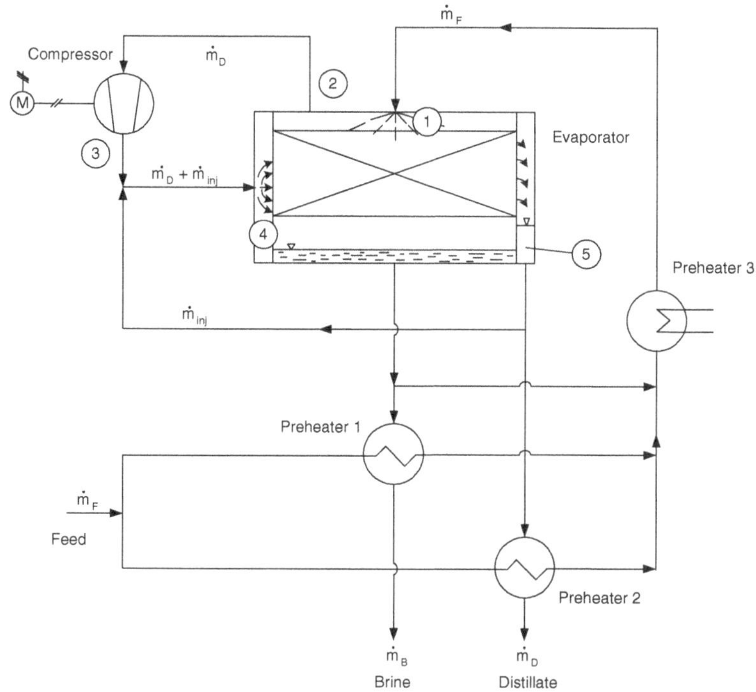

Figure 2.37 Flow sheet and picture of an evaporation plant with mechanical vapour compression (by courtesy of VA TECH WABAG GmbH).

The circulation of a certain amount of water is mostly necessary in order to be safe that the heat exchanger surfaces, here the outside of the tubes, are always wetted. If the film would break up, at these points salt would crystallise. The scaling would not just hinder the heat transfer, but in the end the tube bundle would also become totally blocked.

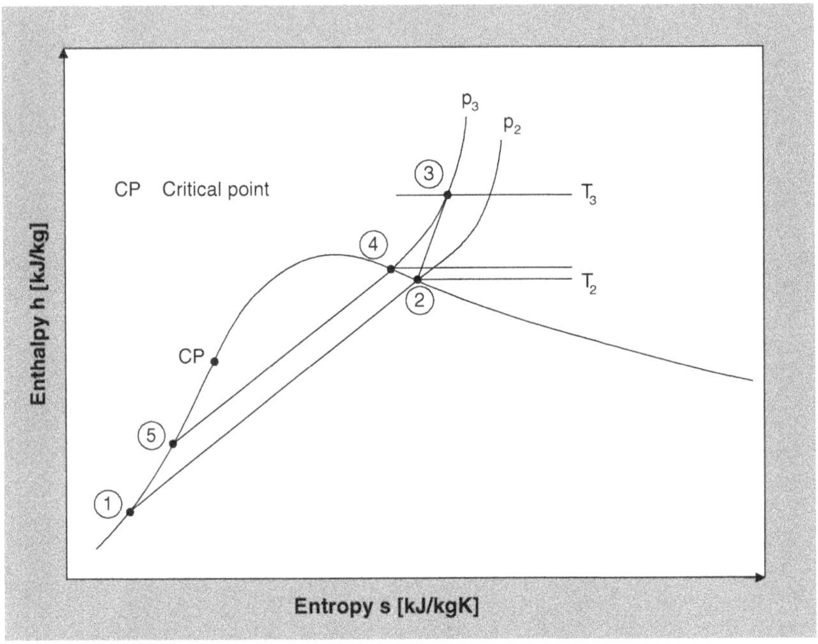

Figure 2.38 Mechanical vapour compression process in an h-s diagram.

According to the fundamentals explained in Chap. 2.1, the mechanical energy introduced into the process is given by the following equation:

$$P_t = \dot{m}_D \cdot (h_3 - h_2) \tag{2.102}$$

Assuming a polytropic change of state in the compressor, the following equations for the temperatures and the pressures at the inlet and outlet of the compressor are valid:

Polytrophic change of state: $p \cdot v^n = const.$ (n = polytropic exponent)

Ideal gas law: $p \cdot v = R \cdot T$

$$\Rightarrow \quad T_3 = T_2 \cdot \left[\frac{p_2}{p_3}\right]^{\frac{n}{n-1}} \tag{2.103}$$

The temperatures and pressures are coupled with the evaporation process as follows:

T_2 Temperature at compressor inlet = evaporation temperature T_V

p_2 Pressure at compressor inlet = evaporation pressure p_V

p_3 Pressure at compressor outlet = condensation pressure p_C

The condensation pressure is given by the condensation temperature according to the boiling pressure curve. The condensation temperature may be derived from the evaporation temperature plus a driving temperature difference necessary for the heat transfer:

$$T_C = T_V + \Delta T_{HT} \tag{2.104}$$

The technical work for a steady-state change of condition may be derived using the 1st rule of thermodynamics. This means:

$$w_{t23} = w_{friction,23} + \int_2^3 vdp + (e_{out2} - e_{out1}) \tag{2.105}$$

Besides friction and changes in external energy, the integral represents the most important part of the technical work. According to Chap. 2.1.1 (Eq. 2.16) the integral for a polytropic change of state may be expressed as follows:

$$\int_2^3 vdp = \frac{n}{n-1} \cdot (p_3 \cdot v_3 - p_2 \cdot v_2) \tag{2.106}$$

With the help of Eq. 2.104 we finally arrive to:

$$\int_2^3 vdp = \frac{n}{n-1} \cdot R \cdot (T_3 - T_2) = \frac{n}{n-1} \cdot R \cdot T_2 \cdot (\frac{T_3}{T_2} - 1)$$

$$\Rightarrow \quad \int_2^3 vdp = \frac{n}{n-1} \cdot R \cdot T_2 \cdot \left[(\frac{p_3}{p_2})^{\frac{n}{n-1}} - 1 \right] \tag{2.107}$$

The polytrophic exponent n can be expressed as follows:

$$\frac{n}{n-1} = \frac{k}{k-1} \cdot \eta_{VC}$$

With

k isentropic exponent (k = 1.326 for water, T < 100°C)
η_{VC} Efficiency of the compressor

Thus Eq. 2.107 becomes:

$$\int_2^3 vdp = \frac{k}{k-1} \cdot \eta_{VC} \cdot R \cdot T_2 \cdot \left[(\frac{p_3}{p_2})^{\frac{k}{k-1} \cdot \eta_{vc}} - 1 \right]$$

If the inner friction in the compressor is taken into account using an efficiency factor η_{VC} and if we ignore the changes in external energies, we obtain the following term for the power demand of the compressor (Eq. 2.102):

$$P_t = \dot{m}_D \cdot (h_3 - h_2)$$

$$P_t = \dot{m}_D \cdot \frac{k}{k-1} \cdot \eta_{VC} \cdot R \cdot T_2 \cdot \left[\left(\frac{p_3}{p_2}\right)^{\frac{k}{k-1} \cdot \eta_{vc}} - 1 \right] \tag{2.108}$$

Hence, in order to calculate the power consumption only the capacity of the plant $\dot{m}_D$ needs to be known, as well as the pressures and temperatures at which the evaporation and the condensation take place. According to Eq. 2.104 this means that a temperature difference ΔT_{HT} must be fixed for the heat transfer. This is an optimisation problem, as this temperature difference appears in two equations: firstly in the heat transfer equation and secondly in the compressor performance. These can be expressed as:

1. $\dot{Q} = k \cdot A \cdot \Delta T_{HT} = \dot{m}_D \cdot \Delta h_{V,T_V}$
2. $P_t = \dot{m}_D \cdot \dfrac{n}{n-1} \cdot R \cdot \Delta T_{HT} \cdot \dfrac{1}{\eta_{VC}}$

The equations above demonstrate that the heat exchanger surface, and thus the costs for the compressor, become smaller the greater the temperature difference and the greater the heat transfer coefficient. When attempting to minimise the investment costs, large temperature differences and large heat transfer coefficients must be aimed for. However, the temperature difference may not simply be selected to have an arbitrarily value for two reasons:

1. The temperature difference for the heat transfer is directly proportional to the compressor power consumption according to the equation listed above. If cheap and simply constructed fans were to be employed, the technical limit for a mechanically driven vapour compressor is a maximum of 10 K. Greater temperature differences can only be achieved using expensive screw-type compressors. However their use is not worthwhile in comparison with the savings in the heat exchange surface.

2. The evaporation of seawater is constrained by the crystallisation of salts (calcium carbonate, calcium sulphate). If bubbles are formed during the evaporation – this process may be observed in any kitchen when water boils -, salt crystals will develop at the edges of the bubbles on the evaporator surface and in time will encrust the whole surface. Boiling with bubbles occurs when the heat flow density (kW per m^2) or the driving temperature difference exceeds a specific value. For water or seawater this is at about 7 K[6]. This limit should on no account be exceeded in seawater desalination.

For the reasons given there are the following constraints on mechanical exhaust vapour compression in practice:

1. Only single-stage process are used for this, as only small temperature differences can be overcome by the compressor (as opposed to thermal compression where greater temperature differences and thus also multiple-effect plants are possible).
2. The costs for a compressor are a constraint on the capacity of the plant, i.e. normally only small plants are equipped with mechanical vapour compression (up to approx. 2,000 m^3/d)

2.3 Performance of Thermal Desalination Processes

2.3.1 Definition of Gained Output Ratio

Regardless of the complexity of the subject "Energy supply", parameters have been established for the assessment of the energy demand of seawater desalination plants which are simple to determine and to apply. However in order to be able to work with these quantities, which in many cases is extremely helpful, requires a thorough understanding of the general prerequisites and the scope of application, which are explained in the following section.

If a query is made as to the energy efficiency of a thermal seawater desalination plant in most cases the gained output ratio, GOR for short, will be given. The GOR is defined as the ratio of fresh water produced to the heating steam and is in line with the above definition of efficiency (benefits to effort):

$$\eta_E = \frac{\textit{fresh water produced}}{\textit{energy demand}} \triangleq \frac{\textit{fresh water produced}}{\textit{heating steam}} = GOR$$

The GOR is thus the relationship of two mass flows, the distillate flow and the heating steam flow, as in the schematic representation of an MED plant in Figure 2.39, which results in the following units.

$$GOR = \frac{\dot{m}_D}{\dot{m}_{HS}} \quad \rightarrow \text{SI-Units:} \quad [GOR] = \left[\frac{kg/s}{kg/s}\right] = [-] \qquad (2.111)$$

Thermal plants with the same GOR therefore require the same amount of heating steam for the production of a certain mass of distillate. In saying this it is simply assumed that this is saturated steam, there is no information on the exact conditions of the heating steam (pressure, temperature). Thus the energy content of the heating steam is not clearly defined. It cannot therefore be concluded that both plants consume exactly the same amount of fuel or that, for instance, the costs for the supply of thermal energy are the same. The GOR is therefore not quite suitable as a scale for a process comparison. For this the energy supply plant and the costs for the fuel must also be taken into consideration[4].

The specific heat demand of thermal seawater desalination plants, related to the number of stages and the thermodynamic data, was determined in Chapter 2.2. As may be seen from Figure 2.39, this heat is equal to the condensation heat of the heating steam. (Subcooling of the condensate is not taken into consideration at this point.) The GOR can therefore be calculated from the specific heat demand of the plant as follows:

$$\frac{\dot{Q}}{\dot{m}_D} = \frac{\dot{m}_{HS} \cdot \Delta h_{V,HS}}{\dot{m}_D} \qquad (2.112)$$

$$\Rightarrow \quad \frac{\dot{Q}}{\dot{m}_D} = \frac{\Delta h_{V,HS}}{GOR} \qquad (2.113)$$

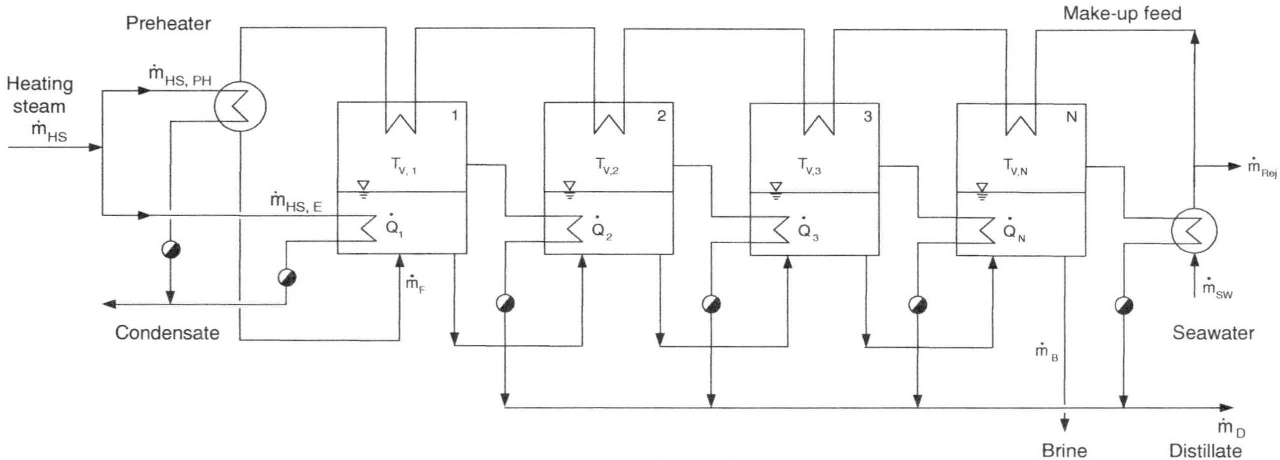

$$GOR = \frac{\text{Distillate}}{\text{Heating steam}} = \frac{\dot{m}_D}{\dot{m}_{HS}} = \frac{\dot{m}_D}{\dot{m}_{HS,PH} + \dot{m}_{HS,E}}$$

.Figure 2.39 Schematic of an MED plant and definition of the GOR.

$$\Rightarrow \quad GOR = \Delta h_{V,HS} \cdot \frac{\dot{m}_D}{\dot{Q}} \tag{2.114}$$

The last term is also known in the literature[1] as the Performance Ratio (PR).

$$PR = \Delta h_{V,HS} \cdot \frac{\dot{m}_D}{\dot{Q}} = \frac{\Delta h_{V,HS}}{\dot{Q}'} \tag{2.115}$$

The heat of evaporation is usually directly entered in this equation as a numerical value with a guideline for the units of the heat and the distillate mass flow, whereby the equation mutates to a numerical value equation. However it is strongly recommended to maintain a distance to such definitions, as they create more confusion than clarity as a scale for the energy demand of a thermal plant.

As an example of the procedure described above the GOR for an MED plant will be determined in the following text. A rough formula for the specific heat demand of an MED plant is (see Eq. 2.54):

$$\frac{\dot{Q}}{\dot{m}_D} = \frac{\Delta h_{V,T_m}}{N^{0.85}}, \tag{2.116}$$

where T_m is the average stage temperature in the MED plant.
From Eq. 2.114 it follows:

$$GOR = \frac{\Delta h_{V,HS}}{\Delta h_{V,T_m}} \cdot N^{0.85} \tag{2.117}$$

If it is assumed that the evaporators in the different stages of an MED plant are all of the same construction, in particular the evaporator in the 1st stage, then the temperature of the heating steam is given by the temperature profile of the plant. Thus for a typical MED plant the ratio of the specific heats of evaporation in Eq. 2.117 can be determined as follows:

Example:

Number of stages: $N = 10$
Vapour temperature 1st effect: $T_{V,1} = 65\ °C$
Vapour temperature last effect: $T_{V,N} = 35\ °C$

Stage decrement: $\Delta T_{stage} = \frac{65-35}{9} K = 3.33\ K$

Heating steam temperature: $T_{V,HS} = 65\ °C + 3.33\ K = 68.3\ °C$

Heat of evaporation (MED-plant): $\Delta h_{V,T_m} = 2,382 \frac{kJ}{kg}$

Heat of evaporation (heating steam): $\Delta h_{V,HS} = 2,338 \frac{kJ}{kg}$

$$\Rightarrow \quad GOR = \frac{2,338}{2,382} \cdot 10^{0.85} = 0.98 \cdot 10^{0.85}$$

$$\Rightarrow \quad GOR = 6.93$$

The 2 % - error (0.98 to 1.00), which results from the fact that different values for the heat of evaporation have to be taken into account in the GOR equation, is comparatively small.

If these points are taken into consideration the gained output ratio can always be used if it is simply a question of obtaining a quick statement on the efficiency of a thermal seawater desalination plant. The GOR is unsuitable as a parameter for a comparative assessment of the energy cost of different seawater desalination processes.

2.3.2 Single Purpose vs. Dual Purpose Plants

Figure 2.40 shows the typical flow sheet for a single purpose plant. Steam is produced in the steam boiler at a relatively high pressure level (HP, 10 to 20 bar). According to the thermodynamic processes in the boiler, the steam is slightly superheated. With the exception of MED plants equipped with thermal vapour compression MSF and MED plants require saturated heating steam at a low temperature and pressure level (LP):

MSF → saturated steam at 2.0 bar / approx. 120 °C
MED → saturated steam at 0.3 bar / approx. 70 °C

The high pressure of the saturated steam must therefore be reduced before it enters the heat exchanger of the evaporator plant (brine heater for MSF, evaporator of the 1st stage for MED). As shown in Figure 2.41, this happens using a throttle valve in

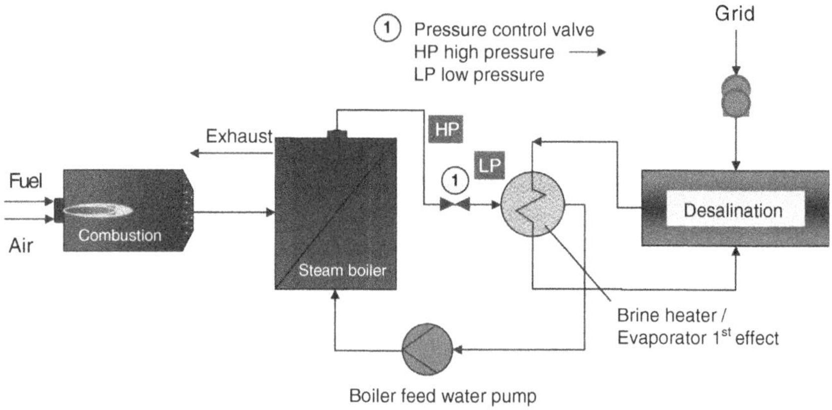

Figure 2.40 Flow sheet of a single purpose plant (steam boiler + desalination).

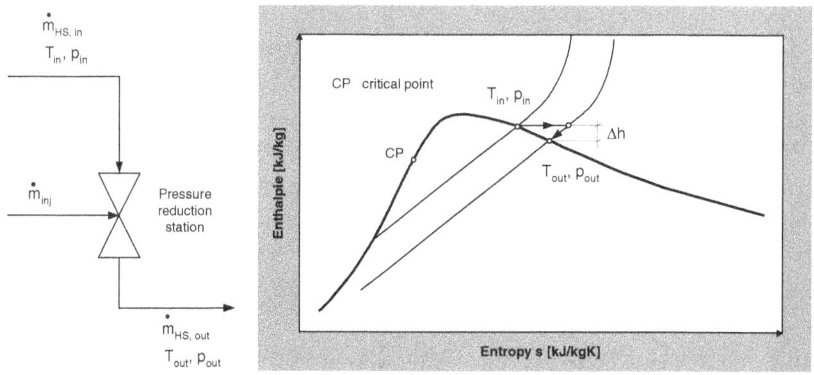

Figure 2.41 Pressure reduction station with water injection for de-superheating and appropriate h-s diagram.

combination with water injection. The latter is necessary in order to transform the steam which is superheated after the throttle into saturated steam. Otherwise superheated steam would enter the heat exchanger. However, as superheated steam acts like a gas, the heat transfer coefficient on the gas side would be very low and the area of the heat exchanger would increased enormously. This makes neither technical nor economic sense. The injection is a simple and effective method of ensuring that saturated steam flows into the heat exchanger, so that the condensation with its corresponding high heat transfer coefficient can immediately start.

As may be seen from the enthalpy – entropy (h-s) - diagram in Figure 2.41 the throttling process is an isenthalp, i.e. the enthalpy h of the flow before and after the throttle stays the same. However, a strong increase in the entropy may be observed, as the change of state is accompanied by large losses, i.e. irreversibilities. This illustrates once more that the supply of a thermal seawater desalination plant using a steam boiler represents a thermodynamically poor way of energy utilisation.

As a result of the water injection the steam mass flow which condenses in the evaporator plant is not equal to the mass flow which was produced by the steam boiler. However, by using a mass and energy balance a calculation can be made to obtain the steam which enters the desalination plant:

Mass balance: $$\dot{m}_{HS,out} = \dot{m}_{HS,in} + \dot{m}_{inj} \tag{2.118}$$

Energy balance: $$\dot{m}_{HS,out} \cdot h_{HS,out} = \dot{m}_{HS,in} \cdot h_{HS,in} + \dot{m}_{inj} \cdot h_{inj} \tag{2.119}$$

Due to the fact, that the heating steam entering the evaporator or the brine heater has to be saturated steam (denoted h'') and that the injected water is the condensate (denoted h') at the same temperature, the following is valid:

$$h_{HS,out} = h''(T_{out})$$

$$h_{inj} = h'(T_{out})$$

In order to obtain the mass flow of the heating steam we have to put these results in the above given mass- and the energy balances. Thus it follows that:

$$\Rightarrow \quad \dot{m}_{HS,out} = \dot{m}_{HS,in} \cdot \frac{h_{HS,in} - h'(T_{out})}{h''(T_{out}) - h'(T_{out})} \tag{2.120}$$

The enthalpies for the saturated steam $h''(T_{out})$ and the condensate $h'(T_{out})$ can be read out from the steam table[4]. Presuming that the mass flow which is produced in the boiler and the enthalpy of the mass flow is given (usually by measuring the temperature and the pressure) the heating steam mass flow can be directly calculated from Eq. 2.120 or using an h-s diagram.

The fuel required for generating the heating steam for the desalination process can be determined using the efficiency of the boiler:

$$\eta_B = \frac{\dot{Q}}{\dot{m}_{Fu} \cdot H_{Fuel}} \tag{2.121}$$

In this equation, H_{Fuel} is the lower calorific value of the fuel (net calorific value).

Following the calculation of the pressure reduction station above, the heat which is transferred is given by:

$$\dot{Q} = \dot{m}_{HS,out} \cdot \Delta h_{V,T_{out}} \tag{2.122}$$

That means that the mass flow of fuel can be expressed as:

$$\dot{m}_{Fu} = \frac{\dot{m}_{HS,out} \cdot \Delta h_{V,T_{out}}}{\eta_B \cdot H_{Fuel}} \tag{2.123}$$

It is now possible to define a modified gained output ratio as the ratio of the fuel mass flow to the distillate produced as:

$$GOR_{Fu} = \frac{\dot{m}_D}{\dot{m}_{Fu}} \tag{2.124}$$

With Eq. 2.123 this gives:

$$GOR_{Fu} = \frac{\dot{m}_D \cdot \eta_B \cdot H_{Fuel}}{\dot{m}_{HS,out} \cdot \Delta h_{V,T_{out}}} \tag{2.125}$$

In Eq. 2.111 the conventional GOR was defined as the ratio of the distillate mass flow to the heating steam mass flow:

$$GOR = \frac{\dot{m}_D}{\dot{m}_{HS,out}}$$

Putting this term into Eq. 2.125 we obtain:

$$GOR_{Fu} = GOR \cdot \frac{\eta_B \cdot H_{Fuel}}{\Delta h_{V,T_{out}}} \tag{2.126}$$

In this definition the GOR_{Fu} can be applied as an efficiency suitable in principle for all thermal processes which are coupled to a steam boiler in single purpose mode.

However what is its practicability like?

Let us continue with the example begun above using the following conditions for the boiler:

Low calorific value: $H_{Fuel} = 39,400 \frac{kJ}{kg_{Fu}}$ (Heavy oil)

Efficiency of the boiler: $\eta_B = 0.75$

Heat of evaporation: $\Delta h_{V,T_{out}} = 2,338 \frac{kJ}{kg}$

As a result we obtain:

$$\Rightarrow \quad GOR_{Fu} = 6.93 \cdot \frac{0.75 \cdot 39,400}{2,338} = 87,68$$

The result makes it obvious that working with an efficiency defined in this way is so arduous:

What do these "87.68" mean at all? Is it good or bad, high or low? An efficiency figure is clear if it is between 0 and 1. How practical is such a figure? It is immediately clear that the good old GOR always succeeds in defending its place for the assessment of energy utilisation in the literature when a direct coupling to the number of stages of the thermal plant is given.

However we would like to stick to this definition for the time being and see if this dimension is of use in comparing dual purpose plants with single purpose-plants.

The simplest and most economic way of coupling a steam turbine power station and a thermal seawater desalination plant is the back pressure turbine, as illustrated in Figure 2.42. In this the steam produced in the boiler is first passed to a turbine which produces electricity. This process should stop at a pressure / temperature level suitable for the requirements of the thermal seawater desalination plant, for example 120°C/2 bar for MSF plants. The seawater desalination plant takes the place of the power station condenser for all intents and purposes.

The main disadvantage of this system is its lack of flexibility: if the turbine has to run at reduced load, then the seawater desalination plant also has to reduce its water production rate (which, depending on the type of plant, is not so simple). For this case and

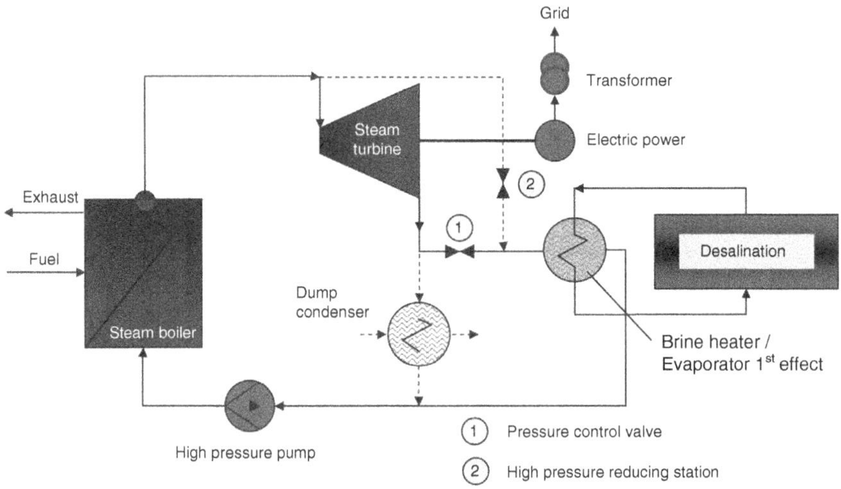

Figure 2.42 Dual purpose plant / Case 1: Back pressure turbine.

also for the case in which the turbine or the seawater desalination plant completely breaks down, a dump condenser and a by-pass around the turbine with a pressure reduction station must be planned. However, this creates extra costs and should be carefully considered during the design phase.

In dual purpose-plants two "products" are generated from the fuel: electrical energy and heating steam. In using such a coupled production the question may be asked as to how the total production costs may be apportioned across the two products.

One possibility, which has the advantage of being justified by thermodynamics, uses the hypothetical approach of non-produced electricity. In this, three systems are compared as represented in Figure 2.43:

1. a conventional condensation power station
2. a steam generator for the supply of the desalination plant
3. a power station with a back pressure turbine for supplying the desalination plant

There are three relevant temperatures which appear:

1. Inlet temperature of the turbine T_{Ta}
2. Outlet temperature of the turbine $T_{T\omega}$
3. Condensation temperature T_C

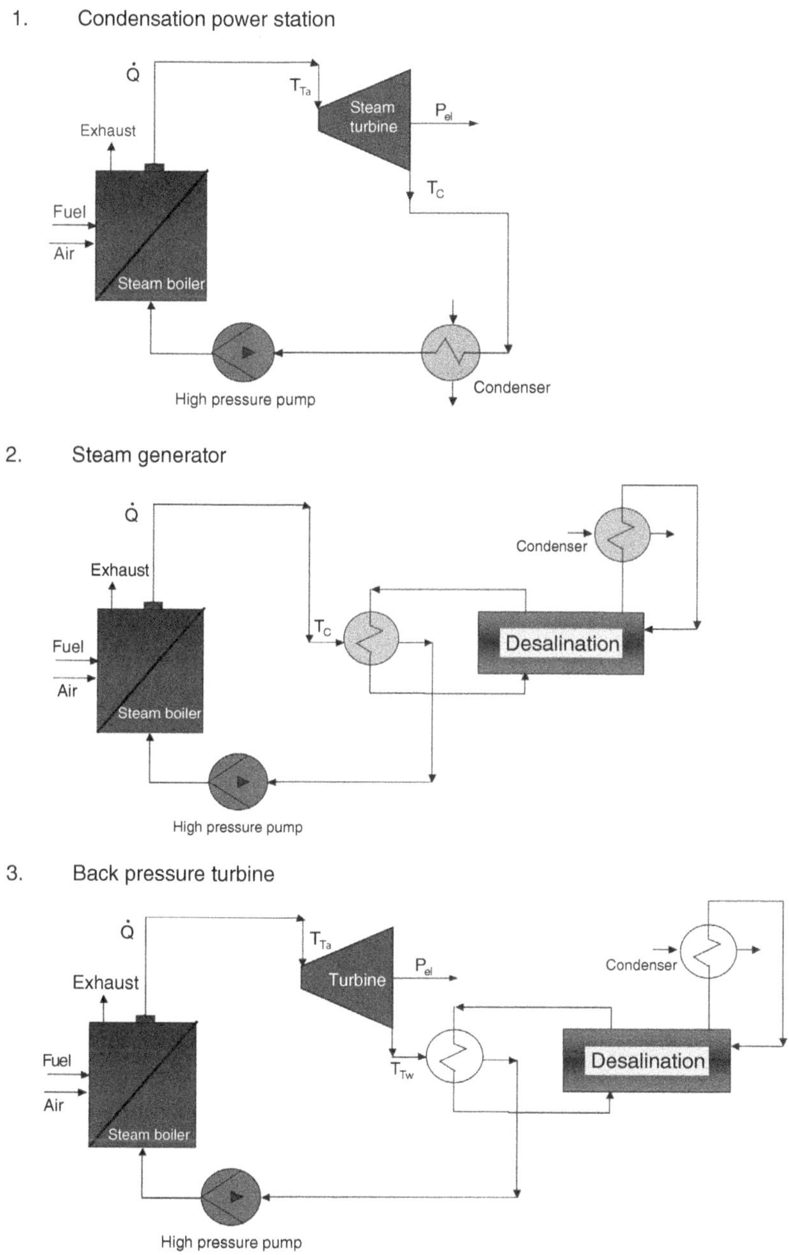

Figure 2.43 1. Condensation power station for electricity generation. 2. Steam generator for heat supply of a desalination plant. 3. Condensation power station with back pressure turbine for electricity generation and heat supply of a desalination plant.

The maximum work which can be performed in an ideal cycle between two temperatures is given by the Carnot Factor, i.e. we can note the following for a condensation power station:

$$\eta_{Carnot,PP} = \frac{work\ output}{heat\ input} = \frac{T_{Ta} - T_C}{T_{Ta}} \tag{2.127}$$

The work performed is pure energy, that means exergy Ex, so that Eq. 2.127 may also be formulated as follows:

$$Ex_T = \frac{T_{Ta} - T_C}{T_{Ta}} \cdot \dot{Q} \tag{2.128}$$

If a seawater desalination plant is connected to this, then less electricity is produced in the turbine, i.e. the Carnot Factor is given as:

$$\tilde{\eta}_{Carnot} = \frac{T_{Ta} - T_{T\omega}}{T_{Ta}} \tag{2.129}$$

or

$$\widetilde{Ex}_T = \frac{T_{Ta} - T_{T\omega}}{T_{Ta}} \cdot \dot{Q} \tag{2.130}$$

The exergy which may be ascribed to the seawater desalination, and which at the same time represents the loss of electricity produced, reads:

$$Ex_{HS} = \frac{T_{T\omega} - T_C}{T_{T\omega}} \cdot \dot{Q} = \eta_{Carnot,HS} \cdot \dot{Q} \tag{2.131}$$

The total fuel supplied to the condensation power station may be theoretically divided into a share which may be ascribed to the electricity generation and a share to the non-produced electricity, in other words the seawater desalination process.

$$\Rightarrow \quad \dot{m}_{Fu,total} = \dot{m}_{Fu,T} + \dot{m}_{Fu,DP} \tag{2.132}$$

The chemical energy stored in a fuel is pure exergy, so that further formulae may be complied:

$$\dot{m}_{Fu,total} \cdot H_{Fuel} = Ex_{PP} = \frac{T_{T\alpha} - T_C}{T_{T\alpha}} \cdot \dot{Q} \tag{2.133}$$

$$\dot{m}_{Fu,DP} \cdot H_{Fuel} = Ex_{HS} = \frac{T_{T\omega} - T_C}{T_{T\omega}} \cdot \dot{Q} \tag{2.134}$$

If these two equations are divided, and Eqs. 2.127 and 2.131 are taken into consideration the following results for the ratio of the two fuel flows:

$$\frac{\dot{m}_{Fu,DP}}{\dot{m}_{Fu,total}} = \frac{Ex_{HS}}{Ex_{PP}} = \frac{\eta_{Carnot,HS}}{\eta_{Carnot,PP}} \tag{2.135}$$

Now it is true that in all three cases (Figure 2.43) the same amount of heat is produced in the boiler and that the same mass of fuel is required:

$$\dot{Q} = \dot{m}_{Fu,total} \cdot H_{Fuel} \cdot \eta_B = \dot{m}_{HS} \cdot \Delta h_{V,HS} \tag{2.136}$$

If this equation is used to replace the total fuel flow in Eq. 2.135, we may obtain for the fuel flow of the desalination plant:

$$\frac{\dot{m}_{Fu,DP}}{\dot{m}_{HS}} = \frac{\Delta h_{V,HS}}{H_{Fuel}} \cdot \frac{\eta_{Carnot,HS}}{\eta_B \cdot \eta_{Carnot,PP}} \tag{2.137}$$

It is helpful to introduce the total efficiency of a power station which comprises the efficiencies for the boiler, the turbine and the Carnot Cycle:

$$\eta_{PP} = \eta_B \cdot \eta_{Carnot,PP} \cdot \eta_T \tag{2.138}$$

Thus Eq. 2.137 becomes:

$$\frac{\dot{m}_{Fu,DP}}{\dot{m}_{HS}} = \frac{\Delta h_{V,HS}}{H_{Fuel}} \cdot \frac{\eta_T \cdot \eta_{Carnot,HS}}{\eta_{PP}} \tag{2.139}$$

This equation has the advantage that the fuel mass flow, ascribable to the seawater desalination plant in dual purpose mode, can be calculated without knowing the relevant temperatures in the power station. It is sufficient to know the conditions of the desalination plant, the efficiency of the power station and the turbine and the type of fuel used.

A Gained Output Ratio based on the fuel may be defined from this equation analogously to the procedure in single purpose mode (Eq. 2.124):

$$GOR_{Fu,DP} = \frac{\dot{m}_D}{\dot{m}_{Fu,DP}} \tag{2.140}$$

Using the conventional definition of the GOR for a thermal desalination plant, and introducing Eq. 2.139, the following is finally obtained for dual purpose mode:

$$GOR = \frac{\dot{m}_D}{\dot{m}_{HS}}$$

$$\Rightarrow \quad GOR_{Fu,DP} = GOR \cdot \frac{\dot{m}_{HS}}{\dot{m}_{Fu,DP}} \tag{2.141}$$

$$\Rightarrow \quad GOR_{Fu,DP} = GOR \cdot \frac{H_{Fuel}}{\Delta h_{V,HS}} \cdot \frac{\eta_{PP}}{\eta_T \cdot \eta_{C,HS}} \tag{2.142}$$

If we compare this term with the result for the single purpose mode,

$$GOR_{Fu,SP} = GOR \cdot \frac{H_{Fuel}}{\Delta h_{V,HS}} \cdot \eta_B$$

we see that the efficiency of the boiler was replaced by a combination of efficiencies. In the case of the single purpose mode an example was calculated above. In the following text a dual purpose plant will be calculated with the values used there. For this it is necessary to give the efficiency rates and the relevant temperatures:

Example:

Efficiency of the power plant $\eta_{PP} = 0.35$

Efficiency of the turbine $\eta_T = 0.7$

Heating steam temperature $T_{V,HS} = 68.3\,°C = T_{T\omega}$

Condensation temperature $T_C = 35°C$

Heat of evaporation z $\Delta h_{V,HS} = 2,338 \frac{kJ}{kg}$

Net calorific value $H_{Fuel} = 39,400 \frac{kJ}{kg_{Fu}}$

Gained Output Ratio $GOR = 6.93$

First of all the Carnot efficiency to be ascribed to the desalination plant must be determined:

$$\eta_{Carnot,HS} = \frac{T_{T\omega} - T_C}{T_{T\omega}} = \frac{68.3 - 35}{68.3 + 273.15} = 0.097$$

Using this and the equations given above, the result is:

$$\Rightarrow \quad GOR_{Fu,DP} = 6.93 \cdot \frac{39,400}{2,338} \cdot \frac{0.35}{0.7 \cdot 0.097} = 602$$

In comparison with the case of a stand-alone steam boiler (single purpose):

$$\Rightarrow \quad GOR_{Fu,SP} = 6.93 \cdot \frac{0.75 \cdot 39,400}{2,338} = 87,68$$

Therefore in this example a dual purpose plant has a GOR (related to the fuel demand) which is almost 7 times bigger than that of a single purpose plant - in other words: the cost for the fuel are 7 times lower in dual purpose mode as in single purpose mode.

Before further technical possibilities for the coupling of a power station with a desalination plant are investigated, a final remark on the efficiencies, in particular to the fuel-related GOR_{Fu} should be allowed. The procedure detailed has shown that anyone working with these formulae is well advised to be clear on the conditions under which these formulae are valid. Both the conventional GOR as well as the fuel-related GOR_{Fu} can lead to quick and sound results when they are applied correctly. If you enter current values for the fuel and the efficiency rates (as in the above example, for instance), then both of the GORs can be relatively simply transferred into each other. The following rule of thumb may be utilised:

Rule of thumb for MSF plants (top brine temperature ≈ 120°C)

Single Purpose Steam generator

$$\Rightarrow \quad GOR_{Fu,SP} = 15 \cdot GOR$$

Dual Purpose Power station with back pressure turbine

$$\Rightarrow \quad GOR_{Fu,DP} = 50 \cdot GOR$$

Rule of thumb for MED and MED/TVC plants (top temperature ≈ 70°C)

Single Purpose Steam generator

$$\Rightarrow \quad GOR_{Fu,SP} = 15 \cdot GOR$$

Dual Purpose Power station with back pressure turbine

$$\Rightarrow \quad GOR_{Fu,DP} = 90 \cdot GOR$$

A general transfer to other possibilities for dual purpose operation, such as for example combined cycle or a block type power station, should be strictly avoided. The fuel demand, and its associated costs, should be determined on a case by case basis using mass and energy balances or even exergy balances.

Figure 2.44 shows the coupling of a desalination plant with a so-called extraction steam turbine. Here a part of the steam is taken out of the turbine at the pressure/temperature level which is necessary for the operation of the seawater desalination plant (e.g. 2 bar/120°C for MSF plants). The remaining steam continues to expand in the turbine producing electricity. With the help of the valves 1, 2, and 3 which control the flow volumes it is theoretically possible to set to any proportion of electricity to water production.

The extraction steam turbine distinguishes itself from the back pressure turbine specifically through this flexibility. However, due to the increased amount of process control systems and the design of the turbine with an extraction port it is more costly to procure.

The larger the power station and the larger the seawater desalination plant, the more varied the alternatives for coupling both, which

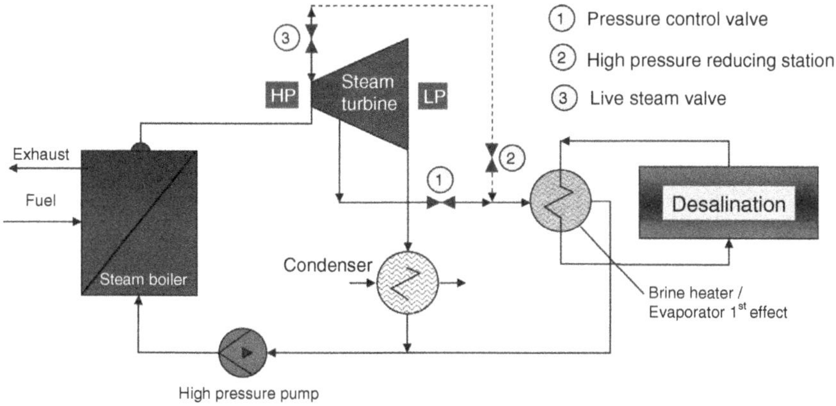

Figure 2.44 Dual purpose plant / Case 2: Controlled extraction steam turbine.

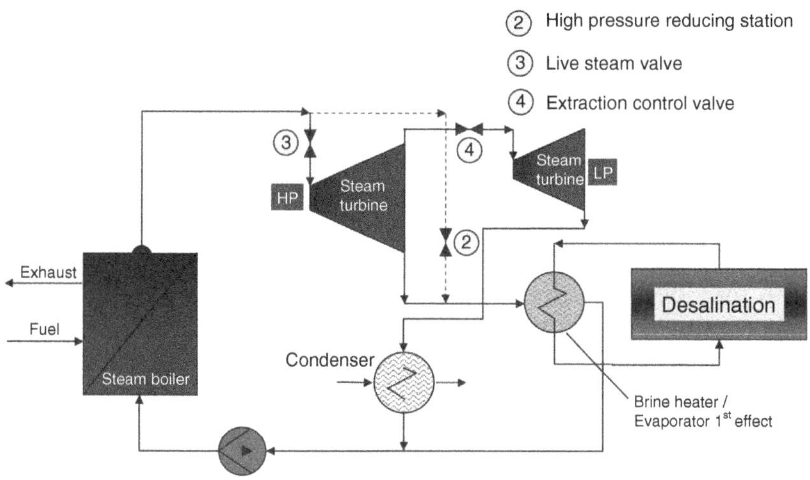

Figure 2.45 Dual purpose plant / Case 3: High pressure turbine + low pressure turbine.

cannot all be dealt with at this point. Figure 2.45 shows, as a further possibility, the constructional segmentation of the turbine into a high-pressure part and a low-pressure part. The desalination plant is situated exactly between these two parts because of the pressure / temperature level. Therefore behind the high-pressure turbine the steam is divided into two parts: one part flows to the desalination plant, the rest of the steam flows to the low-pressure turbine. A control of the ratio of electricity to water production is possible using an extraction control valve.

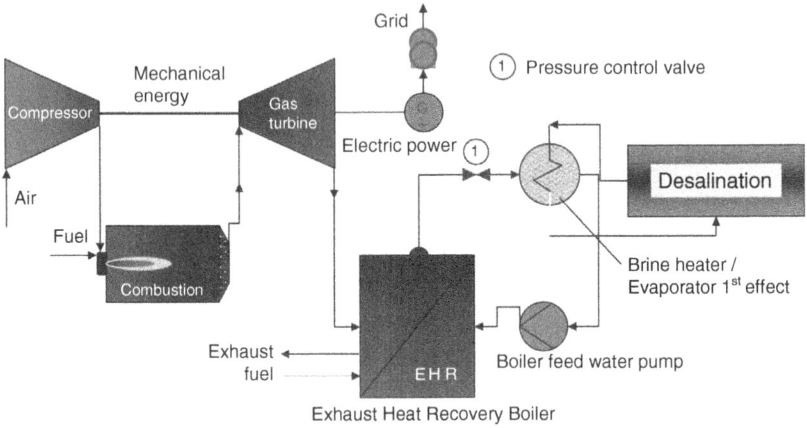

Figure 2.46 Dual purpose plant / Case 4: Gas turbine + exhaust recovery boiler.

Figure 2.46 shows the coupling of a desalination plant with a gas turbine. In this case the exhaust gas from the gas turbine is passed to a steam generator which provides the heating steam for the thermal seawater desalination plant. The operation mode at constant load is extremely simple, as only one pressure control valve is required before the desalination plant. Uneconomic operation may be found on partial load, i.e. the gas turbine no longer delivers the volume of exhaust gas necessary for full load operation of the desalination plant. The loss in capacity can be partially compensated by the installation of an additional burner in the heat recovery boiler, whereby this layout then becomes more and more like a single purpose plant with a steam generator and all its associated negative characteristics. As gas turbine power stations are often not base load power stations it should be checked exactly if and how they can be coupled to a seawater desalination plant.

In principle there are no new aspects which have to be taken into consideration in the combination of a combined cycle with a seawater desalination plant: the exact design of a dual purpose plant can in the end be seen in the context of the electricity and water demand of a particular region or even of the country in question. Above all the aim of this chapter was to make clear that we should prefer the dual purpose mode in thermal seawater desalination in order to optimise the consumption of the chemical energy stored in the fuel.

2.3.3 Specific Primary Energy Consumption

When we speak of primary energy we mean the chemical energy stored in the fuel which can be regarded as pure exergy:

$$Ex = \dot{m}_{Fu} \cdot H_{Fuel} \tag{2.143}$$

In order to determine the primary energy demand a calculation must therefore be made backwards from the energy demand of the plants to the exergy and with this to the fuel. In doing this a difference must be made between the thermal and electrical energy to be supplied to the plants.

The fuel demand for the heat for the thermal plants was already calculated as follows in Chap. 2.3.2 (Eq. 2.123 and Eq. 2.139):

Single purpose

$$\frac{\dot{m}_{Fu,SP}}{\dot{m}_{HS}} = \frac{\Delta h_{V,HS}}{\eta_B \cdot H_{Fuel}} \tag{2.144}$$

Dual purpose

$$\frac{\dot{m}_{Fu,DP}}{\dot{m}_{HS}} = \frac{\Delta h_{V,HS}}{H_{Fuel}} \cdot \frac{\eta_T \cdot \eta_{C,HS}}{\eta_{PP}} \tag{2.145}$$

Using Eq. 2.143 the exergy flow may be determined as:

Single purpose

$$Ex_{SP,th} = \dot{m}_{Fu,SP} \cdot H_{Fuel} = \dot{m}_{HS} \cdot \Delta h_{V,HS} \cdot \frac{1}{\eta_B} \tag{2.146}$$

Dual purpose

$$Ex_{DP,th} = \dot{m}_{Fu,DP} \cdot H_{Fuel} = \dot{m}_{HS} \cdot \Delta h_{V,HS} \cdot \frac{\eta_T \cdot \eta_{C,HS}}{\eta_{PP}} \tag{2.147}$$

Although the heat represents the major part of the primary energy cost for the thermal plants, the demand of electrical energy for pumps and auxiliary equipment must not be forgotten. Without going into the individual items in greater detail, at this point an approximated value for the different plants should be used:

MSF plants (brine recycle mode)

$$\frac{P_{el,MSF}}{\dot{m}_D} = 2.5 \frac{kWh}{m^3} \tag{2.148}$$

MED plants

$$\frac{P_{el,MED}}{\dot{m}_D} = 2.0 \frac{kWh}{m^3} \tag{2.149}$$

The electrical energy is produced in a power station. Therefore the fuel consumption can be directly derived via the total efficiency of the power station:

$$\eta_{PP} = \frac{P_{el}}{\dot{m}_{Fu} \cdot H_{Fu}} \tag{2.150}$$

This results in the following expression for the exergy demand for provision of electrical energy:

MSF plants (brine recycle mode)

$$Ex_{MSF,el} = \dot{m}_{Fu,MSF} \cdot H_{Fu} = \frac{P_{el,MSF}}{\eta_{PP}} \tag{2.151}$$

MED plants

$$Ex_{MED,el} = \dot{m}_{Fu,MED} \cdot H_{Fu} = \frac{P_{el,MED}}{\eta_{PP}} \tag{2.152}$$

The heat demand of an MSF plant and an MED plant can be expressed in a formula as follows, as shown in Chap. 2.2:

MSF (Eq. 2.59):

$$\dot{Q}_{H,MSF} = \dot{m}_D \cdot \frac{\Delta h_{V,T_m}}{N} \cdot (1 + N \cdot \frac{\Delta T_{TTD} + \Delta T_{BPE} + \Delta T_{Losses}}{\Delta T_0})$$

MED (Eq. 2.53):

$$\dot{Q}_{H,MED} = \dot{m}_D \cdot \frac{\Delta h_{V,T_m}}{N} + \frac{CF}{CF-1} \cdot c_P \cdot (\Delta T_{Stage} + \Delta T_{TTD} + \Delta T_{BPE})$$

The heat is provided by the condensation of saturated steam, so that the following terms apply:

$$\dot{Q}_{H,MSF} = \dot{m}_{HS,MSF} \cdot \Delta h_{V,HS} \tag{2.153}$$

$$\dot{Q}_{H,MED} = \dot{m}_{HS,MED} \cdot \Delta h_{V,HS} \tag{2.154}$$

If you now combine Eq. 2.153 respectively Eq. 2.154 with Eq. 2.59 respectively Eq. 2.53 and enter the result into Eq. 2.147 for the dual purpose mode (the single mode may be treated in an analogous manner which does not happen here so that we can retain a clear overview), then, in each case, you obtain an equation for the specific exergy requirement of an MSF or an MED plant:

MSF / dual purpose

$$\frac{Ex_{DP,th,MSF}}{\dot{m}_D} = \left[\frac{\Delta h_{V,T_m}}{N} \cdot (1 + N \cdot \frac{\Delta T_{TTD} + \Delta T_{BPE} + \Delta T_{Losses}}{\Delta T_0}) \right] \cdot \frac{\eta_T \cdot \eta_{C,HS}}{\eta_{PP}} \tag{2.155}$$

MED / dual purpose

$$\frac{Ex_{DP,th,MED}}{\dot{m}_D} = \left[\frac{\Delta h_{V,T_m}}{N} + \frac{CF}{CF-1} \cdot c_P \cdot (\Delta T_{Stage} + \Delta T_{TTD} + \Delta T_{BPE}) \right] \cdot \frac{\eta_T \cdot \eta_{C,HS}}{\eta_{PP}} \tag{2.156}$$

In order to arrive at the total specific exergy demand, and thus the specific primary energy demand of the plants, the share for the electrical energy must be added to the heat. Using Eq. 2.148 and Eq. 2.149 this finally gives:

MSF / dual purpose

$$\frac{Ex_{DP,MSF}}{\dot{m}_D} = \left[\frac{\Delta h_{V,T_m}}{N} \cdot (1 + N \cdot \frac{\Delta T_{TTD} + \Delta T_{BPE} + \Delta T_{Losses}}{\Delta T_0}) \right]$$

$$\cdot \frac{\eta_T \cdot \eta_{C,HS}}{\eta_{PP}} + \frac{P_{el,MSF}}{\eta_{PP} \cdot \dot{m}_D} \tag{2.157}$$

MED / dual purpose

$$\frac{Ex_{DP,MED}}{\dot{m}_D} = \left[\frac{\Delta h_{V,T_m}}{N} + \frac{CF}{CF-1} \cdot c_P \cdot (\Delta T_{Stage} + \Delta T_{TTD} + \Delta T_{BPE}\right]$$

$$\cdot \frac{\eta_T \cdot \eta_{C,HS}}{\eta_{PP}} + \frac{P_{el,MED}}{\eta_{PP} \cdot \dot{m}_D} \tag{2.158}$$

As reverse osmosis only needs electrical energy the determination of the exergy demand is relatively simple. The starting point is the specific electricity demand of an RO plant[4].

$$\frac{P_t}{\dot{m}_P} = \frac{\Delta p_a}{\varphi \cdot \rho_F} \cdot \left(\frac{1}{\eta_P} - (1-\varphi) \cdot \eta_T \cdot \frac{\Delta p_\omega}{\Delta p_a}\right) \tag{2.159}$$

Here, it is assumed that the density of the feed flow is equal to the density of the retentate flow.

As the electricity is produced in a power station the fuel demand can be determined via the total efficiency of the power station. This therefore results in:

$$\frac{P_{el,RO}}{\dot{m}_D} = \frac{1}{\rho} \cdot \left[\frac{\Delta p_{RO} + \Delta p_{LOSS} + b \cdot w_F \cdot CF}{1 - \frac{1}{CF}} \cdot \frac{1}{\eta_P} - \frac{\Delta p_{RO} + b \cdot w_F \cdot CF}{CF-1} \cdot \eta_{ER}\right] \tag{2.160}$$

$$\Rightarrow \quad \frac{Ex_{RO}}{\dot{m}_D} = \frac{1}{\rho} \cdot \left[\frac{\Delta p_{RO} + \Delta p_{LOSS} + b \cdot w_F \cdot CF}{1 - \frac{1}{CF}} \cdot \frac{1}{\eta_P}\right.$$

$$\left. - \frac{\Delta p_{RO} + b \cdot w_F \cdot CF}{CF-1} \cdot \eta_{ER}\right] \cdot \frac{1}{\eta_{PP}} \tag{2.161}$$

If an additional share of electrical energy is also taken into account for the auxiliaries, then we finally obtain for the total specific exergy demand of a reverse osmosis unit:

$$\frac{Ex_{RO}}{\dot{m}_D} = \frac{1}{\rho} \cdot \left[\frac{\Delta p_{RO} + \Delta p_{LOSS} + b \cdot w_F \cdot CF}{1 - \frac{1}{CF}} \cdot \frac{1}{\eta_P} - \frac{\Delta p_{RO} + b \cdot w_F \cdot CF}{CF - 1} \cdot \eta_{ER} \right] \cdot \frac{1}{\eta_{PP}} + \frac{P_{el,MED}}{\eta_{PP} \cdot \dot{m}_D} \tag{2.162}$$

Assuming typical figures for desalination plants according to the state-of-the-art (see Table. 2.2) a graph may be drawn which shows the specific exergy demand and the specific primary energy

Table. 2.2 Typical values of thermal and RO desalination plants as a basis for Figure 2.47

Ex	Primary energy consumption	[kW]
N	Number of stages	[-]
$\dot{m}_D$	Mass flow distillate / permeate	[to/ h]
ΔT_{TTD}	Terminal temperature difference	1.5 K
ΔT_{BPE}	Boiling point elevation	0.7 K
ΔT_{Losses}	Non-equilibrium losses	0.6 K
ΔT_0	Overall temperature difference	80 K
T_T	Temperature of heating steam	393 K (120 °C)
T_{MIN}	Theoretical condensing temperature	313 K (40 °C)
Δh_V	Heat of vaporization (at 120°C)	2200 kJ kg
c_P	Specific heat capacity	4.2 kJ kgK
η_{PP}	Efficiency power plant	0.40
η_T	Efficiency steam turbine	0.90
η_P	Efficiency RO high pressure pump	0.80
η_{ERT}	Efficiency RO energy recovery turbine	0.80
Δp_{RO}	Driving pressure difference	30 bar
Δp_{Losses}	Pressure losses due to friction	4 bar
b	Osmotic pressure coefficient	8 bar/ %TDS
w_F	Feed concentration	4.2 %TDS
CF_{RO}	RO concentration factor	1.5
CF_{MED}	MED concentration factor	2.0
ρ	Density of saline water	1041 kg/ m³
$P_{AUX RO}/m_D$	Specific energy demand for auxiliaries	0.5 kWh/ to
$P_{AUX MSF}/m_D$	Specific energy demand for auxiliaries	2.0 kWh/ to
$P_{AUX MED}/m_D$	Specific energy demand for auxiliaries	2.0 kWh/ to

consumption respectively as a function of the number of stages N. In Figure 2.47 three curves may be identified:

1. MSF plant / dual purpose mode → Eq.2.160
2. MED plant / dual purpose mode → Eq.2.161
3. RO plant → Eq.2.162

The following may be observed:

- The primary energy demand of reverse osmosis amounts to around 15 kWh/m³ independently of the number of stages.
- As was to be expected there is a strong dependency on the number of stages for the thermal plants. MSF plants always require more primary energy than MED plants. However, a state-of-the-art MSF plant (approx. 20 stages, GOR ≈ 8) requires about 30 kWh/m³; state-of-the-art MED plants (approx. 10 stages, GOR ≈ 8) require around 35 kWh/m³.
- Only above a number of stages of N > 40 do MED plants catch up with RO plants, MSF plants only reach values of 35 kWh/m³ at the most, even with a high number of stages.

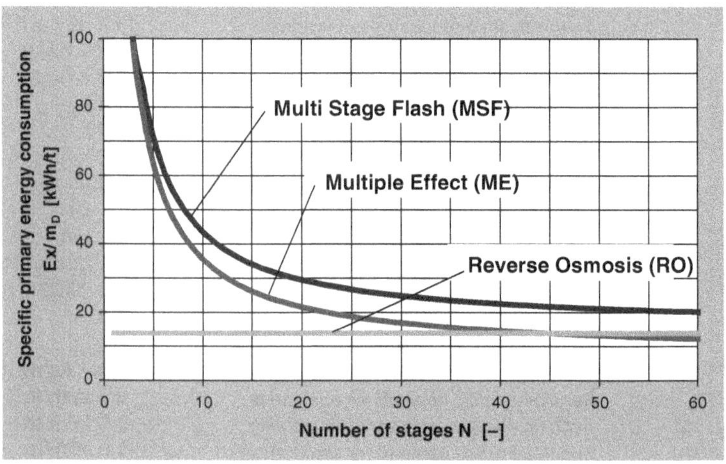

Figure 2.47 Specific exergy and primary energy consumption vs. number of stages (for dual purpose mode).

2.4 Historical Review

All human actions take the path
from the primitive via the complicated to the simple.
Antoine de Saint-Exupery, 1939

Professor Robert Rautenbach placed the quotation from Antoine de Saint-Exupery at the head of his book "Membrane processes – Fundamentals of Module and Plant Design" which was first published in 1996, in the knowledge that it is also true for technical processes and provides the main objective which should be strived for in the development of processes. The development of seawater desalination until now is a fine example for the path which technicians and engineers have taken from the primitive via the complicated to the simple, whereby the question may be directly asked as to whether today we have already reached the simple stage and as to how the next steps in the development of seawater desalination could look.

Every age had different ways of approaching the desalination of water and in each of the ages – to use the language of mathematics – "local maxima" were discovered, i.e. simple and good solutions. In the following, I would like to refer to an excellent summary by James D. Birkett, which has been published in the DESWARE – encyclopaedia[7].

The desalination of seawater is in principle not a complicated matter. Every raindrop is evidence of how simple the principle of nature is: Seawater evaporates due to the energy of the sun's rays, the moist air is further warmed and rises into the cooler air layers where very fine water droplets condense to be visible as clouds and they then fall to the ground. The astonishing thing is that rainwater is "soft", i.e. it contains no salt, although its source is salty seawater. Physically this is due to the different steam pressures of water and salt, in other words: only the water is evaporated by the sunrays, the salt remains dissolved in the water. This was known to man from ancient times, as Aristotle wrote in his volume *Meteorologica*[8].

> *"Salt water becomes sweet when it turns into vapour and the vapour does not form saltwater when it condenses again.."*

Sailors were the first people to copy the natural water cycle as shown in Figure 2.48 in that they boiled seawater in a vessel using

Figure 2.48 Ancient representation of drinking water extraction from the sea[9].

a wood fire and condensed the vapour in a sponge. The water squeezed out of the sponge (virtually salt-free) tasted unbeknownst; however, it could be drunk without giving it a second thought. Depending on your point of view this process may be described as "primitive" or as "simple"; we tend to the word "simple", based on the list of Saint-Exupery, as this process contains all the technical process steps and equipment which are necessary today for the correct and simple application: a container for the seawater, a focussed energy supply, the use of a cold sponge with a large surface as a condenser and at the same time as a collecting point for the desalinated water, the distillate. The technical description today would probably be: a discontinuous, single-stage evaporation process under atmospheric conditions.

The task described, i.e. to desalinate seawater in small volumes for personal use, changed significantly in the 15th century as once again sailors were confronted with the question of the supply of drinking water on their long journeys which now spanned the seas of the world.

A cylindrical vessel may be seen on a commemorative coin to celebrate the works of an Englishman named Fitzgerald (see Figure 2.49)

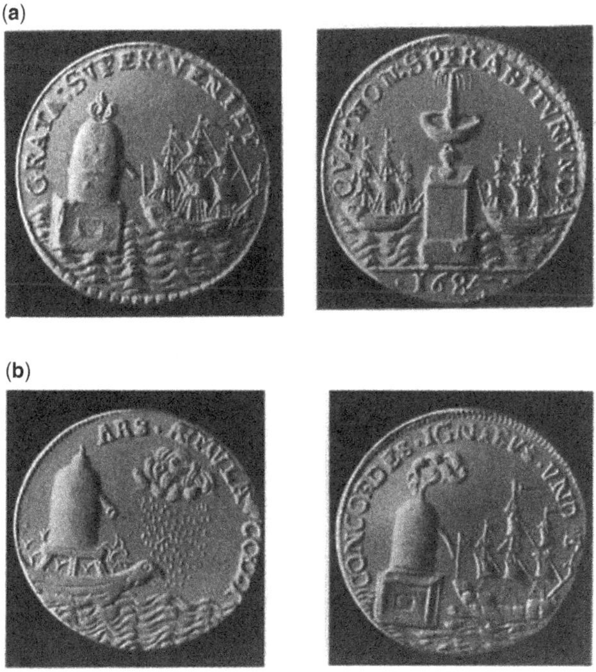

Figure 2.49 Commemorative coin in Fitzgerald's honour (1684)[10].

[10]; its contents, the seawater, are heated over an oven which is fired with either wood or coal. The distillate trickles from the head, where it is likely that the condenser was located. There are other places in literature in which it is reported that a tube has been placed outside the ship's hull in order to condense the vapour with cool seawater.

The solution found by Fitzgerald, and at the same time elsewhere by his contemporary Walcot, was only different from the ancient methods in its use of cast-iron vessels and tubes. In comparison with other technical achievements of the age we would call this "discontinuous, single-stage evaporation process under atmospheric conditions" primitive, although much praise was offered up when conferring the commemorative coin[10]:

> "By him the waters, acid and marine, are purg'd and freed from their destructive brine: The sailor now to farthest shores may go, Since in his road these lasting fountains flow; . . . . . No more with thirst the feav'rish sea-man dyes, The briny waves afford him fresh supply."

The evaporation equipment for which a patent was registered in 1720 by the Frenchman Gautier[11] had, certainly in comparative terms, left the status of the primitive, it must however be classified as complicated, with its hand-driven evaporator drum and its grill firing located on the axis of this hollow drum. However, in its design it showed initial noteworthy progress in comparison with previous equipments.

The water is not boiled in a vessel, but instead a small amount of water is distributed over an undulated heat transfer surface and is partially evaporated from this thin film of water. Gautier seemingly knew that evaporation from a thin film is highly effective and that it could be optimised through the undulated surface of the drum in regard of the distribution of the water as well as the size of the heat transfer surface. The heuristic rule which today is known to every process engineer may be recognised for the first time in this construction: the so-called space-time-yield, i.e. the maximisation of the volume of a product per volume of the apparatus, so that the costs per unit of product could be minimised by low apparatus costs. Even the condenser, in the shape of its triangular profile, follows this construction principle and offers a large surface while the required volume is small.

The era of seawater desalination plants on ships – even the Titanic had several single-stage vessel-like evaporators on board – was marked solely by the task of producing desalinated water at a

Figure 2.50 Gautier's evaporation equipment (1720)[11].

relatively low plant cost, i.e. capital cost. Because of the low water demand on board the question of energy was not of significance. The boilers on the Titanic had to generate many times more energy for the ship's drive using steam turbines in comparison with the demand for the evaporator. Cast iron was used as a material, the high corrosion rates of the seawater were counteracted by thick walls. As a result of the boiling process, crusts of salt were formed on the walls and the heat exchanger tubes. Evaporation meant at that time: heavy, unwieldy boilers with big corrosion and scaling problems and a great demand for maintenance and repair. At that time people were far away from a simple solution, the designs varied, depending on the creativity of the inventor, between "primitive" and "complicated".

2.5 State-of-the-Art

At this point the chapter is concerned with the question as to whether the current "state-of-the-art" has reached a local maximum or an absolute maximum.

2.5.1 Multiple-Effect Distillation (MED)

In the 19th century there was progress in the world of the evaporator: however it was not seawater desalination which led to an upturn in this technology but rather the enormous growth worldwide in this period for pure, white refined sugar. In referring to this we must briefly explain the sugar production process. The sugar cane or sugar beet is initially cut into small pieces in order to obtain good conditions for the subsequent sugar extraction. After the extraction there is a watery sugar solution as well as the remains of the plant. The well-known white sugar is obtained by the slow removal of the water until the solubility of the sugar is exceeded, so that fine crystals are formed. The removal of the water by evaporation requires energy which is introduced to the process in the form of heat by the burning of a fuel. However, different to the small ships' evaporators, suddenly the energy question appears as a result of the amounts needed and the market prices for sugar! The engineers who were given the task of finding a solution to this question immediately suggested burning the plant remains, the so-called begasse. This idea was good, but not good enough, as the energy balance in the sugar factory did not work, i.e. more energy was required than the begasse can provide. This led to additional energy to be purchased and transferred to the sugar factory, which significantly raised the price of the sugar. This price pressure drew from the engineers the idea of multiple-effect or multi-stage evaporation which may well be described as a "quantum leap in evaporator technology" (see Figure 2.51). The Frenchman Rillieux registered a US patent in 1846[12] for a multiple-effect evaporator and commented on it that:

> „In a multi-effect, one pound of steam applied to the apparatus will evaporate as many pounds of water as there are bodies in the set."

and

> *"Suddenly the begasse by itself had sufficient fuel value to maintain the refining process."*

Inspired by the progress in evaporator technology, the sugar industry experienced a boom in the following decades. The technical solutions had a high standard and, looked at today, may be described as "simple" if not "brilliantly simple". The current evaporator plants in the sugar industry only vary slightly from the technology of that time. The most decisive aspect however in connection with seawater desalination which is being discussed here, is that in the sugar industry an essential energy saving concept for all thermal separation processes had been discovered in the multiple-effect evaporation principle.

Up to the beginning of the 1960s seawater desalination restricted itself to smallish plants for supplies to military locations in the

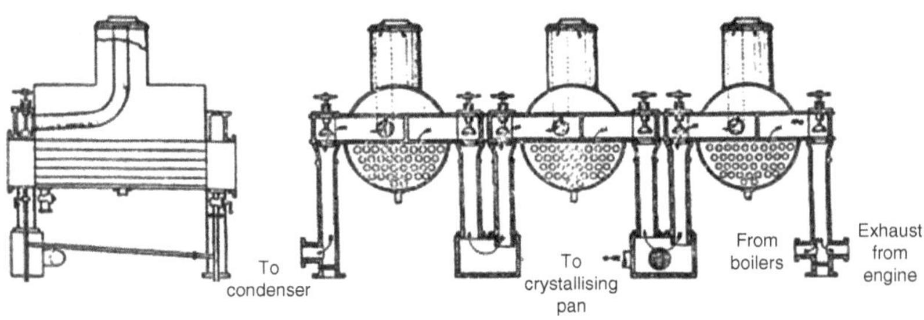

Figure 2.51 Rillieux's multiple-effect evaporator (1846)[12].

Caribbean and in the Mediterranean, for example on Malta. The technology used was a copy of the plants from the sugar industry, as no special "seawater desalination industry" had been established due to a lack of demand.

The task of desalinating seawater is however a different situation compared to the concentrating of sugar syrup. While in the one case the sugar is the product, the evaporated water is the product from seawater desalination; while the crystals are desired in the production of sugar, they are definitely to be avoided in seawater desalination in order to prevent scaling on the heat transfer surfaces. Therefore in a sugar factory a technically complex "sugar evaporator" must be built in any case with the equivalently high capital costs. The application of this complicated equipment in seawater desalination may lead to a plant which somehow works, however inevitably produces higher drinking water costs.

The technology which was used in the 1960s in seawater desalination was unwieldy, difficult to operate due to the ever-present risk of scaling, and expensive to procure due to the high corrosion danger through the salt water - and it had in addition a high energy consumption due to the relatively small number of stages: the production of water from seawater was an exotic and expensive business and was only used where the need was great and money was not of significance, for example for the deployment of troops at strategically important locations.

The largest plant in the world in 1959 was on the Antilles, had 6 stages, submerged evaporators (similar to an immersion heater) and a capacity of 2 MIGD, that is about 7,500 m^3/d[13].

The picture of the plant on a postage stamp (see Figure 2.52) has significant similarities with Rillieux's patent drawing and shows no noteworthy developments.

The main disadvantage of multiple-effect evaporators of the "immersion heater" type - or in the technical jargon "submerged tubes" – is the inevitable scaling on the heat exchanger tubes. However this can be avoided if the evaporation takes place from thin films with low temperature differences, whereby steam bubbles, with consequential local super-saturation, do not occur. The advantages of this principle were recognised by Gautier in 1720 and – how could it be otherwise – also by the sugar industry. The single stage film evaporator, which was in use in 1888 in Philadelphia, has all the characteristics of today's apparatus equipment of the same type (see Figure 2.53 [15] and Figure 2.54 [16])

Figure 2.52 Multiple-effect seawater desalination plant on Aruba (1959)[13].

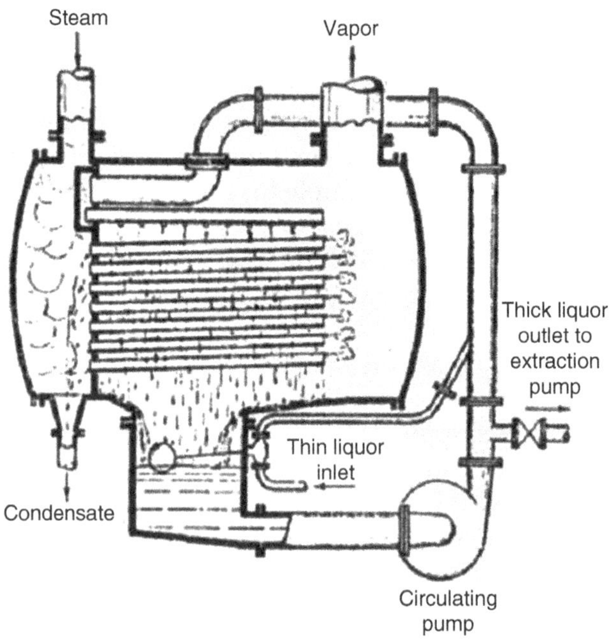

Figure 2.53 Horizontal tube sprayed film evaporator S. Morris Lillie Sugar Manufacturing Company Philadelphia 1888 [15].

Figure 2.54 300 t/d - MVC evaporator during installation in a power plant (by courtesy of Hamworthy Serck Como GmbH, Geesthacht, Germany) [16].

The sugar syrup (or the seawater) is distributed on the outside of horizontal tubes. A continuous water film is then formed which flows downwards from tube to tube (hence the name "falling film evaporator"). The tubes are heated on the inside with condensing steam which transfers its heat to the film and leads to the evaporation of a certain amount of water. The main advantage of this evaporator type is in its high heat transfer coefficient in the film evaporation which leads to a lower heat transfer area being needed. Thus expensive, specifically corrosion resistant materials, such as titanium, can be used, as the low heat transfer area leads to moderate costs for the heat exchanger.

MSF plants demonstrate a low level of energy consumption if they are coupled with a power station. Because of this almost all large MSF plants are operated as so-called dual-purpose plants. However electricity plus water is not required at every location and it is not always worthwhile to build a power station, particularly if the demand for water is not so high.

Multiple-effect plants with falling film evaporators have a high market share for plants with a capacity of 5,000 to 25,000 m³/d per unit. These plants are supplied by a steam boiler installed especially for this purpose.

2.5.2 Multi-Stage-Flash Evaporation (MSF)

Seawater desalination plants would probably look exactly the same today if oil had not been discovered all around the Arab Gulf. The World Health Organisation quotes a volume of 1,000 m^3 per year as the minimum water availability of one person without risking any negative effects on her health. The water resources in the Gulf States are on average below 100 m^3/a·capita. The 1960s therefore brought a new task for the engineers: the production of drinking water from seawater in significantly larger volumes – and in safe, reliable manner and at low costs. It was Professor Silver, typically enough a Scotsman, who in 1965 registered the patent for the so-called Multi-Stage-Flash process – in short MSF – and thus laid the foundation for the current state of the art. The most brilliant aspect of Professor Silver's idea was the technical implementation of the flash process in a kind of simplicity which still cannot be found today. With his concept he completely avoided the "immersion heater principle" (submerged tubes), i.e. no evaporation took place on a heated surface – he simply physically separated the heat input from the evaporation. With this he solved the main difficulty in evaporation – the scaling of the heat exchanger surfaces.

A new age in seawater desalination began with the MSF process. From that moment the installed plant capacity increased continually up to now 40 million m^3/d. The largest plants currently in operation all work according to the MSF principle. Figure 2.53 shows an aerial photograph of the Umm Al Nar East desalination plant in Abu Dhabi comprising three MSF evaporator units with a capacity of 27,240 m^3 per day (6 MIGD)[14]. Each unit has a total of 16 stages and achieves a Gained Output Ratio (GOR) of 6. A characteristic feature of these types of plants is the side by side to a power plant providing the heating steam for the evaporation process (keywords: dual purpose plants or co-generation).

Modern MSF plants comprise about 20 stages at a GOR of 8. Thus they are not the most energy-efficient alternative in seawater desalination and also not the cheapest in relation to the investment. However, the robust construction, the long operating life of 20 to 30 years and the opportunity for use of waste heat from a turbine when coupled to a power station are decisive factors for the high market share of MSF desalination plants.

Figure 2.55 Umm al nar east electricity and water generation plant (by courtesy of Sidem)[14].

2.5.3 Multiple-Effect Distillation with Thermally Driven Vapour Compression

To use the energy from the steam generated at a pressure of between 10 and 20 bar in an optimal manner, the multiple-effect plants work with a thermo-compressor (known as MED – TVC for short) Figure 2.56 shows a typical picture of such an MED-TVC plant. Each of the three units comprises 6 stages or effects or cells. A thermo-compressor (recognisable clearly at the top of the plant) sucks up a part of the vapour produced in the 4th effect (or the final effect according to the tailor-made design). The so-called suction steam is compressed by motive steam from the boiler. The mixture of the steam flows (suction plus motive) provides heating steam for the first effect. Thus the Gained Output Ratio (GOR) of the plant shown in Figure 2.56 rises up to 9. Thermo-compression plants are normally fitted with 4 or 6 effects and achieve GOR values of 7 to 9, comparable with those MSF plants.

The losses through the mixing process in the compressor are relatively high; this is however compensated, like in the MSF process, by the simple construction of the plant and the uncomplicated thermo-compressor. There is usually no staged pre-heating of the seawater; the water is distributed in parallel into the stages, which leads to a simplification of the whole process.

Figure 2.56 MED - TVC plant with 3 units each 13,333 m³/d / Tobruk, Libya (by courtesy of SIDEM (VEOLIA Water Systems) [14].

At this point the following points must be noted. Processes, MSF and MED-TVC, simplify the multiple-effect principle: in the MSF process there is no evaporation, simply pre-heating and then flashing; in the MED-TVC process there is no pre-heating, simply evaporation. In both cases this leads to a simple design for the total plant. All attempts made to combine MSF and MED, i.e. to integrate pre-heater and evaporator in one stage and to operate the whole equipment in a multiple-effect manner, have failed due to the complicated nature of the design.

When speaking of the MED-TVC technique we can also refer to it as a local maximum in relation to its simplicity and reliability, however it should be made clear that this takes place at the cost of a poor utilisation of fuel energy.

2.6 Future Prospects

What does the future of seawater desalination look like? Will a new task appear? And will the state-of-the-art be sufficient to fulfil this task? What will come after MSF, the currently dominant process which is also simple and reliable?

In our opinion the new task has two main aspects. The demand for water will continue to increase in the MENA states, i.e. in the Middle East and in North Africa simply due to the increase in the population and the increasing standard of living and will remain clearly below the WHO value of 1,000 $m^3/d \cdot$capita.

Most likely China will fall below the mark of 2.000 $m^3/d \cdot capita$ in the next 20 years, hence the water supply issue will climb up on the agenda. At the same time the currently most important source of energy in the MENA-state countries, oil and gas, will become scarce commodities and will drive the water production costs upwards. Seawater desalination in the future must therefore be in a position to supply a constantly growing market in a reliable manner and at low cost. In this area the move to alternative forms of energy will play a decisive part.

Extensive investigations were carried out in this area in the past, beginning with the use of solar ponds up to the coupling of seawater desalination plants with nuclear power stations. However, the energy market has changed in recent years and this must be taken into consideration. In addition allowance must be made for the fact that large volumes of water (300,000 m^3/d and more) will have to be provided at one location.

The following may be noted as keywords when speaking of alternative sources of energy:

- Solar energy
- Wind power
- Fuel cells
- Micro-turbines
- Block-type thermal power stations

As well as the energy supply itself the plants must be evidently optimised in such a way that the thermal and mechanical energy consumption is minimised. This can be achieved for the multiple-effect plants (MSF and MED) in the simplest way through an increase in the number of stages. The decisive factor here is the temperature range in which the processes work. At the moment the maximum temperature is restricted to about 110°C, the lowest temperature through the seawater which serves as coolant water. The top limit is given by the solubility limit of calcium sulphate (gypsum), by the steeply increasing vapour pressure of water above 110°C, and the danger of corrosion at high temperatures. In order to move these boundaries upward the following must happen:

1. the precipitation of calcium sulphate must be prevented,
2. evaporator chambers must be constructed which can stand the pressure,
3. corrosion-resistant materials must be developed.

A higher number of stages mean in any case a reduction in the thermal energy consumption of the plant but also the construction of a larger number of stages with the relevant evaporators and preheaters which also have to operate at higher pressures. Therefore an increase in the number of stages is only worthwhile where cheap materials can be used for the heat exchanger and a cheap design can be made. In addition the process-related thermodynamic and hydrodynamic losses increase with the number of stages, so that the energy saving from a certain number of stages is only marginal. A warning comes from an attempt in the 1980s to operate an MED plant with 55 stages, which only achieved a GOR of 37 instead of an ideal GOR of 55, the efficiency was therefore only 65 %.

Research in the development of "high-temperature evaporator plants", for instance in the area of the materials or scaling inhibitors, should therefore definitely be accompanied by a carefully prepared cost-benefit study based on a thermodynamic and hydrodynamic simulation of the process.

As the MED process is superior to the MSF process in terms of thermal efficiency, efforts should be made to optimise the MED process design, whereby the main focus should be on the evaporator as heat exchanger unit. The evaporator type used previously in the MED plants is a horizontally arranged tube bundle made from metal, for instance titanium or aluminium or copper nickel. However, in other fields evaporators are in operation which are made of plastic sheets, stuck together to form blocks and can be completely replaced after a service life of some years – similarly to the modules of a membrane plant.

Low temperature evaporators, able to operate below 70°C with waste heat, could be developed as a counterpart to the high temperature evaporator plants. The installation of a higher number of stages assumes small temperature differences for the heat transfer, which would directly lead to the need for larger heat transfer areas. The evaporators which would have to be developed must therefore consist of cheap materials which are corrosion-resistant and guarantee a high heat transfer coefficient. Thin plastic sheets can well meet these demands; however it has not yet been definitely clarified as to how the wetting of the sheets could be made without scaling occurring. Plastic tubes are a possible alternative, whereby the question may immediately arise regarding the mechanical stability and the wall thickness.

Should it prove possible to develop efficient low temperature evaporators, the next step would be to investigate the optimum coupling with a suitable energy source. As an alternative to the large plants, many, small decentralised plants could be erected which run with fuel cells or with wind turbines. The evaporator plants would be standardised plants, for example in containers, comparable with block-type thermal power stations, which would be placed in series to create larger units.

At this point we would not wish to go into the full breadth of the optimisation potential of reverse osmosis when operating with seawater (operating time of the membranes, improvement in fouling and scaling behaviour for the membranes and the modules, chemical consumption, personnel requirements) – there is much research work in progress in this area. We would rather like to propose as a vision of the future a closer form of interlocking between the membrane technology and thermal seawater desalination plants.

So, for example, the coupling of an MSF plant with a reverse osmosis plant, whereby the brine, the waste from the MSF plant, would serve as the feed for the reverse osmosis plant. This hybrid plant would have the advantage that no conventional pre-treatment of the seawater would be required before the reverse osmosis, as the heated brine from the MSF plant has a greatly reduced tendency to fouling and scaling. The disadvantages of the process are in the need to have to operational membranes and modules at temperatures higher than 40°C and in the increased osmotic pressure and the subsequently reduced yield of the reverse osmosis process.

Intensive research should also be made into the use of micro- and ultrafiltration plants for pre-treatment of the seawater as well as nanofiltration for partial removal of the bivalent ions, in particular of calcium which is responsible for the formation of crystalline calcium carbonate and calcium sulphate and thus for scaling on the membranes.

The continuously growing demand for desalinated seawater poses the urgent question of the effects of an increase in the salt content of the sea through the discharge of the concentrate from the plants. This not only affects the design parameters of the plants, such as the osmotic pressure and the maximum possible yield, but also the living conditions in the sea. A focus of future research activities must address the question of concentrate disposal[17].

References

1. H. E. Hömig, *Seawater and Seawater Desalination* Vulkan-Verlag, Essen 1978, ISBN 3-8027-243-0.
2. R.B. Bird, W.E. Steward and E.N. Lightfoot, *Transport Phenomena* John Wiley & Sons, Inc., New York, London. (1960).
3. F.H. Geoffrey, *Heat Exchanger Design Handbook 2002* (HEDH 2002), ISBN 1-56700-181-5.
4. J. Gebel and S. Yüce, *An Engineer's Guide to Desalination* VGB Powertech Service GmbH, Essen. (2008). ISBN 978-3-86875-000-3.
5. B.P. Robert, *Steam Jet Ejections for the Process Industries,* McGraw-Hill, Inc., (1994). ISBN 0-07-050618-3.
6. VDI Wärmeatlas Berechnungsblätter für den Wärmeübergang A24 VDI-Verlag, Düsseldorf 5. Auflage. (1988). ISBN 3-18-400760-X
7. D.B. James, *The History of Desalination before Large Scale DESWARE-Encyclopedia of Desalination and Water Resources*, EOLSS Publishers Co. Ltd., Oxford. (2000). ISBN 0-953-49442-X http://www.desware.net
8. R.J. Forbes, *A Short History of the Art of Distillation*, Leiden. (1948).
9. J.J. Howarth, *Product Literature for P and B Evaporators,* Aiton Ltd, Derby (1984).
10. G. Nebbia and G.N. Menozzi ,Aspetti storici della dissalzione, *Acqua Ind* **41–42**, 3–20. (1996)
11. J.R. Partington, *History of Chemistry*, Vols 2–3, McMillan, London (1962).
12. N. Deerr, *The History of Sugar*, Vol. II, Chapman and Hall, New York (1950).
13. E.M. Arrnidell and J.D. Birkett, Arubas Early Experience in Desalination The International Desalination and Waste Water Quarterly (D&WR) Vol.12/3, November/December 2002
14. Sidem - Société International de Dessalement, Paris, France http: www.veoliawater.com/services/solutions
15. O. Lyle, Efficient Use of Steam, His Majesty's Stationary Office, London 1946
16. Hamworthy Serck Como GmbH, Geesthacht, Germany. (www.hamworthy.com)
17. S. Lattemann, Hoepner T., Seawater Desalination - Impact of Brine and Chemical Discharge on the Marine Environment Balaban Desalination Publication, L'Aquila, Italy 2003

SECTION III

MEMBRANE PROCESSES

3

The Reverse Osmosis Process

Mark Wilf

Abstract

Reverse osmosis (RO) is membrane based, energy efficient, water desalination process. Application of RO process is usually more expensive than application of conventional water treatment technologies. However, RO is the only process enabling cost effective reduction of seawater or brackish water salinity, down to the potable salinity range. This chapter includes fundamentals of the RO process, technical information on commercial RO products, configuration of membrane units and RO plants, process parameters and monitoring of performance of RO membrane units.

Keywords: Reverse Osmosis, RO, desalination, membranes, membrane modules, membrane system configuration, RO process, RO operating parameters.

3.1 The Reverse Osmosis Process

Osmosis is a natural process involving fluid flows across a semipermeable membrane barrier. It is selective in the sense that the solvent passes through the membrane at a faster rate than the passage of dissolved solids. The difference of passage rate results in solvent - solids separation. The direction of solvent flow is determined by its chemical potential, which is a function of pressure, temperature, and concentration of dissolved solids. Pure water in contact with both sides of an ideal semipermeable membrane at equal pressure and temperature has no net flow across the membrane because the chemical potential is equal on both sides. If a soluble salt is added to water on one side of the membrane, the chemical potential of this salt solution is reduced. Osmotic flow

Jane Kucera (ed.) Desalination , (157–204) 2014 © Scrivener Publishing LLC

from the pure water side across the membrane to the salt solution side will occur until the equilibrium of chemical potential is restored (Figure 3.1a). Equilibrium occurs when the hydrostatic pressure differential resulting from the volume changes on both sides is equal to the osmotic pressure. This is a solution property independent of the membrane. Application of an external pressure to the salt solution side, which is equal to the osmotic pressure, will also cause equilibrium. Additional pressure will raise the chemical potential of the water in the salt solution and cause a solvent flow to the pure water side, because it now has a lower chemical potential. This phenomenon is called reverse osmosis (Figure 3.1b).

The osmotic pressure, P_{osm}, of a solution can be determined indirectly by measuring the concentration of dissolved salts in solution:

$$P_{osm} = R\,(T + 273)\,\Sigma(m_i) \tag{3.1}$$

Where P_{osm} is osmotic pressure (in bar), R is universal gas constant (0.082 l atm/mol °K), T is the temperature (in °C), and $\Sigma(m_i)$ is the sum of molar concentration of all constituents in a solution.

An approximation for P_{osm} can be made by assuming that 1000 ppm concentration of Total Dissolved Solids (TDS) equals about 0.77 bar (11 psi) of osmotic pressure. For example, in RO unit operating at 75% recovery rate, feed salinity is 3,000 ppm TDS and concentrate salinity is about 11,500 ppm TDS. Accordingly, osmotic pressure of the feed is 2.3 bar (33 psi) and the concentrate is 8.7 bar (126 psi). The equation 3.1 only holds for dilute salt solutions and temperatures close to 25 C. At significantly different conditions a

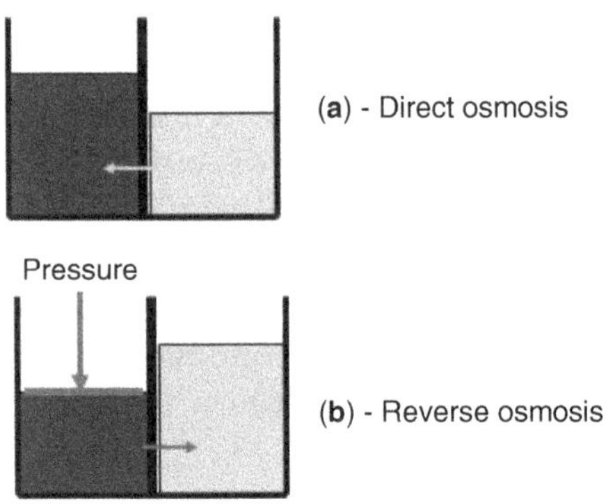

Figure 3.1 The direct and reverse osmosis process.

more rigorous calculations, that takes into consideration ions activities rather than concentrations, has to be applied.

3.2 Permeate Recovery Rate (Conversion Ratio)

Permeate recovery is one of more important parameters in the design and operation of RO systems. Recovery or conversion rate of feed water to product (permeate) is defined by Equation 3.2 and 3.3 and illustrated in Figure 3.2:

$$R = 100\% \, (Q_p / Q_f) \tag{3.2}$$

$$R = 100\% \, (Q_p / (Q_p + Q_c)) \tag{3.3}$$

Where R is recovery rate (in %), Q_p is the product water flow rate, Q_f is the feed water flow rate and Q_c is the concentrate flow rate.

The recovery rate affects salt passage and product flow. As the recovery rate increases, the salt concentration on the feed-brine side of the membrane increases, which causes an increase in salt flow rate across the membrane. Also, a higher salt concentration in the feed-brine solution increases the osmotic pressure, reducing the net driving pressure available and consequently reducing the product water flow rate.

In a multistage RO system recovery rate is defined for each stage and for combined system (Figure 3.3).

3.3 Net Driving Pressure

The net driving pressure (NDP) is the driving force of the water transport through the semipermeable membrane. The value of

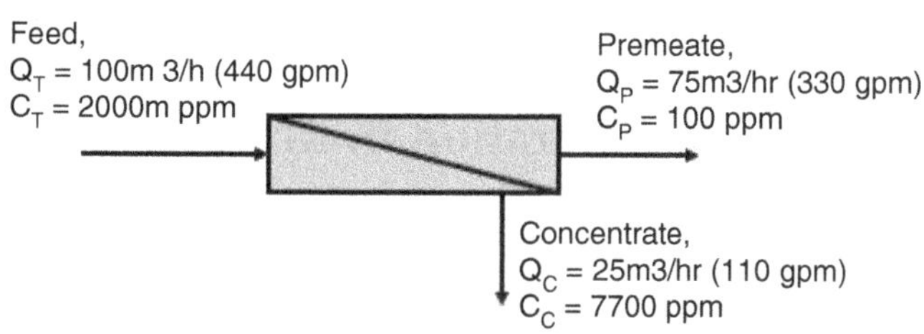

Figure 3.2 Mass balance in reverse osmosis unit. Recovery ration 75%.

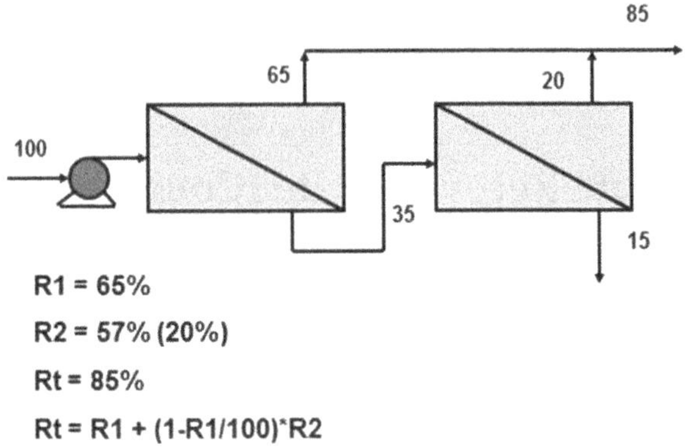

Figure 3.3 Recovery rate in a two stage RO unit.

NDP decreases along the RO unit. Therefore, for the purpose of membrane performance calculations it is defined as an average NDP. The NDP is defined as the fraction of the applied pressure in excess of average osmotic pressure of the feed and any pressure losses in the system according to the following equation:

$$NDP = P_f - P_{os} - P_p - 0.5\,P_d \quad (+\,P_{osp}) \tag{3.4}$$

Where: P_f = feed pressure
P_{os} = average feed osmotic pressure
P_p = permeate pressure
P_d = pressure drop across RO elements
P_{osp} = osmotic pressure of permeate.

In regular RO applications osmotic pressure of permeate is negligible. However, in NF applications, where salt rejection is relatively low, permeate salinity is significant compared to the feed concentration. Therefore, osmotic pressure of permeate has to be considered in calculation of NDP in NF systems.

3.4 Salt - Water Separation in Reverse Osmosis Process

The mechanism of water and salt separation by reverse osmosis is not fully understood. Current scientific thinking suggests two water transport models: porosity and diffusion. That is, transport of water through the membrane may be through physical pores present in the membrane (porosity), or by diffusion from one bonding site to another within the membrane. The theory suggests that the chemical nature of the membrane is such that it will absorb and pass water preferentially to dissolved salts at the solid/liquid interface. This may occur by weak chemical bonding of the water to the membrane surface or by dissolution of the water within the membrane structure.

The chemical and physical nature of the membrane determines its ability to allow for preferential transport of solvent (water) over solute (salt ions). It is also known that part of dissolved species rejection is a result of size discrimination, i.e. larger molecules are better rejected by the RO membranes then the small ones. Another part of rejection process is result of repulsion of dissolved ions due to presence of charges (usually negative) on the membrane surface. Membrane repels ions having the same charges as these present on the membrane surface. Usually ions of multiple negative valency (high ion charges) are better rejected than single valency ions. For example, rejection of sulfate ions is higher than rejection of chloride ions. Due to condition of maintaining electro-neutrality of solutions on both sides of the membrane, repulsion of one type of ion hinders passage of the co-ion and increases overall rejection. Conversely, presence of ion with high passage through the membrane will increase passage of corresponding co-ion. The contribution of this charge depended rejection will vary with ionic composition of solution treated.

3.5 Water Transport

The rate of water passage through a semipermeable membrane is defined in Equation 3.5:

$$Q_w = (\Delta P - \Delta P_{osm})\, K_w\, (S/d) \tag{3.5}$$

Where Q_w is the rate of water flow through the membrane, ΔP is the hydraulic pressure differential across the membrane, ΔP_{osm} is the osmotic pressure differential across the membrane, K_w is the membrane permeability coefficient for water, S is the membrane area, and d is the membrane thickness.

The above equation can be simplified to:

$$Q_w = A\,S\,NDP \tag{3.6}$$

Where A is water transport coefficient and represents a unique constant for each membrane material type, and NDP is the net driving pressure or net driving force for the mass transfer of water across the membrane.

A units are: g/cm2-sec (gallons/foot^2-day-psi)

3.6 Salt Transport

The rate of salt flow through the membrane is defined by Equation 3.7:

$$Q_s = \Delta C\,K_s\,(S/d) \tag{3.7}$$

Where Q_s is the flow rate of salt through the membrane, K_s is the membrane permeability coefficient for salt, ΔC is the salt concentration differential across the membrane, S is the membrane area, and d is the membrane thickness.

K_s units are: cm/sec (ft/sec)

The equation 3.7 can be simplified to:

$$Q_s = B\,S\,(\Delta C) \tag{3.8}$$

Where B is the salt transport coefficient and represents a unique constant for each membrane type, and ΔC is the concentration gradient which is the driving force for the transfer of dissolved ions through the membrane.

Equations 3.6 and 3.8 show that for a given membrane:

a) Rate of water flow through a membrane is proportional to net driving pressure differential (NDP) across the membrane.
b) Rate of salt flow is proportional to the concentration differential across the membrane and is independent of applied pressure.

Accordingly to above relations, the salinity of the permeate, C_p, depends on the relative rates of water and salt transport through reverse osmosis membrane:

$$C_p = Q_s/Q_w \tag{9}$$

The fact that water and salt have different mass transfer rates through a given membrane creates the phenomena of water - salt separation and salt rejection. No membrane is ideal in the sense that it absolutely rejects salts; rather the different transport rates of water and dissolved ions create an apparent rejection. The equations 3.6 – 3.9 explain important design considerations in RO systems. For example, an increase in operating pressure will increase water flow without significantly affecting salt flow, thus resulting in lower permeate salinity. On the other hand higher recovery rate will increase concentration gradient and result in higher permeate salinity.

3.7 Salt Passage and Salt Rejection

Salt passage is defined as the ratio of concentration of salt on the permeate side of the membrane relative to the average feed concentration according to Equation 3.10:

$$SP = 100\% \ (C_p/C_{fm}) \tag{3.10}$$

Where SP is the salt passage (in %), C_p is the salt concentration in the permeate, and C_{fm} is the mean salt concentration in feed stream.

Applying the fundamental equations of water flow and salt flow (Equations 3.5 – 3.9) illustrates some of the basic principles of RO membranes. For example, apparent salt passage is an inverse function of pressure; that is, the salt passage increases as applied pressure decreases. This is because with reduced feed pressure permeate flow rate decreases and hence dilution of salt, on the permeate side of the membrane, decreases as well (the salt flows through the membrane at a constant rate as the rate of flow is independent of pressure).

Salt rejection is the opposite of salt passage, and is defined by Equation 3.11:

$$SR = 100\% - SP \tag{3.11}$$

Where SR is the salt rejection (expressed as %), and SP is the salt passage as defined in Equation 3.10. Salt rejection is important performance parameter of RO membranes, determining suitability of given membrane for various applications.

The above relations for water and salt transport imply constant values of a transport rates. However, salt and water transport rates are strongly affected by temperature, changing at similar rates with temperature fluctuations.

3.8 Temperature Effect on Transport Rate

Feed water temperature effect rate of diffusive flow through the membrane. For RO calculations the following equation is being used to calculate temperature correction factor (TCF), applied for calculation of water permeability:

$$TCF = 1/\exp(C\,(1/(273+t)-1/298)) \quad (3.12)$$

t is temperature C

C is constant, characteristic of membrane barrier material. For polyamide membranes a C values of 2500 – 3000 are being used.

It is customary for RO applications use temperature of 25 C (77 F) as the reference temperature, for which TCF = 1.0. The water and salt transport increases about 3% per degree C. Figure 3.4 shows value of the TCF in the temperature range of 5 – 50 C.

The relation shown in Figure 3.4 implies that due to increased water permeability with temperature increase, the operating feed pressure should be lower at higher temperature. This is indeed the situation in case of processing of low salinity feed (brackish applications). This is also the case for RO seawater applications in the low range of feed water temperatures. However, at feed water temperatures above 30 C the subsequent decrease of required feed pressure levels off. The effect of increased water permeability is reduced

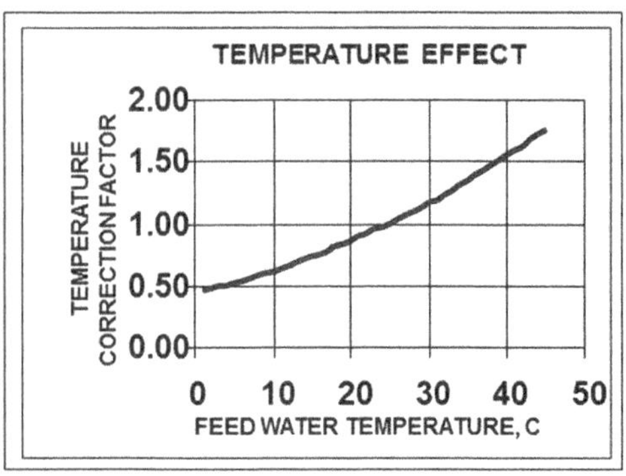

Figure 3.4 Water permeability temperature correction factor.

by increased osmotic pressure of the seawater feed. In addition, increase of salt passage and potential need for partial second pass processing, may actually result in higher overall power consumption at the high end of feed water temperature.

3.9 Average Permeate Flux

Average permeate flux is another important design parameter of the RO process.

APF - average permeate flux is combined permeate flow rate divided by the total membrane area installed in the RO unit. Units: l/m2-hr or gfd (gallon/ft2-day)

$$APF = Qp/(EN\ MA) \tag{3.13}$$

Where

Q_p = permeate flow rate
EN = number of elements in the system
MA = membrane area per element

Example # 1:
RO system produces 400 m3/day (105,700 gallons/day). Membrane array consists of 3 pressure vessels, each housing 6 membrane elements. Each element has 37 m^2 of membrane area (400 ft^2). APF is calculated as follows:

APF = 400,000 l/day/(3X6X37m^2X24) = 25.0 l/m^2-hr
APF = 105,700 gal/day /(3X6X400ft^2) = 14.6 gfd

3.10 Specific Water Permeability of a Membrane

Specific permeability, or specific flux - SF, characterizes the membrane material in terms of water flux rate driven by the gradient of applied net driving pressure.

$$SF = APF/NDP \tag{3.14}$$

Specific permeability depends on resistance of membrane to water flow. This resistance is composite of flow resistance of membrane barrier, support layers and any foulant layer on the membrane surface. It is usually calculated for the feed water temperature of 25 C. Specific flux results at different temperature can be corrected to reference value at 25 C by applying temperature correction factor (TCF), specific for given membrane type.

The specific flux units are: l/m^2-hr-bar (gfd/psi)

Example # 2

RO membrane element is tested at the following test conditions:

Feed salinity = 1500 ppm NaCl

Recovery rate = 15% (0.15)

Feed pressure = 10.3 bar (150 psi)

Pressure drop = 0.2 bar (2.9 psi)

Permeate pressure: 0.1 bar (1.5 psi)

Permeate flow: 41.6 m^3/day (11,000 gpd)

Membrane area: 39.5 m^2 (430 ft^2)

Average permeate flux = 41.6X1000/(24 X 39.5) = 43.9 l/m^2-hr (25.8 gfd)

Average feed salinity = 1500 X 0.5X(1+ 1/(1 – 0.15)) = 1632 ppm NaCl

Average osmotic pressure = 1632/1000X 0.77 = 1.3 bar (19 psi)

Net driving pressure = 10.3 – 1.3 – 0.1 – 0.5X 0.2 = 8.8 bar (128.0 psi)

Specific flux, SF = 43.9/8.8 = 4.99 l/m^2-hr-bar (0.20 gfd/psi)

3.11 Concentration Polarization

As water flows through the membrane and salts are rejected by the membrane, a boundary layer is formed near the membrane surface. In this layer the salt concentration exceeds the salt concentration in the bulk solution. This increase of salt concentration at the membrane surface is called concentration polarization. During the RO process there is a convective flow of water and ions toward the membrane surface. Ions rejected by the membrane, diffuse back to the bulk due to the concentration gradient. The observed effect of concentration polarization is reduction of actual product water flow rate and salt rejection versus theoretical estimates.

3.12 Commercial RO/NF Membrane Technology

The semipermeable membrane for nanofiltration and reverse osmosis applications consists of a film of polymeric material composed of a skin layer several thousands angstroms thick and spongy

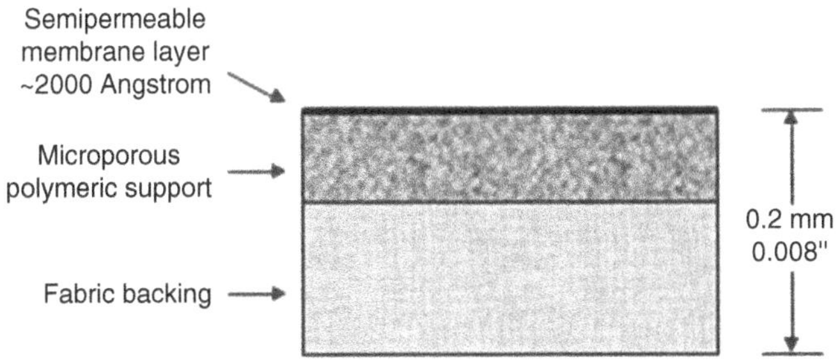

Figure 3.5 Cross section configuration of flat sheet RO membrane.

supporting layer approximately 0.25 - 0.50 mm (0.001 – 0.002") thick cast on a fabric support. The overall thickness of membrane is 0.15 – 0.20 mm (0.06 – 0.08"). The schematic configuration of membrane layers is shown in Figure 3.5. Scanning electron microscopy (SEM) picture of cross section of polyamide membrane, at 500 times magnification is shown in Figure 3.6.

The commercial RO membranes must have high water permeability and a high degree of semipermeability; that is, the rate of water transport must be much higher than the rate of transport of dissolved ions. The membrane material must be stable over a wide range of pH and temperature, and have good mechanical integrity. The stability of membrane performance: permeability and salt rejection, over a period of time at field conditions defines the commercially useful membrane life, which is, for the current commercial membranes, in the range of 5 to 10 years.

There are two major groups of polymeric materials, which are used to produce commercial reverse osmosis membranes: cellulose acetate (CA) and polyamide (PA). Polymer structure (Figure 3.7), chemical tolerance, membrane manufacturing, operating conditions, and performance differ significantly for each group of polymeric material.

3.13 Cellulose Acetate Membranes

The original cellulose acetate membrane, developed in the late 1950's by Loeb and Sourirajan, was made from cellulose diacetate polymer [1]. Current CA membrane is usually made from a blend of cellulose diacetate and triacetate. The membrane is formed by casting a thin film acetone-based solution of cellulose acetate

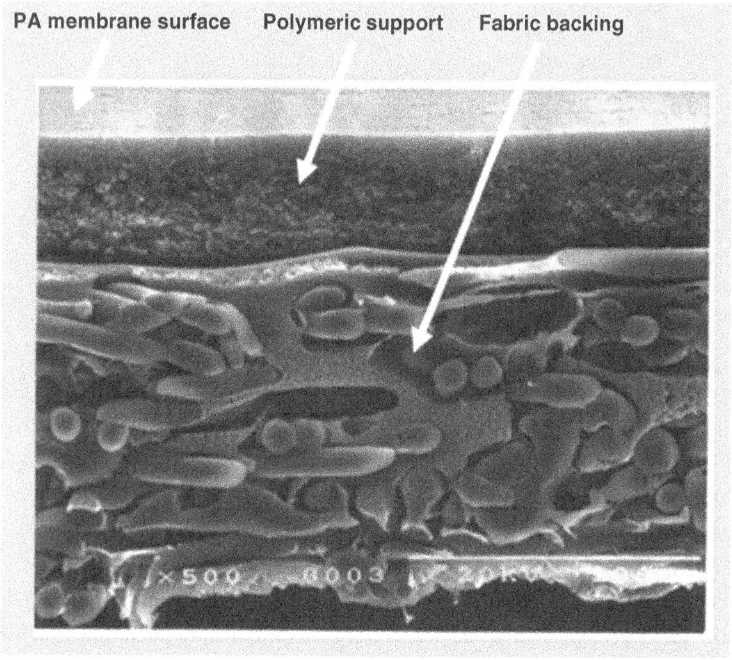

Figure 3.6 Scanning electron microscopy picture of cross section of polyamide membrane.

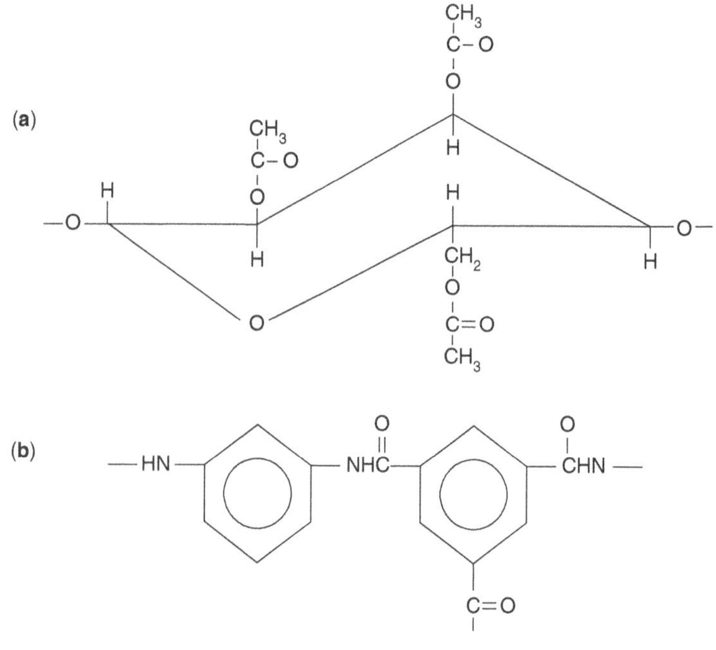

Figure 3.7 Chemical structure of cellulose triacetate (A) and polyamide (B) membrane polymer.

polymer with swelling additives from a trough onto a non-woven polyester fabric. Two additional steps, a cold bath followed by high temperature annealing, complete the casting process. During casting, the solvent is partially removed by evaporation. After the casting step, the membrane is immersed into a cold water bath which removes the remaining acetone and other leacheable compounds. Following the cold bath step, the membrane is annealed in a hot water bath at a temperature of 60 - 90°C. The annealing step improves the semipermeability of the cellulose acetate membrane resulting in a decrease of water transport and a significant decrease of salt passage. After processing, the cellulose membrane has an asymmetric structure with a dense surface layer of about 1000 - 2000 angstrom (0.1 - 0.2 micron) which is responsible for the salt rejection property. The rest of the membrane film is spongy and porous and has high water permeability. Salt rejection and water flux of a cellulose acetate membrane can be controlled by variations in temperature and duration of the annealing step. Description of manufacturing process of cellulose acetate membranes and its properties can be found in number of publications [2, 3, 4, 5].

Cellulose acetate membrane polymer hydrolyzes rapidly at extreme pH. Therefore, the operating feed water pH range for CA membrane is 6 – 8. Accordingly, cellulose acetate membrane elements can be only cleaned in the narrow range of pH close to neutral (pH: 6 – 8). However, CA membrane polymer has sufficient tolerance to free chlorine that enables operation with chlorinated feed water and on line disinfection to control bacterial growth. For this reason cellulose acetate is still membrane of choice for applications where frequent disinfection of RO system with free chlorine is practiced, such as pharmaceutical industries and some food applications. Also, one of membrane manufacturers presently produces capillary RO membranes for seawater desalting using cellulose acetate polymer. Except for the previously mentioned applications, rest of the desalination market is dominated by the composite polyamide membranes in spiral wound configuration.

3.14 Composite Polyamide Membranes

The manufacturing process of composite polyamide membranes consists of two distinct steps. First, a polysulfone support layer is cast onto a non-woven polyester fabric. The polysulfone polymer solution is applied from a trough onto a moving polyester backing fabric. After polysulfone application and formation of UF membrane layer the fabric travels through water bath to remove solvent and is collected on a drum.

The polysulfone layer is very porous and is not semipermeable; i.e. does not have the ability to separate water from dissolved ions solution. However, it has high water permeability In the next process step, the drum with polysulfone membrane is moved to the second machine where interfacial polarization takes place. There, a semipermeable membrane skin is formed on the polysulfone substrate by interfacial polymerization of two monomers, one: metaphenylenediamine (MPD) containing amine groups and the other: trimesoyl chloride (TMC) provides carboxylic acid chloride functional groups. The polymerization reaction is very rapid and takes place on the surface of the polysulfone support forming a barrier, 1000 – 2000 angstrom thick. This barrier is responsible for the semipermeable property: passage of water and rejection of dissolved species. Following polymerization zone, membrane web enters a rinse bath. The membrane is rinsed to remove excess reagents and passed trough the oven to dry [6].

This two-step manufacturing procedure enables independent optimization of the distinct properties of the membrane support and salt rejecting skin. The resulting composite membrane is characterized by higher specific water flux and lower salt passage than cellulose acetate membranes.

Polyamide composite membranes are stable over a wider pH range than the cellulose acetate membranes. However, polyamide membranes are susceptible to oxidative degradation by free chlorine, while cellulose acetate membranes can tolerate limited levels of exposure to free chlorine.

The early composite membranes made of aliphatic polymers [4] were very sensitive to presence of oxidants and suffered from inadequate stability of performance in field conditions. The later generation of composite membranes, based on aromatic

polyamide invented by Cadotte [6, 7] have some tolerance to free chlorine, good stability in wide range of feed pH (2 – 10) and shows excellent long term performance stability with majority of feed water types. This type of membrane material is used today almost exclusively to manufacture commercial RO membrane elements.

The variety of types of membranes made of composite aromatic polyamide includes seawater, brackish and nanofiltration membrane elements. Composite membranes are used in all areas of applications: seawater and brackish water desalting, potable water softening, wastewater reclamation, food processing and other industrial applications.

3.15 Membrane Module Configurations

The reverse osmosis technology started with tubular and plate and frame configurations. Due to low packing density, these initial module configurations were gradually phased out of potable applications and at present are being very infrequently used in conventional reverse osmosis applications. However, new configurations of plate and frame modules are still being used in food processing applications and for treatment of waste streams including land fill leaches. In the past, the two major membrane module configurations used for reverse osmosis applications were hollow fiber and spiral wound. At present, vast majority of RO membrane manufactures offer elements in spiral wound configuration only.

3.16 Spiral Wound Elements

The concept of spiral wound membrane element device was introduced shortly after the invention of the hollow fiber configuration [8]. In a spiral wound configuration two flat sheets of membrane are separated with a permeate collector channel material to form a leaf. The leaf assembly is sealed on three sides with the fourth side left open for permeate to exit (Figure 3.8). A feed/brine spacer material sheet is added to the leaf assembly. A number of these assemblies or leaves are wound around a central plastic permeate tube. The permeate tube is perforated to collect the permeate from the multiple leaf assemblies (Figure 3.8). During the element assembly process membrane leaves are rolled around the permeate tube in a spiral configuration (Figure 3.9). The membrane leaves are kept in this form with a tape wrapped around the element and the outer shell, which is usually made of reinforced fiberglass.

The feed/brine flow through an element is in a straight axial path from the feed end to the opposite brine end, running parallel to the membrane surface. Fraction of the feed permeates through the membrane and flows through the permeate carrier fabrics to the central permeate tube. The remaining fraction of feed water continue to flow through the feed channel and becomes a concentrate (Figure 3.9). The feed channel spacer is in the form of a two level

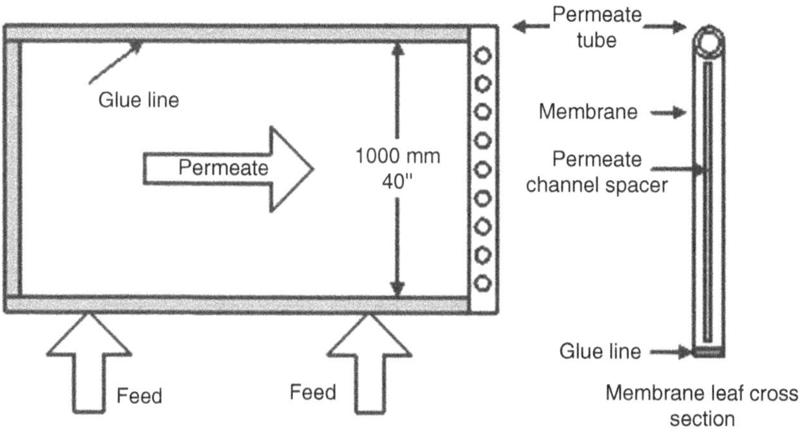

Figure 3.8 Configuration of flat sheet membrane leaf.

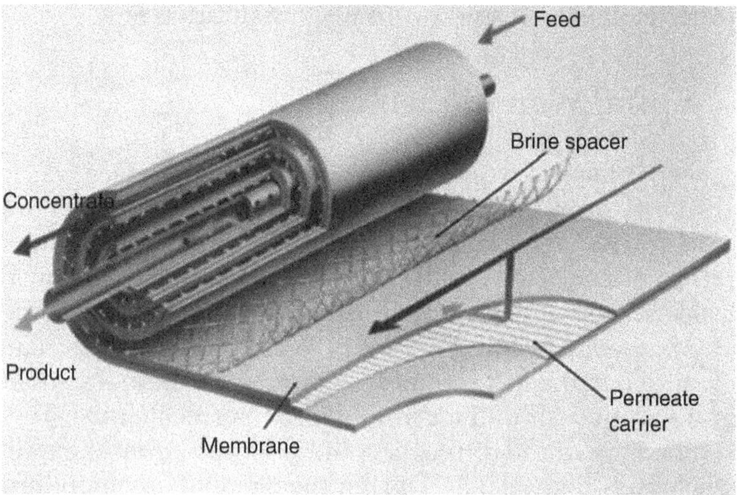

Figure 3.9 Configuration of spiral wound membrane element.

(biplanar) net. This two level net separates membranes from adjacent leaves and induces turbulence in the feed stream to reduce concentration polarization. The thickness of the feed channel is in the range of 0.7 – 09 mm (0.028 – 0.034"). However, the cross section of feed channel open to flow is much smaller, due to the presence of feed spacer.

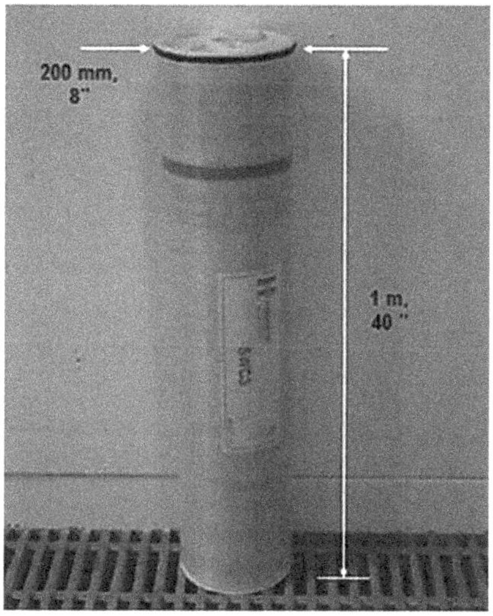

Figure 3.10 Commercial spiral wound membrane element.

The typical commercial spiral wound membrane elements are approximately 100 or 150 cm (40 or 60 inches) long and 10 or 20 cm (4 or 8) inches in diameter (Figure 3.10).

Membrane manufacturers specify concentrate flow rate requirements to control concentration polarization by limiting recovery rate (or conversion) per element to 10 - 20 percent. Therefore, recovery rate is a function of the feed-brine path length. In order to operate at acceptable recoveries, spiral systems are usually staged with three to eight membrane elements connected in series in a pressure tube (Figure 3.11). The concentrate stream from the first element becomes the feed to the following element, and so on for each element within the pressure tube.

Each element contains brine seal, which is in the form of flexible o-ring, usually position at the front end of element. The brine seal seals the space between the element outer wall and inner wall of the pressure tube. Brine seal prevents feed water to bypass the element, which would otherwise result in low flow through element and high recovery rate. Concentrate stream from the last element exits the pressure tube to the next processing stage or to waste. The permeate tubes of each element are connected to adjacent element through permeate interconnector, forming a common permeate tube. The first and the last element in the pressure vessel is connected through an adaptor

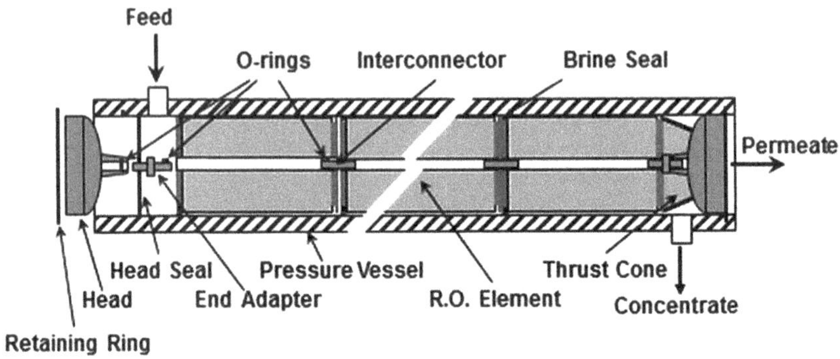

Figure 3.11 Configuration of pressure vessel with membrane elements.

to the pressure vessel permeate outlet (Figure 3.11). Permeate from all elements in the pressure vessel exits the vessel as a common permeate stream. A single pressure vessel with six to eight membrane elements connected in series can be operated at up to 50 - 70 percent recovery under normal design conditions.

The dimensions and geometry of spiral wound membrane elements is highly standardized. Spiral wound membrane elements produced by various manufacturers are of very similar configuration and outer dimensions. They can be operated in the same pressure vessels and are easily interchangeable.

Recently, large diameter (> 200 mm, 8″) elements are available from number of membrane manufacturers The industry standardized on membrane element dimensions of 406 mm diameter by 1016 mm long (16″ Φ by 40″ L). Such elements have about four times the membrane area (and the permeate flow) of the current, 200 mm (8″) diameter element and provide reduction of RO system capital cost. One of the membrane manufactures (Koch Membrane Systems) provides limited offering of 18″ (457 mm) diameter membrane elements, which are used in small to medium size RO units.

3.17 Spiral Wound Element Categories

Large desalination systems utilize membrane elements that are 1 m (40″) long and 200mm (8″) in diameter. Smaller, 100 mm diameter elements are used for small systems (light commercial, small potable) and pilot testing. The spiral wound membranes are used commercially in three major application categories: potable water softening (nanofiltration), brackish water desalting and seawater desalting. They are categorized accordingly as nanofiltration, brackish and seawater elements. Although operated at different feed pressure range, the elements in all categories of applications are of almost identical configuration and utilize the same materials of construction. One of the major differences between spiral elements used in low and high feed pressure applications is feed spacer. The type used in high pressure, seawater elements is of tighter type, reducing membrane embossing. Traditionally, brackish water elements were manufactured with slightly higher (about 10%) membrane area then seawater elements. At present, this difference is still maintained to some extend but membrane area of elements in all categories has increased.

The representative properties and nominal performance of nanofiltration, brackish and seawater elements are listed in Table 3.1. The nominal performances are measured during testing of a single element at nominal test conditions. The nominal test conditions include feed salinity (as NaCl), feed pressure, recovery rate and temperature (25 C). At field conditions, where operating parameters are different then the conditions during the factory tests, the elements are operating at a different performance level. In softening and brackish applications 200 mm (8″) diameter element will produce about 24 m3/day (6400 gallons per day). In seawater applications a single element of the same size will produce about 12 m3/day (3200 gallons per day). In brackish water RO systems the reduction of feed salinity will be about 97% i. e. permeate salinity will be about 3 % of the feed salinity. The corresponding values for seawater systems are about 99.0% reduction of feed salinity or producing permeate salinity of about 1.0 % concentration of the feed salinity. In softening (nanofiltrtaion) applications the actual system salt rejection will very much depend on type of elements selected. It can be as high as 90% for systems designed for salinity and hardness reduction and as low as practically insignificant for systems designed to remove color and dissolved organics only.

The nominal performances of membrane elements listed in Table 3.1 indicate decrease of water permeability and reduced salt transport, as transition from nanofiltration membrane elements through brackish elements to seawater elements category. Accordingly, at field conditions desalination systems equipped with nanofiltration membranes will operate at lowest feed pressure, whereas seawater desalination system will require the highest feed pressure to produce the same flow rate of permeate per unit of membrane area. The required feed pressure is function of

Table 3.1 Representative nominal performance of membrane elements.

Element type	Softening	Brackish	Seawater
Element dimensions	1m long, 200 mm diameter (40″ X 8″)	1m long, 200 mm diameter (40″ X 8″)	1m long, 200 mm diameter (40″ X 8″)
Membrane area, m2 (ft2)	37.2 (400)	39.5 (430)	36.8 (400)
Permeate flow, m^3/d (gpd)	29.5 (7,800)	41.6 (11,000)	24.6 (6,500)
Salt rejection, %	80.0	99.60	99.8
Test feed pressure, bar (psi)	5.2 (75)	10.3 (150)	55.2 (800)
Test feed salinity, mg/l NaCl	500	1500	32,000
Test recovery rate, %	15	15	10
Test flux rate, l/m^2-hr (gfd)	33.2 (19.5)	43.5 (25.6)	27.6 (16.3)
Permeability, l/m^2-hr-bar (gfd/psi)	7.3 (0.29)	4.9 (0.20)	1.0 (0.04)
Relative salt transport value	663	17.4	5.5

membrane permeability and operating parameters of the RO system: feed water temperature, salinity and recovery rate.

Permeate salinity is function of membrane salt transport property (relative salt transport) and system operating parameters listed above.

For the same feed salinity and system operating parameters, RO system equipped with seawater membranes would produce the lowest permeate salinity.

The projected field performances are calculated using membrane manufactures provided computer programs. The results of computer calculations are quite accurate for brackish and seawater applications. At present, the calculated results for softening applications are approximate and manual corrections based on field experience (usually pilot unit operation) have to be applied.

For nanofiltration membranes, due to high salt passage there is strong interaction of various ions during the transport process. In addition, some of the nanofiltration membrane surfaces are strongly charged. Therefore, salt passage for a given mixed ions composition could be significantly different than the nominal salt passage determined using single a salt solution.

3.18 RO System Configuration

The configuration of RO system is affected to some extend by type of feed water being processed. Extensive information on configuration of RO desalination systems can be find in references (16, 17). Figure 3.12 shows configuration diagram of RO system processing well water. This configuration is representative of brackish plant or seawater plants receiving feed water from wells. In brackish RO systems, feed water filtration is usually very limited, including only cartridge filtration. Systems processing surface water from open intake, brackish or seawater, require more extensive filtration of the raw water (Figure 3.13). The pretreatment in RO systems processing surface water may consist of single or two-stage media filtration, combined with flocculation and/or clarification. RO systems in wastewater reclamation plants utilize, almost universally, membrane pretreatment: ultrafiltration or microfiltration (Figure 3.14). Membrane pretreatment is also being used, with increasing frequency in seawater RO desalination systems.

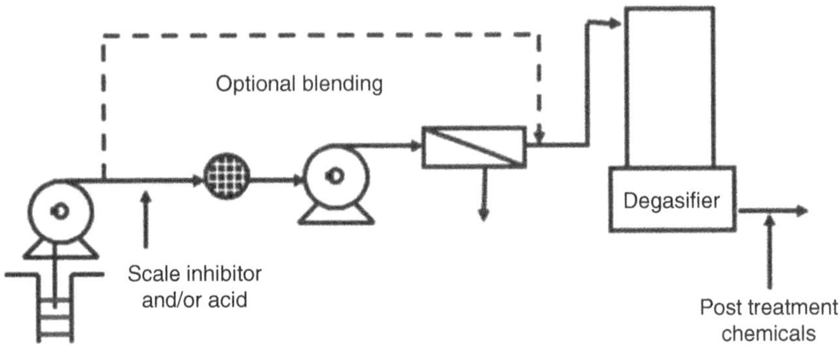

Figure 3.12 Configuration of RO system processing well water.

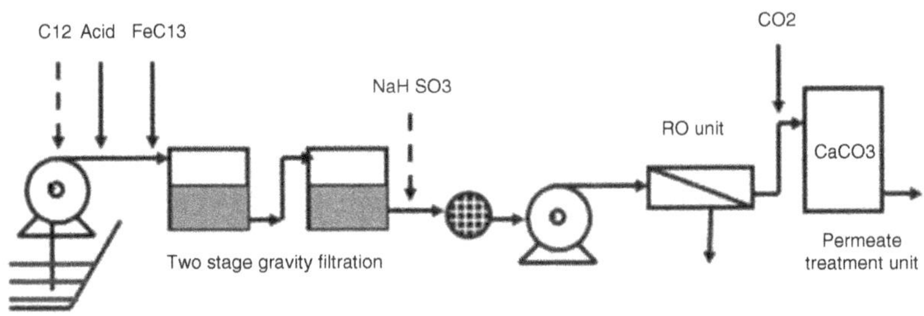

Figure 3.13 Configuration of RO system processing surface water equipped with media filtration pretreatment.

RO desalination systems consist of the following basic equipment components:

- Feed water supply unit
- Pretreatment system
- High pressure pumping unit
- Membrane element assembly unit
- Instrumentation and control system
- Electric power supply system
- Permeate treatment and storage unit
- Cleaning unit

The configuration and operation of all components of RO system are designed to produce and maintain adequate quality of feed water to the membrane elements, maintain stable performance of all system components, produce design permeate flow and quality and maintain design economics of the desalination process.

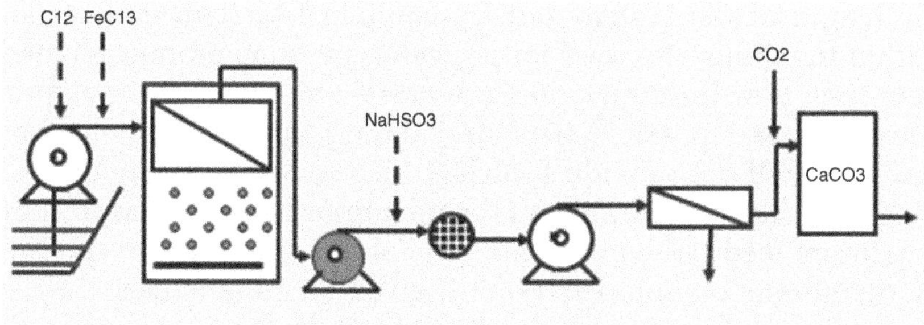

Figure 3.14 Configuration of RO system processing surface water equipped with membrane filtration pretreatment.

3.19 Membrane Assembly Unit

The membrane assembly unit (RO train) is the "heart" of the RO system. This is the unit where the separation between water and dissolved species takes place. It consists of a stand, supporting pressure vessels, interconnecting piping, feed, permeate and concentrate manifolds. RO train also includes an instrumentation panel with local display of flow pressure and conductivities. In some systems permeate sampling panel is also included. This panel is a collecting point for permeate sampling tubing from individual pressure vessels. Membrane elements are installed in the pressure vessels.

The RO system is divided into groups of pressure vessels, called concentrate stages. In each stage, pressure vessels are connected in parallel, with respect to the direction of the feed/concentrate flow. The number of pressure vessels in each subsequent stage decreases in the direction of the feed flow. The configuration is usually in the ratio of 2:1. Thus, one can visualize that the flow of feed water through the pressure vessels of a system resembles a pyramid structure: a high volume of feed water flows in at the base of pyramid, and a relatively small volume of concentrate leaves at the top. The decreasing number of parallel pressure vessels from stage to stage compensates for the decreasing volume of feed flow, which is being progressively converted to the permeate. Permeate from all pressure vessels in each stage, is combined together in a common permeate manifold.

The objective of the taper configuration of pressure vessels is to maintain a similar feed/concentrate flow rate per vessel through the length of the system and to maintain feed/concentrate flow within the limits specified for a given type of membrane element. Very high flow through a pressure vessel will result in a high pressure drop and possible structural damage of the element. Very low flow will not provide sufficient turbulence, and may result in excessive salt concentration at the membrane surface. The limits of maximum feed flow and minimum concentrate flow are specified by membrane manufacturers for a given membrane element type depending mainly on combined height of the feed channels in the element and type of feed spacer net used.

3.20 Concentrate Staging

A commercial RO unit usually consists of single pump and a multistage array of pressure vessels. A simplified block diagram of a two stage RO unit is shown in Figure 3.15. The concentrate from the first stage becomes the feed to the second stage; this is what is meant by the term "concentrate staging." The flows and pressures in the multistage unit are controlled with the feed and concentrate valves. The feed valve, after the high pressure pump, controls feed flow to

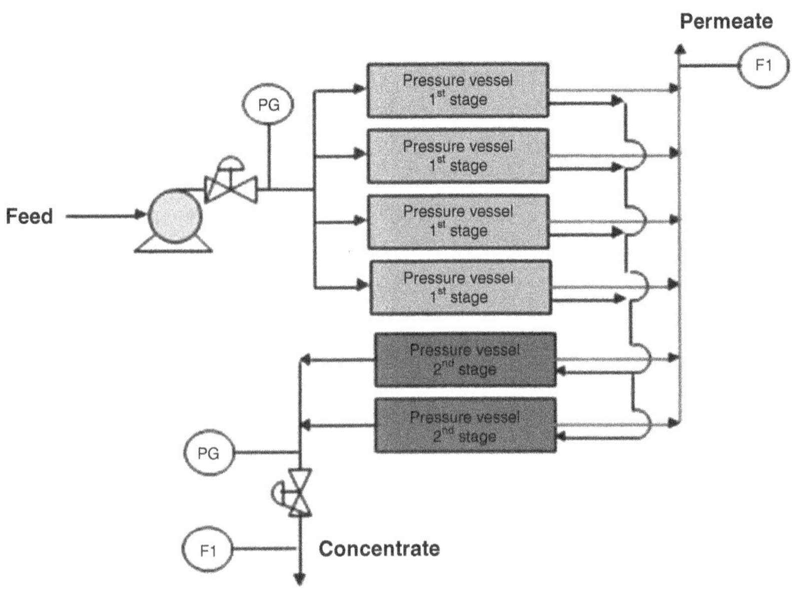

Figure 3.15 Schematic configuration of a two stage RO unit.

the unit. The concentrate valve, at the outlet of RO block, controls feed pressure inside the unit.

For a given RO unit, the number of concentrate stages will depend on the permeate recovery ratio and the number of membrane elements per pressure vessel. In order to avoid excessive concentration polarization at the membrane surface, the recovery rate per individual membrane element should not exceed 18%. It is common engineering practice to design brackish RO systems so that the average recovery rate per 40 inch long membrane element will be about 6 - 8%. Accordingly, the number of concentrate stages for an RO unit having 6 elements per pressure vessel would be two stages for recovery rates over 60%, and three stages for recovery rates over 75%. With pressure vessels containing seven to eight elements, a two stage configuration would be sufficient for recovery rates up to about 85%.

Figure 3.16 shows a picture of commercial, two stage, brackish train. The array is 32:14 pressure vessels with 7 elements per vessel. The picture shows two parallel feed manifold with 4 X 8 pressure vessel connected. Unit configuration is eight vessels high and six vessels wide. This translates into unit dimensions of 4.0 m high, 2.9 m wide and 8.0 m long (13.1′ X 9.5′ X 26′). The first stage

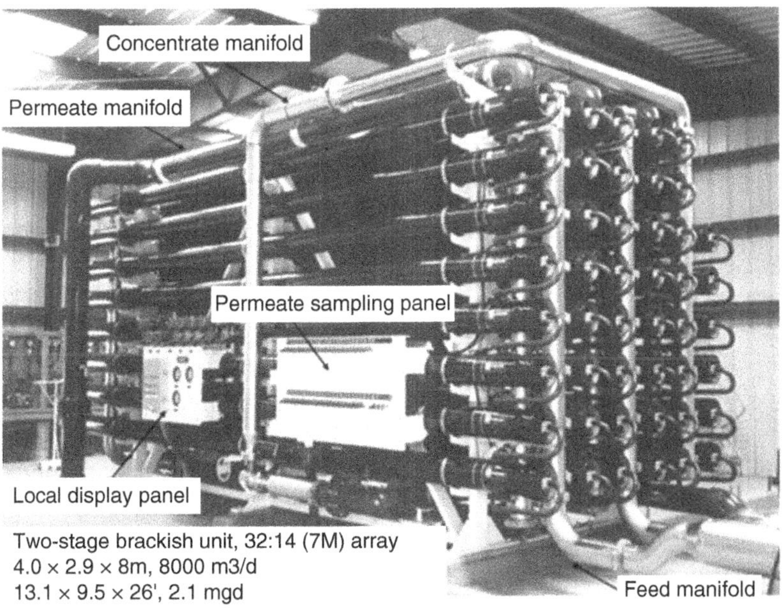

Figure 3.16 Two stage RO unit. Permeate capacity about 8,000 m3/day (2.1 mgd)

concentrate is collected from the 32 first stage pressure vessels and fed to 14 second stage vessels. The concentrate outlets from the second stage vessels are connected to the concentrate manifold. The concentrate throttling valve located on the concentrate pipe is shown as well. The train is equipped with permeate sapling panel, which enables sampling permeate conductivity from individual vessels and local display panel of flow, pressure, feed temperature and conductivities.

The unit contains total of 322 membrane elements, each 200 mm (8") diameter, 1000 mm (40") long, and is capable of about 8,000 m^3/day (2.1 mgd) of permeate at recovery rate of 80%.

3.21 Permeate Staging (Two Pass Systems)

For some applications, the single pass RO system may not be capable of producing permeate water of a required salinity. Such conditions could be encountered in two types of RO applications:

- Seawater RO systems, which operate on a very high salinity feed water, at high recovery ratio and/or at high feed water temperature.
- Brackish RO applications which require very low salinity permeate such as supply of makeup water for pressure boilers or production of rinse water for microelectronics applications.

To achieve an additional reduction in permeate salinity, the permeate water produced in the first pass is processed again in a second RO system. This configuration is called a two pass design, or "permeate staging." Figure 3.17 contains schematic diagram of a two pass system.

Depending on permeate quality requirements, part or all of the first pass permeate volume is desalted again in the second pass unit. The system configuration is known as a complete or partial two pass system according to whether all of the 1st pass permeate or only some fraction is fed to the second pass unit. The first pass permeate is a very clean water. It contains very low concentrations of suspended particles and dissolved salts. Therefore, it does not require any significant pretreatment. The second pass RO unit can operate at a relatively high average permeate flux and high recovery rate without concerns of concentration polarization and scaling. The common

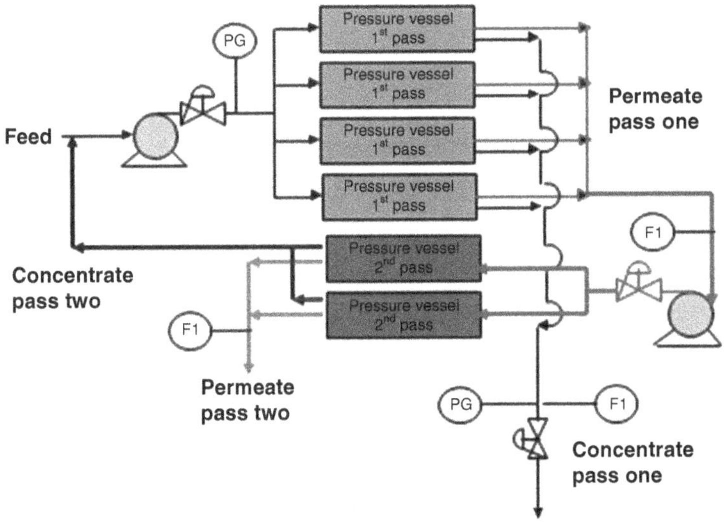

Figure 3.17 Schematic configuration of two pass RO unit.

design parameters for the second pass RO unit are average flux rate of 34 l/m2-hr (20 gfd) and recovery rate of 85% - 90%. In a two pass system, the permeate from the first pass flows through a storage tank or is fed directly to the suction of the second pass high pressure pump. It is a common procedure in a two pass seawater RO systems to return concentrate from the second pass unit to the suction of the high pressure pump of the first pass unit. The dissolved salts concentration in the concentrate from the second pass is usually lower the concentration of the feed to the first pass unit. Therefore, blending feed water to the first pass with small flow rate of the second pass concentrate, reduces slightly the salinity of the feed to the first pass, and increases the overall utilization of the feed water (Figure 3.17).

3.22 Partial Two Pass Configuration

There are number of possible configurations of the two pass RO units. One configuration, which is a partial two pass system, is shown in Figure 3.18. In such configuration the first pass permeate is split into two streams. One stream is processed by the second pass unit, and it is then combined with the unprocessed part of permeate from the first pass. Provided that the partial two pass system can produce the required permeate quality, this configuration results

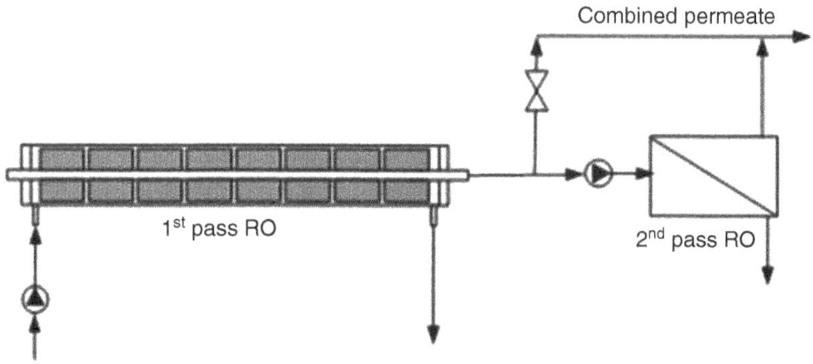

Figure 3.18 Schematic configuration of a partial two pass RO unit.

in smaller second pass unit, therefore lower capital and operating costs, as well as higher combined permeate recovery rate (utilization of the feed water), compared to a complete two pass system.

Another partial two pass configuration, which takes advantage of internal salinity distribution of RO permeate in the pressure vessel, is becoming increasingly popular in seawater systems. This unique concept of partial two pass system, designated as "split partial", has been proposed in the past [8] but only lately is being implemented in large seawater systems [18, 19].

In the RO system each subsequent element, in the direction of feed flow, produces water of increasingly higher salinity. This is due to increasing feed salinity along the pressure vessel and decreasing permeate flux. affected by the decrease of NDP. The NDP decrease results from increasing osmotic pressure and gradual decrease of feed pressure along the pressure vessel, due to friction losses. As the elements are connected together through the permeate tube, the permeate from the individual elements is mixed together and leaves pressure vessel through permeate port at composite salinity.

In a split partial configuration the first pass system is configured as a single stage unit. The permeate is collected from both ends of pressure vessels. The fraction of permeate collected from the feed end is of lower salinity and flows directly to permeate storage tank. The fraction collected at the concentrate end is processed with the second pass RO unit (Figure 3.19).

The split partial process results in reduction of the overall system size and lower power consumption. The smaller the fraction of the first pass permeate that has to be processed with the second pass RO the higher the benefits of "split partial" configuration as compared to the regular partial two pass design. A comparison of conventional two pass and split partial two pass system configuration is provided in Table 3.2. The results illustrate advantage of split partial configuration. It requires smaller

capacity of first and second pass unit and reduced energy demand for the same product water quality and capacity, as compared to the conventional partial two pass

The flow rate processed by the second pass RO unit affects required capacity of the primary RO. With increased capacity of the secondary RO larger fraction of primary RO permeate is discharged as concentrate of the secondary RO. This has to be compensated by increased capacity of the primary RO. Therefore, addition of second pass processing affects both the capital and operating cost (mainly power consumption but also some contribution from chemical consumption and membrane replacement cost).

3.23 Calculation of System Performance

3.23.1 Manual Method of Membrane System Performance Calculations

The common approach to projecting performance of RO system is to calculate permeate flow according to the net driving pressure

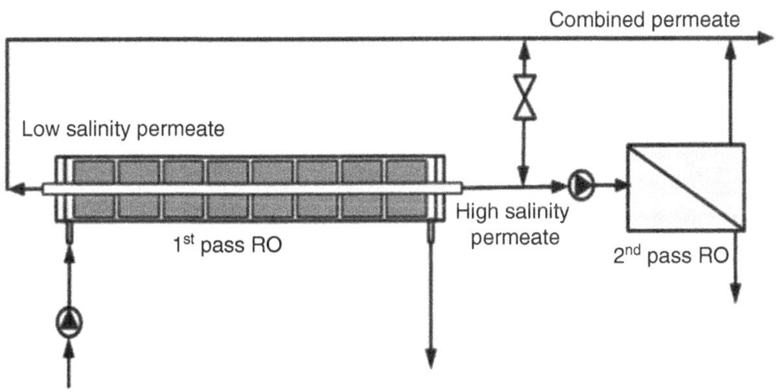

Figure 3.19 Schematic configuration of a "split partial" two pass RO unit.

Table 3.2 Comparison of conventional and split partial two pass system configurations.

	First pass	**Second pass**
Conventional design		
Permeate flow, m3/hr	4,312.5	1958.3
Processed by second pass, %		**45.4**
Recovery ratio, %	50.0	90.0
Feed pressure, bar	65.4	14.1
No. of pressure vessels	1030	190
No. of elements	**8,240**	**1,520**
Power requirement, kWhr	12,525	826
Combined power req., kWhr	**13,351**	
Split partial design		
Permeate flow, m3/hr	4,166	62.5
Processed by second pass, %		**18.9**
Recovery ratio, %	50	90.0
Feed pressure, bar	66.5	13.4
No. of pressure vessels	1,000	60
No. of elements	**8,000**	**480**
Power requirement, kWhr	12,307	321
Combined power req., kWhr	**12,628**	
Difference of no. of elements, (%)	**240 (2.9)**	**1,040 (68)**
Power saving, kWhr(%)	**723 (5.4)**	

model and base calculations of permeate salinity on salinity gradient between feed and permeate as a driving force of the salt transport. The reference conditions are the nominal element performance, as tested at standard test conditions, defined by membrane manufactures. Single point calculations for basic system configuration can be conducted manually (as shown below). However repeated calculations, required for optimization of process design, are conducted using computer programs available from all major membrane manufacturers. The manual calculations process goes through the following steps:

1. According to the type of feed water source select membrane element type and the value of system average permeate flux (APF).
2. Using nominal test conditions and nominal element performance calculate specific permeability of the selected membrane element (SP).
3. Using the above values of APF and SP calculate the required net driving pressure.
4. Based on feed water composition, feed water type or project specifications select system recovery ratio and calculate average feed salinity.
5. Calculate corresponding average feed osmotic pressure (Equation 3.2 or salinity – osmotic pressure relations).
6. Make assumptions regarding system array, pressure drop per stage and permeate back pressure.
7. Calculate required feed pressure.
8. Calculate permeate salinity based on average feed salinity, average system permeate flux, nominal element salt passage and element permeate flux at nominal test conditions.

Example # 3
Brackish two stage RO system. Feed salinity 2500 ppm TDS. Recovery rate: 85%. Average flux rate 27.2 l/m^2-hr (16 gfd). Feed water temperature 25 C.

Calculations of specific element performance

Element type: Brackish, membrane area: 36.8 m^2

Nominal element performance: 34.07 m^3/d @ pressure 10.3 bar

Salt rejection 99.6% @ flux rate 38.6 l/m^2hr (22.7 gfd)

Nominal test conditions: feed salinity 1500 ppm NaCl, recovery rate 15%

Average feed salinity during nominal test: 1500X0.5 (1 + 1/(1- 0.15)) = 1632 ppm NaCl

Average feed osmotic pressure: 1.25 bar (18.1 psi)

Nominal NDP: 10.3 – 1.25 = 9.05 bar (131.2 psi)
Specific permeability: 38.6/9.05 = 4.26 l/m^2-hr-bar (0.17 gfd/psi)

Calculation of system performance
System NDP required: 27.2/4.26 = 6.4 bar (93 psi)
Friction pressure drop per stage 2 bar (29 psi), total for system 4 bar (58 psi).
Permeate back pressure 0.5 bar (7.2 psi).
Feed salinity 2500 ppm TDS, osmotic pressure 1.9 bar (28 psi)
Average feed osmotic pressure: 1.9X0.5(1 + 1/(1 – 0.85)) = 7.3 bar (105.8 psi)
Required system feed pressure: 6.4+7.3 + 4 + 0.5 = 18.2 bar (264 psi)

Permeate salinity:
Average feed salinity = 0.5(2500 + 2500/(1 – 0.85)) = 9583 PPM
Permeate salinity: 9583(1-99.6/100) (38.6/27.2) = 54 PPM

Permeate salinity is function of average feed salinity and the system permeate flux rate as compared to the nominal flux.

Additional corrections that should be applied to these calculations include correction for system configuration (flux distribution), temperature and element age

3.23.2 Calculations of RO Performance Using Computer Programs

For the purpose of RO systems design, the projected performances of RO unit are calculated using computer programs developed by manufacturers of membrane elements.

The calculation process is conducted through the following steps:

1. Entry of feed water composition
2. Adjustment of feed water pH, if required to prevent calcium carbonate scaling
3. Entry of permeate flow rate, system recovery rate, feed water temperature
4. Selection of membrane elements type and array (number of concentrate stages in the RO membrane unit,

number of pressure vessels for each stage and number of elements per vessel).
5. The membrane array is adjusted to arrive to average permeate flux value in the range recommended for a given water source.
6. Computer conducts calculations of feed pressure, permeate composition and scaling indexes of the concentrate
7. Display of system parameters and calculation results

The computer program conducts calculations in a similar way as it has been demonstrated in example #3. The computer calculations starts with nominal performance of elements selected to determine specific permeability and salt transport. However, the calculations are conducted in element by element approach. Using feed water composition, calculation of performance of single element in the system results in composition of concentrate that is used as a feed composition for the next element in pressure vessel.

Computer programs for calculation RO system performances are available for free download from web sites of membrane manufacturers.

3.24 Monitoring of Process Parameters and Equipment Performance in RO System

Monitoring of process parameters and equipment performance is required to satisfy the following objectives:

- Prevents system from operation at conditions that may result in personnel injury or equipment damage
- Maintains required sequence and timing of equipment operation
- Maintains operation of equipment within the design process limits
- Maintains production of the design quantity and quality of water
- Stores and processes operational data, generates reports, displays information

- Enables controlled intervention in system operation
- Enables estimation of maintenance requirement (including membrane cleaning)

Example of listing of process parameters that are being monitored in RO plants is listed below:

- Raw water conductivity
- Raw water temperature
- Raw water flow
- Raw water pump suction and discharged pressure
- Raw water turbidity
- Dosing rates of pretreatment chemicals
- Raw water free (combined) chlorine
- Filtration unit head loss
- Filter effluent turbidity and SDI
- Cartridge filters pressure drop
- High pressure pump suction and discharged pressure
- Feed water pressure
- Feed water pH
- Feed water free (combined) chlorine
- RO permeate flow
- RO permeate pressure
- RO permeate conductivity
- RO permeate temperature
- RO permeate pH
- RO concentrate flow
- RO concentrate pressure
- Dosing rate of post-treatment chemicals
- Product water turbidity
- Product water free (combined) chlorine
- RO permeate storage tank level

Monitoring of the above process parameters provides data required for determination of performance of the RO plant and condition of membrane elements.

The recorded performance of RO membranes have to be adjusted to account for fluctuation of feed water composition, temperature, recovery rate and feed pressure. This approach enables comparison of system performance over time and determination stability of membrane performance.

3.25 Normalization of RO System Performance

Performance of RO/NF system is a result of aggregate performance of individual membrane elements. Each subsequent element in a pressure vessel, connected in series, operates at different values of feed salinity and feed pressure. Along the system feed salinity increases and feed pressure decreases. The performance and operating conditions are significantly different then the corresponding nominal values. In addition, system performances are affected by fluctuation of operating parameters: feed salinity, temperature, recovery rate and feed pressure.

In order to identify intrinsic changes of membrane performance such as permeability or salt passage, at the early stages of membrane deterioration process, system operational data are recorded at frequency at least once per day and normalized performance are calculated. The generic method of RO performance normalization is described in the ASTM procedure [20]. In actual commercial RO applications the following normalization approaches are adopted:

1. Normalization to the reference (initial) operating conditions of the plant.
2. Normalization to the nominal element(s) test conditions
3. Calculation of water transport and salt transport values for the membrane elements in operation.

In the normalization calculations process each set of plant (or desalting stage) flows, pressures and salinities data is initially reduced to the average values. These average values are assumed to be representative for an element positioned somewhere in the middle of the system, on the feed – concentrate cross section line: i.e. element that process an average feed salinity at an average applied feed pressure and produces average permeate flow. The averages are calculated based on feed – concentrate values. Then based on this data the water and salt permeability are calculated. In the normalization approach # 1 every set of the performance data of the system are being recalculated to the initial operating conditions: temperature, average feed salinity and NDP.

Any of the above performance normalization method will provide good presentation of membrane unit performance trend. Some advantage of the first method is that, in addition to normalized permeate flow and salt passage, it usually also provides trend of the pressure drop. Pressure drop is an important indicator of early stage of fouling, which results in blockage of the element feed channels. In the normalization approach # 2 the performance of the system are calculated and presented as a performance of an average element, it would perform, if tested at the nominal test conditions. The normalization approach # 3 is very similar to the first one. In this calculations performance of RO system is reduced to performance of an average element. Then based on this data the water and salt permeability are calculated.

Normalization of salt passage and permeate flow is derived from the salt transport relations. According to these relations the salt passage is function of salinity gradient and quantity of permeate available for dilution (permeate flux rate). Therefore, salt passage at a given operating conditions SP(1) is related to different operating conditions accordingly to the corresponding average permeate flux rate (APF):

$$SP(2) = SP(1)\ APF(2)/APF(1) \tag{14}$$

Permeate flow (Qp) at condition #1 can be related to operating conditions #2 accordingly to corresponding net drawing pressures (ND) and temperature correction factors (TCF)

$$Qp(2) = QP(1)\ NDP(1)/NDP(2)\ TCF(1)/TCF(2) \tag{15}$$

Example # 4
Normalization of system performance to the initial operating conditions

Recovery rate – R
$R = Qp/(Qp + Qr)$
$R(1) = 200/(200 + 50) = 0.80$
$R(2) = 180/(180 +60) = 0.75$

Concentration factor – CF
$CF = \ln(1/(1\text{-}R))/R$
$CF(1) = \ln(1/(1\text{-}0.80))/0.80 = 2.01$
$CF(2) = \ln(1/(1\text{-}0.75))/0.75 = 1.85$

Average feed salinity, ppm – Cfavg
$Cfavg = Cf\ CF$
$Cfavg(1) = 2000X\ 2.01 = 4020$
$Cfavg(2) = 2500\ X\ 1.85 = 4625$

Table 3.3 System configuration and operating parameters

Brackish RO unit. Array 20:10. Number elements per vessel - 7			
		Initial values of operating parameters, conditions 1	Current values of operating parameters, conditions 2
Feed salinity	ppm TDS	2000	2500
Permeate salinity	ppm TDS	30	50
Feed pressure,	bar (psi)	14 (203)	16 (232)
Concentrate pressure	bar (psi)	10.5 (152)	11 (160)
Feed– concentrate pressure drop	Bar (psi)	3.5 (51)	5 (72)
Permeate pressure	bar (psi)	1.5 (22)	1.5 (22)
Feed temperature	C	17	18
Permeate flow	m3/hr (gpm)	200 (880)	180 (792)
Concentrate flow	m3/hr (gpm)	50 (220)	60 (264)
Element type		8040 ESPA2	8040 ESPA2
Number of elements		210	210
Membrane area per element	m2 (ft2)	37 (400)	37 (400)

Average osmotic pressure, bar (psi) – Posmavg
Posmavg = 0.77 Cfavg/1000
Posmavg(1) = 0.77(4020/1000) = 3.1 (45)
Posmavg(2) = 0.77(4625/1000) = 3.5 (51)

Average permeate flux, l/m2-hr (gfd) – APF
APF = Qp 1000/(A EN)
APF(1) = 200X1000/(37X210) = 25.7 (15.1)
APF(2) = 180X1000/(37X210) = 23.2 (13.6)

Temperature correction factor – TCF
TCF = exp(2700(1/(273+t) – 1/298))
TC F(1) = exp(2700(1/(273+22)-1/298) = 1.284
TC F(1) = exp(2700(1/(273+18)-1/298) = 1.243

Net driving pressure bar (psi) – NDP
NDP = Pf – 0.5(Pf-Pc) – Pp - Posmavg
NDP(1) = 14.0 – 0.5(14.0 – 10.5) – 1.5 – 3.1 = 7.65 (111)
NDP(2) = 16.0 – 0.5(16.0 – 11.0) – 1.5 – 3.5 = 8.5 (123)

Specific flux L/m2-hr-bar (gfd/psi) – SF
SF = APF TCF/NDP
SF(1) = 25.7X1.284/7.65 = 4.31 l/m2-hr-bar (0.172)
SF(2) = 23.2X1.243/8.50 = 3.39 l/m2-hr-bar (0.136)
Salt passage, % - SP

SP = 100 Cp/Cfavg
SP(1) = 100X30/4020 = 0.74
SP(2) = 100X50/4625 = 1.08

Normalized salt passage, % - NSP
NSP(2) = SP(2)APF(2)/APF(1)
NSP(2) = 1.08X23.2/25.7 = 0.97%

Average feed flow m3/hr (gpm)
Qfav = (Qf+Qc)/2
Qfav(1) = (200 + 50)/2 = 125 (550)
Qfav(2) = (180 + 60)/2 = 120 (528)

Normalized pressure drop, bar (psi)
NPD(2)= DP1 (Qfav1/Qfav2)^1.4
NPD(2) = 5 (125/120)^1.4 = 5.29 (77)

Membrane elements fouling and performance restoration

Table 3.4 Summary operating parameters normalized to initial performances

Normalized value	Initial values	Current values	Difference, %
Specific flux, l/m²-hr-bar (gfd/psi)	4.31 (0.172)	3.39 (0.136)	-21
Salt passage, %	0.74	0.97	+31
Pressure drop, bar (psi)	3.5 (51)	5.29 (77)	+51

3.26 Membrane Elements Fouling Process

Membrane fouling affects membrane performance and can occur with any type of feed water and membrane types. Fouling is even encounter in secondary RO units processing RO permeate. Fouling affect is moderate in its initial stages, accelerating fast, and resulting in significant performance deterioration if not addressed on time. Summary of major fouling phenomena is presented in Table 3.5. Each one is presented as a separate category, however, in field conditions combined fouling processes are occurring either in parallel or one fouling process could be precursor to another one.

Correction of fouling conditions is a three step process:

1. Early detection of fouling process
2. Identification of fouling conditions and their mitigation
3. Correction of membrane performance

The most effective way of early detection of fouling process is through periodic evaluation of trends of normalized performance: product flow or water permeability, salt passage or salt transport and pressure drop. Any changes of normalized performance values, beyond the initial decline of water and salt transport are indicative of fouling process. The usual process design assumptions are that due to fouling, permeability will decrease and salt passage will increase by 5% – 10% per year. Performance deterioration at higher rate can only be tolerated if additional safety margin was applied during process design. Otherwise, system performance will be outside project specifications.

Table 3.5 Summary of major membrane fouling categories

Fouling factor	Initial fouling stage effect	Advanced fouling stage effect	Potential membrane damage
Exposure to free chlorine	Some permeability and salt passage decline. Initially in the lead element(s)	Increase of permeability and significant increase of salt passage	Irreversible damage of membrane barrier
Colloidal matter	Some increase of pressure drop. Initially in the lead element(s).	Significant increase of pressure drop, some decline of permeability and increase of salt passage	Element telescoping and extrusion of brine spacer
Dissolved natural organic matter (NOM)	Some permeability and salt passage decline.	Moderate decline of permeability, same salt passage decline	None
Biological matter	Some increase of pressure drop. Some permeability and salt passage decline. Initially in the lead element(s).	Severe increase of pressure drop. Some permeability and salt passage decline.	Element telescoping and extrusion of brine spacer
Inorganic scale	Some increase of pressure drop. Some permeability decline. Initially in the tail element(s).	Severe increase of pressure drop. Some permeability decline and salt passage increase.	Severe blockage of feed channels
Petroleum products	Significant permeability decline in the lead element(s). Small effect on salt passage	Severe decrease of permeability. Small effect on salt passage	None at low concentration. At high concentrations barrier integrity damage
Composite foulants (organics + colloids)	Some increase of pressure drop. Some permeability decline. Initially in the lead element(s).	Severe increase of pressure drop. Severe permeability decline and some salt passage increase.	Severe blockage of feed channels. Element telescoping

The identification of fouling conditions is a complex step wise process. It always starts with evaluation of updated composition and quality of the feed water. As mentioned previously, it is quite common that the process design of a new RO plants is based on approximate composition of raw water as feed water sources are usually being developed during system construction. The actual feed water should be analyzed in respect of concentration of scaling constituents (mainly in case of brackish sources) and RO feed water quality indicators such as turbidity, SDI and TOC. In instances that raw water has high fouling potential, useful information about nature of the foulants can be obtained by analyzing deposits on the SDI filter pad using scanning electron microscopy (SEM) and energy dispersive x-ray analysis (EDX). These tests may provide some indication about effectiveness of pretreatment and help to determine what type of foulants are arriving to membrane elements with the feed water. An example of comprehensive analysis of fouled membrane elements is included in reference 13.

Fouling process in RO system is usually identified through results of performance normalization. In brackish water multistage systems more insight into fouling phenomena is obtained through normalization of performance of individual membrane stages rather then normalization of total unit performance. Higher rate of performance decline in the first stage are indicative of foulants arriving with feed water. Typical fouling phenomena observed in the first stage are pressure drop increase due to blockage of the feed channels by colloids or biological growth. Another process, observed preferentially in the first stage at the beginning of the fouling process, is flux decline due to adsorption of organics. If the elements in the last stage are more affected then it is likely that fouling process is result of high concentration of fouling constituents due to an excessive recovery rate. The most common fouling phenomena, observed in the last stage, is formation of inorganic scale: either carbonate or sulfate. Less frequently scale of polymerized silica precipitates from the concentrate stream. Sometimes a mixed layer, containing both inorganic and organic foulants, is formed on the membrane surface. The mixed foulant layers are quite common as precipitation of one component, usually serves as precipitation centers and initiates precipitation of other constituents, which are at saturation concentrations.

Once the presence of fouling process in the RO system has been established through performance normalization, the next step is to remove elements for evaluation. For meaningful evaluation a lead and tail elements are required. Sometimes, a full load of a single pressure vessel is removed for testing. The first step is to examine elements appearance and determine their weight. Presence of slime on element outside surface indicates biological fouling. Accumulation of particles on the inlet element area indicates inadequate operation of the filtration system. Reddish-brown deposit usually indicates carryover of iron flocculant from the pretreatment system.

The next step in the evaluation process is to test membrane element performance at nominal test conditions and compare results with ex-factory test results.

The final step in the membrane elements examination process is evaluation of effectiveness of cleaning procedures. Based on the results of the above described elements examination, cleaning procedures are selected and tested initially on single elements. Membrane elements for conducting of the initial cleaning tests should be selected from positions in systems that have been most affected by fouling. The elements performances should be tested prior to cleaning and after the cleaning procedure is completed. If more than one cleaning solution is being evaluated, it is recommend to test element performance after each cleaning solution has been applied, at least during the initial cleaning attempts. The cleaning procedures that were found most effective in single elements cleaning tests are applied to restore performance of elements in the commercial trains. Extensive discussion of membrane evaluation process and interpretation of results is included in references 10 and 11.

3.27 Performance Restoration

3.27.1 Chemical Cleaning

Cleaning of elements in RO train is conducted using the cleaning unit. The configuration of cleaning unit is shown in Figure 3.20. It consists of cleaning tank, heater, recirculation pump, cartridge filter and connecting piping. Larger cleaning units also include separate tank for dissolving and mixing of cleaning solutions. Materials of construction of the cleaning unit should be selected to withstand low and high pH cleaning solutions (pH 2 – 11) at temperatures up

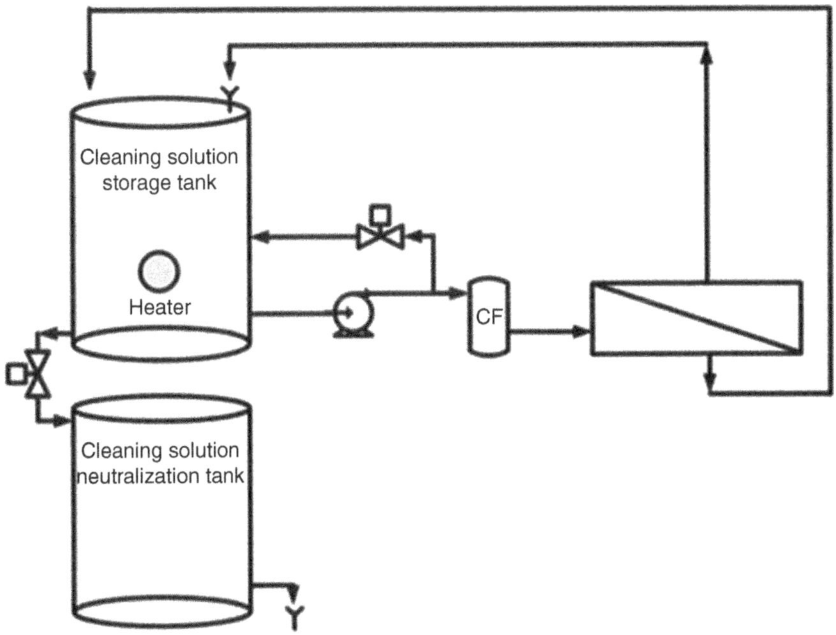

Figure 3.20 Configuration of membrane cleaning unit.

to 50 C. The size of cleaning tank and capacity of cleaning pump is determined by the number of pressure vessels that will be cleaned at one time. During cleaning operation the flow rate of cleaning solution per vessel should be close to 7 - 9 m^3/hr (~ 30 – 40 gpm). The cleaning tank volume should hold enough cleaning solution volume to provide at least 5 min of pump capacity. If, for example, 48 pressure vessels will be cleaned at one time, the operational volume of the cleaning tank should be about 19 m^3 (5,000 gallons).

In large RO systems the connecting piping of the cleaning unit is permanently attached to all trains. Valves or removable piping segments are used to connect/disconnect given train or train segment to the cleaning unit.

Cleaning operation sequence includes:

1. Flushing RO train with permeate water.
2. Connecting train or train segment to the cleaning unit.
3. Preparing cleaning solution in the cleaning unit.
4. Recirculating cleaning solution for 1 – 4 hr through the RO train.

5. Flushing cleaning solution
6. Repeating steps 2 – 5 with next cleaning formulation or reconnecting cleaned train to high pressure pump and restoring normal operation.

Membrane cleaning, like any other dispersive process, is more effective at elevated temperature. Cleaning should be conducted at temperature of cleaning solution in the range of 35 – 40 C. Cleaning solutions can be purchased from specialized suppliers or generic cleaning formulations can be used. Composition of generic cleaning formulations can be obtained from all major membrane manufacturers.

One of the generic low pH cleaning formulation, frequently used, is 2% solution of citric acid. pH of such solution is about 2.5. Citric acid cleaning solution is very effective in removal of deposits of metal hydroxides and dissolving of carbonate scale. If it has been established that fouling deposit contains mainly calcium carbonate or metal hydroxides, temporary operation with feed water acidified to low pH (pH = 4.5 – 5) with mineral acid (H2SO4 or HCl), may be sufficient to restore membrane performance. Cleaning, through operation at low feed pH, is only possible if discharge of low pH concentrate is allowed by local regulation at a given site.

The generic high pH cleaning formulations consist of solutions of NaOH in combination with EDTA or SDBS (surfactant). These cleaning solutions have pH of 10 – 11 and are effective in removal deposits of organic matter from membrane surface. It has been found (62) that EDTA or surfactants are essential components of high pH cleaning solutions and their presence contributes to improved removal of surface deposits that contain Ca ions imbedded in the organic fouling layer. In majority of cases fouling layer is of a mixed nature, it contains a mixture of inorganic and organic matter. The effective cleaning sequence is to apply low pH cleaning followed by application of high pH formulation.

RO systems operating on well water feed seldom need membrane cleaning. Cleaning frequency is usually less than one cleaning per 2 – 3 years of operation. RO systems treating surface water feed require more frequent membrane cleaning. In well designed and operated seawater RO systems cleaning is conducted at 1 – 2 years intervals. In seawater systems with inadequate feed water quality, required cleaning frequency could be much higher. For the purpose of operating cost estimation, budget for cleaning operation is usually based on two cleaning events per year. If more frequent cleanings are required, then it is an indication of inadequate pre-treatment process.

Example #5
Calculation of annual cleaning cost.
System permeate capacity: 100,000 m3/day (26.4 mgd),
RO unit configuration: 8 trains, 120 PV per train, 8 elements per vessel
Train segment size for a single cleaning: 60 pressure vessels.
Annual cleaning frequency: 2
Cleaning procedure: low pH cleaning followed by high pH

Free volume of pressure vessels: 60 X 8 X 0.025m^3 = 12 m^3 (3200 gallons)
Volume of manifolds (10% of PV): 0.1 x 12 m^3 = 1.2 m^3 (320 gallons)
Volume of connecting piping (50% of PV): 0.5 x 12 m^3 = 6 m^3 (1600 gallons)
120 PV X 8 m3/hr X 3min/60 min = 24 m3 (6,300 gallons)
Total volume of cleaning solution: 19.2 m^3 + 24 m3 = 43.2 m^3 (11,400 gallons)

Chemicals quantity for annual cleaning operation:
Solution 1 – citric acid
2% Citric acid: 0.02 x 43.2 = 0.864 t/cleaning
8 train X 2 segments X 2 cleanings/year X 0.864= 27.6 t
Solution 2 – NaOH + SDBS
0.2% NaOH: 0.002 x 43.2 = 0.086 t/cleaning
8 train X 2 segments X 2 cleanings/year X 0.086 = 2.8 t
0.2% SDBS: 0.002 x 43.2 = 0.086 t/cleaning
8 train X 2 segments X 2 cleanings/year X 0.086 = 2.8 t

Annual cost of cleaning chemicals

Citric acid:	27.6t X $2500/t =	$69,000
SDBS:	2.8t X $3000/t =	$8,400
NaOH:	2.8t X $250/t =	$700
Total cost of cleaning chemicals per year		$78,100

Cost per water produced: $0.002/m^3 ($0.008/kgallon)

As indicated by above example cost of generic cleaning chemicals is not significant if cleaning frequency is limited to two cleanings per year. Additional cost, associated with cleaning operation, that should be considered is loss of production capacity. The system off line time required for cleaning is in the range of 1 – 1.5 days, which is corresponds to about 0.4% of availability The major expense related to cleaning, at some locations, could be the disposal cost of spent cleaning solutions.

In majority of cases cleaning operation is capable to restore some of the lost permeability and reduce pressure drop. Very seldom salt rejection is improved. Usually, it remains the same or can even decline after cleaning. This is because foulant layer plugs imperfections and damaged areas in membrane barrier and effective cleaning opens them again to salt passage.

If cleaning attempts does not result in sufficient performance improvement, membrane element replacement is the only practical solution available for additional performance correction. Usually a considerable fraction of elements in the system have to be replaced to achieve noticeable performance improvement. Number of elements that require replacement can be reduced if elements with worst performance can be identified in the RO system.

3.27.2 Direct Osmosis Cleaning

Direct osmosis is phenomena frequently observed in RO systems treating high salinity feed water. When RO units shuts down, there is a reverse flow of permeate through the membrane to the feed side. This process is utilized in new cleaning method called Direct Osmosis – High Salinity (DO – HS) membrane performance restoration [21, 22]. The DO – HS method constitutes injection of high salinity solution to the suction of high pressure pump, while the RO unit is in operation. The injection lasts number of seconds. The high pulse flows through the membrane elements from the feed to the concentrate. During the flow of high salinity wave through the feed channels of RO elements, the flow trough the membrane is reversed. The reverse permeate flow lifts the foulants from the membrane surface. The continued feed - concentrate flow sweeps lifted foulant particles from the membrane elements, out of the membrane unit.

This method of membrane performance restoration is especially effective in controlling membrane biofouling and removing colloidal deposits.

References

1. S. Loeb and S. Sourirajan, *Adv. Chem. Ser.* **38**,117 (1962).
2. O.K. Buros, R.B. Cox, I. Nusbaum, A.M. El-Nashar, and R. Bakish, The USAID Desalination Manual, IDEA (1981).
3. Reverse Osmosis and Synthetic Membranes, Theory - Technology - Engineering, S. Sourirajan (Ed.), National Research Council Canada Publications, Ottawa (1977).
4. R. Rautenbach and R. Albrecht, *Membrane Processes,* John Willey and Sons, New York (1989).
5. S. Parekh (Ed.), *Reverse Osmosis Technology Application for High - Purity - water Production*, Marcel Dekker Inc., New York (1988).
6. R. Peterson, J. Cadotte, and J. Buttner, Final report of development of FT-30 membrane in spiral wound modules, Contract No 14-34-0001-8547 for US Department of Interior, (October 1882).
7. Material Science of Synthetic Membranes, J. E. Cadote and D. R. Lloyd (Eds), *Amer. Chem. Soc.* Symposium Series, Washington DC, pp. 273–294, (1985).
8. D.T. Bray and H.F. Menzel, Design Study of Reverse Osmosis Pilot Plant, Office of Saline Water, Research and Development Progress Report No. 176 (1966).
9. H. Iwahori, M. Ando, R. Nakahara, M. Furuichi, S. Tawata, and T. Yamazato, Seven years operation and environmental aspects of 40,000 m3/day seawater RO plant at Okinawa, Japan, Proceedings of IDA Congress, Bahamas (2003).
10. K. Tasaka, H. Iwahori, and Y. Kamiyama, Fouled membrane analysis of the operated elements from more than 80 plants in Japan, Proceedings IDA Conference, Madrid (1997).
11. Donald T. Bray, U.S. Patent 4,046,685 (1977).
12. P. Glueckstern, M. Wilf, J. Etgar, and J. Ricklis, "Use of Micro-processors for Control, Data Acquisition and on Line Performance Evaluation in a Reverse Osmosis Plant". Proceedings of the Second World Congress on Desalination and Water Reuse. Bermuda (November 1985).
13. R. Schneider, L. Ferreira, P. Binder, and J. Ramos, Analysis of foulant layer in all elements of an RO train. *J. Membr. Sci.* **261**, 152-162 (2005).
14. Z. Amjad, Reverse Osmosis failures; cause and prevention, Ultrapure Water, (December 1991).
15. Lynn Stevens, Jeff Kowal, Ken Herd, Dr. Mark Wilf, and Wayne Bates, Tampa Bay seawater desalination facility: Start to finish, IDA Conference, Bahamas (2003).
16. The Guidebook to Membrane Desalination Technology. Reverse Osmosis, Nanofiltration and Hybrid Systems. Process, Design and Applications by M. Wilf with chapters by C. Bartels, L. Awerbuch,

M. Mickley, G. Pearce, and N. Voutchkov, Balaban Desalination Publications (2006).

17. The Guidebook to Membrane Technology for Wastewater Reclamation. Wastewater Treatment, Pollutants, Membrane Filtration, Membrane Bioreactors, Reverse Osmosis, Fouling, UV Oxidation, Process Control, Implementation, Economics, Commercial Plants Design by M. Wilf with chapters by C. Bartels, D. Bloxom, J. Christopher, A. Festger, K. Khoo, V. Frenkel, J. Hudkins, J. Muller, G. Pearce, R. Reardon, and A. Royce, Balaban Desalination Publications (January 2010).
18. V.P. Caracciolo, N.W. Rosenblatt, and V.J. Tomsic, DuPont hollow fibers membranes, in *Reverse Osmosis and Synthetic Membranes*, S. Sourirajan (Ed.), National research Council of Canada Publications, Ottawa (1977).
19. J.W. McCutchan, S. Loeb, P.A. Buckingham, and A.W. Ayers, Preliminary economic study U.C.L.A. reverse osmosis process for brackish water desalination, UCLA Engineering Department Report 63–62, (1963).
20. Standard practice for standardizing reverse osmosis performance data, ASTM D 4516 – 00 (2010).
21. I. Liberman, RO membrane cleaning method, US Patent Application 20070181497 (2007).
22. A. Sagiv, N. Avraham, C. Dosoretz, and R. Semiat, Osmotic backwash mechanism of reverse osmosis membranes, *J. Membr. Sci.* **322**, 225–233. (2008).

4

Nanofiltration – Theory and Application

Christopher Bellona

Abstract

Nanofiltration (NF) represents a relatively recent development in membrane technology with characteristics that fall between ultrafiltration (UF) and reverse osmosis (RO). NF membranes capable of ion and small uncharged solute separation (i.e., loose NF) represent a unique membrane process that can be used for a variety of applications. With effective pore sizes in the range of 1 nm, solute removal is believed to be a function of steric hindrance, and Donnan and dielectric exclusion which, allows for manipulation of the NF process to achieve preferential removal of multi-valent ions, separation of uncharged solutes based on size, and demineralization of process streams containing valuable constituents. Significant recent interest in the use of NF for a wide variety of separation processes indicates that further membrane development will lead to increased use of NF in the near future. This chapter provides an overview of NF technology including history, theory and application.

Keywords: Nanofiltration, ultrafiltration, reverse osmosis, membrane, desalination, water treatment, water reuse, industrial applications, rejection, separation

Jane Kucera (ed.) Desalination, (205–254) 2014 © Scrivener Publishing LLC

4.1 Introduction

Nanofiltration (NF) represents a relatively recent development in membrane technology with characteristics that fall between ultrafiltration (UF) and reverse osmosis (RO). NF has been engineered to provide selective separation of solutes and now consists of a wide range of membrane materials and configurations that can be used for a variety of applications. While RO membranes dominate the seawater desalination industry, NF is employed in a variety of water treatment and industrial applications for the selective removal of ions and organic substances, and also includes certain seawater desalination applications. Significant recent interest in the use of NF for a wide variety of separation processes indicates that further membrane development will lead to increased use of NF in the near future. The following sections present a brief introduction into NF technology including theory and application.

4.2 Defining Nanofiltration

Since the development of the first reverse osmosis (RO) membranes by Loeb and Sourirajan, researchers have continually developed and reformulated membrane materials giving rise to membrane products with a range of filtration characteristics and fluxes [1, 2]. While the major goal of early RO membrane development research was to produce potable water from seawater, there was also interest in developing membranes tailored for specific applications, particularly those requiring less than 99% sodium chloride removal and lower operating pressures [3]. This desired range of filtration was occupied by membrane materials capable of moderate to high divalent ion rejection, rather high passage of monovalent salts and high rejection of low-molecular weight organic compounds. Additionally, it was observed that many of these membranes could operate at significantly lower pressures than RO membranes. Such membranes were initially categorized as 'open RO', 'loose RO', 'intermediate RO/UF', 'selective RO' or 'tight UF' [2].

The term nanofiltration (NF) appears to have been first used commercially by the Filmtec Corporation in the mid-1980s to describe a new line of commercially available membranes with characteristics between ultrafiltration (UF) and RO [4]. These membranes were described in several Filmtec Corporation and Dow Chemical Company patents [5, 6] as possessing:

> '…*certain unusual properties. These membranes exhibit a significant water flux at relatively low transmembrane pressures and also are ion selective. The patented membranes show much greater rejection of multivalent anion salts than of monovalent anion salts*'.

Guilhem [7] further characterized the newly commercialized NF membranes produced by Filmtec Corporation as those that:

> '… *fall between reverse osmosis and ultrafiltration membranes in terms of the size of the species which will be passed through the membrane. As used herein, the term "nanofiltration membrane" refers to having micropores or openings between chains in a swollen polymer network, which micropores or openings are estimated to have an average diameter on the order of about 10 angstrom (i.e., one nanometer).*'

Because the physical and operational differences between RO and UF are rather large, a rigorous and universally accepted definition of NF has largely proven elusive. For example, researchers have defined NF membranes based on various characteristics including: a molecular weight cut-off (MWCO) between 100 and 500 Daltons [1, 3, 8]; a MWCO between 200 and 10,000 Daltons [9]; magnesium sulfate rejection between 90 and 98% and sodium chloride rejection less than 50% [4]; selective rejection of solutes with molecular weights of 300 to 400 Daltons and exclusion of ions according to their charge densities [10]; sodium chloride passage proportional to the sodium chloride concentration to a power of 0.4 [11]; pore size between 0.5 and 2 nm [12]. In addition, the International Union of Pure and Applied Chemistry (IUPAC) defines nanofiltration as a pressure-drive membrane process in which particles and dissolved macromolecules smaller than 2 nm are rejected.

As a further complication, newer generation RO membranes including low-pressure RO (LPRO) and ultra-LPRO (ULPRO) have similar pressure requirements as commercially available 'NF' membranes with salt rejection less than RO membranes developed for seawater and brackish water desalination. As a result, the line between NF, and UF and RO membranes is often blurred which has made NF as a standalone technology somewhat controversial to some researchers [13]. To illustrate the difficulty in defining NF, a summary of the characteristics of commercially available NF membranes is provided in Table 4.1. Commercially available NF

Table 4.1 List of commercially available NF membranes with select properties (all properties came from manufacturer unless otherwise noted).

Membrane	Manufacturer	Type	Material	NaCl Rej.	$MgSO_4$ Rej.	$CaCl_2$ Rej.	MWCO (Da)
TFC-SR2	Koch[a]	NF	TFC/Polyamide	10 – 30	95	NA[b]	300 – 400
TFC-S		RO/NF[d]	TFC/Polyamide	85	99	NA[b]	180[b]
TFC-SR3D		NF	TFC/Polyamide	NA[b]	>99	NA[b]	200
TFC-SR100		NF	TFC/Polyamide	NA[b]	>99	NA[b]	>200
ESNA1-LF-LD	Hydranautics[e]	NF	TFC/Polyamide	NA[b]	NA[b]	NA[b]	NA[b]
NANO-BW		NF	TFC/Polyamide	NA[b]	99.7	NA[b]	NA[b]
NF90	Dow Filmtec[f]	NF	TFC/Polyamide	85 – 90	>97	NA[b]	~100[g]
NF270		NF	TFC/Polypiperazine	NA[b]	97	40 – 60	~180[g]
NE-70	CSM[h]	NF	TFC/Polyamide	40 – 70	97	40 – 70	NA[b]
NE-90		NF	TFC/Polyamide	85 – 95	97	90 – 95	NA[b]
Desal-5-DK	Osmonics[i]	NF	TFC/Polyamide	NA[b]	96 – 98	NA[b]	150 – 300

Membrane	Manufacturer	Type	Material	NaCl Rej.	$MgSO_4$ Rej.	$CaCl_2$ Rej.	MWCO (Da)
Membralox-NF	Pall[j]	NF	Ceramic/Titania	NA[b]	NA[b]	NA[b]	1000 – 5000
Ionopor nano	Rauschert[k]	NF	Ceramic/Titania	NA[b]	NA[b]	NA[b]	450 – 750
N30F	Nadir[l]	SR-NF[m]	Polyethersulfone	NA[b]	NA[b]	NA[b]	400[n]

[a]Koch Membrane Systems, Wilmington, MA, USA

[b]Data not available

[c]Data from Sharma

[d]Previously listed as NF, now listed as softening RO membrane

[e]Nitto Denko Company, Oceanside, CA, USA

[f]Dow Filmtec, Midland, MI, USA

[g]Data from Bellona (unpublished)

[h]CSM Woongjin Chemical Co, Jung-gu, Seoul, South Korea

[i]GE Osmonics, Vista, CA, USA

[j]Pall Corporation, White Plains, NY, USA

[k]Rauschert GmbH, Germany

[l]Nadir Filtration GmbH, Weisbaden, Germany

[m]Solvent resistant NF

[n]Data from Van der Bruggen [67]

membranes can span a range of molecular weight cut-offs (MWCO) of 100 all the way up to 1,000 – 5000 Daltons, and NaCl rejection between 10 and 95%, but tend to have a relatively narrow range of $MgSO_4$ rejection.

Because commercially available NF membranes span a range of characteristics, researchers have adopted the terms 'tight' and 'loose' to differentiate between the two major classes of polymeric cross-linked NF membranes. Solute removal by tight NF is believed to be fundamentally similar to LPRO in that non-selective solute removal occurs due to hindered diffusion in the membrane polymer and, to a lesser extent, through charge (or Donnan) exclusion resulting from interactions between ions and ionic functional groups in the membrane polymer. Loose NF membranes however, are believed to rely more on charge exclusion to restrict ion transport and show selective passage of ions with low valence. It is worth noting that controversy exists regarding whether or not loose NF, and to a lesser extent RO and tight NF, have a defined pore structure, although is now commonly reported that mass transfer in NF is through distinct pores. As pointed out by Linder and Kedem [2], both porous and non-porous modeling approaches can describe mass transport in NF systems indicating that the NF truly lies between RO and UF membrane processes.

In light of the previous discussion, NF as a truly standalone technology (from RO and UF) is primarily limited to loose NF membranes that can provide selective passage of ions and uncharged solutes based on charge and size. Discussion of NF below focuses on loose NF unless otherwise noted.

4.3 History of Nanofiltration

Cellulose acetate membranes with characteristics of NF were essentially developed along side early asymmetric cellulose acetate (CA) RO membranes in the 1960s. As pointed out in the excellent review of the history of NF by Linder and Kedem, the 1964 US patent of Loeb and Sourirajan included data indicating that a range of NaCl passage (~5 – 82%) and improved flux could be obtained by cellulose acetate (CA) membranes by varying the casting solution composition, evaporation period and annealing [2, 14]. Thus, asymmetric CA membranes with rejection properties between UF and RO were commercially available from the early 1970s although the term NF wasn't coined until later (see previous section) [2, 4].

Due to the intermediate behavior of these membranes and the initial application for divalent cation removal from surface water, the term 'softening membranes', became a common and still used moniker for these membranes. However, researchers began evaluating these membranes for use in a variety of applications including separating pharmaceutical fermentation liquors, separating and concentrating components in the food and dairy industry, and color and hardness removal in the water treatment industry. In one of the first reported water treatment applications, a special softening cellulose diacetate membrane (marketed by Fluid Systems (ROGA-8150)) with a monovalent rejection rate of 47% was evaluated as early as 1976 in Florida as a substitute for lime softening [8]. According to Conlon and McClellan [8], CDA RO membranes were hydrolyzed in-situ using caustic solution to yield the desired membrane properties. From many accounts, these membranes covered the range of NF in terms of rejection but had several issues limiting their applicability including low initial fluxes, poor chemical and biological stability, and continual changes in rejection and flux during operation [2]. Due to these issues, various other polymeric materials were investigated to make asymmetric NF membranes (e.g., polysulfone, polyelectrolyte complexes, other cellulosic formulations) however, optimum flux and stability performance was never fully realized and by the 1970s new methods for producing both RO and NF membranes were being evaluated.

Watson and Hornburg [1] presented a brief summary of membrane development through the 1980's and put the emergence of commercially available thin-film aromatic polyamide and polypiperazineamide low-pressure NF membranes around 1986. These NF membranes were produced by the Filmtec Corporation and included the NF-50 and NF-70 which could provide high $MgSO_4$ rejection and NaCl passage. Around the same time (1985), researchers from Toray Industries reported on the development of 'super low pressure' polypiperazineamide reverse osmosis membranes with similar sodium chloride rejection performance (50 – 70%) [15]. These membranes were evaluated for organic compound removal and had MWCOs of approximately 200 g/mol based on glucose rejection. According to Linder and Kedem [2], further developments of NF membrane technologies included chemical modifications of aromatic polyamide RO membranes to produce desired flux and rejection performance, and NF developed from a variety of different polymers including polyvinyl alcohol, polyethyleneimine, and polysulfone.

Commonly applied NF membranes are now composed of proprietary aromatic polyamide and polypiperazineamide polymers. Manufacturers of these membranes included Dow/Filmtec, GE/Osmonics, Hydranautics/Nitto Denko, Koch, Toray, CSM/Saehan, and other smaller companies (Table 4.1). However, other commercially available products based on alternative materials exist. Sulfonated polysulfone based membranes have been marketed by several companies due to high stability over a wide range of pH and temperature, and polyvinyl derivative NF membranes are marketed by several companies for concentration of maple sap for maple syrup production. Additionally, inorganic membranes (i.e., ceramic membranes) with the designation of NF are becoming more common although these membranes have relatively high MWCOs putting them in the filtration range of tight UF.

A recent breakthrough in NF technology has been the development of commercially available organic solvent resistant NF membranes (often termed organic solvent NF (OSN)). These membranes have found numerous applications in industry with first large-scale OSN system used by Exxon-Mobil for lubricant dewaxing [16]. While researchers have been developing OSN membranes for a variety of purposes since the 1990s, recent breakthroughs have been reported including enhanced stability in a wider variety of solvents, greater control over membrane properties, and development of membranes spanning a range of MWCOs [17].

By the late 1980s, NF was viewed as a legitimate membrane treatment option for a variety of applications. As a result of continuous development of NF membrane materials, research into NF has increased significantly since 1990. A summary of scholarly articles published on NF is presented in Figure 4.1 which demonstrates the increased interest in NF for a variety of applications.

4.4 Theory

NF has unique characteristics that allows for selective filtration of dissolved salts and organic compounds at relatively low operating pressures. As a result, NF has been employed in a variety of water treatment and industrial applications to meet a wide range

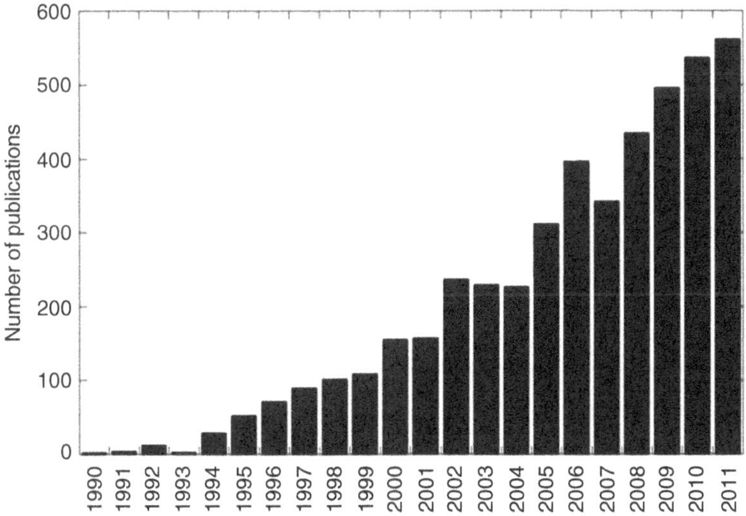

Figure 4.1 Number of publications per year since 1990 with 'nanofiltration' listed in the topic (from ISI Web of Knowledge (Thomas Reuters)).

of objectives. Considerable work has been performed to better understand NF filtration mechanisms, model NF performance, engineer NF membranes for certain applications, and alleviate issues related to NF operation such as fouling. These fundamental areas are reviewed in the sections below.

4.4.1 Mechanisms of Solute Removal

As opposed to RO membranes, NF membranes are generally (although somewhat controversially) accepted to have a pore structure with pore diameters larger than most dissolved salts [18]. Because NF membranes carry a surface charge (usually negatively charged at neutral pH), solute rejection is attributed to three mechanisms: size exclusion or steric hindrance, electrostatic (or Donnan) exclusion, and dielectric exclusion [12, 19]. Size exclusion refers to either the total exclusion of a solute from the membrane polymer or pore structure, whereas steric hindrance refers to hindered transport within the membrane relative to water. Donnan exclusion is important for charged solutes and occurs due to the attraction of counter-ions to the charged membrane surface which results in exclusion of co-ions. Dielectric exclusion refers to the phenomenon whereby the aqueous solution and polymeric membrane matrix have different dielectric constants making it less favorable for charged solutes to enter the membrane polymer [20].

4.4.1.1 *Ion Rejection*

The rejection of ions (e.g., salts) by NF is a complex process dependent upon ion characteristics (i.e., size and valence), membrane characteristics (i.e., effective pore size, thickness, porosity, charge density), feed water composition (i.e., ion concentration, ion composition, pH), temperature and operating conditions (i.e., flux, recovery). Compared to size exclusion/steric hindrance effects which can be estimated by the size of an ion relative to the effective membrane pore size, ion rejection resulting from electrostatic effects is relatively complicated. These complications arise because membrane surface charge is a function of pH, and is influenced by the types and concentrations of ions in solution. Additionally, ion-partitioning between the feed solution and the membrane is influenced by other ions in solution.

For binary salt solutions (e.g., NaCl only) at fixed pH, rejection primarily depends on the effective membrane pore size (due to steric effects), the ratio of membrane charge density to salt concentration, and permeate flux [21, 22]. In general, ion rejection increases with increasing ion size and decreasing effective pore radius (steric effect), increasing salt valence (Donnan exclusion), decreasing salt concentration for a fixed membrane surface charge (Donnan exclusion, Figure 4.2), and increasing permeate flux. Solution pH influences membrane surface charge due to ionization of functional groups in the active layer and for most NF membranes, an increase in pH results a greater negative surface charge and increased ion rejection (Figure 4.2). It is worth noting that several researchers have shown that certain ion pairs (e.g., $CaCl_2$) may not follow this expected trend due to interactions between salts and the membrane surface [23, 24].

For multicomponent salt solutions, which are common in real applications, rejection of individual ions is complex and may never be fully understood [20]. While the three mechanisms are still valid, ion-partitioning between the bulk solution and membrane is dependent upon the other ions in solution and requirement that electroneutrality be maintained across the membrane. Because certain ions are more permeable than others, their transport across a membrane must be balanced by the transport of an

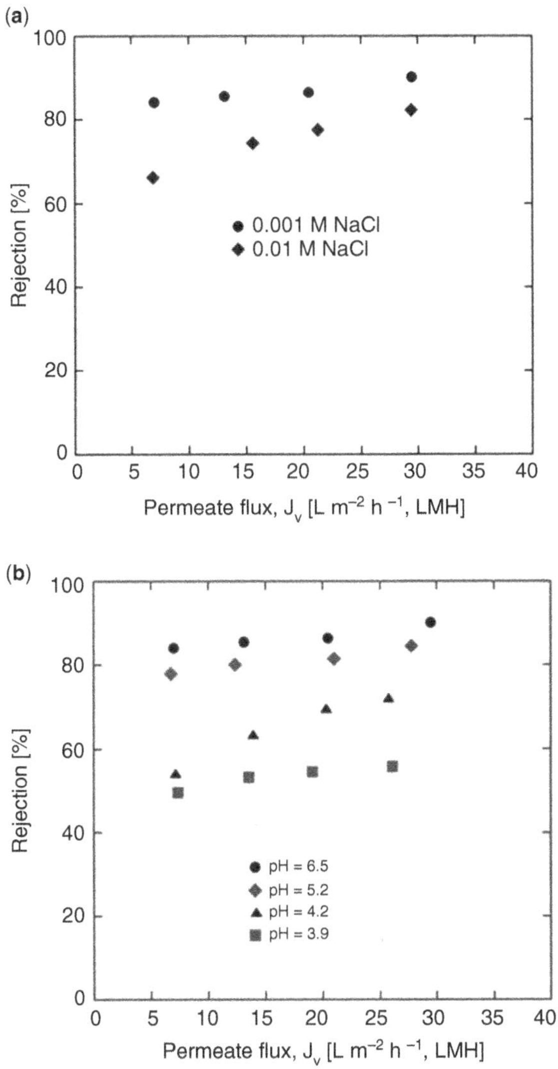

Figure 4.2 NaCl rejection by an NF membrane for a NaCl solution as a function of permeate flux and concentration (top) and permeate flux and pH (bottom, 0.001 M NaCl).

ion of opposite charge [25]. This transport coupling between certain ion pairs can have a significant overall effect on rejection and is difficult to describe quantitatively although several models have been developed [22, 26–28]. However, it is this phenomena which makes the case of negative rejection possible, for example, in the case of chloride in quaternary solutions of NaCl and Na_2SO_4 when chloride concentration is low relative to sodium and sulfate.

As a result of aforementioned ion rejection mechanisms - NF membranes, with pore sizes in the range of 1 nm and relatively low charge densities, have only limited rejection of small monovalent ions (e.g., potassium, sodium, bicarbonate, chloride) compared to multi-valent ions (e.g., calcium, magnesium, carbonate, sulfate). While this limits NF to certain treatment applications, it results in lowered pressure requirements and lower concentrations of monovalent salts in the concentrate stream compared to RO, and the possibility for preferential separation of ions based on their size and valence. Researchers have used the properties of NF membranes and the influence of flux, pH and ionic strength on rejection to develop processes for purifying and fractionating various process streams in different industries. These applications will be discussed further below.

4.4.1.2 Organic Solute Rejection

NF has been evaluated for applications where removal of organic matter and organic contaminants is desired. Similar to ion rejection, organic solute rejection can be a complex phenomenon dependent upon membrane and solute properties, feed solution chemistry, temperature, and operating conditions. Due to recent interest in NF for a variety of applications, there has been a significant amount of research into the removal of organic solutes by different NF membranes. One of the main outcomes of this research has been the observation that rejection of organic solutes is due to three main mechanisms: size exclusion/steric hindrance, electrostatic effects (Donnan and dielectric exclusion), and solute-membrane interactions.

For nonionic or uncharged organic solutes without significant solute-membrane interaction, rejection is mainly a function of the ratio between solute size and the effective pore size of the membrane, or ratio between solute's molecular weight (MW) and the MWCO of the membrane [29]. However, a number of researchers have demonstrated that solute size can be difficult to characterize [30–32] and that MW is often a poor parameter to characterize rejection performance. As such, the Stokes radius (r_s) is commonly used to characterize molecular size which can be calculated from Eqn. (4.1):

$$r_S = \frac{kT}{6\pi\eta D_s} \tag{4.1}$$

where k is the Boltzmann constant, T is temperature, η is the dynamic viscosity of water, and D_s is the solute liquid-phase diffusion coefficient. The use of Eqn. (4.1) requires the calculation of a solute's diffusion coefficient (D_s) which can be experimentally measured or calculated using several methods one of which, is using the Wilke-Chang correlation (Eqn. (4.2)):

$$D_S = \frac{7.4x10^{-8}(\varphi M)^{0.5}T}{\eta V_S^{0.6}} \tag{4.2}$$

where φ is an association factor (usually 2.26 for aqueous solutions), M is molar mass of the solvent and V_s is the molar volume of the solute. A solute's volume can be experimentally determined or calculated using several approaches based on additive molar volume methods such as the LaBas approach where a solute's molar volume is the sum of the volume of all atoms in the molecule according to Eqn. (4.3):

$$V_S = \sum_i n_i V_{B,i} \tag{4.3}$$

where $V_{B,i}$ is the LaBas molar volume of each atom, ring or substituent group in the molecule. Values for $V_{B,I}$ can be found in other references such as [33]. The rejection of 12 nonionic organic solutes by an NF membrane as a function of r_s is presented in Figure 4.3. As demonstrated in Figure 4.3, NF can be used to separate solutes based on molecular size which has been used in variety of industries.

For a wide variety of nonionic organic solutes, common molecular size descriptors have been shown to only marginally correlate with rejection [32, 34]. As a result, researchers have attempted to develop novel molecular size descriptors [35–37] or explain the discrepancy due to other molecular properties that may affect rejection including hydrophobicity, polarity and ability to form hydrogen bonds [32, 38–40].

For certain nonionic solutes, rejection is often lower than anticipated based on molecular size due to interactions between a solute and membrane. Several studies have demonstrated that compounds can adsorb to and partition through membrane materials which is dependent upon the properties of the solute and membrane [41, 42]. Williams et al. [43] proposed that solutes adsorb through two main mechanisms: specific adsorption where solutes can form hydrogen

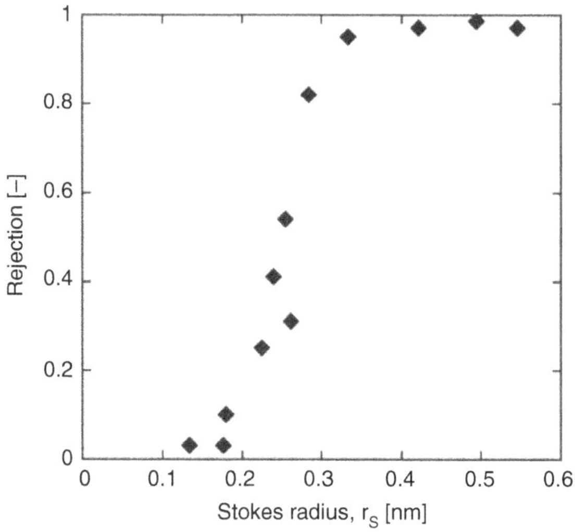

Figure 4.3 Rejection of nonionic organic solutes by a NF membrane as a function of Stokes radius at a permeate flux of 20 LMH.

bonds with the membrane polymer; and non-specific adsorption where hydrophobic or non-polar solutes have greater affinity for the membrane polymer compared to water. These compounds often exhibit high initial rejection during the adsorption stage followed by significant decreased rejection as a function of filtration time [44]. Bellona et al. [45] recently found that this phenomenon is more significant for NF membranes compared to RO and is difficult to predict based on solute properties. Researchers have attempted to predict the compounds likely to exhibit this behavior using octanol/water partitioning coefficients (Log K_{ow}), acid dissociation constants (pK_a), Taft and Hammett constants, and dipole moment with only limited success. This adsorption phenomena has been reported for phenols [41, 43], steroidal hormones [46], trihalomethanes [47], perfluorinated compounds [44] and other compounds.

Because most ionic organic solutes, for which both electrostatic and steric effects are important, are relatively large compared to salts, they are generally well removed by most NF membranes regardless of size [48]. Exceptions include certain small organic acids (e.g., acetic acid, certain amino acids) and aliphatic amines (e.g., triethylamine). To illustrate, the rejection of negatively and positively charged organic solutes by an NF as a function of Stokes radius is presented in Figure 4.4. With the

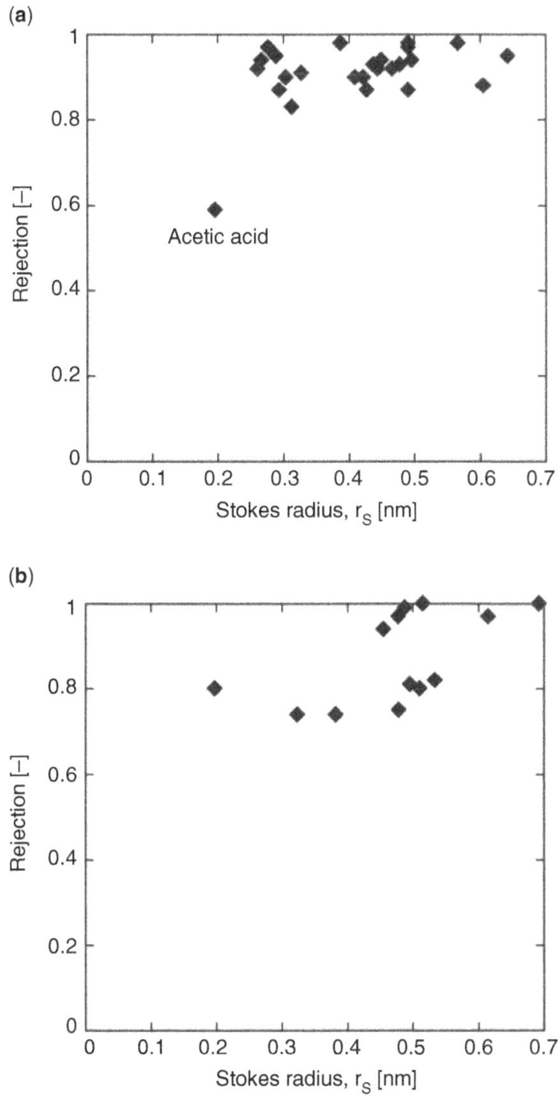

Figure 4.4 Rejection of negatively charged (top) and positively charged (bottom) organic solutes by a NF membrane as a function of Stokes radius at a permeate flux of 20 LMH and pH of 6.5.

exception of acetic acid, all negatively charged organic solutes exhibited rejection above 80 percent. Rejection of positively charged organic compounds showed more variability, which has been reported in the past [49].

4.4.2 Modeling NF Separations

A considerable amount of effort has been put towards developing quantitative approaches to describe mass transport in membrane systems. Early modeling work that laid the foundation for recently developed approaches include those based on irreversible thermodynamics [50, 51] and mass transport through charged porous membranes described by the extended Nernst-Planck equation [9, 28, 52]. Several research groups have expanded on early approaches using the extended Nernst-Planck to include descriptions solute partitioning due to electrostatic (Donnan), dielectric and steric mechanisms [18, 19, 22]. Other models based on force balances around a solute permeating a membrane pore (finely porous model, and surface-force pore model) and modified forms of the solution-diffusion model have also been applied to describe NF systems [53, 54]. This effort has resulted in a number of models that can be applied to NF membrane systems, and the choice of one model over another is often dependent on the system being evaluated and model complexity. The following sections present a brief summary of modeling approaches that can be applied to NF systems.

4.4.2.1 Donnan Steric Pore Model

As previously mentioned, modeling approaches based on the extended Nernst-Planck equation have been developed by several research groups and are commonly applied in studies to better understand and optimize NF separations. One of these approaches, termed the Donnan steric pore model (DSPM), describes hindered solute transport through a membrane due to steric and electrostatic partitioning between the membrane and bulk liquid phase. Solute flux (J_i) can be described by:

$$J_i = -K_{i,d} D_{i,\infty} \frac{dc_i^m}{dx} + K_{i,c} c_i^m J_v - \frac{z_i c_i^m K_{i,d} D_{i,\infty}}{RT} F \frac{d\psi^m}{dx} \tag{4.4}$$

where $D_{i,\infty}$ is bulk phase diffusivity, c_i^m is the membrane phase ion concentration, z_i is the ion valence, R is the universal gas constant, T is absolute temperature, F is the Farraday constant, $d\Psi^m/dx$ is the electric potential gradient across the membrane, J_v is the permeate flux, and $K_{i,d}$ and $K_{i,c}$ are the hindrance factors for convention and diffusion, respectively. As described by Eqn. (4.4), solute flux through a membrane is a combination of diffusive, convective and electrical transport phenomena.

Eqn. (4.4) applies to ions however, for nonionic solutes, the last term on the right side of Eqn. (4.4) is zero. The hindrance factors, $K_{i,d}$ and $K_{i,c}$, account for hydrodynamic effects caused by interactions between the solute and pore walls (steric hindrance) and are usually derived based on a cylindrical pore structure and empirical equations relating hindrance factors to the ratio between the solute radius and pore radius, λ ($\lambda = r_s/r_p$). Common expressions for ion partitioning in the membrane include steric, electrostatic (Donnan) and dielectric terms, for example:

$$\frac{c_i^m}{C_i} = \Phi_i exp\left(\frac{-z_i F}{RT}\Delta\psi\right) exp\left(\frac{-\Delta W_i}{kT}\right) \tag{4.5}$$

where Φ_i is a steric partitioning term ($\Phi = (1-\lambda)^2$), C_i is the liquid phase ion concentration, k is the Boltzmann constant, and ΔW_i is the solvation energy barrier due to differences between the liquid and membrane dielectric constant.

For ions, the application of the DSPM requires an iterative approach to solve the differential equations describing the concentration and electric potential gradients across the membrane. The model has four main fitting parameters including the membrane solvent permeability coefficient (L_p), membrane charge density (X_d), membrane pore radius (r_p) and membrane thickness to porosity ratio ($\Delta x/A_k$). The iterative solution procedure and number of fitting parameters has mainly limited the DSPM to binary or ternary ion solutions and model water solutions. However, the advantage of the DSPM model is that rejection performance can be analyzed as a function of structural and electrical properties of the NF membrane which, is both useful from a theoretical and membrane characterization stand-point. The DSPM has been applied to describe separation of ions in the textile and pharmaceutical industry [55–57], as well as predict the removal of organic contaminants [32, 46, 58] by NF membranes.

4.4.2.2 *Irreversible Thermodynamic or Phenomenological Model*

Based on thermodynamics of irreversible processes, the phenomenological model has been applied to describe separation performance of both nonporous and porous membranes. Solute rejection can be calculated as:

$$Rej. = 1 - \frac{C_p}{Cf} = \frac{\sigma(1-F)}{1-\sigma F} \tag{4.6}$$

where σ is the reflection coefficient that describes the case of limiting rejection and F is calculated by:

$$F = exp\left(-\frac{1-\sigma}{P}J_v\right) \tag{4.7}$$

where P is the solute permeability coefficient and is analogous to solute diffusion in the membrane. This approach is often described as a 'black-box' model as the two model coefficients (σ and P) encompass the structural and electrical properties of the membrane and solute that influence rejection. As such, the phenomenological model is difficult to use for both predictive and NF membrane characterization purposes. For example, describing salt rejection by NF requires characterization of σ and P through experimentation and model fitting with the water matrix and membrane of interest.

The major advantage of the phenomenological model is that it is simple to apply and has been found to describe the rejection behavior of a variety of solutes [59]. Several researchers have attempted to improve upon the non-specific nature of the phenomenological model by correlating σ and P with ion concentration, descriptors of molecular size and membrane properties [21, 29, 32]. Recently, Sharma and Chellam [21] used the phenomenological model to predict the removal of organic matter and salts by NF treatment of surface water at high system recoveries.

4.4.2.3 *Other Modeling Approaches*

As previously mentioned, researchers have developed various modeling approaches that have been used to describe the rejection performance of NF treating different water types. Derivations of the solution-diffusion model that includes convection and terms to correct for high recovery have been applied to describe solute rejection at pilot- and full-scale NF applications by several researchers

[33, 54, 60, 61]. One example is the homogenous solution model (HSDM) modified to include film theory:

$$C_p = \frac{C_f K_S e^{\frac{J_v}{K_b}}}{K_W \left(\Delta P - \Delta \Pi\right)\left(\frac{2-2R}{2-R}\right) + K_S e^{\frac{J_v}{K_B}}} \tag{4.8}$$

where C_p is the permeate concentration, C_f is the feed concentration, K_b is the back-diffusion mass transfer coefficient at the membrane, K_s and K_w are the membrane mass transport coefficients for the solute and water, and R is recovery. Zhao et al. [61] applied this model to predict salt rejection by tight NF membranes during pilot-scale applications and developed an approach to correlate the solute mass transport coefficient to the salt type.

Several approaches have been developed to describe the previously discussed phenomenon of solute-membrane affinity. In the case of solutes with strong membrane affinity, rejection has been observed to decrease with increasing permeate flux which, cannot be described by the previously mentioned modeling approaches. Several models can describe this behavior including the surface-force pore model (SFPM) and versions of the finely porous model. The SFPM is based on hindered solute transport in a pore and includes terms for steric hindrance, electrostatic exclusion and adsorption [53, 62]. Mehdizadeh and Dickson [63] refined the model to include diffusion and applied it to describe the rejection of salt by porous RO membranes. Similar to the DSPM, the major difficulty of the SFPM is related to complexity including solving the model equations, calculating the numerous fitting parameters, and application to multi-component solutions.

4.4.3 Membrane Fouling

Membrane fouling presents one of the biggest challenges to the application of membrane technologies. Here, fouling is defined as loss of membrane permeability due to deposition or accumulation of material at the membrane interface including microbes, inorganic and organic colloids, particulates, organic matter and precipitated salts [64]. Membrane fouling can also impact rejection performance due to enhanced concentration polarization within the layer of accumulated material [65, 66]. Membrane fouling has been studied extensively and shown to be a function of feed water composition and chemistry, membrane properties, and operating conditions [67]. Because NF membranes cannot be backwashed, fouling control is of critical importance and is usually achieved through a variety of measures including pretreatment, feed water conditioning (e.g., antiscalant chemicals, pH adjustment, chloramines), and proper operation (e.g., conservative flux and recovery). A variety of cleaning strategies have been developed with the most successful dependent upon the feed water composition, deposited material and membrane [67].

For specific applications, NF is often preferred over RO due to lower pressure requirements, energy demands and operating costs. Because fouling requires an increase in driving pressure to meet a predetermined permeate flux, fouling can reduce the cost benefits of NF, as well as, deteriorate separation efficiency. Although pretreatment and operating conditions can be optimized to minimize fouling, researchers have demonstrated that certain membranes foul and experience greater flux decline than others although the mechanisms are not entirely clear [48, 68–70]. Membrane hydrophobicity, surface charge and roughness are all believed to play a role in membrane fouling propensity although researchers have reported contradictory effects of these membrane properties on flux decline [70–72]. These contradictory results are likely due to the variety of foulants evaluated and different methods used to study fouling. In general, all membrane processes are prone to fouling and care should be taken when selecting both pretreatment measures and NF membranes for specific applications.

Schaefer et al. [64] recently provided a comprehensive, 70 page review that highlights the large volume of work that has been devoted to membrane fouling. The authors demonstrate that the mechanisms of fouling are complicated and dependent upon the foulant type and membrane. The primary causes of flux decline are considered to be due to organic fouling, colloidal fouling, biofouling, and scale formation.

Organic matter fouling is largely attributed to adsorption which has been shown to be dependent upon solute concentration and characteristics (e.g., hydrophobicity, dipole moment, charge, molecular size), membrane properties (e.g., surface charge, hydrophobicity), and operating conditions (e.g., flux, recovery). Because flux decline is often positively correlated with strength of adsorption which, is largely dependent upon hydrophobic interactions between a solutes and a membrane, hydrophobic membranes have been demonstrated to foul to greater degrees than hydrophilic membranes. While researchers have postulated that high negative membrane surface charge should repulse negatively charged organic matter leading to lower fouling, this phenomenon is dependent on permeate flux and can be offset by solution chemistry (e.g., presence of divalent cations) [73].

Colloidal fouling of NF and RO membranes is believed to be a function of colloidal properties (e.g., size, shape, charge, hydrophobicity), membrane properties (e.g., porosity, roughness, permeability, hydrophobicity), and operating conditions (e.g., permeate flux, cross-flow velocity). Fundamental studies have demonstrated that initial colloidal deposition is a function of particle transport to the membrane, interactions between the colloid and membrane, and shear enhanced back diffusion. Additionally, colloidal fouling has been shown to be dependent upon membrane roughness with rougher membranes exhibiting greater flux decline [67].

Biofouling refers to colonization and growth of microbes on membrane surfaces which, can lead to severe flux decline with the development of biofilms and production of extracellular substances [64]. Biofouling can be problematic both from an operational standpoint and a cleaning stand-point because well developed biofilms can be difficult to completely remove. Depending on the level of pretreatment and feed water composition, control of biological growth is often achieved through dosing of chemical oxidants at low concentrations such as chloramines.

Scale formation is caused by sparingly soluble salts exceeding their solubility limit as they are rejected by the membrane and concentrated along a membrane element. Scale formation is dependent upon the feed composition, rejection of sparingly soluble species, and system recovery. If required, scale formation can be managed by limiting system recovery and/or through feed water conditioning to increase solubility (pH adjustment) or crystal induction time (antiscalants). Compared to RO, for certain applications, NF can operate a higher recovery due to partial passage of sparingly soluble salts [69].

4.5 Application

The characteristics of NF, particularly selective separation of salts, good organic removal and relatively low pressure requirements, have lent to its use in a wide variety of applications including water treatment, and in the dairy, chemical, beverage, food, pharmaceutical, pulp and paper, and textile industry. The following sections summarize recent applications of NF in these areas.

4.5.1 Water and Wastewater Treatment Industry

4.5.1.1 Water Treatment

The advent of commercially available NF membranes provided drinking water utilities, particularly in Florida, with a new and economically viable approach for water softening and the removal of color and disinfection by-product (DBP) precursors from source waters. For most of these applications, rejection of monovalent salts was of minimal importance, and membranes were sought that operated at low-pressure and could provide high removal of hardness, color and trihalomethane precursors due to recently enacted drinking water regulations [3, 8, 74]. In many cases, NF was preferred over RO for these applications, not only due to reduced energy requirements but also due to a more dilute concentrate waste stream, and a product water requiring less stabilization to minimize distribution system corrosion [75]. As a result, several utilities in Florida began testing early cellulose diacetate (CDA) NF membranes to treat groundwater as early as 1976 as an alternative to lime softening [8]. Advantages of NF over lime softening include a small footprint, reduced chemical requirements, reduced chemical storage, increased organic matter removal, and no sludge production [75].

Increased water demand and the need to use alternative sources for drinking water has resulted in the construction of a number of NF facilities in Florida for treating shallow groundwater including one of the world's largest NF plants in Boca Raton (Table 4.2). This facility, the 151 megaliter per day NF system at the Glades Road Water Treatment Plant in Boca Raton, reportedly operates without antiscalant, acid addition or post-treatment pH adjustment which saves the utility approximately $927,000 per year [76]. Additionally, NF is currently employed by utilities in the US for the removal of fluoride, iron, organic compounds, radon and TDS at several facilities in California, Illinois, and Nevada (Table 4.2).

One of the largest applications of NF for drinking water treatment has been at the Méry-sur-Oise treatment facility in Northern Paris, France since 1999 [77, 78]. Prior to the addition of NF, the facility relied on coagulation/flocculation, sedimentation, slow-sand filtration, ozonation and granular activated carbon for the removal of organic matter from the Oise River source. Because organic removal could be problematic under certain conditions, a parallel NF plant was built with a capacity of approximately 140 megaliters per day. Treatment objectives for the NF included bulk organic matter and pesticide removal as well as providing an additional barrier towards microbial pathogens. While effective for the treatment objectives, the NF membranes have exhibited significant seasonal fouling events due to microbial activity and changes in organic matter properties [79]. Several pretreatment steps and stages of treatment have been employed to reduce the fouling propensity of the water including storage in a raw water reservoir, ozonation, dual-media filtration, and microfiltration.

Beyond employing NF for removal of hardness and bulk organic matter from water supplies, NF has been evaluated for the removal of a number of different constituents including arsenic [80–88], DBPs [42, 89–91], fluoride [87, 88, 92–95], heavy metals [96–100], inorganic carbon [10, 87, 101], nitrate [101–110], pesticides [36, 38, 102, 111–119], oxyanions (e.g., bromate, perchlorate, phosphate, sulfate) [120–125], and various organic emerging organic contaminants [38, 45, 46, 49, 112, 126–132].

4.5.1.2 *Wastewater Treatment and Reuse*

RO is commonly employed in potable wastewater reuse applications (i.e., indirect potable reuse) in Australia, Singapore and the U.S. (i.e., Arizona, California) due to effectiveness as a barrier to inorganic ions, metals and organic compounds [48]. Potable reuse applications are configured as a multiple barrier treatment train approach that includes an integrated membrane system consisting of MF or UF pretreatment followed by RO, and commonly, an advanced oxidation process (usually ultraviolet light with peroxide (UV-AOP)). Several researchers have proposed and/or evaluated the use of NF for reuse applications to reduce the energy requirements associated with RO, and selectively remove organic contaminants while allowing for passage of salts [48, 69, 133, 134].

Bellona et al. [48] evaluated a variety of spiral-wound RO, LPRO and NF membranes for the treatment of microfiltered wastewater effluent and reported that while all membranes tested could provide high organic compound rejection, the major differences

Table 4.2 List of water treatment facilities employing NF in the U.S. (data obtained from the American Membrane Technology Association).

State	Utility	Start Year	Source Water	Capacity (megaliters/ day)	Treatment Purpose
CA	Port Hueneme Water Agency	1999	GW	2.9	Hardness and TDS
CA	Deep Aquifer Treatment System	2002	GW	30.3	Color
FL	City of Boca Raton	2005	GW	151.4	Softening, color, DBPFP
FL	Boynton Beach West WTP	1994	GW	30.3	Softening
FL	Town of Jupiter	2010	GW	54.9	Hardness and organics
FL	City of Sunrise	2002	GW	45.4	Hardness
FL	Hollywood	1995	GW	136.3	Color
FL	Palm Beach	1996	GW	35.2	Hardness
FL	Cooper City	1998	GW	11.4	Color and DBPFP
FL	City of Dunedin	1992	GW	36	Hardness and iron
FL	City of Fort Myers	1992	SW	45.4	DBPFP
FL	Gulf Utilities (Corkscrew)	1991	GW	1.9	Hardness and iron

State	Utility	Start Year	Source Water	Capacity (megaliters/ day)	Treatment Purpose
FL	Palm Beach Co. 9 WTP - Sandlefoot Cove	2001	GW	87.1	NA
FL	Palm Beach Co. WTP 3	1996	GW	35.2	NA
FL	Palm Beach Park of Commerce	1988	GW	0.7	Hardness
FL	Palm Coast WTP 2	1992	GW	24.1	Chloride and TDS
FL	Palm Coast WTP 3	NA	NA	11.4	NA
FL	St Lucie West Service District	1998	GW	3.8	Color, iron, hardness and TOC
FL	Tropical Farms WTP	1996	GW	5.7	NA
FL	City of Wauchula	1990	GW	1.2	Sulfate and TDS
FL	City of Wellington	1990	GW	23.9	DBPFP
IL	Dupage County	1998	GW	5.8	Hardness and iron
IL	City of Chenoa	1992	GW	1.3	Fluoride and radium
IL	City of Itasca	1997	GW	0.5	Hardness and iron
NV	Nevada Lake Mead Echo Bay	2000	GW	0.8	TDS and DBPFP
NV	Nevada Lake Mead Overton Beach	2000	SW	0.4	TDS and DBPFP

Abbreviations: GW - groundwater, SW - surface water, NA - not available, TDS - total dissolved solids, DBPFP - disinfection byproduct formation potential between membranes included the initial specific flux (prior to fouling), fouling propensity, and the rejection of monovalent ions including nitrate and ammonia (Figure 4.5). During pilot testing of a tight NF with high permeability and rejection (NF-90, Dow/Filmtec), significant flux decline due to fouling was observed which, resulted in only marginal energy savings compared with conventional RO membranes. Subsequent pilot testing with two loose and low fouling NF membranes demonstrated that significant cost savings could be attained however, the major disadvantages were found to be poor rejection of nitrate, and greater number of organic contaminant detections in NF permeates compared with the RO membrane [69, 133]. However, for wastewater treatment plants practicing biological removal of nitrogen, and potable reuse applications employing UV-AOP for contaminant destruction, certain NF membranes may offer significant advantages over RO, e.g., lower energy requirements, lower potential for inorganic scaling due to partial passage of divalent salts, and less concentrated waste stream [69].

Due to potential advantages of using loose NF for wastewater reclamation, the Los Angeles County Sanitation District of Southern California recently conducted pilot testing of a dual NF/RO treatment process for indirect potable reuse applications using surface

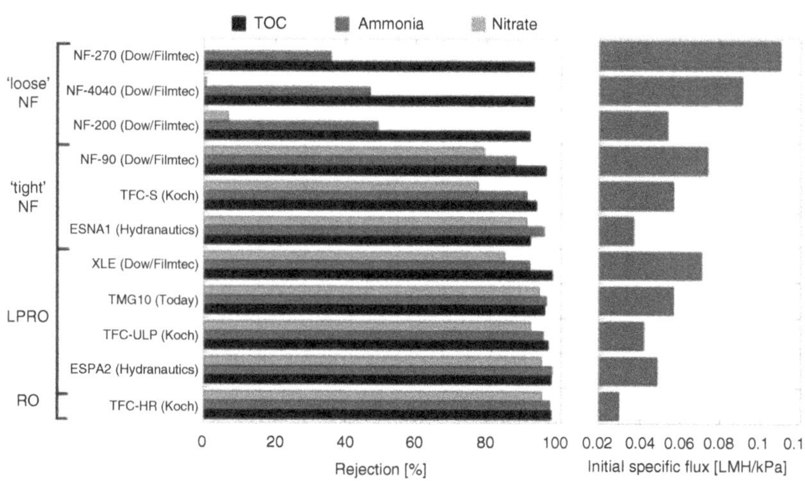

Figure 4.5 Comparison of rejection performance (left) and pressure requirements (right, specific flux for 20 LMH permeate flux) between various RO, LPRO and NF membranes when treating microfiltered secondary effluent.

spreading for groundwater recharge [135]. The proposed process utilized MF followed by 'loose' NF to produce permeate for the groundwater spreading operation with the concentrate treated by a secondary RO system to increase overall recovery. A desktop study evaluating this concept indicated that the system could operate at a recovery over 92% whereas typical reuse membrane systems operate at recoveries less than 85% [136].

Several alternate NF configurations have been developed/evaluated for treatment of wastewater and wastewater effluent including tubular polymeric NF, tubular ceramic NF, and flat-sheet NF operated as banks or in stacks with sheer inducing rotors. Tubular configurations have the advantage of operation at high cross-flow velocity to minimize fouling however, low packing densities can be an issue and ceramic NF have relatively high MWCOs compared with polymeric NF [137]. Other configurations such as NF stacks with sheer inducing rotors can be operated with high-fouling feed water however, packing densities are usually rather low.

Currently, the application of NF in wastewater treatment requires significant pretreatment generally including pre-filtration with MF or UF membranes to remove suspended solids and dissolved macromolecules. For applications without pre-filtration, there is a need to develop NF membranes with low fouling propensity, better fouling mitigation strategies (e.g., backwashing), resistance to oxidation (e.g., chlorine tolerance to combat biofouling), and good rejection of dissolved constituents such as organic contaminants. While several spiral-wound NF products have relatively low fouling propensity and good rejection characteristics, these membranes have low chlorine tolerance and cannot be backwashed. Although several companies have developed unique products (e.g., vibratory sheer enhanced processing (VSEP) and vortex enhanced filtration) to minimize flux decline, more research is necessary to develop processes that can efficiently integrate NF technology into wastewater treatment and reuse facilities.

4.5.1.3 Desalination

Although thermal processes (e.g., multi-stage flash (MSF)) and RO are the dominant technologies used for seawater and brackish water desalination, NF has been proposed for certain niche applications in the desalination industry. These applications include a dual NF system that was extensively piloted by the Long Beach

Water Department (LBWD) in Southern California, and NF as a pretreatment to RO or MSF for seawater desalination.

The dual-staged NF system (termed NF^2 by the researchers [138]) concept tested by LBWD is essentially a double pass tight NF process where permeate from the first stage is treated by a second stage of NF membranes to reduce the TDS concentration. The system was designed to recycle concentrate from the second stage and permeate from the first stage (optional) to reduce the TDS concentration and osmotic pressure of the feed stream. Because NF operates at lower pressures than RO, it was expected that lower energy requirements could be achieved while producing low TDS permeate through second stage polishing. Several tight NF membranes were initially tested including the NF-90 (Dow/Filmtec), NE-90 (Saehan) and TS-80 (Trisep) due to high permeability and relatively high monovalent salt rejection [139]. The process was pilot tested with the NF-90 and NE-90 membranes to evaluate the main factors influencing energy requirements and final permeate quality. While acceptable permeate TDS concentrations (< 400 mg/L) could be produced by the NF-90 and NE-90 membranes at energy requirements reportedly on the low-end of RO based desalination technologies, overall system recoveries were on the order of 30 – 40% which, are lower than typical SWRO systems (~50%) [138]. Additionally, energy consumption was found to be proportional to the permeate TDS concentration, with energy costs comparable to SWRO required to achieve permeate TDS concentrations below 200 mg/L. The researchers at LBWD as well as AlTaee and Shariff [140] have attempted to alleviate the low recovery and elevated permeate TDS issues by replacing second stage membranes with RO membranes to varying success. Through desktop simulations, AlTaee and Shariff [140] reported that a NF/BWRO system could produce high quality permeate at lower operating costs than typical SWRO facilities.

Hassan et al. [141] proposed a desalination process using NF as a pretreatment step to SWRO or MSF which has subsequently been pilot and demonstration tested at the Umm Lujj facility in Saudi Arabia. The advantage of NF pretreatment is reported to be the reduction in sparingly soluble salts, microorganisms, organic foulants, turbidity and TDS which allows the SWRO to operate at lower energy consumption and higher recovery (70 %), and the MSF at a higher brine temperature and recovery (80 %). Subsequent pilot testing indicated that hybridization of RO and

MSF desalination processes by introducing NF could reduce typical SWRO desalination costs by approximately 30% [142]. Based on these results, the Umm Lujj desalination facility was retrofitted with NF pretreatment for long-term demonstration testing of the NF/SWRO process [143]. Although the NF membranes used had organic fouling issues, process modifications (i.e., addition of more NF elements to reduce the flux of lead elements) have reportedly solved any issues [143]. Macedonio et al. [144] conducted a cost and energy analysis on several integrated desalination processes (RO, MSF, membrane crystallization (MCr) that included NF pretreatment and reported that while NF can significantly increase the recovery of desalination systems, the cost savings are only marginal when energy recovery devices are used.

4.5.2 Food Industry

4.5.2.1 Dairy Industry

Pressure driven membrane technologies have been used in various applications by the dairy industry for over thirty years [12, 145]. While early applications mainly involved treatment of whey solutions produced through cheese production, the dairy industry has adopted MF, UF, NF and RO membranes for different aspects of the dairy industry including removal of bacteria and particles (MF and UF), fractionation and recovery of constituents from process streams (MF, UF, NF), demineralization (NF), and concentration of various solution components and solutions (MF, UF, NF and RO) [146]. NF is becoming an important component of the dairy industry for demineralization, water reuse and waste minimization. In many of these applications, NF is employed in an integrated membrane system and used in conjunction with MF and UF membrane processes.

Whey, which contains more than 93% water, is the predominant by-product of cheese production and contains approximately half of the original nutrients of milk including calcium, lactose and protein [147]. Whey can be processed to extract amino acids, lactose, lipids, and proteins for various products, or concentrated and/or evaporated to produce animal feed or whey powders for use in a variety of end-products [145, 148]. Because production of animal feed and whey powder requires concentration and partial demineralization of the various types of whey (e.g., acid whey, salty whey, sweet whey), NF has been employed for whey processing due to selective passage of monovalent ions and high organic rejection [12, 148–150]. Past work has demonstrated that NF operated in diafiltration mode can achieve similar whey concentration and demineralization as commonly employed methods that use RO or evaporation followed by electrodialysis or ion-exchange, but at lower capital and operating costs [148, 151]. Bargeman et al. [148] cautions that proper membrane selection is a necessity as lactic acid rejection can vary widely among membranes with similar salt rejection and that biofouling and membrane scaling can be issues when desalting certain whey streams using NF. Yorgun et al. [152] evaluated several NF and RO membranes for whey treatment and reported that although several NF membranes had poor lactose rejection, adding a subsequent RO step to treat NF permeate allowed for protein concentration by NF, lactose concentration by RO and a high quality permeate stream that could be discharged or reused.

NF is also being evaluated for the treatment of UF-whey permeate streams for the production of lactose, minimization of volume prior to disposal, and production of water for other process streams [153]. In proposed processes for lactose production, UF is first used concentrate proteins for whey powder production, followed by an NF step for demineralization and lactose concentration, and a crystallization or spray-drying step for production of lactose [148]. Bargeman et al. [148] reports that the benefits of using NF as opposed to evaporation is the reduction of concentration costs, the removal of ions that impact the crystallization step, reduction of post-crystallization lactose processing costs, and higher lactose yields. The NF process can be operated in several modes including concentration (i.e., permeate is continually discharged and feed volume decreases) and continuous diafiltration (i.e., permeate is continually discharged while water is added to the feed to maintain constant volume) mode and several recent studies have shown that a combination of concentration and dialfiltration modes can result in relatively high demineralization with minimal lactose losses [154]. Several researchers have indicated the fouling and scaling can be a problem with NF treatment of UF-whey permeate when high concentration factors are desired [152, 153].

Both unprocessed whey and UF-whey permeate is commonly considered a waste stream requiring disposal (e.g., sent to wastewater treatment plant). Additional waste streams in dairy operations can include various cleaning (acids, alkalines, detergents) and sanitization streams, diluted milk, and various process streams from unit processes. Although the quality of dairy wastewater can vary significantly, it is often characterized by high concentrations of organic matter as well as relatively high suspended solids, nutrients, and total dissolved solids. Several treatment schemes using NF have been proposed for the reclamation of dairy waste water due to lower pressure requirements compared to RO. These include using integrated UF/NF [155], rotational shear enhanced flat-sheet NF [156, 157], and vibratory sheer enhanced NF [158] to treat raw dairy wastewater as well as, NF [159] and dual-stage NF/RO [160, 161] to treat dairy wastewater treatment plant effluent. NF has also been reportedly used to reclaim sodium hydroxide, nitric acid and detergents from cleaning solutions for reuse [148, 162].

4.5.2.2 Sugar and Beverage Industry

NF has been investigated for a variety of food and beverage processing applications to replace expensive or energy intensive processes currently used for fractionation, isolation and concentration. This includes the fruit juice industry where researchers have evaluated NF for both the concentration of juices prior to packaging [163–165] and recovery of bioactive compounds such as polyphenols [166, 167]. NF is a promising technology for these applications due to lower pressure requirements compared to RO, better separation of low molecular weight components compared to UF and RO, and capability of operation at ambient temperature (compared with thermal processes for concentration) which prevents thermal degradation of isolated constituents [166]. Significant recent interest has also been given to NF in the wine- and beer-making industry for sugar control and dealcholization due to high ethanol passage and rejection of aroma and taste compounds [168–170].

Several researchers have evaluated NF for the fractionation of saccharide solutions and isolation of oligosaccharides from various process liquors (e.g., from hydrolysis or bioreactor processing of a variety of feedstocks) which are being increasingly used as prebiotics and low-calorie sweeteners in food products [171, 172]. For these applications fractionation is based on MW differences between monsaccharides and oligosaccharides and a NF membrane with low rejection of monosaccharides (e.g., glucose, lactose, sucrose, xylose) and high rejection of oligosaccharides is desired which has reportedly requires NF membranes with MWCOs somewhere between 300–400 (i.e., MWs of lactose and sucrose are approximately 342 Daltons) and 1,000 Daltons [171, 173]. Studies have demonstrated that efficient isolation of oligosaccharides by carefully chosen NF membranes can be achieved when operated in continuous diafiltration or variable volume diafiltration modes [174]. Similar approaches using NF for isolation and concentration of alternative sweeteners such as stevioside, lactic acid from fermentation broths and starch from pasta from pasta blanch water have also been reported [148].

4.5.3 Chemical Processing Industry

Current or proposed NF applications in the chemical processing industry are too numerous to completely summarize here, however, a comprehensive review was recently provided by Kyburz and Meindersma [16]. NF has been implemented for metal and solvent recovery, acid purification, process water treatment and reuse, and fractionation and concentration in a variety of chemical industries. Two of the most mature applications of NF include the pharmaceutical and dye-processing/textile industry.

4.5.3.1 Pharmaceutical Industry

In the pharmaceutical industry, NF is often integrated into process trains designed to produce, isolate, concentrate and purify pharmaceutical substances using a variety of technologies such as chemical and enzymatic reactors, solvent extraction, evaporation, washing, centrifugation, precipitation, and crystallization. Similar to the dairy and sweetener industry, NF can be used for fractionation, purification, demineralization and concentration, and is often integrated with other membrane processes such as MF and UF. Specific applications include polymeric NF diafiltration for demineralization and purification of quaternary ammonium salt solutions [175], demineralization of amino acid solutions [16], and recovery of pharmaceutically active products [55, 176–180]. OSN membranes are also being used for applications in the pharmaceutical industry, primarily to remove post-reaction impurities at ambient temperatures prior to crystallization [17]. A group of researchers recently evaluated spiral-wound OSN for purification of pharmaceutical ingredients from a tetrohydrofuran solution for a period of 120 days and reported stable flux and consistent separation performance over time [17]. Other researchers have evaluated OSN for separation of pharmaceutical products from catalysts [181, 182] and genotoxic impurities [183, 184].

4.5.3.2 *Textile industry*

The textile industry generates a large amount of wastewater which is characterized by high concentrations of salts, dyes and chemical oxygen demand (COD) [185]. Dye wastewater is difficult to treat due to non-biodegradable dyes, high ionic strength and presence of other pollutants. NF has been studied as a way to separate dyes, COD and other pollutants from dye-bath effluents for waste volume reduction and water reuse [186–191]. Because dyes have large molecular weights, and various salts are often used at high concentrations in the dyeing process, NF is often preferred over other membrane processes due to high dye rejection and high passage of salts which, reduces the osmotic pressure differential and energy requirements compared to RO [188, 192]. Additionally, both the NF brine with concentrated dyes and salt rich permeate could be recycled to reduce chemical demand [188, 193].

Because the composition of dye-facility wastewater can vary significantly, researchers have reported variable success using NF as a treatment technology. For example, due to the presence of low MW dyes or dyes at high concentration in wastewater, dye rejection by NF may not be high enough to produce a colorless effluent [185]. High dye and COD concentrations, and charge neutralization caused by significant ionic strength can result in severe fouling and flux decline, higher energy requirements and frequent chemical cleanings [194, 195]. Because salt rejection by NF is a complex phenomena that depends on salt concentration, valence of the salts in solution, membrane properties and operating conditions, it is often extremely difficult to predict NF rejection and flux during operation. Several researchers have reported conditions where enhanced salt rejection was observed which significantly increased the osmotic pressure and energy requirements of NF [196]. As a result of these issues, recent studies have evaluated adsorption processes [192], coagulation-flocculation [197, 198], ozonation [185] and UF [189] as pretreatment to reduce NF fouling as well as RO and oxidation post-NF, to remove remaining dye and COD prior to reuse [185]. It is also worth noting that because NF is a separation process, treatment or disposal of NF concentrated textile wastewater is a major issue that requires further research and development efforts to solve.

4.6 Conclusions

NF is a relatively recent development in membrane technology that has found a variety of uses in water treatment and industrial processes. Although commercially available NF membranes span a range of characteristics falling between UF and RO, NF membranes capable of ion and small uncharged solute separation (i.e., loose NF) represent a unique membrane process that can be used for a variety of applications. With effective pore sizes in the range of 1 nm, solute removal is believed to be a function of steric hindrance and Donnan and dielectric exclusion which, allows for manipulation of the NF process to achieve preferential removal of multi-valent ions, separation of uncharged solutes based on size, and demineralization of process streams containing valuable constituents. Similar to other membrane technologies, there is a need to develop more selective and robust membrane materials to provide better separation control and combat fouling, respectively. The increased interest in NF over the past 20-years indicates that with further membrane development research, the application of NF in water treatment and industrial processes will increase significantly in the near future.

References

1. B.M. Watson and C.D. Hornburg, Low-energy membrane nanofiltration for removal of color, organics and hardness from drinking water supplies. *Desalination* **72**, 11–22 (1987).
2. C. Linder and O. Kedem, History of nanofiltration membrane 1960 to 1990, in *Nanofiltration - Theory and Application*, A.I. Schaefer, A.G. Fane, and T.D. Waite (Eds), pp. 1–31, Elsevier Advanced Technology, Oxford, UK (2005).
3. J. Cadotte, R. Forester, M. Kim, R. Petersen, and T. Stocker, Nanofiltration membranes broaden the use of membrane separation technology. *Desalination* **70**, 77–88 (1988).
4. P. Eriksson, Nanofiltration extends the range of membrane filtration. *Environ. Prog.* **7**, 58–62 (1988).
5. R.F. Fibiger, J.-y. Koo, D.J. Forgach, R.J. Petersen, D.L. Schmidt, R.A. Wessling, and T. Stocker, Novel polyamide reverse osmosis

membranes, US Patent 4859384, assigned to Filmtec Corporation - Dow Chemical Company (1989).

6. J.E. Cadotte, Reverse osmosis membrane, US Patent 4259183, assigned to Midwest Research Institute (1981).
7. M. Guilhem, Combined membrane and sorption process for selective ion removal, US Patent 4806244, assigned to The Dow Chemical Company (1989).
8. W.J. Conlon and S.A. McClellan, Membrane softening: A treatment process comes of age. *J. Am. Water Work Assoc.* 47–51 (1989).
9. X.-L. Wang, T. Tsuru, S.-i. Nakao, and S. Kimura, Electrolyte transport through nanofiltration membranes by the space-charge model and the comparison with Teorell-Meyer-Sievers model. *J. Membrane Sci.* **103**, 117–133 (1995).
10. A.E. Simpson, C.A. Kerr, and C.A. Buckley, The effect of pH on the nanofiltration of the carbonate system in solution. *Desalination* **64**, 305–319 (1987).
11. P. Eriksson, M. Kyburz, and W. Pergande, NF membrane characteristics and evaluation for sea water processing applications. *Desalination* **184**, 281–294 (2005).
12. J.M.K. Timmer, *Properties of nanofiltration membranes; model development and industrial application*, Eindhoven University of Technology, (2001).
13. A.I. Schaefer, A.G. Fane, and T.D. Waite, Introduction, in *Nanofiltration: Principles and applications*, A.I. Schaefer, A.G. Fane, T.D. Waite (Eds), pp. 1–4, Elsevier, Oxford, UK (2005).
14. S. Loeb and S. Sourirajan, High flow porous membranes for separating water from saline solutions, US Patent 3133132, assigned to The Reagents of the University of California, Berkeley, CA (1964).
15. K. Kurihara, T. Uemura, Y. Nakagawa, and T. Tonomura, The thin-film composite low pressure reverse osmosis membrane. *Desalination* **54**, 75–88 (1985).
16. M. Kyburz and G.W. Meindersma, Nanofiltration in the chemical processing industry, in *Nanofiltration: Principles and applications*, A.I. Schaefer, A.G. Fane, and T.D. Waite (Eds), pp. 329–361, Elsevier, Oxford, UK (2005).
17. I. Sereewatthanawut, F.W. Lim, Y.S. Bhole, D. Ormerod, A. Horvath, A.T. Boam, and A.G. Livingston, Demonstration of molecular purification in polar aprotic solvents by organic solvent nanofiltration. *Org. Process Res. Dev.* **14**, 600–611 (2010).
18. W.R. Bowen, A.W. Mohammad, and N. Hilal, Characterisation of nanofiltration membranes for predictive purposes -- use of salts, uncharged solutes and atomic force microscopy. *J. Membrane Sci.* **126**, 91–105 (1997).
19. S. Bandini and D. Vezzani, Nanofiltration modeling: The role of dielectric exclusion in membrane characterization. *Chem. Eng. Sci.* **58**, 3303–3326 (2003).

20. A.E. Yaroshchuk, Recent progress in the transport characterisation of nanofiltration membranes. *Desalination* **149**, 423–428 (2002).
21. R.R. Sharma and S. Chellam, Solute rejection by porous thin film composite nanofiltration membranes at high feed water recoveries. *J. Colloid Interface Sci.* **328**, 353–366 (2008).
22. W.R. Bowen, J.S. Welfoot, and P.M. Williams, Linearized transport model for nanofiltration: Development and assessment. *AICHE J.* **48**, 760–773 (2002).
23. S. Bandini and C. Mazzoni, Modelling the amphoteric behaviour of polyamide nanofiltration membranes. *Desalination* **184**, 327–336 (2005).
24. S. Bandini, J. Drei, and D. Vezzani, The role of pH and concentration on the ion rejection in polyamide nanofiltration membranes. *J. Membrane Sci.* **264**, 65–74 (2005).
25. F.G. Donnan, The theory of membrane equilibria. *Chem. Rev.* **1**, 73–90 (1924).
26. J. Schaep, C. Vandecasteele, A. Wahab Mohammad, and W. Richard Bowen, Modelling the retention of ionic components for different nanofiltration membranes. *Separ. Purif. Technol.* **22**(23), 169–179 (2001).
27. G. Hagmeyer and R. Gimbel, Modelling the salt rejection of nanofiltration membranes for ternary ion mixtures and for single salts at different pH values. *Desalination*, **117**, 247–256 (1998).
28. X.-L. Wang, T. Tsuru, S.-i. Nakao, and S. Kimura, The electrostatic and steric-hindrance model for the transport of charged solutes through nanofiltration membranes. *J. Membrane Sci.* **135**, 19–32 (1997).
29. B. Van der Bruggen and C. Vandecasteele, Modelling the retention of uncharged molecules with nanofiltration. *Water Res.* **36**, 1360–1368 (2002).
30. Y. Kiso, K. Muroshige, T. Oguchi, M. Hirose, T. Ohara, and T. Shintani, Pore radius estimation based on organic solute molecular shape and effects of pressure on pore radius for a reverse osmosis membrane. *J. Membrane Sci.* **369**, 290–298 (2011).
31. Y. Kiso, T. Kitao, K. Jinno, and M. Miyagi, The effects of molecular width on permeation of organic solute through cellulose acetate reverse osmosis membranes, *J. Membrane Sci.* **74**, 95–103 (1992).
32. Rolf U. Haden, Comparing the phenomenological and hydrodynamic modeling approaches for describing the rejection of emerging nonionic organic contaminants by a nanofiltration membrane, in *Contaminants of Emerging Concern in the Environment: Ecological and Human Health Considerations*, pp. 397–420, American Chemical Society, Washington, DC (2010)
33. J.A.M.H. Hofman, A.J. Gijsbertsen, and E. Cornelissen, Nanofiltration retention models for organic contaminants, Water Research Foundation and KIWA Final Research Report, (2007).

34. B. Van der Bruggen, J. Schaep, D. Wilms, and C. Vandecasteele, Influence of molecular size, polarity and charge on the retention of organic molecules by nanofiltration. *J. Membrane Sci.* **156**, 29–41 (1999).
35. Y. Kiso, K. Muroshige, T. Oguchi, T. Yamada, M. Hhirose, T. Ohara, and T. Shintani, Effect of molecular shape on rejection of uncharged organic compounds by nanofiltration membranes and on calculated pore radii. J. Membrane Sci. **358**, 101–113 (2010).
36. Y. Kiso, Y. Sugiura, T. Kitao, and K. Nishimura, Effects of hydrophobicity and molecular size on rejection of aromatic pesticides with nanofiltration membranes. J. Membrane Sci. **192**, 1–10 (2001).
37. F. Zheng, Z. Zhang, C. Li, and Q. Yuan, A comparative study of suitability on different molecular size descriptors with the consideration of molecular geometry in nanofiltration. *J. Membrane Sci.* **332**, 12–23 (2009).
38. C. Bellona, J.E. Drewes, P. Xu, and G. Amy, Factors affecting the rejection of organic solutes during NF/RO treatment--a literature review. *Water Res.* **38**, 2795–2809 (2004).
39. T. Matsuura and S. Sourirajan, Physicochemical criteria for reverse osmosis separation of alcohols, phenols, and monocarboxylic acids in aqueous solutions using porous cellulose acetate membranes. *J. Appl. Polym. Sci.* **15**, 2905–2927 (1971).
40. L. Braeken, R. Ramaekers, Y. Zhang, G. Maes, B.V.d. Bruggen, and C. Vandecasteele, Influence of hydrophobicity on retention in nanofiltration of aqueous solutions containing organic compounds. *J. Membrane Sci.* **252**, 195–203 (2005).
41. K. Kimura, G. Amy, J. Drewes, and Y. Watanabe, Adsorption of hydrophobic compounds onto NF/RO membranes: an artifact leading to overestimation of rejection, *J. Membrane Sci.* **221**, 89–101 (2003).
42. T.-U. Kim, J.r.E. Drewes, R. Scott Summers, and G.L. Amy, Solute transport model for trace organic neutral and charged compounds through nanofiltration and reverse osmosis membranes. *Water Res.* **41**, 3977–3988 (2007).
43. M.E. Williams, J.A. Hestekin, C.N. Smothers, and D. Bhattacharyya, Separation of organic pollutants by reverse osmosis and nanofiltration membranes: Mathematical models and experimental verification. *Ind. Eng. Chem. Res.* **38**, 3683–3695 (1999).
44. E. Steinle-Darling, E. Litwiller, and M. Reinhard, Effects of sorption on the rejection of trace organic contaminants during nanofiltration. *Environ. Sci. Technol.* **44**, 2592–2598 (2010).
45. C. Bellona, K. Budgell, D. Ball, J.E. Drewes, and S. Chellam, Models to predict organic contaminant removal by RO and NF membranes. *IDA J.* **3**, 40–44 (2011).
46. L.D. Nghiem, A.I. Schaefer, and M. Elimelech, Removal of natural hormones by nanofiltration membranes: Measurement, modeling, and mechanisms, *Environ. Sci. Technol.* **38**, 1888–1896 (2004).

47. R. Cheng, J. Glater, J.B. Neethling, and M.K. Stenstrom, The effects of small halocarbons on RO membrane performance. *Desalination* **85**, 33–44 (1991).
48. C. Bellona, J.E. Drewes, G. Oelker, J. Luna, G. Filteau, and G. Amy, Comparing nanofiltration and reverse osmosis for drinking water augmentation, *J. Am. Waterworks Assoc.* **100**, 102–116 (2008).
49. A.R.D. Verliefde, E.R. Cornelissen, S.G.J. Heijman, J.Q.J.C. Verberk, G.L. Amy, B. Van der Bruggen, and J.C. van Dijk, The role of electrostatic interactions on the rejection of organic solutes in aqueous solutions with nanofiltration. *J. Membrane Sci.* **322**, 52–66 (2008).
50. K.S. Spiegler and O. Kedem, Thermodynamics of hyperfiltration (reverse osmosis): Criteria for efficient membranes. *Desalination* **1**, 311–326 (1966).
51. O. Kedem and A. Katchalsky, Thermodynamic analysis of the permeability of biological membranes to non-electrolytes. *Biochim. Biophys. Acta.* **27**, 229–246 (1958).
52. L. Dresner, Some remarks on the integration of the extended Nernst-Planck equations in the hyperfiltration of multicomponent solutions. *Desalination* **10**, 27–46 (1972).
53. T. Matsuura and S. Sourirajan, Reverse osmosis transport through capillary pores under the influence of surface forces. *Ind. Eng. Chem. Process Des. Dev.* **20**, 273–282 (1981).
54. Y. Zhao and J.S. Taylor, Incorporation of osmotic pressure in an integrated incremental model for predicting RO or NF permeate concentration. *Desalination* **174**, 145–159 (2005).
55. W.R. Bowen, B. Cassey, P. Jones, and D.L. Oatley, Modelling the performance of membrane nanofiltration—application to an industrially relevant separation. *J. Membrane Sci.* **242**, 211–220 (2004).
56. D.L. Oatley, B. Cassey, P. Jones, and W. Richard Bowen, Modelling the performance of membrane nanofiltration--recovery of a high-value product from a process waste stream, *Chem. Eng. Sci.* **60**, 1953–1964 (2005).
57. W.R. Bowen and W.A. Mohammad, A theoretical basis for specifying nanofiltration membranes -- Dye/salt/water streams. *Desalination* **117**, 257–264 (1998).
58. A.R.D. Verliefde, E.R. Cornelissen, S.G.J. Heijman, J.Q.J.C. Verberk, G.L. Amy, B. Van der Bruggen, and J.C. van Dijk, Construction and validation of a full-scale model for rejection of organic micropollutants by NF membranes. *J. Membrance Sci.* **339**, 10–20 (2009).
59. E.A. Mason and H.K. Lonsdale, Statistical-mechanical theory of membrane transport. *J. Membrance Sci.* **51**, 1–81 (1990).
60. Y. Zhao, J.S. Taylor, and S. Chellam, Predicting RO/NF water quality by modified solution diffusion model and artificial neural networks. *J. Membrance Sci.* **263**, 38–46 (2005).

61. Y. Zhao, J. Taylor, and S. Hong, Combined influence of membrane surface properties and feed water qualities on RO/NF mass transfer, a pilot study. *Water Res.* **39**, 1233–1244 (2005).
62. S. Jain and S.K. Gupta, Analysis of modified surface force pore flow model with concentration polarization and comparison with Spiegler–Kedem model in reverse osmosis systems. *J. Membrance Sci.* **232**, 45–62 (2004).
63. H. Mehdizadeh and J.M. Dickson, Theoretical modifications of the surface force-pore flow model for reverse osmosis transport. *J. Membrance Sci.* **42**, 119–145 (1989).
64. A.I. Schaefer, N. Andritsos, A.J. Karabelas, M.V. Hoek, R.P. Schneider, and M. Nystrom, Fouling in nanofiltration, in *Nanofiltration: Principles and application*, A.I. Schaefer, A.G. Fane, T.D. Waite (Eds.), pp. 170–240 Elsevier, Oxford, UK (2005).
65. H.Y. Ng and M. Elimelech, Influence of colloidal fouling on rejection of trace organic contaminants by reverse osmosis. *J. Membrance Sci.* **244**, 215–226 (2004).
66. S. Lee, J. Cho, and M. Elimelech, Influence of colloidal fouling and feed water recovery on salt rejection of RO and NF membranes. *Desalination* **160**, 1–12 (2004).
67. B. Van der Bruggen, M. Mänttäri and M. Nyström, Drawbacks of applying nanofiltration and how to avoid them: A review. *Separ. Purif. Technol.* **63**, 251–263 (2008).
68. P. Xu, C. Bellona and J.E. Drewes, Fouling of nanofiltration and reverse osmosis membranes during municipal wastewater reclamation: Membrane autopsy results from pilot-scale investigations. *J. Membrance Sci.* **353**, 111–121 (2010).
69. C. Bellona, D. Heil, C. Yu, P. Fu, and J.E. Drewes, The Pros And Cons Of Using Nanofiltration In Lieu Of Reverse Osmosis For Indirect Potable Reuse Applications. *Separ. Purif. Technol.* **85**, 69–76 (2012).
70. W. Peng, I.C. Escobar, and D.B. White, Effects of water chemistries and properties of membrane on the performance and fouling--a model development study. *J. Membrance Sci.* **238**, 33–46 (2004).
71. M. Manttari, T. Pekuri, and M. Nystrom, NF270, a new membrane having promising characteristics and being suitable for treatment of dilute effluents from the paper industry. *J. Membrance Sci.* **242**, 107–116 (2004).
72. N. Hilal, H. Al-Zoubi, N.A. Darwish, A.W. Mohamma, and M. Abu Arabi, A comprehensive review of nanofiltration membranes:Treatment, pretreatment, modelling, and atomic force microscopy. *Desalination* **170**, 281–308 (2004).
73. S. Hong and M. Elimelech, Chemical and physical aspects of natural organic matter (NOM) fouling of nanofiltration membranes. *J. Membrance Sci.* **132**, 159–181 (1997).

74. W.J. Conlon, C.D. Hornburg, B.M. Watson, and C.A. Kiefer, Membrane softening: The concept and its application to municipal water supply. *Desalination* **78**, 157–176 (1990).
75. S.S. Beardsley and S.A. McClellan, Membrane softening: An emerging technology helping Florida communities meet the increased regulation for quality potable water, in *American Waterworks Association Membrane Technology Conference*, Reno, NV, (1995).
76. C.A. Kiefer, F. Brinson and B. Suratt, Process optimization at new nanofiltration plant meets water-quality goals, saves money for Boca Raton. *Fla Water Resour. J.* 48–56 (2006).
77. C. Ventresque, V. Gisclon, G. Bablon, and G. Chagneau, An outstanding feat of modern technology: The Mery-sur-Oise nanofiltration treatment plant (340,000 m3/d). *Desalination* **131**, 1–16 (2000).
78. B. Cyna, G. Chagneau, G. Bablon, and N. Tanghe, Two years of nanofiltration at the Mery-sur-Oise plant, France. *Desalination* **147**, 69–75 (2002).
79. N. Her, G. Amy, A. Plottu-Pecheux, and Y. Yoon, Identification of nanofiltration membrane foulants. *Water Res.* **41**, 3936–3947 (2007).
80. Y. Sato, M. Kang, T. Kamei, and Y. Magara, Performance of nanofiltration for arsenic removal. *Water Res.* **36**, 3371–3377 (2002).
81. P. Brandhuber and G. Amy, Alternative methods for membrane filtration of arsenic from drinking water. *Desalination* **117**, 1–10 (1998).
82. H. Saitua, R. Gil, and A.P. Padilla, Experimental investigation on arsenic removal with a nanofiltration pilot plant from naturally contaminated groundwater. *Desalination* **274**, 1–6 (2011).
83. R.S. Harisha, K.M. Hosamani, R.S. Keri, S.K. Nataraj, and T.M. Aminabhavi, Arsenic removal from drinking water using thin film composite nanofiltration membrane. *Desalination* **252**, 75–80 (2010).
84. A. Figoli, A. Cassano, A. Criscuoli, M.S.I. Mozumder, M.T. Uddin, M.A. Islam, and E. Drioli, Influence of operating parameters on the arsenic removal by nanofiltration. *Water Res.* **44**, 97–104 (2010).
85. C.M. Nguyen, S. Bang, J. Cho, and K.-W. Kim, Performance and mechanism of arsenic removal from water by a nanofiltration membrane. *Desalination* **245**, 82–94 (2009).
86. E.M. Vrijenhoek and J.J. Waypa, Arsenic removal from drinking water by a ‚Äúloose‚Äù nanofiltration membrane. *Desalination* **130**, 265–277 (2000).
87. A.P. Padilla and H. Saitua, Performance of simultaneous arsenic, fluoride and alkalinity (bicarbonate) rejection by pilot-scale nanofiltration. *Desalination* **257**, 16–21 (2010).
88. L.A. Richards, B.S. Richards, H.M.A. Rossiter, and A.I. Sch√§fer, Impact of speciation on fluoride, arsenic and magnesium retention by nanofiltration/reverse osmosis in remote Australian communities. *Desalination* **248**, 177–183 (2009).

89. I. Sentana, R.D.S. Puche, E. Sentana, and D. Prats, Reduction of chlorination byproducts in surface water using ceramic nanofiltration membranes. *Desalination* **277**, 147–155 (2011).
90. R. Chalatip, R. Chawalit, and R. Nopawan, Removal of haloacetic acids by nanofiltration. *J. Environ. Sci.* **21**, 96–100 (2009).
91. V. Uyak, I. Koyuncu, I. Oktem, M. Cakmakci, and I. Toroz, Removal of trihalomethanes from drinking water by nanofiltration membranes. *J. Hazard Mater.* **152**, 789–794 (2008).
92. F. Elazhar, M. Tahaikt, A. Achatei, F. Elmidaoui, M. Taky, F. El Hannouni, I. Laaziz, S. Jariri, M. El Amrani, and A. Elmidaoui, Economical evaluation of the fluoride removal by nanofiltration. *Desalination* **249**, 154–157 (2009).
93. M. Tahaikt, A. Ait Haddou, R. El Habbani, Z. Amor, F. Elhannouni, M. Taky, M. Kharif, A. Boughriba, M. Hafsi, and A. Elmidaoui, Comparison of the performances of three commercial membranes in fluoride removal by nanofiltration. Continuous operations. *Desalination* **225**, 209–219 (2008).
94. M. Tahaikt, R. El Habbani, A. Ait Haddou, I. Achary, Z. Amor, M. Taky, A. Alami, A. Boughriba, M. Hafsi, and A. Elmidaoui, Fluoride removal from groundwater by nanofiltration. *Desalination* **212**, 46–53 (2007).
95. K. Hu and J.M. Dickson, Nanofiltration membrane performance on fluoride removal from water. *J. Membrance Sci.* **279**, 529–538 (2006).
96. Z.V.P. Murthy and A. Choudhary, Application of nanofiltration to treat rare earth element (neodymium) containing water. *J. Rare Earth.* **29**, 974–978 (2011).
97. Z.V.P. Murthy and L.B. Chaudhari, Separation of binary heavy metals from aqueous solutions by nanofiltration and characterization of the membrane using Spiegler,ÄìKedem model. *Chem. Eng. J.* **150**, 181–187 (2009).
98. A. Taleb, M., T. Chaabane, S. Taha, and R. Maachi, Treatment of heavy metals by nanofiltration present in the lake Reghaia. *Desalination* **221**, 277–283 (2008).
99. S. Bouranene, P. Fievet, A. Szymczyk, M. El-Hadi Samar, and A. Vidonne, Influence of operating conditions on the rejection of cobalt and lead ions in aqueous solutions by a nanofiltration polyamide membrane. *J. Membrance Sci.* **325**, 150–157 (2008).
100. M. Nystrom, L. Kaipia, and S. Luque, Fouling and retention of nanofiltration membranes. *J. Membrance Sci.* **98** 249–262 (1995).
101. A. Santafe-Moros and J.M. Gozalvez-Zafrilla, Nanofiltration study of the interaction between bicarbonate and nitrate ions. *Desalination* **250**, 773–777 (2010).
102. B. Van der Bruggen, K. Everaert, D. Wilms, and C. Vandecasteele, Application of nanofiltration for removal of pesticides, nitrate and

hardness from ground water: Rejection properties and economic evaluation. *J. Membrance Sci.* **193**, 239–248 (2001).

103. K. Hayrynen, E. Pongracz, V. Vaisanen, N. Pap, M. Manttari, J. Langwaldt, and R.L. Keiski, Concentration of ammonium and nitrate from mine water by reverse osmosis and nanofiltration. *Desalination* **240**, 280–289 (2009).
104. A. Santafe-Moros, J.M. Gozalvez-Zafrilla, and J. Lora-Garcia, Nitrate removal from ternary ionic solutions by a tight nanofiltration membrane. *Desalination* **204**, 63–71 (2007).
105. F. Garcia, D. Ciceron, A. Saboni, and S. Alexandrova, Nitrate ions elimination from drinking water by nanofiltration: Membrane choice. *Separ. Purif. Technol.* **52**, 196–200 (2006).
106. A. Santafe-Moros, J.M. Gozalvez-Zafrilla, J. Lora-Garcia, and J.C. Garcia-Diaz, Mixture design applied to describe the influence of ionic composition on the removal of nitrate ions using nanofiltration. *Desalination* **185**, 289–296 (2005).
107. A. Santafe-Moros, J.M. Gozalvez-Zafrilla, and J. Lora-Garcia, Performance of commercial nanofiltration membranes in the removal of nitrate ions. *Desalination* **185**, 281–287 (2005).
108. L. Paugam, S. Taha, G. Dorange, P. Jaouen, and F. Quemeneur, Mechanism of nitrate ions transfer in nanofiltration depending on pressure, pH, concentration and medium composition. *J. Membrance Sci.* **231**, 37–46 (2004).
109. L. Paugam, S. Taha, J. Cabon, and G. Dorange, Elimination of nitrate ions in drinking waters by nanofiltration. *Desalination* **152**, 271–274 (2003).
110. J. Bohdziewicz, M. Bodzek, and E. Wasik, The application of reverse osmosis and nanofiltration to the removal of nitrates from groundwater. *Desalination* **121**, 139–147 (1999).
111. A. Caus, S. Vanderhaegen, L. Braeken, and B. Van der Bruggen, Integrated nanofiltration cascades with low salt rejection for complete removal of pesticides in drinking water production. *Desalination* **241**, 111–117 (2009).
112. A. Verliefde, E. Cornelissen, G. Amy, B. Van der Bruggen, and H. van Dijk, Priority organic micropollutants in water sources in Flanders and the Netherlands and assessment of removal possibilities with nanofiltration. *Environ. Pollut.* **146**, 281–289 (2007).
113. K. Kosutic, L. Furac, L. Sipos, and B. Kunst, Removal of arsenic and pesticides from drinking water by nanofiltration membranes. *Separ. Purif. Technol.* **42**, 137–144 (2005).
114. Y. Zhang, B. Van der Bruggen, G.X. Chen, L. Braeken, and C. Vandecasteele, Removal of pesticides by nanofiltration: effect of the water matrix. *Separ. Purif. Technol.* **38**, 163–172 (2004).

115. B. Van der Bruggen and C. Vandecasteele, Removal of pollutants from surface water and groundwater by nanofiltration: Overview of possible applications in the drinking water industry. *Environ. Pollut.* **122**, 435–445 (2003).
116. R. Boussahel, A. Montiel, and M. Baudu, Effects of organic and inorganic matter on pesticide rejection by nanofiltration. *Desalination* **145**, 109–114 (2002).
117. Y. Kiso, K. Nishimura, T. Kitao, and K. Nishimura, Rejection properties of non-phenylic pesticides with nanofiltration membranes. *J. Membrance Sci.* **171**, 229–237 (2000).
118. B. Van der Bruggen, J. Schaep, W. Maes, D. Wilms, and C. Vandecasteele, Nanofiltration as a treatment method for the removal of pesticides from ground waters. *Desalination* **117**, 139–147 (1998).
119. P. Berg, G. Hagmeyer, and R. Gimbel, Removal of pesticides and other micropollutants by nanofiltration. *Desalination* **113**, 205–208 (1997).
120. K. Kosutic, I. Novak, L. Sipos, and B. Kunst, Removal of sulfates and other inorganics from potable water by nanofiltration membranes of characterized porosity. *Separ. Purif. Technol.* **37**, 177–185 (2004).
121. S. Bertrand, I. Lemaitre, and E. Wittmann, Performance of a nanofiltration plant on hard and highly sulphated water during two years of operation. *Desalination* **113**, 277–281 (1997).
122. K. Listiarini, J.T. Tor, D.D. Sun, and J.O. Leckie, Hybrid coagulation-nanofiltration membrane for removal of bromate and humic acid in water. *J. Membrance Sci.* **365**, 154–159 (2010).
123. G.T. Ballet, A. Hafiane, and M. Dhahbi, Influence of operating conditions on the retention of phosphate in water by nanofiltration. *J. Membrance Sci.* **290**, 164–172 (2007).
124. Y. Yoon, G. Amy, and J. Yoon, Effect of pH and conductivity on hindered diffusion of perchlorate ions during transport through negatively charged nanofiltration and ultrafiltration membranes. *Desalination* **177**, 217–227 (2005).
125. Y. Yoon, G. Amy, J. Cho, N. Her, and J. Pellegrino, Transport of perchlorate (ClO4-) through NF and UF membranes. *Desalination* **147**, 11–17 (2002).
126. C. Bellona and J.E. Drewes, The role of membrane surface charge and solute physico-chemical properties in the rejection of organic acids by NF membranes. *J. Membrance Sci.* **249**, 227–234 (2005).
127. A.M. Comerton, R.C. Andrews, and D.M. Bagley, The influence of natural organic matter and cations on the rejection of endocrine disrupting and pharmaceutically active compounds by nanofiltration. *Water Res.* **43**, 613–622 (2009).
128. A.M. Comerton, R.C. Andrews, D.M. Bagley, and C. Hao, The rejection of endocrine disrupting and pharmaceutically active compounds

by NF and RO membranes as a function of compound and water matrix properties. *J. Membrance Sci.* **313**, 323–335. (2008)

129. K. Kimura, G. Amy, J.E. Drewes, T. Heberer, T.-U. Kim, and Y. Watanabe, Rejection of organic micropollutants (disinfection by-products, endocrine disrupting compounds, and pharmaceutically active compounds) by NF/RO membranes. *J. Membrance Sci.* **227**, 113–121 (2003).
130. L.D. Nghiem and P.J. Coleman, NF/RO filtration of the hydrophobic ionogenic compound triclosan: Transport mechanisms and the influence of membrane fouling. *Separ. Purif. Technol.* **62**, 709–716 (2008).
131. L.D. Nghiem, A.I. Schaefer, and M. Elimelech, Pharmaceutical retention mechanisms by nanofiltration membranes. *Environ. Sci. Technol.* **39**, 7698–7705 (2005).
132. S. Hajibabania, A. Verliefde, J.A. McDonald, S.J. Khan, and P. Le-Clech, Fate of trace organic compounds during treatment by nanofiltration. *J. Membrance Sci.* **373**, 130–139 (2011).
133. C. Bellona and J.E. Drewes, Viability of a low-pressure nanofilter in treating recycled water for water reuse applications: A pilot-scale study. *Water Res.* **41**, 3948–3958 (2007).
134. V. Yangali-Quintanilla, S.K. Maeng, T. Fujioka, M. Kennedy, and G. Amy, Proposing nanofiltration as acceptable barrier for organic contaminants in water reuse. *J. Membrance Sci.* **362**, 334–345 (2010).
135. B. Mansell, Pilot-scale testing of a high recovery NF/RO integrated treatment system for indirect potable reuse, in *WEFTEC*, Los Angeles, CA, (2011).
136. C. Yu, J.E. Drewes, B. C., J. Min, S. Li, and P. Fu, Maximizing recovery of recycled water for groundwater recharge employing an integrated membrane system, in WateReuse Research Foundation Final Report, WRRF-08-010, pp. 94 (2011).
137. H. Zhou, and D.W. Smith, Advanced technologies in water and wastewater treatment, *J. Environ. Eng.* **1**, 247–264 (2002).
138. Y.A. Le Gouellec, R.C. Cheng, and T.J. Tseng, A novel approach to seawater desalination using dual-staged nanofiltration, in *Water Research Foundation*, AWWARF, Denver, CO (2006).
139. C.J. Harrison, Y.A. Le Gouellec, R.C. Cheng, and A.E. Childress, Bench-Scale testing of nanofiltration for seawater desalination. *J. Environ. Eng.* **133**, 1004–1014 (2007).
140. A. AlTaee and A.O. Sharif, Alternative design to dual stage NF seawater desalination using high rejection brackish water membranes. *Desalination* **273**, 391–397 (2011).
141. A.M. Hassan, M.A.K. Al-Sofi, A.S. Al-Amoudi, A.T.M. Jamaluddin, A.M. Farooque, A. Rowaili, A.G.I. Dalvi, N.M. Kither, G.M. Mustafa, and I.A.R. Al-Tisan, A new approach to membrane and thermal

seawater desalination processes using nanofiltration membranes (Part 1). *Desalination* **118**, 35–51 (1998).

142. M.A.K. Al-Sofi, A.M. Hassan, O.A. Hamed, A.G.I. Dalvi, M.N.M. Kither, G.M. Mustafa, and K. Bamardouf, Optimization of hybridized seawater desalination process. *Desalination* **131**, 147–156 (2000).
143. A.S. Al-Amoudi and A.M. Farooque, Performance restoration and autopsy of NF membranes used in seawater pretreatment. *Desalination* **178**, 261–271 (2005).
144. F. Macedonio, E. Curcio, and E. Drioli, Integrated membrane systems for seawater desalination: Energetic and exergetic analysis, economic evaluation, experimental study. *Desalination* **203**, 260–276 (2007).
145. M. Rosenberg, Current and future applications for membrane processes in the dairy industry. *Trends. Food. Sci. Tech.* **6**, 12–19 (1995).
146. P.A. Marcelo and S.S.H. Rizvi, Applications of membrane technology in the dairy industry, in *Hanbook of Membrane Separations*, A.K. Pabby, S.S.H. Rizvi, A.M. Sastre (Eds.), pp. 635–670,CRC Press, Boca Raton, FL (2008).
147. USDA Foreign Agriculture Service (FAS), U.S. Whey Exports, (1999). www.fas.usda.gov/dlp2/circular/1999/99-12dairy/uswhey.html
148. G. Bargeman, M. Timmer, and C. van der Horst, Nanofiltration in the food industry, in *Nanofiltration: Principles and Applications*, A.I. Schaefer, A.G. Fane, T.D. Waite (Eds.), pp. 305–328,Elsevier, Oxford, UK, (2005)
149. H.S. Alkhatim, M.I. Alcaina, E. Soriano, M.I. Iborra, J. Lora, and J. Arnal, Treatment of whey effluents from dairy industries by nanofiltration membranes. *Desalination* **119**, 177–183 (1998).
150. K. Pan, Q. Song, L. Wang, and B. Cao, A study of demineralization of whey by nanofiltration membrane. *Desalination* **267**, 217–221 (2011).
151. H.C. van der Horst, J.M.K. Timmer, T. Robbertsen, and J. Leenders, Use of nanofiltration for concentration and demineralization in the dairy industry: Model for mass transport. *J. Membrance Sci.* **104**, 205–218 (1995).
152. M.S. Yorgun, I.A. Balcioglu, and O. Saygin, Performance comparison of ultrafiltration, nanofiltration and reverse osmosis on whey treatment. *Desalination* **229**, 204–216 (2008).
153. G. Rice, S. Kentish, A. O'Connor, G. Stevens, N. Lawrence, and A. Barber, Fouling behaviour during the nanofiltration of dairy ultrafiltration permeate. *Desalination* **199**, 239–241 (2006).
154. B. Cuartas-Uribe, M.I. Alcaina-Miranda, E. Soriano-Costa, J.A. Mendoza-Roca, M.I. Iborra-Clar, and J. Lora-Garcia, A study of the separation of lactose from whey ultrafiltration permeate using nanofiltration. *Desalination* **241**, 244–255 (2009).
155. J. Luo, L. Ding, B. Qi, M.Y. Jaffrin, and Y. Wan, A two-stage ultrafiltration and nanofiltration process for recycling dairy wastewater. *Bioresource. Technol.* **102**, 7437–7442 (2011).

156. J. Luo, L. Ding, Y. Wan, P. Paullier, and M.Y. Jaffrin, Application of NF-RDM (nanofiltration rotating disk membrane) module under extreme hydraulic conditions for the treatment of dairy wastewater. *Chem. Eng. J.* **163**, 307–316 (2010).
157. J. Luo, L. Ding, Y. Wan, P. Paullier, and M.Y. Jaffrin, Fouling behavior of dairy wastewater treatment by nanofiltration under shear-enhanced extreme hydraulic conditions. *Separ. Purif. Technol.* **88**, 79–86 (2012).
158. O. Akoum, M.Y. Jaffrin, L.H. Ding, and M. Frappart, Treatment of dairy process waters using a vibrating filtration system and NF and RO membranes. *J. Membrance Sci.* **235**, 111–122 (2004).
159. T. Mustafa, Influence of filtration conditions on the performance of nanofiltration and reverse osmosis membranes in dairy wastewater treatment. *Desalination* **170**, 83–90 (2004).
160. M. Vourch, B.a. Balannec, B. Chaufer, and G.r. Dorange, Nanofiltration and reverse osmosis of model process waters fromthe dairy industry to produce water for reuse. *Desalination* **172**, 245–256 (2005).
161. B. Balannec, M. Vourch, M. Rabiller-Baudry, and B. Chaufer, Comparative study of different nanofiltration and reverse osmosis membranes for dairy effluent treatment by dead-end filtration. *Separ. Purif. Technol.* **42**, 195–200 (2005).
162. P. Fernandez, F.A. Riera, R. Ålvarez, and S. Ålvarez, Nanofiltration regeneration of contaminated single-phase detergents used in the dairy industry. *J. Food Eng.* **97**, 319–328 (2010).
163. L.F. Sotoft, K.V. Christensen, R. Andresen, and B. Norddahl, Full scale plant with membrane based concentration of blackcurrant juice on the basis of laboratory and pilot scale tests. *Chem. Eng. Process. : Process Intensification* **54**, 12–21 (2012).
164. J. Warczok, M. Ferrando, F. Lopez, and C. Guell, Concentration of apple and pear juices by nanofiltration at low pressures. *J. Food Eng.* **63**, 63–70 (2004).
165. S. Banvolgyi, S. Horvath, E. Bekassy-Molnar, and G. Vatai, Concentration of blackcurrant (Ribes nigrum L.) juice with nanofiltration. *Desalination* **200**, 535–536 (2006).
166. M. Cisse, F. Vaillant, D. Pallet, and M. Dornier, Selecting ultrafiltration and nanofiltration membranes to concentrate anthocyanins from roselle extract (Hibiscus sabdariffa L.). *J. Food Eng.* **44**, 2607–2614 (2011).
167. C. Conidi, A. Cassano, and E. Drioli, A membrane-based study for the recovery of polyphenols from bergamot juice. *J. Membrance Sci.* **375**, 182–190 (2011).
168. M. Catarino and A. Mendes, Dealcoholizing wine by membrane separation processes. *Innov. Food Sci & Emerg. Technol*, **12**, 330–337 (2011).

169. N. Garcia-Martin, L. Palacio, P. Pradanos, A. Hernandez, M. Ortega-Heras, S. Perez-Magarino, and D.C. Gonzalez-Huerta, Evaluation of several ultra- and nanofiltration membranes for sugar control in winemaking. *Desalination* **245**, 554–558 (2009).
170. N. Garcia-Martin, S. Perez-Magarino, M. Ortega-Heras, C. Gonzalez-Huerta, M. Mihnea, M.L. Gonzalez-Sanjos, L. Palacio, P. Pradanos, and A. Hernandez, Sugar reduction in musts with nanofiltration membranes to obtain low alcohol-content wines. *Separ. Purif. Technol.* **76**, 158–170 (2010).
171. R. Vegas, S. Luque, J.R.n. Alvarez, J.L. Alonso, H. Dominguez, and J.C. Parajo, Membrane-Assisted Processing of Xylooligosaccharide-Containing Liquors. Journal of Agricultural and Food Chemistry, 54 (2006) 5430–5436.
172. A.K. Goulas, A.S. Grandison, R.A. Rastall, Fractionation of oligosaccharides by nanofiltration, Journal of the Science of Food and Agriculture, 83 (2003) 675–680.
173. A.K. Goulas, P.G. Kapasakalidis, H.R. Sinclair, R.A. Rastall, A.S. Grandison, Purification of oligosaccharides by nanofiltration, *J. Membrance Sci.*, 209 (2002) 321–335.
174. W. Li, J. Li, T. Chen, C. Chen, Study on nanofiltration for purifying fructo-oligosaccharides: I. Operation modes, *J. Membrance Sci.*, 245 (2004) 123–129.
175. N. Capelle, P. Moulin, F. Charbit, R. Gallo, Purification of heterocyclic drug derivatives from concentrated saline solution by nanofiltration, *J. Membrance Sci.*, 196 (2002) 125–141.
176. M.B. Martinez, B. Van der Bruggen, Z.R. Negrin, P. Luis Alconero, Separation of a high-value pharmaceutical compound from waste ethanol by nanofiltration, Journal of Industrial and Engineering Chemistry.
177. A.I. Cavaco Morao, A. Szymczyk, P. Fievet, A.M. Brites Alves, Modelling the separation by nanofiltration of a multi-ionic solution relevant to an industrial process, *J. Membrance Sci.*, 322 (2008) 320–330.
178. D.L. Oatley, B. Cassey, P. Jones, W. Richard Bowen, Modelling the performance of membrane nanofiltration - recovery of a high-value product from a process waste stream, Chemical Engineering Science, 60 (2005) 1953–1964.
179. K.Y. Wang, T.-S. Chung, The characterization of flat composite nanofiltration membranes and their applications in the separation of Cephalexin, *J. Membrance Sci.*, 247 (2005) 37–50.
180. A. Zhu, W. Zhu, Z. Wu, Y. Jing, Recovery of clindamycin from fermentation wastewater with nanofiltration membranes, *Water Res.*, 37 (2003) 3718–3732.

181. C.J. Pink, H.-t. Wong, F.C. Ferreira, A.G. Livingston, Organic Solvent Nanofiltration and Adsorbents; A Hybrid Approach to Achieve Ultra Low Palladium Contamination of Post Coupling Reaction Products, Organic Process Research & Development, 12 (2008) 589–595.
182. P. van der Gryp, A. Barnard, J.-P. Cronje, D. de Vlieger, S. Marx, H.C.M. Vosloo, Separation of different metathesis Grubbs-type catalysts using organic solvent nanofiltration, *J. Membrance Sci.*, 353 (2010) 70–77.
183. G. Szekely, J. Bandarra, W. Heggie, B. Sellergren, F.C. Ferreira, Organic solvent nanofiltration: A platform for removal of genotoxins from active pharmaceutical ingredients, *J. Membrance Sci.*, 381 (2011) 21–33.
184. G. Szekely, J. Bandarra, W. Heggie, B. Sellergren, F.C. Ferreira, A hybrid approach to reach stringent low genotoxic impurity contents in active pharmaceutical ingredients: Combining molecularly imprinted polymers and organic solvent nanofiltration for removal of 1,3-diisopropylurea, *Separ. Purif. Technol.* 86 (2012) 79–87.
185. C. Tang, V. Chen, Nanofiltration of textile dye effluent, in: A.I. Schaefer, A.G. Fane, T.D. Waite (Eds.) Nanofiltration: Principles and applications, Elsevier, Oxford, UK, 2005, pp. 380 – 393.
186. R. Jiraratananon, A. Sungpet, P. Luangsowan, Performance evaluation of nanofiltration membranes for treatment of effluents containing reactive dye and salt, *Desalination* 130 (2000) 177–183.
187. B. Van der Bruggen, E. Curcio, E. Drioli, Process intensification in the textile industry: the role of membrane technology, J Environ Manage, 73 (2004) 267–274.
188. L. Shu, T.D. Waite, P.J. Bliss, A. Fane, V. Jegatheesan, Nanofiltration for the possible reuse of water and recovery of sodium chloride salt from textile effluent, *Desalination* 172 (2005) 235–243.
189. C. Fersi, M. Dhahbi, Treatment of textile plant effluent by ultrafiltration and/or nanofiltration for water reuse, *Desalination* 222 (2008) 263–271.
190. M.I. Alcaina-Miranda, S. Barredo-Damas, A. Bes-Pia, M.I. Iborra-Clar, A. Iborra-Clar, J.A. Mendoza-Roca, Nanofiltration as a final step towards textile wastewater reclamation, *Desalination* 240 (2009) 290–297.
191. N.B. Amar, N. Kechaou, J. Palmeri, A. Deratani, A. Sghaier, Comparison of tertiary treatment by nanofiltration and reverse osmosis for water reuse in denim textile industry, J Hazard Mater, 170 (2009) 111–117.
192. S. Chakraborty, S. De, J.K. Basu, S. DasGupta, Treatment of a textile effluent: application of a combination method involving adsorption and nanofiltration, *Desalination* 174 (2005) 73–85.

193. A. Sungpet, R. Jiraratananon, P. Luangsowan, Treatment of effluents from textile-rinsing operations by thermally stable nanofiltration membranes, *Desalination* 160 (2004) 75–81.
194. E. Ellouze, N. Tahri, R.B. Amar, Enhancement of textile wastewater treatment process using Nanofiltration, *Desalination* 286 (2012) 16–23.
195. C.Y. Tang, T.H. Chong, A.G. Fane, Colloidal interactions and fouling of NF and RO membranes: A review, Adv Colloid Interface Sci, 164 (2011) 126–143.
196. B. Van der Bruggen, B. Daems, D. Wilms, C. Vandecasteele, Mechanisms of retention and flux decline for the nanofiltration of dye baths from the textile industry, *Separ. Purif. Technol.* 22–23 (2001) 519–528.
197. M. Riera-Torres, C. Gutierrez-Bouzan, M. Crespi, Combination of coagulation-ìflocculation and nanofiltration techniques for dye removal and water reuse in textile effluents, *Desalination* 252 (2010) 53–59.
198. A. Bes-Pia, M.I. Iborra-Clar, A. Iborra-Clar, J.A. Mendoza-Roca, B. Cuartas-Uribe, M.I. Alcaina-Miranda, Nanofiltration of textile industry wastewater using a physicochemical process as a pre-treatment, *Desalination* 178 (2005) 343–349.

5

Forward Osmosis

Jeffrey McCutcheon and Nhu-Ngoc Bui

University of Connecticut Department of Chemical & Biomolecular Engineering
Center for Environmental Sciences and Engineering

Abstract
Forward osmosis (FO) is an emerging desalination technology that has garnered an increasing amount of attention in recent years. In FO, water is driven across a semi-permeable membrane by an osmotic pressure gradient that is generated by a draw solution. The membrane rejects dissolved contaminants, much like they are in reverse osmosis. The draw solution is comprised of highly soluble solutes that are easily removed and reused in the process. When designed with an appropriate membrane and draw solution, FO promises to enable low cost desalination with improved recovery and fouling resistance. The road to commercialization still contains a number of technical hurdles. Membranes must be designed specifically for FO while retaining the high permselectivity of conventional reverse osmosis membranes. Draw solutes must be designed for high solubility, easy removal, and low toxicity. These challenges, while considerable, have not deterred a substantial worldwide research effort on forward osmosis. Commercialization of FO hinges on continued work in these area while eventual successful demonstration on the pilot scale will secure FO in its place among conventional desalination technologies.

Keywords: Forward osmosis, membrane, draw solution, low cost desalination

5.1 The Limitations of Conventional Desalination

Desalting seawater or brackish water has in recent years become a commoditized industry. Reverse osmosis (RO) now comprises 50%

Jane Kucera (ed.) Desalination , (255–286) 2014 © Scrivener Publishing LLC

of worldwide desalination capacity of 3,500,000 million gallons per day [1]. This increasing reliance on RO for a growing worldwide desalination industry is largely due to the remarkable improvements in RO membrane technology in recent decades. The thin film composite (TFC) membrane replaced the cellulose acetate membrane, exhibiting qualities of ultra-high permeability and selectivity [2–4]. The power requirement of RO dropped considerably, dramatically improving the commercial prospects for large scale RO desalination [5].

A number of companies offer RO membranes with similar chemistries and performance. These membranes rely on conventional polyamide chemistry and do not exhibit dramatic performance improvements from one year to the next. Incremental improvements of membrane performance (specifically permeance or selectivity) year to year have not led to dramatic reductions in RO energy use or cost. In recent years, there have been a number of innovative mixed matrix membrane designs that have been proposed, including biomimicry with aquaporin [6], aligned carbon nanotubes [7], zeolite nanocomposite membranes [8], and graphene [9]. These approaches, while very novel, have limited commercial viability and have yet to be demonstrated to have dramatically better performance over existing RO membrane technology under real conditions. Moreover, these membranes also fail to address the key performance limitation in RO: thermodynamics. Increasing permeance or selectivity of a membrane does not address the thermodynamic limitations of RO [10–12]. These limitations stem from the osmotic pressure of the dissolved solutes present in saline water.

5.1.1 Osmotic Pressure

Osmotic pressure is a physical manifestation of chemical potential difference between two solutions separated by a semi-permeable membrane. In essence, this pressure represents a solution's "thermodynamic desire" to expand its volume and allow the solutions dissolved solute molecules to move as far from each other as possible (similar to gas molecules in a fixed volume exerting pressure on their container). The osmotic pressure, π, is often described mathematically using the Van't Hoff equation.

$$\pi = iCRT \tag{5.1}$$

where i is the solute dissociation constant, C is the concentration of the solute, R is the gas constant, and T is the temperature. This equation indicates that the osmotic pressure is proportional to the concentration of the dissolved solute molecules or ions. Solutes that dissociate, such as inorganic salts, have higher osmotic pressures since they form multiple ions in solution. This equation is valid only for "ideal" solutions. While this definition is vague, it generally means that the solution is dilute.

Osmotic pressure plays a significant role in RO system performance and economics. We can see this in the following basic equation for RO membrane performance.

$$J_W = A(\Delta P - \Delta \pi) \tag{5.2}$$

Where J_W is the water flux, A is the water permeance, ΔP is the transmembrane hydraulic pressure, and $\Delta\pi$ is the transmembrane osmotic pressure. For highly selective processes, the permeate osmotic pressure is often assumed negligible and therefore $\Delta\pi$ is sometimes referred to simply as the feed solution osmotic pressure.

This equation allows us to see that increasing the osmotic pressure of a solution will result in a lower flux unless hydraulic pressure is increased. For example, seawater has a higher osmotic pressure (around 25 bars) than brackish water (typically 2–5 bars) and therefore RO must be operated at a higher transmembrane hydraulic pressure. More importantly, though, is the limitations on recovery that are imposed by osmotic pressure. As more water is removed, the osmotic pressure increases nonlinearly and restricts flux [13]. If seawater has an osmotic pressure of 25 bar and 50% of the water is recovered (as would be in a conventional RO plant), the osmotic pressure of the retentate is about 50 bars. If you recover another half of the water (total recovery of 75%), then the osmotic pressure is about 100 bar. Since typical RO plants are operated around 55–60 bar, one can see why recovery is limited to about 50% in RO. Operating at higher pressures would be prohibitively expensive and energy intensive while also requiring changes in membrane housing design and more expensive instrumentation and fittings.

What this ultimately means is that as osmotic pressure increases with increasing recovery, the permeance plays a diminishing role in overall performance and cost. Therefore, developing new, possibly expensive, high permeance membranes will not result in a substantial change in RO cost. This is especially true given the already excellent permeance of existing RO membranes. What if, however, we could stop trying to fight osmotic pressure and instead use it to our advantage?

5.2 Forward Osmosis

Forward osmosis (FO) is a process that utilizes *osmotic* pressure to drive water from a contaminated or saline water feed solution across a semi-permeable membrane that retains the dissolved solutes. This flux is driven by osmotic pressure generated by a draw solution or osmotic agent. This osmotic separation requires little energy (only the pumping of fluids to the membrane element) as it occurs spontaneously due to the tendency toward thermodynamic equilibrium. The draw solution contains dissolved solutes that are also, in ideal circumstances, retained by the membrane. After osmosis, this diluted solution is sent to a secondary separation process that recovers the solute and recycles it while liberating clean, drinkable water from the draw solute. It is this secondary process that requires some energy. The choice of draw solute, which will be discussed in more detail later, determines the process required and the amount and type of energy needed to run it. A schematic diagram of the process is shown in Figure 5.1.

Ultimately, the process economics is determined by the performance of the membrane and the draw solute recovery. The former has been the focus of most of the academic research at the time of this books publishing, while the latter is largely determined by the type of draw solute chosen.

5.2.1 History of FO

Osmosis is a physical phenomenon that was first observed in the mid-18th century and then conceived in 1854 by Abbe Nollet, a Scotland chemist. The earliest studies of osmotic phenomenon involved observation of osmosis through natural materials (e.g. animal bladders, plant cells, collodion (nitrocellulose), rubber, or porcelain [13]). The first coining of the term "forward osmosis" came in the mid-1960s, when a patent by Batchelder [14] described an inexpensive process to demineralize saline water without using

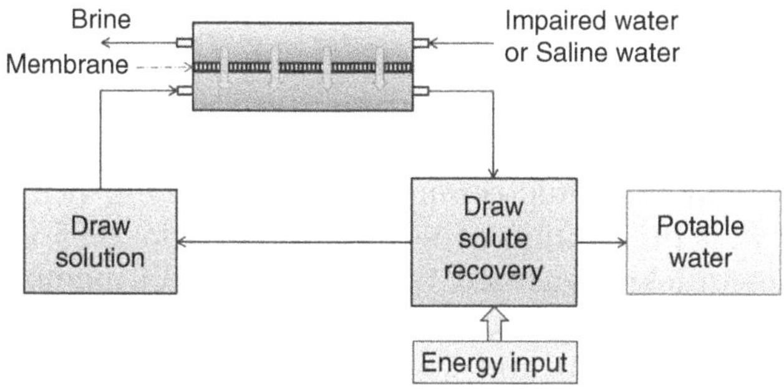

Figure 5.1 - Schematic of the generalized forward osmosis desalination process

a large amount of external energy like other technologies such as distillation, evaporation, freezing or electrodialysis. At this early stage, several processes had been engineered and proposed as potential applications for natural osmosis. Osmotic process were subsequently considered for liquid recovery [15, 16], solution concentration [16], desalination [17–19], power generation [20–22], nutritious drink production out of non-potable liquids [23], waste solution purification or solute extraction from solutions or suspensions [24]. However, in all studies that presented data, very low fluxes were observed. The primary reason behind this poor performance were the membranes chosen. These osmotic processes required the use of a salt rejecting membrane, and at the time of many of these studies, the only such membranes available were designed for RO. Available RO membranes during the early years of forward osmosis included flat-sheet RO membranes from Eastman, cellulose acetate hollow fiber membrane from Dow [25], B-9 (flat-sheet) and B-10 Permasep (hollow fiber) from Dupont [26, 27], or Loeb-Sourirajan type CA-3000 from Toray [28]. Because of the ease of use, most of these early studies were limited to flat sheet membranes. Their poor performance was attributed to their asymmetric structures meant for use under hydraulic pressure operation. RO membranes employed a thick porous supporting layer that exhibits severe mass transfer resistance, known as internal concentration polarization, which was found to be the single greatest impediment to good water flux during osmosis. This will be discussed later section of this chapter [29, 30]. In the 1990s, Osmotek Inc. (Albany, Oregon) (currently Hydration Technology Innovation (HTI)) pioneered a manufacturable osmotic membrane designed specifically for FO. This membrane was integrated into a number of commercial products used in industrial water purification, emergency relief, military water purification, and recreational purposes [31–47]. This membrane became the gold standard of the industrial and academic community interested in FO, especially after the first publication of its performance in 2005 [48]. Other companies quickly emerged that produced FO membranes and processes. These include Oasys Water (Boston, MA), Catalyx Inc. (Anaheim, California) and Porifera (Hayward, CA). Hydration Technology Innovations dominated this field in the 2000s, providing membrane free of charge to academic groups from around the world for use in academic research on osmotic processes. They now boast that over 100 Ph.D. level scientists have been trained in osmotic processes using their membranes [49] . This academic research has led to a multitude of membrane designs emerging into the field. Laboratory scale flat-sheet [8, 50–63] or hollow fiber [60, 64-70] membranes quickly emerged with properties tailored to FO and thus exhibited superior flux and selectivity performance. Meanwhile, investigations on suitable osmotic agents which are inexpensive, abundant, soluble, and recoverable have also occured [19, 38, 48, 71–79]. Recent improvement in membrane performance and draw solute efficiency has led to a number of reports on the potential of osmotic process in various applications [32, 35, 36, 38–41, 47, 80, 81]. These efforts will be discussed throughout this chapter.

5.2.2 Benefits of Forward Osmosis

It is important to note that FO is not seen as a replacement to RO technology. In fact, in certain configurations, FO could feasibly be used in coordination with RO. Some of the advantages of FO are described below.

Low Energy Cost. Often FO is described as a technology that uses less energy than reverse osmosis. This is an inaccurate statement. FO is not intended to use less energy than any process, but rather use less *expensive* energy. For example, RO uses electricity, which is inherently a high value energy. Some versions of FO, however, can utilize extremely low grade thermal energy which has no other beneficial use [82]. The energy type, quality and quantity needed are entirely dependent on the choice of draw solute, which makes conducting an economic comparison between FO and RO difficult. In RO, there is only one way to generate driving force: hydraulic pressure. The primary means of generating this force is using an electrically driven pump. In FO, the osmotic driving force can be generated using an infinite number of solutes or their combination, but the removal and reuse of those solutes are dependent on a secondary separation process which requires a specific type of energy. Nevertheless, if the solute can be tailored to be regenerated using a low cost or free energy source, there is a distinct advantage over electrically driven processes. This will be discussed in greater detail below.

Reduced Impact of Thermodynamic Restriction. As recovery increases, osmotic pressure increases nonlinearly. If high recovery desalination is desired (for either increased productivity or reduced brine discharge volume), the osmotic pressure limitations must be overcome. In RO, one can simply increase the pressure, but this will result in higher cost and possibly require the redesign of the membrane or element to tolerate higher pressure. In FO, one only needs to increase the concentration of the draw solution, which is straightforward if the solute has a high solubility. This may however increase the energy load on the draw solution recovery system.

Low Fouling Propensity. Fouling is one of the most critical problems in membrane processes today. Organics, inorganics, biological organisms, and colloidal matter can deposit onto membranes and severely reduce flux performance of membranes. Several recent studies have demonstrated that FO has a lower fouling propensity [42, 83, 84] when compared to hydraulically driven processes. This largely attributed to the lack of compaction of the fouling layers which normally occurs in hydraulically driven processes. Compaction of a fouling layer reduces the porosity of the layer, making them tougher and more resistant to cleaning agents. Fouling does occur in FO, though it is perceived as a less dense layer which is easier to remove with conventional cleaning techniques.

Lower Cost. Overall, each of the above benefits can lead to a lower cost separation process. Whether it is devised as a hybrid with existing RO processes [85, 86] or a standalone process, the use of osmosis can generate many of these benefits which can reduce energy cost or increase membrane lifetime through reduced fouling.

Economic analysis of the FO desalination process has yet to be fully developed, and only one pilot plant study with an assessment has been published (produced water treatment, not drinking water). With a number of companies taking interest in FO as of 2013, one can expect growth in this area in the near future.

5.3 The Draw Solution

While any solute that dissolves in water could be considered a draw solute, not all solutes will lead to an economically viable FO process. The solute chosen must meet a number of criteria to ensure reliable and economic performance. [71]. Major criteria for an ideal draw solute are: (1) high osmotic efficiency (i.e. high solubility and relatively low molecular weight due to the colligative property of osmotic pressure); (2) minimal reverse diffusion to maintain the driving force, avoid contaminating the feed solution, and limit the need for replenishment; (3) chemically compatible with FO membrane materials; (4) non-toxic; (5) economically recoverable; and (6) low cost [13, 48, 71, 73, 75, 78]. Conventional draw solutes are inorganic salts (of monovalent and divalent cations) and sugars (sucrose, fructose, glucose). Since 2005, a variety of new draw solutes have been reported such as thermolytic NH_4HCO_3 salt [48, 72, 73], functionalized magnetic nanoparticles [76, 77, 87, 88], 2-methylimidazole based charged organic compounds [74], fertilizers [38], stimuli-responsive polymeric hydrogels [89], polyelectrolytes [79], switchable solvent [78], or hexavalent phosphazene (a hybrid organic-inorganic material) [75]. We review various classifications of draw solutes below.

5.3.1 Inorganic solutes

Inorganic salts have a distinct advantage over many draw solutes because of their ability to dissociate into more than one ion. Many early FO processes considered mixtures of water with inorganic solutes including soluble gases (e.g. sulfur dioxide) [14], precipitable salts (e.g. aluminum sulfate) [17], seawater [25] or sugar [23, 24] (e.g. fructose, sucrose, glucose) as a source of osmotic driving force. These solutes were either removed by heating, air-stripping, precipitating or used directly as a nutritious drink [48]. Sodium chloride

(NaCl) has routinely been used as a draw solute due to its low cost, high solubility, and low toxicity, abundance, and ease of regeneration using other conventional desalination processes (e.g. distillation or RO) without risk of scaling [34, 42, 71, 84]. The drawback of using conventional desalination technologies for draw solute regeneration is that there is no energy benefit since the regeneration process must use as much energy as the single step process. Thermal regeneration of draw solutes emerged in 2002 as a new method of recovering the draw solution. McGinnis introduced a two-stage FO process for seawater desalination in which osmotically efficient draw solutes (e.g. heated solution of saturated KNO_3 for the first stage and SO_2 draw solution for the second stage) and a recycle loop were combined to increase draw solute recovery [19]. McCutcheon and McGinnis *et al.* discovered that a draw solution comprising of two highly soluble gases – ammonia (NH_3) and carbon dioxide (CO_2) – satisfies many of the criteria for an ideal draw solution [48, 72, 73]. By dissolving ammonium bicarbonate salt (NH_4HCO_3) in water and adjusting the ratios of NH_3, CO_2, and the salt, a draw solution with a very high osmotic efficiency (> 250 atm) exceeding that of seawater can be obtained. Separation of the product water from the draw solution can be achieved by moderately heating the solution (~ 60°C) to decompose NH_4HCO_3 into NH_3 and CO_2 gases. These gases can then be removed by a separation process (e.g. distillation, or membrane-based process). Achilli and coworkers have developed a protocol for selecting the optimal draw solute for different FO applications by systematically combining desktop screening process with laboratory data, modeling analyses and cost consideration. As a result, a group of seven draw solutions with different characteristics was found to be the most suitable inorganic solutes including $CaCl_2$, $KHCO_3$, $MgCl_2$, $MgSO_4$, $NaHCO_3$, NaCl and Na_2SO_4 [71].

5.3.2 Nanomaterials

Suspended nanomaterials are a recent development in draw solute choice. Their large size relative to dissolved solutes makes their regeneration less costly and reduces the likelihood of solute crossover to the feed side. Adding surface functionality containing dissociable groups offers a means of increasing their osmotic pressure and enhancing their stability. Surface modification can also increase the effective size of NPs which reduces the amount of draw solutes leaking to the feed side. This was considered to be an important performance criteria given the cost of replenishing the nanoparticles lost across the membrane

Probably the most interesting nanoparticle developed as a draw solute is based on magnetoferritin. After the FO step, these magnetic materials are separated from aqueous streams by a magnetic field [76]. However, magnetic fields may be unable to capture all of the smallest NPs. Also, reusing NPs may be complicated by agglomeration after magnetic removal. Irreversible aggregation can be mitigated in part by using thermo-sensitive magnetic NPs which can be separated in a low strength magnetic field at a temperature above the lowest critical solution temperature of the coating polymers [87]. Sonication is also an option, but this would add an energy cost. Membrane-based processes (e.g. ultrafiltration) can also be effectively used for regenerating these draw solutes.

5.3.3 Organic Solutes

Organic solutes offer a highly soluble option for draw solutions while also enabling a number of different regeneration schemes. An interesting study by Kim and coworkers analyzed 4058 compounds as possible draw solutes and found that 5 of the 7 total suitable solutes were organic (methanol, ethanol, 2-butanone, 2-propanol and methyl acetate) [90]. While these screening methods may use dubious criteria, studies like this suggest that we should not dismiss organic compounds as draw solutes, even if they do not dissociate.

Mark and coworkers [78] have recently attempted to apply switchable polarity solvents (SPSs) as draw solutes for FO. SPSs can switch back and forth between water-miscible and water-immiscible phases dependent on the presence of carbon dioxide at ambient pressures. The separated non-polar phase (i.e. immiscible in water) can then be regenerated to a full strength draw and reused in the presence of CO_2 [78]. Furthermore, SPS traces in the separated water can simply be removed by reverse osmosis. Although these highly soluble compounds generate remarkable osmotic pressures, these solutions are solvents and thus require a solvent tolerant membrane. Commercial cellulose acetate FO membranes were found to degrade during the test with SPSs. Either solvent tolerant membranes may have to be developed or alternative SPS chemistry will be required.

Hybrid organic-inorganic multi-valent salts derived from phosphazene have also been synthesized and characterized as FO draw solutes [75]. The advantages of this draw system are hydrolytic stability and versatile chemistry [75]. This draw solute was shown to give high degrees of dissociation which suggests a high osmotic potential. However, hydrolysis of cellulose acetate membranes were also observed.

Ge *et al.* reported the use of highly soluble polyelectrolytes of a series of polyacrylic acid sodium salts (PAA-Na) as draw solutes for FO [79]. Low solute leakage, easy recycle and structural flexibility are the advantages of this solute system. However, the high viscosity of this draw solution may require more energy to pump during the FO process. Concentration polarization is also enhanced in high viscosity fluids.

5.4 The Membrane

Membrane performance has long limited the advancement of forward osmosis. Membranes designed for RO were originally used to evaluate the performance of FO and found to perform very poorly. In 2005, however, a new membrane was found to perform better than commercial RO mebranes. These data was reported in *Desalination* [48] and reprinted in Figure 5.2.

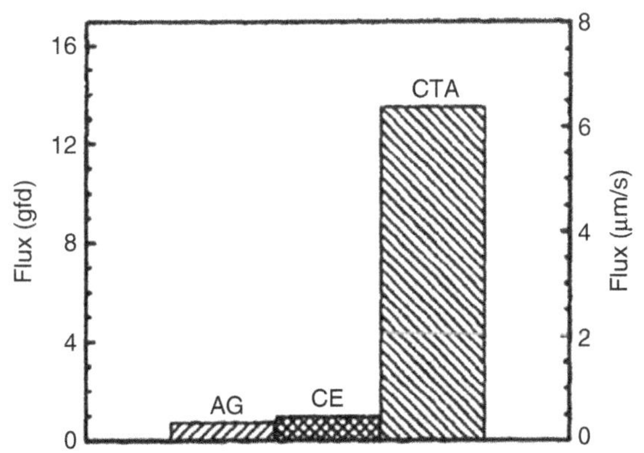

Figure 5.2 – Flux performance of 2 commercially available RO membranes from GE Water (AG, CE) and a cellulose triacetate forward osmosis membranes from Hydration Technology Innovations (CTA). The draw solution used was a 6M ammonia-carbon dioxide solution and the feed solution was a 0.5M sodium chloride solution. The temperature of the test was 50°C. Reprinted from [48]. Tests were done in a custom built crossflow benchtop FO testing system.

This data was among the first data to compare a commercially available FO membrane with RO membranes in an osmotic flux test. This FO membrane was found to have an order of magnitude higher osmotic flux performance than its more selective and permeable RO membrane counterparts.

We can explain this result by looking at the differences between these two membranes. The AG membrane is a conventional thin film composite (TFC) membrane designed for RO while the CTA (commonly referred to as CA in the literature today) is an asymmetric cellulose triacetate membrane. Figure 5.3 shows cross sectional SEM images of a representative TFC membrane and FO membrane

Figure 5.3 shows that a typical RO membrane has a 3 tiered structure employing a highly permselective but exceedingly fragile aromatic polyamide layer. The inherent fragility of this layer necessitates the use of support layers that provide mechanical strength under pressure. The key difference between this membrane and the HTI FO membrane is the lack of much of the support layer, having instead derived strength from an embedded mesh thus allowing it to be made much thinner than a typical RO membrane.

So why do these support layers matter in FO? It is all a question of mass transfer resistance and will be discussed in the next section.

5.4.1 Mass Transfer Limitations in Forward Osmosis

Consider for a moment the process of osmosis. For osmosis to occur, a selective barrier must exist between two solutions with different concentrations of dissolved solutes. This barrier must retain at

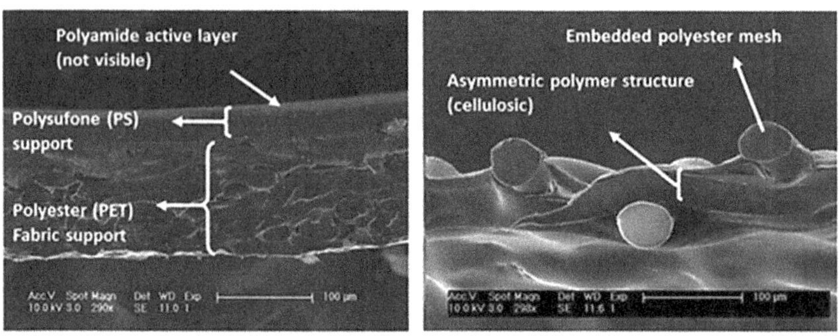

Figure 5.3 - Scanning electron micrographs of a representative RO TFC membrane (left) and the HTI FO membrane (right).

least some of the solutes on either side of the membrane in order to induce an osmotic flow. Now, as water moves across the membrane, solutes in the dilute solution will be retained against the membrane surface, while solutes on the draw side of the membrane will be diluted by the permeating water. These effects are called concentrative and dilutive concentration polarization (CP), respectively, and have been discussed in detail in a number of publications on FO and other processes [29, 52, 54, 55, 91–94].

Now, imagine that on one side of that selective barrier exists a supporting structure that is porous, but non-selective for these dissolved solutes. Therefore, the solutes must diffuse through this porous supporting structure just to reach the selective barrier and "exert" their osmotic pressure. Then, as water permeates the membrane, these solutes will be diluted *within this structure*, where no mixing occurs and replenishment by diffusion is hindered by the low porosity and high tortuosity of the material. This phenomenon is known as *internal* concentration polarization and causes a reduction in osmotic pressure at the downstream interface of the membrane selective layer. Internal CP is illustrated in Figure 5.4.

The mathematical models describing this phenomenon can be found elsewhere [29]. However, one will note that any CP phenomenon reduces the effective osmotic pressure driving force and thus flux. If flux is to be improved, the CP phenomenon must be minimized.

CP affects every membrane process. In RO, for instance, CP causes a concentration rise at the membrane interface and a corresponding increase in osmotic pressure which reduces flux. This can be controlled by either reducing flux, or increasing the mixing at the surface of the membrane. The same approach can be used for FO, but only on the feed side of the membrane. Internal CP cannot be mitigated by mixing because it exists within the protective confines of the membrane support layer. One option is to use a draw solute with a higher diffusivity. However, if you are restricted to a particular draw solute, the only way to improve mass transfer in this region is to change the properties of this support layer by making it thinner and more open.

The academic literature quantifies the membrane specific structural resistance to diffusion using a term known as the structural parameter, S. This term is related to the intrinsic structural characteristics of the membrane support layer.

$$S = \frac{t\tau}{\varepsilon} \tag{5.3}$$

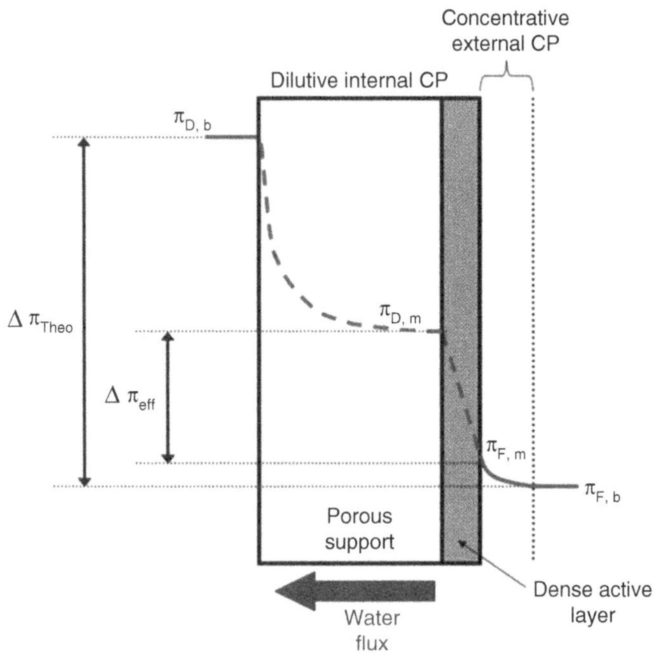

Figure 5.4 – Illustration of concentration polarization phenomenon near the active (selective) layer of the membrane. π refers to the osmotic pressure of the solution. $\Delta\pi$ is indicative of the osmotic driving force across the membrane while the subscripts "Theo" and "eff" refer to the theoretical and effective driving force, respectively. Subscripts D, F, m, and b refer to the "draw", "feed", "selective layer interface", and "bulk" osmotic pressures, respectively

Where ε is the porosity, τ is the tortuosity, and t is the thickness. This term can also be seen as an average distance that a solute molecule must diffuse before reaching the selective layer. So, a membrane that has a support layer that is 100 μm thick with a porosity of 50 % and a tortuosity of 2 will have an S value of 400 μm.

The key to reducing internal CP and maximizing flux is by reducing the membrane structural parameter. This is done by either making the support layer thinner, more porous, or less tortuous.

5.4.2 Tailored membranes for FO

There are a number of criteria that must be met for tailoring membranes specifically for FO. These include the following:

Superior permselectivity. The permeability and selectivity of any FO membrane should match or exceed that of existing RO membranes. This is critical not only for high water flux and salt rejection, but also for retention of the draw solute.

Chemical resistance and thermal stability. The membrane may be exposed to exotic chemistries as new draw solutes are developed. Moreover, these draw solutions may have been thermally regenerated and therefore may come into contact with the membrane while heated. These membranes must be able to tolerate the environment within the system.

Mechanically strong. While these membranes are not exposed to high pressures, the membrane must be fabricated on a large scale and be placed into a module. This requires the membrane to have a modest amount of strength and resist pinhole formation.

Easy and inexpensive to manufacture. These new membranes should not involve complex materials or processing that make large scale manufacturing difficult or expensive.

Tolerate modest pressures. In the full-scale FO process, there will be pressure drop across the module. This may lead to a transmembrane pressure that could impact the membrane performance. This means that these membranes must tolerate some pressure, though it will be far less that of RO. These issues may be resolved with the use of properly designed spacers on either side of the membrane, but this is an area that is still underdeveloped in FO.

Thin, porous, and non-tortuous support layers. As described above, the key to making a good FO membrane lies in the support layer design. Support layers must be made to be highly porous, non-tortuous, and thin (*i.e.* have a low S parameter) [13, 52, 53, 56, 95]. The challenge, however, is to do this while retaining the above described characteristics. A number of membranes have been developed that address these criteria. They will be discussed in greater detail below.

Hydrophillic chemistry. Hydrophilicity is critical for good water flux through the selective layer of a membrane. However, it is equally if not more important for transport in the support layer [30, 52, 56]. Without proper wetting of this layer, transport can be inhibited since transport can only occur in the wetted porosity of the support layer. Without proper wetting, internal CP can be greatly enhanced. It is important that the membrane not experience plasticization in the presence of water, however.

If these criteria are met for a given membrane-draw solution combination, then the membrane should function well. A number of such membranes have been proposed by the academic community in the past 5 years. These include both flat sheet and hollow fiber membranes.

5.4.2.1 Flat sheet

HTI's cellulose triacetate membrane (CTA-HTI) is the first commercial flat-sheet FO membrane that was manufactured on a continuous line. Shown in Figure 5.3, this membrane is very different than a typical RO thin film composite membrane. It has a thickness of about 50 μm and has been shown to exhibit superior performance when compared to RO membranes in FO testing [13, 29, 48]. The mechanical support of the CTA-HTI is provided by an embedded polyester mesh instead of a fabric backing, which is common in RO membrane. However, these membranes still are inhibited by internal CP and a number of alternative structures and chemistries have been proposed. Starting around 2010, there was a substantial increase in the number of studies on redesigning the structure of TFC membranes for FO by focusing either on improving permselectivity and longevity of the selective layer [8, 57, 96, 97] or modifying the membrane support properties [52, 53, 55, 56, 95, 98]. For example, using commercially available RO membranes, Arena and coworkers modified the chemistry of the support layer using polydopamine and increased the hydrophilicity to improve wetting. This had a dramatic effect on osmotic water flux [55]. Another study used an intrinsically hydrophilic nylon-6,6 microfiltration membrane as a support for a TFC FO membrane [95]. A variety of polymers have been selected as materials for forming FO membrane supports, such as cellulose-derived materials [99–101], polysulfone (PSu) [53, 63], polyethersulfone (PES) [102], sulphonated copolymer made of PES and polyphenylsulfone [50], polyacrylonitrile [56], Torlon® polyamide-imide [103], and blends thereof [56]. Membrane supports having finger-like porous structure were studied by different groups to understand how a support having a tortuosity $\tau \approx 1$ affects FO performance [50, 53, 63]. While Yip and coworkers [53] used different solvents (NMP and DMF) in casting mechanism to create pores, Wei *et al.* incorporated poly (vinylpyrrolidone) (PVP) and LiCl into the casting dope [63].

Double-skin layer membranes have been investigated as another approach for novel FO membranes [57, 62]. The second skin layer was designed to reduce internal CP and fouling in the support layer. Wang *et al.* [62] developed double dense-layer cellulose acetate membrane by manipulating the mechanism of phase inversion casting method and annealing post-treatment. Qi and coworkers designed double-skinned FO membranes based on layer-by-layer (LbL) assembly in which oppositely charged polyelectrolyte layers were alternatively deposited at both sides of the porous supports [57]. The membrane was stabilized via cross-linking treatment and was reported to give high flux and antifouling capability. In these studies, minimization of internal CP, structural parameter (S) and fouling propensity (of the support layer) were reported. However, osmotic flux performance of these double-skin membranes was low compared to other membranes. This may be due to the fact that while a thin skin layer was formed at the bottom of the membrane to reduce ICP, the external concentration polarization at its surface actually increased because the support layer surface was no longer porous. Moreover, the second skin layer reduced the permeance of the membrane because water moving across the membrane now had to diffuse across two skin layers.

Nanofiber-supported TFC membranes have also emerged as a potentially new FO membrane platform. Possessing a high porosity, low tortuosity and an interconnected porous structure, nanofibrous mats exhibit exceptionally low structural parameters [52, 56, 104]. Bui *et al.* [52,56] and Song *et al.* [104] have independently shown that nanofiber is a promising mid-layer support material for high-flux TFC membranes. Hoover *et al.* reported an active role of polyester nanofiber as a backing layer which, instead of a conventional polyester nonwoven, provides support for a cast polysulfone mid-layer which in turn supported a polyamide selective layer [98].

Very recently, some novel approaches have been used to make FO membranes. Nguyen and coworkers deposited a poly amino acid 3-(3,4-Dihydroxyphenyl)-L-alanine (L-DOPA), a zwitterionic polymer, on the HTI-CA membrane surface to enhance the fouling resistance of this membrane by 30% [96]. Dumée *et al.* [7] developed a TFC membrane on a self-supporting bucky-papers (BPs) made of hydroxyl-functionalized carbon nanotubes. In its early stage, this support showed a low contact angle of < 20°, a high water uptake capacity of 17 wt% and a large porosity of > 90%. However, a major challenge of this type of support lies in the high compressibility of BPs which may increase the tortuosity and decrease the mechanical stability of the membranes during use [7]. Another interesting membrane design comes from Jensen and coworkers who use biomimetic membranes for water extraction from liquid aqueous media by FO [105]. This aquaporin containing liquid membrane system comprises aquaporin water channels in a dispersion of amphiphilic molecules, preferably comprising vesicles in the form of proteoliposomes. Many of these novel membranes may never make it to commercial scale manufacturing because of their cost or complexity, but they serve as useful platforms to study new membrane designs and fundamentals of membrane transport.

5.4.2.2 *Hollow Fiber*

Hollow fibers are emerging as a preferable membrane platform for a number of membrane separations due to their high packing density relative to flat sheet membranes. While hollow fibers have been commercialized for applications in dialysis and microfiltration, they had not been considered for FO until very recently. To produce high-performance hollow fibers, proper selection of the membrane material is critically important because it determines (1) the spinnability and mechanical integrity, (2) the hydrophilicity / hydrophobicity and fouling tendency for water-related applications, (3) the bio-compatibility (for medical uses), and (4) the chemical stability for applications in harsh conditions [106]. Once an adequate material is selected, desirable characteristics of the fiber (e.g. pore size and distribution, selective layer thickness, or porous structure) can be tuned via the phase inversion process. For FO, a relatively thin, defect-free, yet highly porous substructure with a permselective skin layer is critical for a high-performance membrane.

Initially, polybenzimidazole (PBI) and cellulose acetate were selected to form FO hollow fiber membranes. These integral self-supported membranes were thermally treated or chemically cross-linked to increase selectivity or mitigate defects. PBI nanofiltration (NF) membranes with adjusted pore sizes were first considered for use in forward osmosis by Wang and coworkers [68, 69]. PBI was selected due to its robust mechanical strength, excellent chemical stability and hydrophilicity. However, this type of membrane had a relatively low selectivity to monovalent ions. Therefore, the draw solutions used for this membrane were limited to divalent salts to ensure a reasonably low salt crossover. Such a membrane would have limited capabilities in desalination applications outside of water softening. This group later reported on an effective use of dual-layer polybenzimidazole – polyethersulfone / polyvinylpyrrolidone (PBI-PES/PVP) hollow fiber nanofiltration membranes generated by co-extrusion technology. The use of this membrane was to enrich lysozyme solution without denaturing or changing the conformation of the component of interest [107]. Sun *et al.*, from the same group, investigated hyperbranched polyethyleneimine (PEI) induced cross-linking of polyamide-imide NF hollow fiber membranes for effective removal of ciprofloxacin. By inducing polyelectrolyte PEI via cross-linking, the membrane pore size was significantly reduced. The membrane surface became more hydrophilic and positively charged and, as a result, the membrane obtained better rejection, low fouling propensity and effectively removed ciprofloxacin from water. Size exclusion, charge repulsion and solute-membrane affinity were believed to be the mechanism causing this remarkable change in NF membrane performance. Yet these membranes were ultimately used for osmotic concentration and not for forward osmosis.

Wang and coworkers pioneered a thin film composite FO hollow fiber membrane using a PES support [65, 66, 70]. This membrane employs a 300–600nm-thick selective layer that was formed by in-situ polymerization of polyamide. This is similar to RO membrane chemistry and could be deposited on either the outer surface (shell) or inner surface (lumen) of a porous PES hollow fiber substrate. The membrane performed well, exhibiting higher water flux with lower salt leakage compared to previous studies [70]. Meanwhile, Setiawan [65] from the same group developed a membrane with a positively charged nanofiltration-like selective layer using asymmetric microporous hollow fiber substrate made of Torlon® polyamide-imide. Despite the high pure water permeability coefficient (about 2.2 L/m^2.hr.bar), the membrane showed a relatively low osmotic water flux. Furthermore, this membrane exhibited a low selectivity to NaCl (a rejection of 49 % was observed). In another study, FO hollow fiber membranes with an antifouling NF-like selective layer was developed by Setiawan *et al.* [66]. Porous PAI ultrafiltration hollow fiber was first spun by phase inversion. After that, polyelectrolyte post-treatments were carried out using positively charged polyethyleneimine (PEI) for cross-linking and negatively charged polystyrene sulfonate sodium salt (PSS) for depositing. The membrane was able to maintain a steady water flux of 11 L/m^{2}hr after 4 hrs when 1000ppm bovine serum albumin and 2000 ppm Na_2SO_4 were used as the feed and 0.5M Na_2SO_4 was used as the draw. These results indicate the potential of this membrane to resist fouling in that given time frame.

In general, each of these new membrane platforms exhibited impressive performance. However, it is not easy to compare the relative performances of these membranes due to the different testing conditions (cross-flow velocity, draw solute, draw and feed concentrations, temperature, and so on). The performances of the above discussed membranes amongst others are therefore presented in Table 5.1. In this table, osmotic water flux is normalized by the theoretical

Table 5.1 Comparison of properties and FO performance (at 20–25 °C) of flat-sheet and hollow fiber FO membranes having been reported in literature [8, 50, 53–55, 57, 61, 63, 65, 95, 101, 104, 108–114]. Pure water permeance, A, and salt permeability coefficient, B were obtained from reverse osmosis test whereas osmotic water flux, J_w, and structural parameter, S were determined from forward osmosis test.

	TFC membrane structure	Draw *vs.* Feed	$\Delta\pi_{theo}$** (bar)	A (LMH/bar)	B (LMH)	J_w (LMH)	$J_w/\Delta\pi_{theo}$ (LMH/bar)	S (μm)	Ref.
Commercial HTI	CTA	1.5M NaCl *vs.* DI water	75.82	0.683±0.025	0.340±0.039	12.41±0.47	**0.164±0.006**	578±16	Bui et al., ES&T 2013
Flat-sheet	PA - PES nanofiber	0.5M NaCl *vs.* DI water	24.38	1.65±0.14	N/A	33.9*	**1.390***	106±8	Song et al. Adv. Mater. 2011
	LbL membranes – PAN casting	0.5M $MgCl_2$ *vs.* DI water	36.95	1.99±0.13	N/A	~35	**0.947**	N/A	Qi *et al.* JMS 2012
	PA - PSf casting	1.0M NaCl *vs.* DI water	48.78	1.9±0.30	0.33±0.19	25.0±4.1	**0.513±0.086**	312±72	Tiraferri et al. JMS 2011
	PA – PSf casting	1.0M NaCl *vs.* DI water	48.78	1.63±0.18	0.84±0.19	20.5±3.8	**0.420±0.079**	389±150	Tiraferri et al. JMS-2011
	PA – PAN nanofiber	1.5M NaCl *vs.* DI water	75.82	2.036±0.949	1.572±1.161	29.3±3.6	**0.384±0.047**	291±53	Bui et al., ES&T 2013
	Zeolite-embedded PA – PSf casting	1M NaCl *vs.* DI water	49.23	0.873±0.185	N/A	17.5±2.0	**0.355±0.041**	N/A	Ma et al. JMS 2012
	PA - PAN/CA 8/2 nanofiber	1.5M NaCl *vs.* DI water	75.82	1.799±1.137	0.577±0.218	26.7±2.2	**0.350±0.029**	311±63	Bui et al., ES&T 2013
	Polydopamine-coated SWXLE	1.5M NaCl *vs.* DI water	72.94	2.498	N/A	~21.0±2.0	**0.288±0.027**	N/A	Arena et al., JMS 2011
	PA – PSf casting – PET nanofiber	1.0M NaCl *vs.* DI water	47.94	1.13±0.12	0.232±0.093	13.0±0.1	**0.271±0.003**	651±40	Hoover et al. Desal. 2012

	PA – PES/sulfonated PSf casting	2.0M NaCl *vs.* DI water	108.03	0.77*	0.11*	26*	**0.241***	238*	Wang et al. AIChE 2012
	PA – PSf casting	1.5M NaCl *vs.* DI water	75.82	1.15±0.06	N/A	18.2±1.0	**0.240±0.013**	492±38	Yip et al. ES&T 2010
	PA – PSf casting	2.0M NaCl *vs.* 10mM NaCl	107.53	1.78±0.23	0.20±0.03	25.1±4.0	**0.233±0.037**	670±170	Wei et al. JMS -2011
	PA / PESU-co-sPPSU casting	2.0M NaCl *vs.* DI water	108.03	0.73	0.25	21.0	**0.194**	324	Widjojo et al. JMS 2011
	PA - CAP-II casting	1.5M NaCl *vs.* DI water	75.82	1.42*	0.132*	14.5±2.5	**0.192±0.329**	695	Li et al. I&EC res.-2012
	PA – Nylon 6,6 casting of 3M	1.5M NaCl *vs.* DI water	72.20	0.917	0.300	~ 6.0±0.5	**0.083±0.007**	1940	Huang et al. JMS 2013
Hollow fiber	PA-PES	1M NaCl *vs.* DI water	49.23	3.45*	0.22*	~ 23*	**0.467***	N/A	Chou *et al.* Desal 2010
	PEI – Torlon® PAI	0.5M $MgCl_2$ *vs.* DI water	36.95	2.25*	0.113*	9.7*	**0.264***	N/A	Setiawan *et al.* JMS 2011
	Crosslinked PBI/*p*-xylylenedichloride	1M $MgCl_2$ *vs.* DI water	93.2	N/A	N/A	~ 9*	**0.097***	N/A	Wang *et al.* Chem.Eng.Sci.2009
	Dual-layer PBI-PES/PVP	5M $MgCl_2$ *vs.* DI water	1148.7	N/A	N/A	24.2*	**0.021***	N/A	Yang *et al.* ES&T 2009
	CA	2M $MgCl_2$ *vs.* DI water	258.3	0.47*	N/A	5.0*	**0.019***	N/A	Su *et al.* JMS 2010

* - Reproducible data from independent samples were not reported in these studies.

**- Osmotic pressure was obtained using modified Van't Hoff equation (for dilute solution) and reference [115] (for non-ideal solution).

Acronyms describing polymers used to make membranes in these studies: (CTA – cellulose triacetate, PA – polyamide, PES – polyethersulfone, PAN – polyacrylonitrile, PSf – polysulfone, CA – cellulose acetate, PET – polyester, PESU-co-sPPSU – copolymer of polyethersulfone and polyphenylsulfone , CAP – cellulose acetate propionate, PEI – polyethyleneimine, PAI – polyamide-imide, PBI – polybenzimidazole, PVP – polyvinylpyrrolidone, SWXLE – sea water reverse osmosis membrane made by Dow Chemicals).

osmotic driving force, $J_w/\Delta\pi_{theo}$. This effective osmotic permeance can be used as a standard parameter to evaluate the performances of different membrane structures and chemistry in forward osmosis. It can be seen that membrane with higher normalized water flux is desired to improve forward osmosis performance.

5.5 Process Design and Desalination Applications

While the academic community has been busy developing novel membranes and draw solutions, the desalination industry has largely kept out of the FO business. HTI has long been a leader in osmotic separations given their proprietary membrane design, but desalination was not the original intended use of that membrane. It was intended for use in an osmotic dilution process where a concentrated sugar-electrolyte drink was used as a draw solution to draw water out of a contaminated water source. The resulting diluted draw solution would be drinkable and contain calories and electrolytes. However, this membrane sparked interest in using FO for desalination and since its widespread use in academic research, HTI has developed new technologies in produced water treatment for the oil and gas industry and a landfill leachate concentration process [116]. While neither is a desalination process for producing drinking water per se, the technology has proven itself valuable for other saline water separation processes.

A number of small startup companies are commercializing FO technology. Two of the most promising are Oasys Water™ and Modern Water™. Oasys licenses the patents on the commonly discussed ammonia-carbon dioxide draw solution and has developed a pilot scale plant for concentrating oilfield produced water [117]. Modern Water advertises two operating seawater desalination plants that use FO with a third being commissioned at the time of this writing, but they offer few details as to the operation and cost of these systems [118]. Other small companies are also dabbling in the technology, but as of mid 2013, no large membrane company had developed FO technology that is ready for large scale commercialization.

5.6 Future Directions

Forward osmosis is a dynamic field that is undergoing a number of stresses that could ultimately see its demise, wild success, or somewhere in between. Many believe that FO will never be a viable large scale desalination option and will rather be relegated to niche separation applications. The academic community is quick to tout the benefits of this emerging technology, as they have found a new, relatively uncharted area of separation science to explore. While this has led to an explosion of publications in the area of osmotic separations, it has not pushed the technology closer to commercialization. The future of FO lies with the small companies pushing for its use in a number of critical applications such as in the oil and gas industry, wastewater reuse, or in hybridization with existing RO systems. As of early 2013, these companies are in the precarious position of having to meet investor targets while battling a deeply entrenched desalination industry with an already enormous capital investment in RO desalination capacity around the world. As they navigate the "valley of death" common in new technology commercialization, the FO community eagerly awaits one of the larger membrane companies to throw its hat in the ring and bring its resources and expertise to bear against the challenges that are preventing FO from reaching its full potential.

5.7 Acknowledgements

The authors acknowledge financial support from the National Science Foundation (CBET #1067564) and the US EPA (R834872).

References

1. V.S. Frenkel, Seawater Desalination: Trends and Technologies, *ISBN: 978-953-307-311-8*, in *Desalination: Trends and Technologies*, M. Schorr (Ed.), InTech, (2011)
2. J.E. Cadotte, Interfacially synthesized reverse osmosis membrane, *US Patent 4,277,344*, year 1981.
3. J.E. Cadotte, Reverse osmosis membrane, *U.S. patent documents* year 1977: U.S.
4. J.E. Cadotte, Reverse Osmosis Membrane, *US Patent 4,259,183*, year 1981.
5. M. Elimelech and W.A. Phillip, The Future of seawater desalination: Energy, technology, and the environment. *Sci.* **333**(6043), 712–717. (2011).
6. P.H. Jensen, J.S. Hansen, T. Vissing, M.E. Perry and C.H. Nielsen, Biomimetic Membranes Uses and Thereof, *US 2012/0080377 A1* (2012).
7. L. Dumée, J. Lee, K. Sears, B. Tardy, M. Duke and S.Gray, Fabrication of thin film composite poly(amide)-carbon-nanotube supported

membranes for enhanced performance in osmotically driven desalination systems. *J. Membr. Sci.* **427**, 422–430 (2013).
8. N.Ma, J. Wei, R. Liao and C.Y. Tang, Zeolite-polyamide thin film nanocomposite membranes: Towards enhanced performance for forward osmosis. *J. Membr. Sci.* **405–406**, 149–157 (2012).
9. B.M. Ganesh, A.M. Isloor and A.F. Ismail, Enhanced hydrophilicity and salt rejection study of graphene oxide-polysulfone mixed matrix membrane. *Desalination* **313**, 199–207 (2013).
10. L. Song, J.Y. Hu, S.L. Ong, W.J. Ng, M. Elimelech and M. Wilf, Performance limitations of the full-scale reverse osmosis process. *J. Membr. Sci* **214**, 239–244 (2003).
11. A. Zhu, P.D. Christofides and Y. Cohen, Effect of thermodynamic restriction on energy cost optimization of RO membrane water sesalination. *Ind. Eng. Chem. Res* **48**, 6010–6021 (2009).
12. L. Song, J.Y. Hu, S.L. Ong, W.J. Ng, M. Elimelech and M. Wilf, Emergence of thermodynamic restriction and its implications for full-scale reverse osmosis processes. *Desalination* **155**, 213–228 (2003).
13. T.Y. Cath, A.E. Childress and M. Elimelech, Forward osmosis: Principles, applications, and recent developments. *J. Membr. Sci.* **281**, 70–87 (2006).
14. G.W. Batchelder, Process for the demineralization of water, United States Patent Office US 3,171,799 (1965).
15. D.N. Glew, Process for liquid recovery and solution concentration, United States Patent Office, M. Dow Chemical Company, Mich. a corporation of Delaware, Editor US 3,216,930 (1965).
16. W.T. Hough, Process for extracting solvent from a solution, in United States Patent Office,US 3,532,621 (1970).
17. B.S. Frank, Desalination of sea water, United States Patent US 3,670,897 (1972).
18. J.O. Kessler and C.D. Moody, Drinking water from sea water by forward osmosis. *Desalination* **18**, 297–306 (1976).
19. R.L. McGinnis, Osmotic desalination process, United States Patent US 6,391,205 B1 (2002).
20. M.H. Weingarten, Power Generation Means, United States Patent US 3,587,227 (1971).
21. H.H.G. Jellinek, Osmosis Process for Producing Energy, United States Patent US 3,978,344 (1974).
22. S. Loeb, Method and apparatus for generating power utilizing pressure-retarded-osmosis, United States Patent US 3,906,250 (1975).
23. K. Stache, Apparatus for transforming sea water, brackish water, polluted water or the like into a nutritious drink by means of osmosis, United States Patent US 4,879,030 (1989).
24. J. Yaeli, Method and apparatus for processing liquid solutions of suspensions particularly useful in the desalination of saline water, United States Patent US 5,098,575 (1992).

25. R.E. Kravath and J.A. Davis, Desalination of seawater by direct osmosis. *Desalination* **16**, 151–155 (1975).
26. G.D. Mehta. and S. Loeb, Internal polarization in the porous substructure of a semi-permeable membrane under pressure-retarded osmosis. *J. Membr. Sci* **4**, 261 (1978).
27. G.D. Mehta. and S. Loeb, Performance of permasep B-9 and B-10 membranes in various osmotic regions and at high osmotic pressures. *J. Membr. Sci* **4**, 335–349 (1979).
28. S. Loeb, L. Titelman, E. Korngold and J. Freiman, Effect of porous support fabric on osmosis through a Loeb-Sourirajan type asymmetric membrane. *J. Membr. Sci* **129**, 243–249 (1997).
29. J.R. McCutcheon and M. Elimelech, Influence of concentrative and dilutive internal concentration polarization on flux behavior in forward osmosis. *J. Membr. Sci* **284**, 237–247 (2006).
30. J.R. McCutcheon and M. Elimelech, Influence of membrane support lauer hydrophobicity on water flux in osmotically driven membrane processes. *J. Membr. Sci* **318**, 458–466 (2008).
31. T.Y. Cath, N.T. Hancock, C.D. Lundin, C. Hoppe-Jones and J.E. Drewes, A multi-barrier osmotic dilution process for simultaneous desalination and purification of impaired water. *J. Membr. Sci* **362**, 417–426 (2010).
32. Z. Hanmin, M. Yanjie, J. Tao, Z. Guangyi and Y. Fenglin, Influence of activated sludge properties on flux behavior in osmosis membrane bioreactor (OMBR). J. Membr. Sci **390-391**, 270–276 (2012).
33. E.G. Beaudry and K.A. Lampi, Membrane technology for direct osmosis concentration of fruit juices. *Food Technol* **44**, 121 (1990).
34. T.Y. Cath, D. Adams and A.Y. Childress, Membrane contactor processes for wastewater reclamation in space. I. direct osmotic concentration as pretreatment for reverse osmosis. *J Membr Sci* **257**, 85–98 (2005).
35. S. Zou, Y. Gu, D. Xiao and C.Y. Tang, The role of physical and chemical parameters on forward osmosis membrane fouling during algae separation. *J Membr Sci* **366**, 356–362 (2011).
36. K.L Hickenbottom, N.T. Handcock, N.R. Hutchings,E.W. Appleton, E.G. Beaudry, P. Xu and T.Y. Cath, Forward osmosis treatment of drilling mud and fracturing wastewater from oil and gas separation. *Desalination* **312**, 60–66 (2013).
37. E.M. Garcia-Castello, J.R. McCutcheon and M. Elimelech, Performance evaluation of sucrose concentration using forward osmosis. *J. Membr. Sci* **338**, 61–66 (2009).
38. S. Phuntsho, H.K. Shon, S. Hong, S. Lee and S. Vigneswaran, A novel low energy fertilizer driven forward osmosis desalination for direct fertigation: Evaluating the performance of fertilizer draw solutions. *J. Membr. Sci* **375**, 172–181 (2011).

39. H. Zhu, L. Zhang, X. Wen and X. Huang, Feasibility of applying forward osmosis to the simultaneous thickening, digestion, and direct dewatering of waste activated sludge. *Bioresour. Technol* **113**, 207–213 (2012).
40. M. Xie, W.E. Price and L.D. Nghiem, Rejection of pharmaceutically active compounds by forward osmosis: Role of solution pH and membrane orientation. *Sep Purif Technol* **93**, 107–114 (2012).
41. X. Jin, J. Shan, C. Wang, J. Wei and C.Y. Tang, Rejection of pharmaceuticals by forward osmosis membranes. *J. Hazard. Mater* **227**, 55–61 (2012).
42. Holloway, R.W, A.E. Childress, K.E. Denneth and T.Y. Cath, Forward osmosis for concentration of anaerobic digester centrate. *Water Res* **41**, 4005–4014 (2007).
43. C.R. Martinetti, A.E. Childress and T.Y. Cath, High receovery of concentrated RO brines using forward osmosis and membrane distillation. *J. Membr. Sci* **331**, 31–39 (2009).
44. V. Yangali-Quintanilla, Z. Li, R. Valladares, Q. Li and G. Amy, Indirect desalination of Red Sea water with forward osmosis and low pressure reverse osmosis for water reuse. *Desalination* **280**, 160–166 (2011).
45. D.L Shaffer, N.Y.Yip, J. Gilron and M. Elimelech, Seawater desalination for agriculture by integrated forward and reverse osmosis: Improved product water quality for potentially less energy. *J. Membr. Sci* **415–416**, 1–8 (2012).
46. V. Sant'Anna, L.D.F. Marczak and I.C. Tessaro, Membrane concentration of liquid foods by forward osmosis: Process and quality view. *J. Food Eng* **111**, 483–489 (2012).
47. X. Jin, Q. She, X. Ang and C.Y.Tang, Removal of boron and arsenic by forward osmosis membrane: Influence of membrane orientation and organic fouling. J. Membr. Sci **389**, 182–187 (2012).
48. J.R. McCutcheon, R.L. McGinnis and M. Elimelech, A novel ammonia-carbon dioxide forward (direct) osmosis desalination process. Desalination **174**, 1–11 (2005).
49. Focus on forward osmosis, in *Water Desalination Report*, T. Pankratz (Ed.), Vol. 49, Issue 14, April 15, 2013, United States, www.desalination.com, Related websites - DesalData.com, IDA Desalination Yearbook, Global Water Intelligence, American Water Intelligence
50. N. Widjojo, T.-S. Chung and M. Weber, The role of sulphonated polymer and macrovoid-free structure in the support layer for thin-film composite (TFC) forward osmosis (FO) membranes. J. Membr. Sci **383**, 214–223 (2011).
51. R.C. Ong and T.-S. Chung, Fabrication and positron annihilation spectroscopy (PAS) characterization of cellulose triacetate membranes for forward osmosis. *J. Membr. Sci* **394–395**, 230–240 (2012).
52. N.-N. Bui, M.L. Lind, E.M.V. Hoek and J.R. McCutcheon, Electrospun nanofiber supported thin film composite membranes for engineered osmosis. *J. Membr. Sci* **385–386**, 10–19 (2011).

53. N.Y. Yip, A. Tiraferri, W.A. Phillp, J.D. Schiffman and M. Elimelech, High performance thin-film composite forward osmosis membrane. *Environ. Sci. Technol* **44**, 3812–3818 (2010).
54. A. Tiraferri, A, N.Y.Yip, W.A.Phillip, J.D. Schiffman and M. Elimelech, Relating performance of thin-film composite forward osmosis membranes to support layer formation and structure. J. Membr. Sci **367**, 340–352 (2011).
55. J.T. Arena, S.S. Manickam, K.K. Reimund, B.D. McCloskey, B.D. Freeman and J.R. McCutcheon, Surface modification of thin film composite membrane support layers with polydopamine: Enabling use of reverse osmosis membranes in pressure retarded osmosis. J. Membr. Sci **375**, 55–62 (2011).
56. N.-N. Bui and J.R. McCutcheon, Hydrophilic nanofibers as new supports for thin film composite membranes for engineered osmosis. *Environ. Sci. Technol* **47**, 1761–1769 (2013).
57. S. Qi, Q.C. Qiu, Y. Zhao and C.Y. Tang, Double-skinned forward osmosis membranes based on layer-by-layer assembly-FO performance and fouling behavior. J. Membr. Sci **405–406**, 20–29 (2012).
58. C. Qiu, S. Qi and C.Y. Tang, Synthesis of high flux forward osmosis membranes by chemically crosslinked layer-by-layer polyelectrolytes. *J. Membr. Sci* **381**, 74–80 (2011).
59. M. Sairam, E. Sereewatthanawut, K. Li, A. Bismarck and A.G. Livingston, Method for the preparation of cellulose acetate flat sheet composite membranes for forward osmosis - Desalination using $MgSO_4$ draw solution. *Desalination* **273**, 299–307 (2011).
60. J. Su, T.-S. Chung, B.J. Helmer and J.S. de Wit, Enhanced double-skinned FO membranes with inner dense layer for wastewater treatment and macromolecule recycle using Sucrose as draw solute. *J. Membr. Sci* **396,** 92–100 (2012).
61. K.Y. Wang and T.-S. Chung, Developing thin-film-composite forward osmosis membranes on the PES/SPSf substrate through interfacial polymerization. *AIChE J* **58**(3), 770–781 (2012).
62. K..Y. Wang, R.C. Ong and T.-S. Chung, Double-skinned forward osmosis membranes for reducing internal concentration polarization within the porous sublayer. *Ind. Eng. Chem. Res* **49**, 4824–4831 (2012).
63. J. Wei, C. Qui, C.Y. Tang, R. Wang and A.G. Fane, Synthesis and characterization of flat-sheet thin film composite forward osmosis membranes. *J. Membr. Sci* **372**, 292–302 (2011).
64. W. Fang, R. Wang, S. Chou, L. Setiawan and A.G. Fane, Composite forward osmosis hollow fiber membranes: Integration of RO- and NF-like selective layers to enhance membrane properties of anti-scaling and anti-internal concentration polarization. *J. Membr. Sci* **394–395**, 140–150 (2012).

65. L. Setiawan, R. Wang, K. Li and A.G. Fane, Fabrication of novel poly(amide-imide) forward osmosis hollow fiber membranes with a positively charged nanofiltration-like selective layer. J. Membr. Sci **369**, 196–205 (2011).
66. L. Setiawan, R. Wang, K. Li and A.G. Fane, Fabrication and characterization of forward osmosis hollow fiber membranes with antifouling NF-like selective layer. *J. Membr. Sci* **394–395**, 80–88 (2012).
67. L. Shi, S.R. Chou, R. Wang, W.X. Fang, C.Y. Tang and A.G. Fane, Effect of substrate structure on the performance of thin-film composite forward osmosis hollow fiber membranes. *J. Membr. Sci* **382**, 116–123 (2011).
68. K.Y. Wang, T.-S. Chung and J.-J. Qin, Polybenzimidazole (PBI) nanofiltration hollow fiber membranes applied in forward osmosis process. *J. Membr. Sci* **300**, 6–12 (2007).
69. K.Y. Wang, Q. Wang, T.-S. Chung and R. Rajagopalan, Enhanced forward osmosis from chemically modified polybenzimidazole (PBI) nanofiltration hollow fiber membranes with a thin wall. *Chem. Eng. Sci* **64**, 1577–1584 (2009).
70. R. Wang, L. Shi, C.Y. Tang, S. Chou, C. Qui and A.G. Fane, Characterization of novel forward osmosis hollow fiber membranes. *J. Membr. Sci* **355**, 158–167 (2010).
71. A. Achilli, T.Y. Cath and A.E. Childress, Selection of inorganic-based draw solutions for forward osmosis applications. *J. Membr. Sci* **364**(1–2), 233–241 (2010).
72. J.R. McCutcheon, R.L. McGinnis and M. Elimelech, The ammonia-carbon dioxide forward osmosis desalination process. *Water Cond. Purif* (2006), October 2006.
73. J.R. McCutcheon, R.L. McGinnis and M. Elimelech, Desalination by ammonia-carbon dioxide forward osmosis: Influence of draw and feed solution concentrations on process performance. *J. Membr. Sci* **278**, 114–123 (2006).
74. S.K. Yen, F.M. Haja N, M. Su, K.Y. Wang and T.-S. Chung, Study of draw solutes using 2-methylimidazole-based compounds in forward osmosis. J. *Membr. Sci* **364**, 242–252 (2010).
75. M.L. Stone, A.D. Wilson, M.H. Harrup and F.F. Stewart, An initial study of hexavalent phosphazene salts as draw solutes in forward osmosis. *Desalination* **312**, 130–136 (2013).
76. M.M Ling and T.-S. Chung, Desalination process using super hydrophilic nanoparticles via forward osmosis integrated with ultrafiltration regeneration. *Desalination* **278**, 194–202 (2011).
77. M.M. Ling and T.-S. Chung, Novel dual-stage FO system for sustainable protein enrichment using nanoparticles as intermediate draw solutes. J. Membr. Sci **372**, 201–209 (2011).

78. M.L. Stone, C. Rae, F.F. Stewart and A.D. Wilson, Switchable polarity solvents as draw solutes for forward osmosis. *Desalination* **312**, 124–129 (2013).
79. Q. Ge, J. Su, G.L. Amy and T.-S. Chung, Exploration of polyelectrolytes as draw solutes in forward osmosis processes. *Water Res* **46**, 1318–1326 (2012).
80. E.M. Garcia-Castello and J.R. McCutcheon, Dewatering press liquor derived from orange production by forward osmosis. J. Membr. Sci **372**, 97–101 (2011).
81. X. Jin, C.Y. Tang, Y. Gu, Q. She and S. Qi, Boric acid permeation in forward osmosis membrane processes: Modeling, experiments and implications. *Environ. Sci. Technol* **45**(6), 2323–2330 (2011).
82. M. Elimelech and R. McGinnis, Forward osmosis separation processes, *EP2303436 A2*, (2011).
83. S. Lee, C. Boo, M. Elimelech and S. Hong, Comparison of fouling behavior in forward osmosis (FO) and reverse osmosis (RO). *J. Membr. Sci* **365**, 34–39 (2010).
84. A. Achilli, T.Y. Cath, E.A. Marchland and A.E. Childress, The forward osmosis membrane bioreactor: A low fouling alternative to MBR processes. *Desalination* **239**, 10–21 (2009).
85. Y.-J. Choi, J.-S. Choi, H.-J. Oh, S. Lee, D.R. Yang and J.H. Kim, Toward a combined system of forward osmosis and reverse osmosis for seawater desalination. *Desalination* **247**, 239–246 (2009).
86. O.A. Bamaga, A. Yokochi, B. Zabara and A.S. Babaqi, Hybrid FO/RO desalination system: Preliminary assessment of osmotic energy recovery and designs of new FO membrane module configurations. *Desalination* **268**, 163–169 (2011).
87. M.M. Ling, T.-S. Chung and X. Lu, Facile synthesis of thermosensitive magnetic nanoparticles as "Smart" draw solutes in forward osmosis. *Chem. Comm* **47**, 10788–10790 (2011).
88. M.M. Ling and T.-S. Chung, Surface-dissociated nanoparticle draw solutions in forward osmosis and the regeneration in an integrated electric field and nanofiltration system. *Ind. Eng. Chem. Res* **51**, 15463–15471 (2012).
89. D. Li, X. Zhang, J. Yao, G.P. Simon and H. Wang, Stimuli-responsive polymer hydrogels as a new class of draw agent for forward osmosis desalination. *ChemComm* **47**, 1710–1712 (2011).
90. T.-W. Kim, Y. Kim, C. Yun, H. Jang, Y. Kim and S. Park, Systematic Approach for Draw Solute Selection and Optimal System Design for Forward Osmosis Desalination. *Desalination* **284**, 253–260 (2012).
91. J. R. McCutcheon and M. Elimelech, Influence of membrane support layer hydrophobicity on water flux in osmotically driven membrane processes. *J. Membr. Sci* **318**, 458–166 (2008).

92. N.Y. Yip, A. Tiraferri, W.A. Phillip, J.D. Schiffman and M. Elimelech, High performance thin-film composite forward osmosis membrane. *Environ. Sci. Technol* **44**, 3812–3818 (2010).
93. G.T. Gray, J.R. McCutcheon and M. Elimelech, Internal concentration polarization in forward osmosis: role of membrane orientation. *Desalination* **197**, 1–8 (2006).
94. M.F. Gruber, C.J. Johnson, C.Y. Tang, M.H. Jensen, L. Yde and C. Hélix-Nielsena, Computational fluid dynamics simulations of flow and concentration polarization in forward osmosis membrane systems. *J. Membr. Sci* **379**, 488–495 (2011).
95. L. Huang, N.-N. Bui, M.Y. Meyering, T.J. Hamlin and J.R. McCutcheon, Novel hydrophilic nylon 6,6 microfiltration membrane supported thin film composite membranes for engineered osmosis. *J. Membr. Sci* **437**, 141–149 (2013).
96. A. Nguyen, S. Azari and L. Zou, Coating zwitterionic amino acid $_L$-DOPA to increase fouling resistance of forward osmosis membrane. *Desalination* **312**, 82–87 (2013).
97. J. Wei, X. Liu, C. Qui, R. Wang and C.Y. Tanf, Influence of monomer concentration on the performance of polyamide-based thin film composite forward osmosis membranes. J. Membr. Sci **381**, 110–117 (2011).
98. L.A. Hoover, J.D. Schiffman and M. Elimelech, Nanofibers in thin-film composite membrane support layers: Enabling expanded application of forward osmosis and pressure retarded osmosis. *Desalination* **308**, 73–81 (2013).
99. S. Zhang, K.Y. Wang, T.-S. Chung, H. Chen, Y.C. Jean and G. Amy, Well-constructed cellulose acetate membranes for forward osmosis: Minimized internal concentration polarization with an ultra-thin selective layer. *J. Membr. Sci* **360**, 522–535 (2010).
100. R.C. Ong, T.-S. Chung, B.J. Helmer and J.S. de Wit, Novel cellulose esters for forward osmosis membranes. *Ind. Eng. Chem. Res* **51**, 16135–16145 (2012).
101. X. Li, K.Y.Wang, B. Helmer and T.-S. Chung, Thin-film composite membranes and formation mechanism of thin-film layers on hydrophilic cellulose acetate propionate substrates for forward osmosis processes. *Ind. Eng. Chem. Res* **51**, 10039–10050 (2012).
102. Y. Yu, S. Seo, I.-C.Kim and S. Lee, Nanoporous polyethersulfone (PES) membrane with enhanced flux applied in forward osmosis process. *J. Membr. Sci* **375**, 63–68 (2011).
103. C. Qiu, L. Setiawan, R. Wang, C.Y. Tang and A.G. Fane, High performance flat sheet forward osmosis membrane with an NF-like selective layer on a woven fabric embedded substrate. *Desalination* **287**, 266–270 (2012).
104. X. Song, Z. Liu and D.D. Sun, Nano gives the answer: Breaking the bottleneck of internal concentration polarization with a nanofiber

composite forward osmosis membrane for a high water production rate. *Adv. Mater* **23,** 3256–3260 (2011).

105. P.H. Jensen, J.S. Hansen, C.H. Nielson, M.E. Perry and T. Vissing, Biomimetic membranes uses and thereof, US 2012/0080377 A1. (2012)
106. N. Peng, T.-S. Chung, J.-Y.Lai, G.G. Lipscomb, P. Sukitpaneenit and M.M. Teoh, Evolution of polymeric hollow fibers as sustainable technologies: Past, present, and future. *Prog. Polym. Sci* **37**, 1401 (2012).
107. Q. Yang, K.Y. Wang and T.-S. Chung, A novel dual-layer forward osmosis membrane for protein enrichment and concentration. *Sep. Purif. Technol* **69**, 269–274 (2009).
108. K.Y. Wang, M.M. Teoh, A. Nugroho, T.-S.Chung, Integrated forward osmosis-membrane distillation (FO-MD) hybrid system for the concentration of protein solutions. *Chem. Eng. Sci* **66**, 2421–2430(2011).
109. X. Li, K.Y. Wang, B. Helmer and T.-S. Chung, Thin-film composite membranes and formation mechanism of thin-film layers on hydrophilic cellulose acetate propionate substrates for forward osmosis processes. *Ind. Eng. Chem. Res* **51**, 10039–10050 (2012).
110. S. Chou, R. Wang, L. Shi, Q. She, C. Tang and A.G. Fane, Thin-film composite hollow fiber membranes for pressure retarded osmosis (PRO) process with high power density. *J. Membr. Sci* **389**, 25–33 (2012).
111. S. Chou, L. shi, R. Wang, C.Y. Tang, C. Qui and A.G. Fane, Characteristics and potential applications of a novel forward osmosis hollow fiber membrane. *Desalination* **261**, 365–372 (2010).
112. J. Su, Q. Yang, J.F. Teo and T.-S. Chung, Cellulose acetate nanofiltration hollow fiber membranes for forward osmosis processes. J. Membr. Sci **355**, 36–44 (2010).
113. Q. Yang, K.Y. Wang and T.-S. Chung, Dual-layer hollow fibers with enhanced flux as novel forward osmosis membranes for water production. *Environ. Sci. Technol* **43**, 2800–2805 (2009).
114. C.H. Tan and H.Y. Ng, Modified models to predict flux behavior in forward osmosis in consideration of external and internal concentration polarizations. *J. Membr. Sci* **324**, 209–219 (2008).
115. Hydration technology innovations, http://www.htiwater.com/
116. *FO Process Concentrates Oilfield Brine*, in *Water Desalination Report*October (2012).
117. Modern water, http://www.modernwater.co.uk/.

6

Electrodialysis Desalination

Hong-Joo Lee[a], Seung-Hyeon Moon[b]

[a]*Chonnam National University, Gwangju, Korea*
[b]*Gwangju Institute of Science and Technology, Gwangju, Korea*

Abstract
Desalination processes involve the separation of salt-free fresh water from seawater or brackish water, which is classified as thermal or membrane processes. As one of membrane desalination process where the electrically charged ions are separated through ion exchange membranes, electrodialysis (ED) has been considered as one of the promising salt removal processes in many areas, such as desalination from brackish water, demineralization of wastewater, and purification of boiler feed water. ED has been exploited successfully for decades as an efficient desalination process, and its technical and economical feasibility has been established especially for salt production from seawater and desalination of brackish water.

Keywords: *Desalination; Electrodialysis; Ion exchange membrane; Operating condition; Process design; Process integration; Fouling*

6.1 Principles of Electrodialysis

The 21st century has been dubbed called the "century of water" because of the coming water crisis due to the global population increase and environmental destruction. Fresh water from rivers, lakes, and groundwater totals only 0.01% of the planet's total water resources [1]. Therefore, efficient technologies for removing salts need to be developed to produce affordable drinking water from seawater or brackish water, which comprises 98% of Earth's water.

Desalination processes involve the separation of nearly salt-free fresh water from seawater or brackish water, where the salts are

Jane Kucera (ed.) Desalination , (287–326) 2014 © Scrivener Publishing LLC

concentrated in the rejected brine stream and can be classified with thermal or membrane processes [2]. Thermal separation processes include two main categories: evaporation followed by condensation, and freezing followed by melting of the formed water ice crystals. Thermal separations include multi-stage flash distillation (MSF), multiple-effect distillation (MED), and mechanical vapor compression (MVC). Desalination membrane processes include electrodialysis (ED), reverse osmosis (RO), nanofiltration (NF), forward osmosis (FO), and membrane distillation (MD). Table 6.1 compares advantages and disadvantages for the various desalination membrane processes [3].

ED is an ion exchange membrane process that uses electrical potential as a driving force. Its system typically consists of a cell arrangement with a series of alternating anion and cation exchange membranes between an anode and a cathode to form individual cells having a volume with two adjacent membranes. The cell pair is a repeating unit in an ED stack, a device composed of individual cells in alternating series with electrodes on both ends.

When an aqueous salt solution is circulated in the cell under an electrical potential, the positively charged cations migrate through a cation exchange membrane toward the cathode and the negatively charged anions through an anion exchange membrane toward the anode. The overall result is a potential drop across the cell pairs as well as a change in the ion concentration in alternate compartments. ED consists of compartments for the depleted solution (diluate) and concentrate solution (concentrate), and two contiguous anion and cation exchange membranes in the stack. Figure 6.1 shows a schematic diagram of a typical arrangement in the ED process.

ED was first developed for the desalination of brackish water to produce potable water and operated in large-scale applications for water desalination. A large-scale application of ED is the preconcentration of seawater for the production of table salt. In Japan, after ED concentrates the salt in seawater to about 18–20% solids the brine is further concentrated by evaporation and the salt is recovered by crystallization. ED has become a mature technology due to the development of process and ion exchange membranes. It is mainly installed in small-to-medium size plants with capacities of less than several 100 m^3/d to over 20,000 m^3/d, with a brackish water salinity of 1,000 to 5,000 mg/L total dissolved solids (TDS) [4].

Among desalination membranes processes, RO consists of the largest portion in the membrane desalination processes, which

Table 6.1 Advantages and disadvantages of membrane desalination processes [3]

Desalination type	Principles	Usage	Advantages	Disadvantages
Electrodialysis(ED)/ Electrodialysis reversal (EDR)	Electrochemical separation process that removes ions and other charged species from water and other fluids.	Widely used, since early 1960s	Long membrane lifetime and high efficiency (up to 94% water recovery, usually around 80%)	High capital and operational costs
Reverse osmosis (RO)	Pressure-driven separation process through a membrane, which retains the solute on one side and allows the pure solvent to pass to the other side.	Widely used, first plant installed in Saudi Arabia in 1979	In water purification, effectively removes all types of contaminants to some extent	Requires more pretreatment of the seawater and more maintenance than MSF plants
Nanofiltration (NF)	Pressure-driven separation process using membranes having a pore size in the order of nanometers	Emerging technology	Very high efficiency	High capital cost, unknown lifetime of membrane, no large-scale plant built yet
Forward osmosis (FO)	Separation process using osmotic process through a semi-permeable membrane to effect separation of water from dissolved solutes.	Emerging technology	Low or no hydraulic pressures and no energy needed for separation	Cannot produce pure water, only concentrated solutions
Membrane distillation (MD)	Separation process using difference in vapor pressure of water across the membrane as driving force for desalination	Widely used	simple operation, low fouling	high energy consumption, low flux

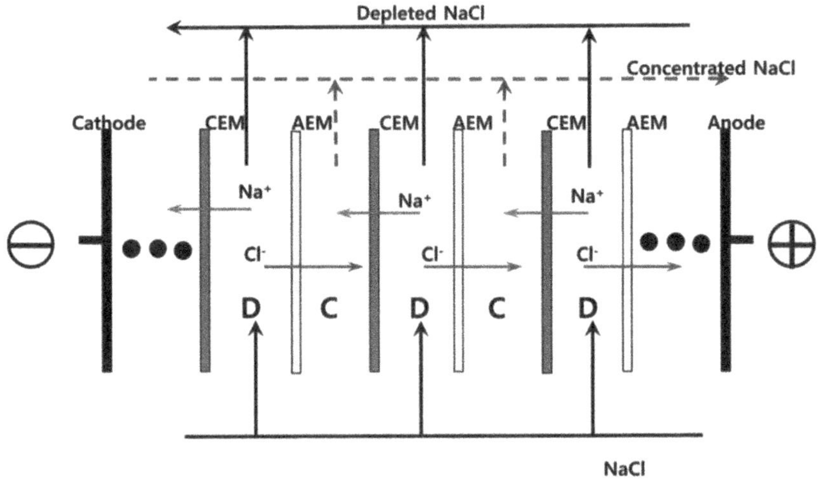

Figure 6.1 Schematic diagram of the typical arrangement in an ED process (CEM: Cation exchange membrane, AEM: Anion exchange membrane, D: Diluate compartment, C: Concentrate compartment).

drives water through the membrane by using the mechanical energy (or the pressure difference across the membrane). In RO, fresh water permeates under high pressure through semi-permeable membranes leaving behind highly concentrated brine solution. Generally, RO is considered to have an economic advantage for the desalination of water with TDS in excess of 10,000 mg/L [5].

In fact, ED is competing directly with RO in the brackish water desalination. In a certain range of feed water salt composition, however, ED has an economic advantage over other desalination processes. The advantages of ED compared to RO include high water recovery rates, long operation life of membranes due to higher chemical and mechanical stability, less membrane fouling or scale, less raw water pretreatment, operation at elevated temperatures, and flexible nature of the process to varied feed water quality [6, 7].

6.2 Preparation and Characterization of Ion Exchange Membranes

6.2.1 Preparation of Ion Exchange Membranes

ED has been motivated by the development of new ion exchange membranes with better selectivity, lower electrical resistance, and improved thermal, chemical, and mechanical properties. Its applications have recently gained a broader interest, especially in the areas of food, drug, chemical process, and biotechnology [8]. The technical feasibility of ED as a separation process is mainly determined by the ion exchange membranes used in the system. The most desired properties of ion exchange membranes are high permselectivity, low electrical resistance, good mechanical stability, and high chemical and thermal stability [9].

Ion exchange membranes are classified into anion and cation exchange membranes depending on the types of ionic or charged groups attached to the membrane matrix. The fixed charged groups partially or completely exclude ions of the same charge (co-ions) from the membrane. An anion exchange membrane with fixed positively charged groups excludes positive ions but is freely permeable to negatively charged ions, referred to as counter-ions. It has positively charged groups, *i.e.*, $-NH_3^+$, $-NRH_2^+$, $-NR_2H^+$, $-NR_3^+$, $-PR_3^+$, $-SR_2^+$, etc., fixed to the membrane structure [10]. A cation exchange membrane with fixed negatively charged groups, containing negatively charged groups, i.e., $-SO_3^-$, $-COO^-$, $-PO_3^{2-}$, $-PO_3H^-$, $-C_6H_4O^-$, etc., fixed to the membrane structure, is freely permeable to positively charged ions. Figure 6.2 shows a schematic diagram of the structure of an anion exchange membrane structure.

According to the preparation method, ion exchange membranes can be divided into homogeneous and heterogeneous membranes. Homogeneous membranes are prepared either by

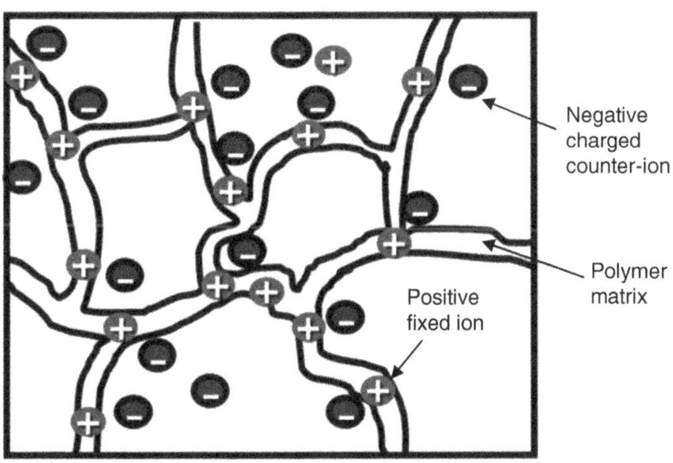

Figure 6.2 Schematic diagram of the structure of an anion exchange membrane structure.

polymerization of monomers having functional groups or by additional functionalization of a polymer film consisting of structures such as inter-polymers, graft-, block-, and co-polymers. Various approaches to prepare homogeneous ion exchange membranes are available by the introduction of ionic groups. These approaches can be classified into three categories based on the starting materials: (a) with a monomer containing a moiety of ionic exchange groups, (b) with a polymer film modified by introducing ionic characters either directly by grafting of a functional monomer or indirectly by grafting non-functional monomer followed by the functionalization reaction, and (c) with polymer blends by introducing ionic moieties, followed by the dissolving of polymer and casting it into a film [11–13].

In previous studies, many different methods have been reported for the preparation of ion exchange membranes with proper properties for special applications. For example, monovalent ion permselective membranes were developed to produce table salt from seawater by concentrating monovalent ions (e.g., Na^+ and Cl^-). A fluorocarbon-type membrane with excellent chemical and thermal stability has been developed for the application in environment of oxidizing agents and elevated temperature. Anion exchange membranes with high proton retention and low proton permeability have been commercialized. In addition, significant efforts have especially concentrated on the development of anion exchange membranes with low fouling tendencies because they are more sensitive to fouling.

Generally, heterogeneous membranes consist of fine colloidal ion exchange resins embedded an inert binder, such as polyethylene, phenolic resins, or polyvinylchloride. They can be produced by calendering ion exchange resins into an inert plastic film. Another procedure is the dry molding of inert film-forming polymers and ion exchange resins, followed by the milling of the mold stock. Ion exchange resins are dispersed in a solution containing a film-forming binder, and then the solvent is evaporated to form ion exchange membranes. Also, heterogeneous membranes can be prepared by dispersion of ion exchange resins in the solution containing a partially polymerized binder polymer [10].

Apart from homogeneous or heterogeneous membranes, ion exchange membranes can also be prepared from inorganic material, such as zeolites, bentonite or phosphate salts. Recently, interests in inorganic-organic composite materials have increased due to their extraordinary properties for special applications. Various membranes including hybrid inorganic-organic ion exchange membranes, amphoteric ion exchange membranes, and mosaic ion exchange membranes have been developed until now [12].

6.2.2 Characterization of Ion Exchange Membranes

The characteristic properties of ion exchange membranes are determined by different parameters, such as the density of the polymer network, the hydrophobic or hydrophilic character of the matrix polymer, the type and concentration of the fixed charges in the polymer, and the morphology of the membrane itself. Mainly, the properties of ion exchange membranes are determined by the basic polymer matrix and the nature and concentration of fixed ionic groups [14, 15]. The basic polymer matrix affects the mechanical, chemical, and thermal stability of the membrane to a large extent. Meanwhile, the concentration of fixed ionic groups determines the permselectivity and electrical resistance of the membrane as well as the electrochemical properties. Microscopic examination gives information on the characteristic values of the membrane properties, which is closely related to the preparation methods of membrane and the improvement of the process efficiency in ion exchange membrane processes.

Ion exchange membranes used in the ED process can be characterized in terms of electrical resistance in different electrolyte solutions, permselectivity of the membrane for different ions of the opposite and same polarity, mechanical properties, ion exchange capacity, transport number of co-ions and counter-ions, water swelling, transport rate of neutral components (especially water), chemical stability in the presence of oxidizing agents and at extreme pH values, hydrophobicity of the surface, diffusion coefficient, and the type and density of fixed charges and their distribution in the membrane matrix [16].

The electric resistance of an ion exchange membrane is one of the factors determining the energy requirements of ED process, which depends on the ion exchange capacity and the mobility of ions within membrane matrix. Electrical resistance is determined by the conductivity measurement in a cell consisting of two well-stirred chambers separated by examined membrane using the alternating current. Practically, the area resistance can be calculated from the difference between the combined area resistance of solution and membrane and the area resistance for just the solution. In recent years, electrical impedance spectroscopy has been used as a reliable method of measuring electrical resistance with alternating current. The frequency of the alternating current is changed and the response to the changing frequency is determined by the impedance spectroscopy [16].

The transport number of an ion exchange membrane is related to the transport of electric charges by specific counter-ions to the total transport of electrical charge through an ion exchange membrane. The degree of the transport number depends on the concentration of electrolyte in the membrane and is related to on the exchange capacity and the cross-linking density. When a membrane separates diluate and concentrate, a concentration gradient is created across the membrane. The apparent transport number of the counter-ion in the membranes is determined by measurement of membrane potential using Ag/AgCl electrodes. The transport number, t_m, for each membrane is calculated by the equation:

$$E_m = \frac{R'T}{F}(2t_m - 1)\ln\frac{a_1}{a_2} \tag{6.1}$$

where E_m is the concentration cell potential, a_1 and a_2 the electrolyte activity in the cells, F the Faraday constant, and R' and T is the gas constant and temperature, respectively. The permselectivity of an ion exchange membrane relates the transport of electric charge by specific counter-ions to the total transport of electrical charge through the membrane. The degree of permselectivity depends on the concentration of electrolytes in the membrane, ion exchange capacity, and the cross-linking density.

The exchange capacity of ion exchange membrane is one of the important characteristic properties related to the functional groups in the membrane structure. It can be determined by titration of the fixed ions of ion exchange membranes; e.g., $–SO_3^-$ or $–R_4N^+$ groups with NaOH or HCl, respectively. The characterized membrane is soaked for 24 hours in 0.1 N NaOH or HCl solution. After the surface water is removed, remaining NaOH or HCl concentration is measured by titration with HCl or NaOH solution. Ion exchange capacity of each membrane is expressed as meq/g-dried membrane based on the consumed H^+ or OH^- concentration from anion or cation exchange membrane, respectively. In general, the ion capacity of the fixed ion in commercial membranes is in the range between 1 and 3 meq/g [10].

In addition, a degree of the hydrophobicity is measured by the contact angle of each membrane using a contact angle meter after an examined membrane is dried. A higher contact angle indicates higher hydrophobicity. Furthermore, electrochemical properties of ion exchange membrane can be characterized in terms of electrokinetic properties and electrochemical properties including current-voltage relationship, chronopotentiometry, which are described in detail elsewhere [16].

Mechanical properties of an ion exchange membrane are affected by the thickness, the swelling capacity, the dimensional stability, the tensile strength, and the hydraulic permeability. The tensile strength and the information concerning the plastic or elastic deformation of a membrane are obtained from a stress versus strain diagram. The swelling capacity, which represents the total water uptake of the membrane in equilibrium with an electrolyte solution, can be determined by measuring the weight difference between a membrane in the wet and dried state. **Tables 6.2 and 6.3** list the properties of commercially available homogeneous and heterogeneous ion exchange membranes, respectively.

6.2.3 Concentration Polarization and the Limiting Current Density

Generally, a concentration polarization takes place at the surface of an ion exchange membrane operating with a certain applied current density and a certain flow velocity of the solutions across the membrane surface at both sides. Even at high turbulence conditions in the bulk solution, the concentration polarization occurs on the ion exchange membrane surface because of a laminar boundary layer existing at the surface of the membrane. It is generally accepted that the concentration polarization decreases the process efficiency mainly due to increase in electrical resistance and the energy consumption, which is accepted as one of unavoidable phenomena in all membrane separation processes. The magnitude of the concentration polarization in an ED process is a function of various parameters such as the applied current density, the feed flow velocity parallel to the membrane surface, the cell design, and the membrane properties [17, 18].

Table 6.2 Characteristic properties of commercial homogeneous ion exchange membranes
(a) Cation exchange membranes [5, 10,12]

Membrane	Company	Electrics resistance (Ωcm^2)	Transport number	Water content (%)	Thickness (mm)	Ion exchange capacity (meq/g)
Aciplex K-192	Astom	1.5-2.0	0.99	35-40	0.13-0.17	-
Aciplex K-501	Astom	2-3	0.99	30-40	0.18-0.20	-
Aciplex K-521	Astom	2-3	0.99	25-30	0.14-0.18	-
Selemion CMV	Asahi Glass	2.0-3.5			0.13-0.15	2.4
Fumasep FKS	Fuma Tech	8	0.96		0.11-0.13	1
Fumasep FKB	Fuma Tech	4	0.98		0.08-0.1	0.9-1.0
Fumasep FK-40	Fuma Tech	1			0.035-0.045	1.2
Fumasep FKD	Fuma Tech	1			0.04-0.06	1.0
Fumasep FKE	Fuma Tech	3	0.98		0.05-0.07	1.0
Fumasep FKL	Fuma Tech	4	0.92		0.11-0.12	-

Nepton CR61-CZL	Ionic	6-9	0.95	40-45	0.6-0.7	2.2-2.5
Nepton CR67-HMR	Ionic	7-11			0.53-0.65	2.1-2.5
Neosepta CM-1	Astom	2.2-3.0			0.15-0.17	2.0-2.5
Neosepta CM-2	Astom	1.5-1.8	0.98	35-40	0.15-0.17	2.0-2.5
Neosepta CMX	Astom	2.5-3.5	0.98	25-35	0.17-0.2	1.5-1.8
Neosepta CMS	Astom	1.5-3.5	0.98	35-40	0.14-0.17	2.0-2.5
Neosepta CMB	Astom	4.0-5.0			0.18-0.24	-
R 1010	Pall RAI	2-4			0.1	1.2
R 5010-L	Pall RAI	2-4	0.90	40-45	0.2-0.3	1.5
R 5010-H	Pall RAI	8-10	0.95	20-25	0.2-0.3	0.9
Morgane CDS	Solvay	1.1-1.8			0.16	-
Morgane CRA	Solvay	1.3-3			0.16	-

(Continued)

Table 6.2 *(Cont.)*
(b) Anion exchange membranes [5, 10,12]

Membrane	Company	Electrics resistance (Ωcm^2)	Transport number	Water content (%)	Thickness (mm)	Ion exchange capacity (meq/g)
Aciplex A-192	Astom	1.8-2.0	0.99	25-35	0.12-0.15	-
Aciplex A-201	Astom	3.6-4.2	0.99	25-28	0.20-0.26	-
Aciplex A-221	Astom	1.4-1.7	0.99	24-30	0.17-0.2	-
Aciplex A-700	Astom	3.5-5.5	0.99	13-18	0.10-0.13	-
Selemion AMV	Asahi Glass	1.5-3.0			0.11-0.15	1.9
Selemion ASV	Asahi Glass	2.3-3.5			0.11-0.15	2.1
Fumasep FAS	Fuma Tech	3	0.95		0.13-0.14	1.0
Fumasep FAB	Fuma Tech	1	0.96		0.10-0.13	1.3
Fumasep FAN	Fuma Tech	2-4			0.09-0.11	0.8
Fumasep FAA	Fuma Tech	2	0.92		0.13-0.15	1.2
Fumasep FAD	Fuma Tech	0.8	0.91		0.08-0.1	1.5
Fumasep FAP	Fuma Tech	1.5	0.92		0.09-0.1	-
AR103QDR	Ionic	14.5			0.56-0.69	1.95-2.20
AR204SZRA	Ionic	6.2-9.3			0.48-0.66	2.3-2.7
AR112-B	Ionic	20-28			0.48-0.66	1.3-1.8

Neosepta AM-1	Astom	1.3-1.8	0.98	25-35	0.12-0.16	1.8-2.2
Neosepta AM-3	Astom	2.0-3.0	0.98	15-25	0.12-0.16	1.8-2.2
Neosepta AMX	Astom	2.0-3.5	0.98	25-35	0.16-0.18	1.4-1.7
Neosepta ACS	Astom	1.5-2.0	0.98	20-30	0.10-0.12	1.4-2.0
Neosepta AFN	Astom	0.4-1.5			0.15-0.20	2.0-3.5
Neosepta AHA	Astom	3.0-3.5			0.18-0.24	-
Neosepta AFX	Astom	0.7-1.5			0.14-0.17	1.5-2.0
R 1030	Pall RAI	0.1			0.04	1.0
R 4030	Pall RAI	1.7			0.07	-
R 5030-L	Pall RAI	4-7	0.85	30-35	0.2-0.3	-
R 5030-H	Pall RAI	11-15	0.90	20.25	0.2-0.3	-
ADP	Solvay	1.8-3.5			0.16	
PC 100D	PCA	5			0.08-0.10	1.2
PC 200D	PCA	2			0.08-0.10	1.3
PC 400D	PCA	10			0.08-0.10	-
PC Acid 35	PCA	-			0.08-0.10	1.0
PC Acid 70	PCA	-			0.08-0.10	1.1
PC Acid 100	PCA	-			0.08-0.10	0.57

Table 6.3 Characteristic properties of commercial heterogeneous ion exchange membranes [5, 10]

Type	Company	Electrics resistance (Ωcm^2)	Transport number	Water content (%)	Thickness (mm)	Ion exchange capacity (meq/g)
Fumasep FTCM (Cation)	Fuma Tech	10	0.95		0.5-0.6	2.2
Fumasep FTAM (Anion)	Fuma Tech	8	0.92		0.5-0.6	1.7
LinAn C (Cation)	LinAn	15	0.90	40-45	0.42	2.0
LinAn A (Anion)	LinAn	15	0.88	30-45	0.42	1.8
Ionac MC 3470 (Cation)	Sybron	10	0.96	25-35	0.38	1.4
Ionac MA 3475 (Anion)	Sybron	25	0.99	35-40	0.41	0.9
Ionac MA 7500 (Anion)	Sybron	10	0.99	25-35	0.46	1.1
Permaplex C 20 (Cation)	Permutit			30-40	0.8	3.0
Permaplex A 20 (Anion)	Permutit			30-40	0.8	2.0
CMI-7000 (Cation)	Membrane International	30	0.94		0.45	1.6
AMI-7001 (Anion)	Membrane International	40	0.90		0.5	1.3

The transport of charged species to the anode or cathode through a set of ion exchange membranes leads to a concentration decrease of counter-ions in the laminar boundary layer at the membrane surface facing the diluate cell and an increase at the surface facing the concentrate cell. When the ion concentration at the surface of cation and/or anion exchange membrane in the diluate becomes nearly zero, the current density will approach the maximum value in the process, as shown in Figure 6.3.

In general, an ED process shows higher electrical resistance or lower current utilization when operated at above LCD. When the LCD is exceeded in an ED operation, the electrical resistance in the diluate drastically increases because of the depletion of ions in the laminar boundary layer at the membrane surface. Therefore, LCD should be considered as one of the significant designing parameters since it determines the efficiency of process in designing of an ED plant [19].

The LCD is affected by variable parameters such as solution properties, diluate concentration, and the hydrodynamics of the solution flow related to the flow channel geometry, the spacer design, the flow velocity, as well as membrane properties. Therefore, the LCD, i_{lim}, can be expressed as the relationship between solution properties, as follows:

$$i_{\text{lim}} = aC_s u^b \tag{6.2}$$

$$\ln\left(\frac{i_{\text{lim}}}{C_s}\right) = \ln a + b \ln u \tag{6.3}$$

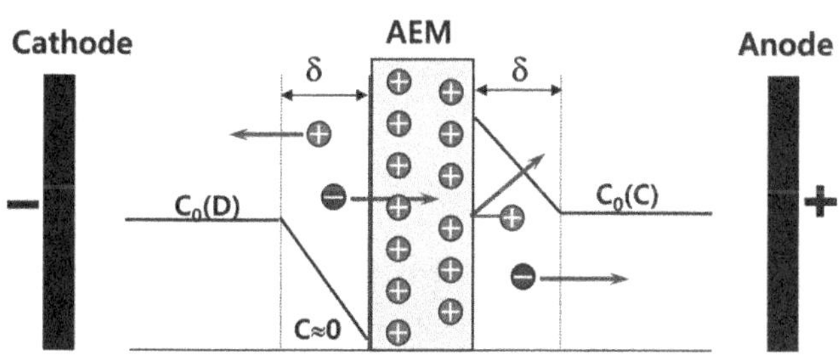

Figure 6.3 Concentration polarization occurring on the surface of an anion exchange membrane (Co(D): Concentration in the bulk solution of diluate compartment, Co(C): Concentration in the bulk solution of concentrate compartment).

where a and b are constants estimated by measurements, C_s the concentration of the solution, and u the linear flow velocity.

For a certain electrolyte, the respective LCD value can be determined by measuring the current as a function of the applied voltage across an ion exchange membrane. The slope of the curve in the current-voltage is inversely proportional to the electrical resistance of the solutions in the boundary layers at the membrane surfaces. The reason for the drastic increase in the boundary layer resistance is the complete depletion of the salt at the membrane surface facing the diluate solution [20]. The LCD value is determined at the point of the notable increase in the boundary layer resistance, as illustrated in Figure 6.4.

In addition, the LCD value can be determined by measuring the total resistance of a cell pair and the pH value in the diluate cell as a function of the current density, i. When the total resistance of a cell pair is plotted versus $1/i$, a minimum is determined to the LCD value. Also, the pH change is considered for the LCD determination since the pH change drastically occurs on the membrane surface at

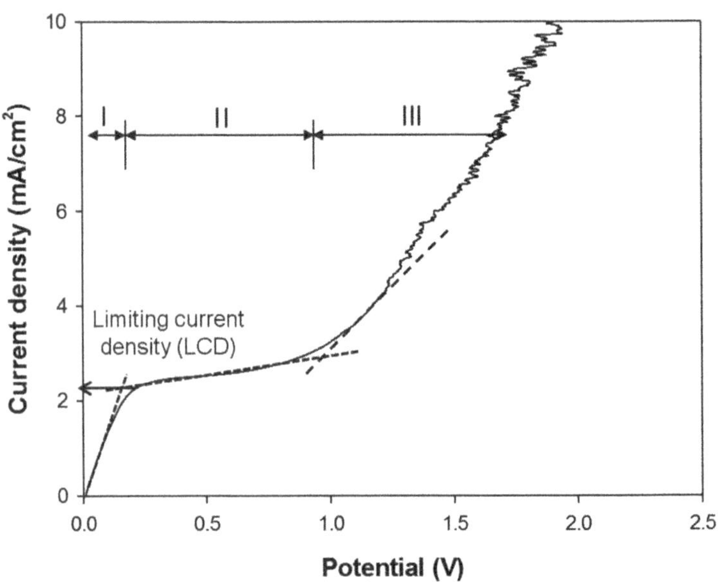

Figure 6.4 Determination of the limiting current density in a graph between supplied potential and corresponding current. (I: Electrical resistance increased due to Ohm's law, II: Boundary layer resistance increase, III: Over-limiting current density).

above LCD due to the water dissociation. A sharp decrease in the pH value is noted in the plot of the pH value against *1/i* when the LCD is exceeded.

The optimum operating current density in an ED process depends to a very large extent on the equipment and, especially, the membrane cost and membrane life. In practical, the operating current density is about 80% of the LCD, even though the current density to be applied in ED is determined by economical considerations [5].

6.3 ED Equipment Design and Desalination Process

6.3.1 ED Stack Design

Successful design of the ED process relies on several crucial parameters, such as the characteristics of solutions (feed, product (or diluate), and concentrate), desired process flow diagram, materials of construction, pretreatment of feed, pump and pressure drops, storage tanks, power supply, controls, limiting current density, current leakage, back-diffusion, solution properties related to scale formation such as Langelier index and calcium sulfate saturation, and membrane stack design [5].

In general, a trade-off occurs between thickness and membrane physical properties. A thicker membrane usually has a greater strength, high mechanical strength, and longer life time, while a thinner membrane has lower electrical resistance and reduced electrical energy requirements. Typically, the thickness of commercially available membranes is about 0.1–0.6 mm [21].

In designing an ED stack, various process parameters should be controlled, which include the feed solution concentration and the desired product and brine concentrations. A proper stack design should provide a maximum effective membrane area per unit stack volume and ensure equal and uniform flow distribution in each compartment. Any leakage between the diluate, concentrate, and electrode cells should be prevented. The distance between the membrane sheets (*i.e.*, cell thickness) should be as small as possible to keep the energy consumption due to electrical resistance of the solution [10].

Gaskets around the perimeter of the cell confine the solutions within cells, which not only separate and seal membranes but also contain manifolds to distribute the process fluids in the different compartments. The supply ducts for the diluate and the concentrate are formed by matching holes in the gaskets, the membranes, and the electrode cells. Ion exchange membranes are separated by a spacer screen that supports the membranes and facilitates turbulent mixing of the solutions.

A spacer separates the membrane and provides a pathway in the cell for the water flow. Cells are made up of two membranes with a spacer in between and stacked with alternating concentrate and diluate compartments to form an ED stage. It should provide a maximum of mixing of the solutions at the membrane surfaces and should cause a minimum in pressure loss [19, 20]. In the effect of the spacer on the current density and cell volume, the so-called shadow effect of the spacer can be estimated by considering the spacer construction or be determined experimentally by comparing the operation of a cell pair with and without a spacer. Generally, the effect of the spacer on the current density can be estimated within an error range of about 5% [10].

Typically, there are two basic types of spacers, a sheet flow stack and tortuous path flow stack, according to different types of spacer in an ED stack. A sheet flow stack with manifolds distributes the flow streams, where the solution flows in a straight path from the entrance to the exit ports located on the opposite sides in the gasket. Meanwhile, the membrane spacer gaskets in the tortuous path flow stack have a long serpentine cutout which defines a long narrow channel for the fluid path. Compared with the sheet flow design, the tortuous path flow design can operate a substantially higher current density in the stack due to high flow velocities in the channels. Operation range of flow velocities in tortuous path flow velocities are 15–30 cm/s, while those in sheet-flow stacks are typically 3–10 cm/s [5].

A pair of electrodes is required for each electrical stage in an ED stack. Electrodes are generally constructed of niobium or titanium with a platinum coating or graphite, stainless steel, etc. Hydrogen ions and oxygen and/or chlorine gas are formed at the anode surface, while hydrogen gas and hydroxyl ions are formed at the cathode surface. A typical voltage drop across a single cell pair is in the range of 1–2 V and the normal current flow is 40 mA/cm^2. For a 200 cell pair stack containing 1 m^2 of membrane, the total voltage is about 200–400 V and the current about 400 A per stack [4]. In most cases, the electrode streams are kept apart and the precipitation of calcium or magnesium salts and the corrosion of electrode surface should be prevented. Normally Na_2SO_4 is used as an electrode rinse solution.

6.3.2 ED Process Design

The degree of desalination is a function of the solution concentration, the applied current density, and the residence time of the solution in the stack. Equipment and process design requirements are quite different according to proper ED applications. In general, operating variables are classified into independent and dependent variables. Independent operating variables include feed concentration of diluate and concentrate streams, electrode stream, operating temperature, and flow rate. Meanwhile, current density, product stream concentration, membrane area, operating voltage, pumping power, and membrane life are examples of dependent variables in the ED process design.

Depending on the feed solution composition and the product requirements, ED can be operated in batch, continuous, feed-and-bleed mode with partial recycling of the diluate and concentrate streams. The flow streams through a stack can be counter- or co-current. The velocities of the flow streams determine the required membrane area for a given capacity desalination plant and the degree of desalination and concentration, and affect the concentration polarization and LCD value [10].

In the batch type ED system, the saline or brackish feed water is recirculated from a feed tank through the demineralization spacers of a single stack until the desired final purity is obtained. The production rate is dependent on the concentration of dissolved minerals in the raw feed water and on the degree of demineralization desired.

In a feed-and-bleed system, a portion of the product solution is supplied and blended with the raw feed solution. The blended solution then is supplied to the ED stack. A feed-and-bleed system is used when large variations in the concentration of the feed solution are encountered and a continuous flow product is desired or when the desired degree of demineralization is low.

In many cases, the desired desalination or concentration of a feed solution cannot be obtained in a single pass through the ED stack. Therefore, two or more stacks are placed in series to achieve a higher degree of desalination or concentration. Staging in an ED system is necessary to provide sufficient membrane area and residence time to remove a specified fraction of salt from the demineralized stream. There are two types of staging, *i,e,* hydraulic staging and electrical staging. A stack with one hydraulic and one electrical stage makes the solution pass across the membrane surface between one pair of electrodes and exit. Hydraulic and electrical multi-staging enables water flow in multiple parallel paths across the membrane surface. Electrical staging is accomplished by inserting additional electrode pairs into a membrane stack, thus increasing flexibility in system design and providing maximum salt removal rates [10, 21].

One of the most important operating variables in ED is the feed temperature. As the operating temperature increases, the resistance of aqueous solution decreases, resulting in a reduction in power consumption. In addition, an elevated temperature operation can improve salt transport through membranes. It is known that each 0.5 °C increase in temperature improves the salt removal per stage by approximately 1% [22]. However, the high temperature operation may decrease the life time of membrane and permselectivity through polymeric membranes.

Many other factors depending on many site-specific factors should be considered carefully for ED desalination process design, which include concentration of organic and inorganic materials in the incoming feed water and the desired quality of the treated water. Prior to desalination, the availability of energy and chemicals and the disposal of waste concentrates should be considered to determine the level of pretreatment.

6.3.3 ED Operation and Maintenance

In the ED process, the direct current provides the driving force for ion migration through ion exchange membranes, and ions migrate between two electrodes in the electric field. Then, ions are removed in the diluate compartment and concentrated in the concentrate compartment simultaneously. Therefore, the process performance in practical applications is a function of membrane properties and feed solution composition, and is determined by several process design parameters such as the stack construction (i.e., the cell geometry and the spacer configuration), the feed flow velocities and mode of operation (i.e., batch or continuous operation with co- or counter-current flow streams). The parameters affect the cost of the processes by determining the investment as well as operating costs [10].

Certain process parameters can be determined by the feed and the product solution properties. Other parameters may be varied in a certain range, which is utilized to optimize the process. These parameters include the current density, the applied voltage, and the stream velocities. Many parameters are interdependent and often counteracting in terms of overall process efficiency. In general, the stack and its design and the mode of operation determine the technical feasibility and the economics of the process in various applications.

To maintain the ED process performance, the membrane stack must be cleaned periodically by cleaning-in-place (CIP) to remove scale and other fouling matters. Generally a decrease in purity of the product water indicates the need for cleaning. Mostly, chemical cleaning solutions (acids or bases) are circulated through the ED stack for 30–60 min for CIP to remove scale and inorganics. Organic contaminants are removed by a subsequent flushing with a base solution. The chemical cleaning is followed by the distilled water rinse. In most cases, CIP may be required at intervals of 1–4 weeks, depending on the quality of the feed water. When membranes are severely fouled, stack disassembly is necessary to restore the process performance. The frequency of stack disassembly and manual cleaning depends on feed quality and operating conditions, varying from a monthly to a semiannual operation [22].

Approximately 10% of total ion exchange membranes will require yearly replacement, whose rate depends on the quality of feed and product. Generally, little maintenance is required by the feed and circulating pumps when suitable construction materials are used. The membranes located at nearest electrodes are affected more significantly where more severe temperature and chemical environments are encountered [21].

6.3.4 Design Parameters in Desalting ED

An ED facility requires a supply of pressurized feed water generally at about 4 bar for pumping the feed/product water, concentrate, and the electrode rinse stream. The pressure drop in the cells is a function of the geometry of the spacers, cell thickness, and geometry of cell distribution [10]. The energy to reduce the salinity of water consists of the DC power supplied to operate the ED stack and the energy required to pump the process water through the plant. Between them, the former primarily predominates in ED process.

To estimate the practical energy requirement, two factors should be considered; the occurrence of electrical resistance through the solution and membranes and increase in the value of membrane potential. The energy required of ED operation is expressed as follows:

$$E = I^2 R t \tag{6.4}$$

where E is the energy requirement, I is the current, R is the electrical resistance, and t is time. The current required to decrease the concentration in dilute solution can be expressed by the equation:

$$I = \frac{zF\Delta C}{tN\eta} \tag{6.5}$$

where z is the electrochemical valence, ΔC the difference in concentration in diluate, N the number of cell pairs, and η the current density. The current density is affected by membrane selectivity, current leakage, back diffusion, diluate and concentrate leakage between membranes, and water transport by electroosmosis and osmosis.

The energy for the actual desalting process (i.e., the ion transfer from the feed solution to the concentrate) is directly proportional to the amount of ionic species to be removed. The pumping energy is independent of the concentration of feed solution. However, it depends on the feed water recovery rate and temperature. The total pumping energy will be about 0.5–1.1 kWh/m^3 product water, indicating that the cost of pumping the solution through stack becomes significant at low feed water salt concentration. Energy consumption in the membrane stack is about 0.7 kWh/m^3 product water per 1,000 mg of TDS removed [21].

As a rule of thumb, the power loss during the operation is about 5% of total energy consumption. Total energy consumption depends on the concentration of the feed water and the total amount of desalination required. To obtain the product water of 500 mg/L TDS, the typical energy consumption will be around 1.5–4 kWh/m^3 for feed water of 1,500 ~ 3,500 mg/L. For higher salinity water, the energy consumption rises significantly to 7-10 kWh/m^3 [21]. For a certain plant capacity, the required membrane area is directly related to the feed water concentration. For the typical brackish water (about 3,000 mg/L TDS), the required membrane area for a plant capacity of 1 m^3/day is about 0.4 m^2 of cation and anion exchange membranes each at an average current density of 12 mA/cm^2 [10].

In the desalting ED process design for batch operation, the total electrical energy consumption for both the constant-current and constant-voltage periods, E, can be calculated as follows [23]:

$$R_{CP} = \frac{a}{C} + b \tag{6.6}$$

$$E = \int_0^{t_{sw}} NI^2 R_{CP} dt + \int_{t_{sw}}^{t_f} \frac{(N\Psi_{CP})^2}{N} \frac{1}{R_{CP}} dt \tag{6.7}$$

$$= NI^2 \int_0^{t_{sw}} (\frac{a'}{C} + b')dt + N\Psi_{CP}^2 \int_{t_{sw}}^{t_f} (\frac{a'}{C} + b')^{-1} dt \tag{6.8}$$

Here, R_{CP} is the single-cell-pair resistance, C, concentration in feed, a' and b', constant determined by experiments, Ψ_{CP}, the voltage applied across each cell pair, and t_{sw}, the time required to reach the switching point. Generally, the effective membrane area of ion exchange membranes is proportional to the amount of ionic species removed from a given feed solution, which can be expressed as follows [5]:

$$A = \frac{zFQ\Delta CN}{i\xi} \tag{6.9}$$

where A is the effective cell pair area, Q the volumetric flow rate of the product stream, i the current density, which is about 80% of the limiting current density, and ξ the current utilization (≤ 1).

In the continuous desalting ED, the actual membrane area is often more than the effective membrane area because of the so-called shadow effect of spacers [10]. For the estimation of the practical membrane area, A_{prac}, the following relationship can be considered using the correction factor accounting for the shadow effect, β (≤ 1):

$$A_{prac} = \frac{A}{\beta} \tag{6.10}$$

The energy, E, in the continuous operation is given by the following equation:

$$E = \frac{N}{A_{prac}Q}\left(\frac{\delta \ln \frac{C_{id}}{C_{ic}}\frac{C_{oc}}{C_{od}}}{\Lambda_s(C_{id} - C_{od})} + R_{CP}\right)\left(\frac{QF(C_{id} - C_{od})}{N\xi}\right)^2 \tag{6.11}$$

where δ is cell thickness and Λ_s, equivalent conductivity of solution. The subscript i,o,d,c, represents inlet, outlet, diluate, concentrate, respectively. In the continuous desalting ED operation, the practically required path length, L_{prac}, is obtained as follows [10]:

$$L_{prac} = \frac{\left[\ln \frac{C_{oc}C_{id}}{C_{ic}C_{od}} + \frac{\Lambda_s R_{CP} + (C_{id} - C_{od})}{\delta}\right] FC_{od}u\delta a}{\left[\frac{C_{od}}{C_{ic}} + 1 + \frac{\Lambda_s C_{od}}{\delta} R_{CP}\right] si_{\lim}\beta\xi} \tag{6.12}$$

where α is the correction factor and s the safety factor.

6.3.5 Economics of the ED Process

The desalination of brackish water and seawater has become a reliable method for water supply all over the world. It has already been practiced successfully for many decades and the technical and economical feasibility have been demonstrated. However, the common processes for desalination such as multi-effect distillation (MED), multi-stage flash (MSF), RO, and ED for treatment of brackish water and seawater requires large quantities of energy to a sustainable method for the drinking water supply [24–26]. Table 6.4 shows process selection guidelines for desalting. In general, ED is accepted as a favorable desalination process when feed water concentration is lower than 3,000 mg/L TDS. In the relatively large scale desalination process, the annual cost of the ED desalting of brackish water (3,000 mg/L TDS) is estimated to be lower compared with different desalting process, SWRO (seawater reverse osmosis), and distillation [27, 28]. The cost estimation results for the desalting processes at large capacity plant are shown in Table 6.5.

Table 6.4 Summary of guideline for desalting selection [27]

Desalination process	Feed water characteristics favorable to economic applications		Normal product quality	
	TDS (mg/L)	Hardness	Temp (°C)	TDS (mg/L)
Distillation	30,000-60,000	Any	35	0.5-25
SWRO (Seawater RO)	15,000-45,000	Any	45	<500
BWRO (Brackish water RO)	500-15,000	Any	45	<500
ED	500-3,000	Any	43	100-500

Table 6.5 Cost estimation of desalination processes [27]

	ED	SWRO	Distillation
Estimation conditions			
Operation per year	350 day	350 day	350 day
Feed water (TDS)	3000 mg/L	Seawater	Seawater
Treated water (TDS)	<500	<500	<500
Capacity	10,000 m^3/d	20,000 m^3/d	91,500 m^3/d
Membrane replacement	10% per year	20% per year	-
Plant life time	20 years	20 years	20 years
Cost estimation results ($/$m^3$)			
Operation and maintenance	0.27	0.50	1.16
Capital cost	0.53	0.60	0.58
Membrane replacement	0.04	0.03	-
Annual cost (total)	0.83	1.12	1.74

As shown before, feed flow velocities, cell and spacer construction, and stack design affect the performance of ED. They directly affect the cost of process indirectly by means of the limiting current density and the current utilization. Process design and economics are closely related in the ED operation since the total cost is the sum of fixed charges associated with amortization of the plant capital cost and operating costs, such as energy and labor costs. The economics of the process is mainly determined by the energy consumption and the investment costs for plants. It is considered that the process economics are related to the membrane properties used in the process, and various design parameters such as cell dimensions, feed flow velocity, and pressure drop of the feed solution in the cell [5].

Capital costs include depreciable items, such as the ED stacks, pumps, electrical equipment, membrane, and non-depreciable items, such as land and working capital. The capital costs of an ED plant strongly depend on the total membrane area required for a certain plant capacity. The membrane area required for ED operation is proportional to the amount of ion species removed from a given feed solution, operating current density, and current utilization. The current utilization determines the portion of the total current passing through an ED stack that is actually used to transfer ions from a feed solution. Generally, the current utilization is less than 100% because of membrane selectivity, water transport, and current passing through the stack.

For desalination of brackish water, the total capital cost for a plant with a capacity of 1000 m^3/day will be in the range of US$200–300/$m^3$/day. Generally, other items such as pumps, piping, and tanks are independent of the concentration in feed water. The membrane replacement cost is often regarded as a separate item because of the relatively short life of the membrane. The cost of the actual membrane is less than 30% of the total capital cost. In general, a useful life of 5–7 years for many brackish water applications is assumed in the cost estimation of an ED system [10].

The operating costs are mainly determined by the required energy, which is determined by the electrical energy required for the actual desalting process and the energy for pumping the solution through the stack. The practical energy requirements and the technological factors governing process and equipment design should be evaluated together according to an economic basis to provide an accurate assessment of the potential of ED for desalination applications.

The energy cost increases with increasing current density, while the required membrane area and the membrane investment cost decreases with increasing current density. The total product costs consisting of the sum of energy costs, maintenance cost, and amortization costs can be varied by the current density and reach minimum value at a certain density. Figure 6.5 shows schematic diagram of ED process costs as a function of the applied current density.

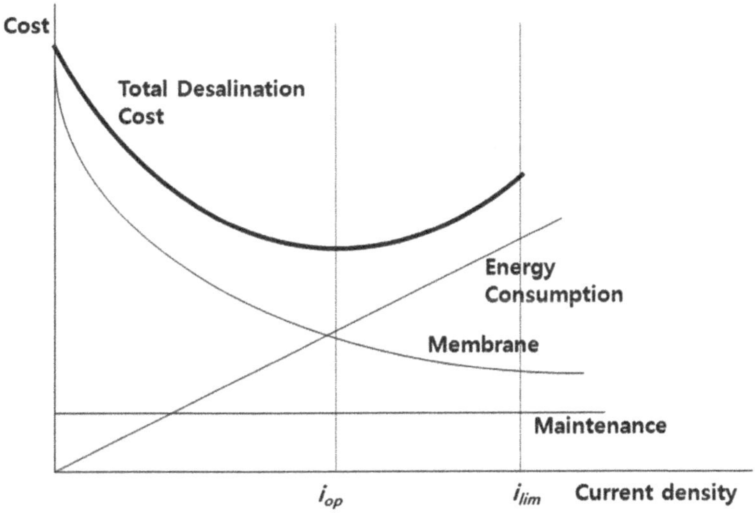

Figure 6.5 ED desalination cost as a function of the current density (i_{op}: operating current density, i_{lim}: limiting current density).

Considering the water desalination costs as a function of the feed solution concentration, ion exchange is the most economical process at very low concentrations. However, the desalination production cost increases sharply as the salt concentration increases in ion exchange process. In the range of 500–5,000 mg/L NaCl, ED becomes the more economical process. At higher salt concentration, however, RO is considered to be a less costly process. Distillation is the most economical process at very high feed solution [29].

6.4 Control of Fouling in an ED Desalination Process

6.4.1 Fouling Mechanism

In the ED desalination process, there are a number of significant considerations including concentration polarization, current efficiency, scaling potential, fouling potential, energy consumption, and process operating variables. Among them, fouling of ion exchange membranes is one of the most important limitations in the design and operation of the ED process [15, 30]. Fouling by the deposition of foulants on the membrane surface decreases the flux and increases the resistance, resulting in decreasing the ED process performance significantly and decreasing the economics of the process.

Table 6.6 illustrates the description, charge properties, and methods for prevention of fouling and/or cleaning for foulants [30]. Among foulants, scale is significant in ED desalination because inorganic fouling occurs by the precipitation (scaling) of slightly soluble inorganic compounds (such as $CaSO_4$ and $CaCO_3$) in the concentrate compartment and by the fixation of multivalent cations on the surface of the cation exchange membrane [31]. To prevent or mitigate inorganic fouling that occurs by scale and colloids, methods can be considered such as maintenance of lower recovery, pH adjustment, cleaning with chemicals using an acidic solution (e.g., citric acid or HCl) or an EDTA-based solution. Fouling mechanism due to organic foulants depends on molecular weight and electric mobility. Organic foulants deposit on the membrane surface due to electric interactions between the surface and their charged groups. As a result, fouling occurs when the accumulated concentration on the surface increases [30]. A base solution is usually used as a cleaning solution to minimize organic fouling.

Table 6.6 Candidate foulants in an ED process [30]

Foulant category		**Description**	**Charge properties**	**Representative foulants**
Inorganics	Scale	Precipitates of sparingly soluble salts in solution	Non-charged	$CaCO_3$ $CaSO_4 2H_2O$ $BaSO_4$ $SrSO_4$,
	Colloids	Agglomeration of suspended matters on membrane surfaces	Negative	SiO_2 $Fe(OH)_3$ $Al(OH)_3$ $FeSiO_3$ $Cr(OH)_3$
Organics		Attachment of organic species to membrane surfaces	Negative	Macromolecules Proteins Polyelectrolytes, Humate Surfactants Alginate

6.4.2 Fouling Potential

The prediction of the fouling potential is important because different fouling behaviors allow the quantitative prediction of the optimized parameters of membrane processes. Many parameters have been used to describe fouling phenomena in pressure-driven membrane processes, such as the silt density index (SDI), the modified fouling index (MFI), and the multi-plugging factor index (MPFI) [32]. Even though some physical parameters were used to describe the fouling potential in the ED process, the fouling indices have a limit to apply to practical processes.

The membrane fouling index as a measure of fouling potential in ED system was suggested based on a fouling layer formation model on the surface of an ion exchange membrane. In the fouling layer formation model, it is assumed that the concentration polarization does not occur and then the boundary layer resistance is negligible when permeable ions exist in the boundary layer facing the diluate solution [30]. For the constant current operation, the membrane fouling index in ED process can be obtained from the slope of the plot of $E(t)/I^2$ vs time:

$$\frac{E(t)}{I^2} = \frac{R_m}{I} + \frac{KC_b r_c}{C_g A^2} t \tag{6.13}$$

$$\frac{E(t)}{I^2} = \frac{R_m}{I} + MFI_{ED} \times t \tag{6.14}$$

where $E(t)$ is the voltage drop, I is the operating current, R_m is the membrane resistance, K is a constant, C_g is the foulant concentration in the fouling gel layer, C_b is the foulant concentration in the bulk solution, r_c is the specific resistance of the fouling layer, and MFI_{ED} is the defined membrane fouling index in ED process.

For the quantitative analysis of the fouling potentials, the membrane fouling index of ED can be obtained based on fouling experimental data under a constant current condition. The decrease in the resistance at the first period is attributed to ion transport through ion exchange membranes in the electric field. As electrical charges accumulate, foulants move to the surface of an anion exchange membrane with a low electric mobility, increasing the foulant concentration near the anion exchange membrane surface. Therefore, three distinct regions can be observed in the plot of the membrane fouling index of ED as shown in Figure 6.6: (i) the acclimation period, when resistance decreases and the foulants migrate to the boundary layer, (ii) the period for the fouling layer formation, when resistance increases due to foulant deposition, and (iii) the period for the concentration depletion, when resistance increases very sharply. It is considered that a fouling layer forms on the membrane surface in the second period when the foulant concentration reaches at a certain critical value [30, 33].

6.4.3 Fouling Mitigation

Many feed solutions for the desalination contain foulants, such as organics with various molecular weights, suspended and colloidal matter or insoluble salts to their saturation level. Therefore, pretreatments of the feed solution to remove and lower the concentration level are required to decrease fouling potential for the effective ED operation.

In general, filtration of the feed solution is applied in almost all ED systems to remove particulate materials. However, additional

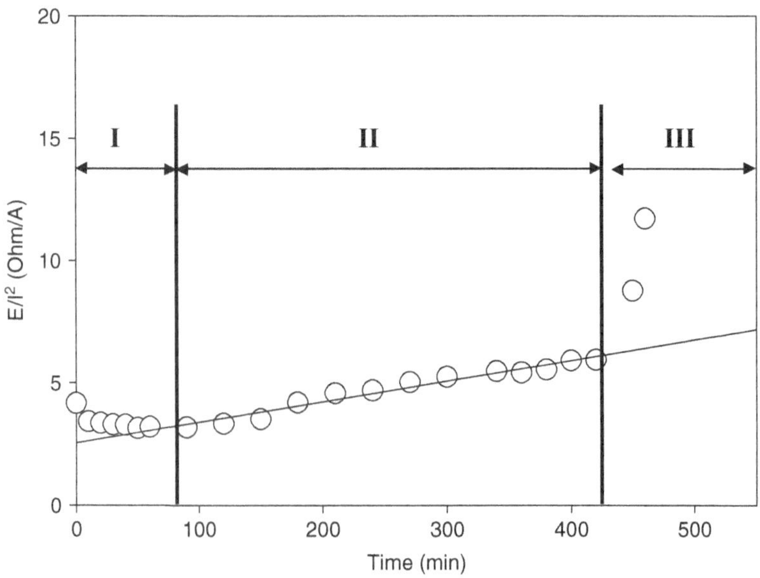

Figure 6.6 Graph for the membrane fouling in ED process in the presence of an organic foulant (I: the period for the acclimation of foulant, II: the period for the fouling layer formation, III: the period for the concentration depletion).

procedures such as flocculation, precipitation of metal hydroxides, and ion exchange may be necessary depending on the properties of the feed solution and water quality of the product. For example, feed solution containing significant quantities of calcium carbonates and bicarbonates, which may precipitate in the concentrate stream, can be prevented by acidification. Especially in the presence of calcium sulfate, a softening step is required before the ED desalination process. Sometimes, certain post-treatment procedures may be necessary before the product water of an ED system can be used as potable or industrial process water.

Many approaches have been studied to minimize fouling during ED process, which include use of turbulence in the compartments and the optimization of process conditions, such as the solution pH and flow rate. Although methods can reduce fouling to some extent, CIP is still needed in practical processes. Several cleaning methods including hydraulic, mechanical, and chemical cleaning methods have been used effectively. Fouling control methods depend on the separation processes and the configuration of the module. Among cleaning methods, chemical cleaning is the accepted as the most effective CIP procedure during ED operation [10].

Fouling can be controlled using the polarity change in ED processes. When an electric field is applied to a feed solution, negatively charged particles or large organic anions migrate to anion exchange membranes and deposit on the membrane surface. The negatively charged species will migrate away from ion exchange membranes back into the feed stream if the polarity is reversed, as shown in shown in Figure 6.7. Electrodialysis reversal (EDR) is designed to produce demineralized water continuously without

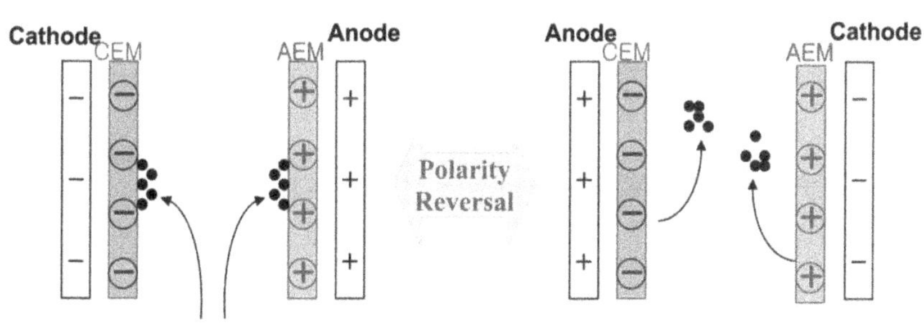

Figure 6.7 Principle of the fouling mitigation by the periodical change of the polarity.

constant chemical addition during normal operation and to eliminate significant problems such as fouling, concentration polarization, and etc [34].

In the EDR operation, the polarity of the electrodes is automatically reversed periodically (about three to four times per hour) and the diluate and concentrate streams are interchanged. There is always a certain amount of the product lost to the waste stream, the so-called off-spec product. Because of the polarity reversal, no flow compartment in the stack is exposed to high solution concentrations for more than 15–20 min [4]. Therefore, the polarity reversal of ED process may not be acceptable in certain applications which require high quality products.

Even though fouling mitigation methods has the effectiveness to decrease fouling rate at some degree, still they have difficulty in reducing the fouling rate significantly and recovering process performance in the ED process. A modified DC power supply, the power sources with electric pulse, was suggested as CIP and fouling mitigation method [35]. The use of an electric pulse enhances the mobility of the charged particles in the fouling layer, thus decreasing the electric resistance of the ED system. Previous studies showed that the pulsation of electric field with an optimal frequency decreases the fouling potential of the already fouled membrane systems. It was understood by that the electric pulse enhances the electrophoretic mobilities of the charged particles in the fouling layer near on the membrane surface, thus decreasing fouling potentials [30, 35].

6.5 Prospects for ED Desalination

6.5.1 Integration with ED for the Desalination

Recently, the importance of the water reuse in industrial processes has increased since there has been a substantial decrease in the quality of raw water for industrial use. Often ED can be employed very effectively and economically in recycling industrial process water due to high recovery rates and stable operation at wide pH range and somewhat high temperature by the integration with other membrane processes [10].

The production of potable water from brackish water is the largest single application of ED due to high water recovery rates and long operation life time and flexible operation compared to RO [36].

The integration of ED processes is found for the water desalination to utilize the effectiveness of ion exchange membrane processes, whose applications can be determined by economic considerations.

As one of the integrated processes, the ED and RO process was examined in the desalination of seawater with salt contents in the range of 35,000–45,000 mg/L. The salt concentration of the sea water is reduced to about 2,000–4,000 mg/L using RO in the first stage, and then to less than 500 mg/L using ED in the second stage [37, 38]. An example of the hybrid approach for economical zero liquid discharge is the installation of EDR plus RO along with evaporation and crystallization. The integrated process with EDR and RO to pre-concentrate and reclaim plant wastewaters resulted in downsizing of the evaporator system and reduction of overall capital costs [39].

For high recovery of product water without compromising the quality, ED unit can be coupled with RO to treat the feed water and bring down its salinity prior to RO, with recycle of RO reject in ED feed. Generally, RO units become not as effective when the salinity of feed water increases, leading to deterioration in the recovery as well as quality. The concentrate stream from RO in the integrated ED-RO process has comparable or lower salinity than the feed water [40, 41]. Figure 6.8 illustrates a schematic diagram for the integrated ED process with pretreatment and RO. Here, RO is utilized to produce potable water after post-treatment. Meanwhile,

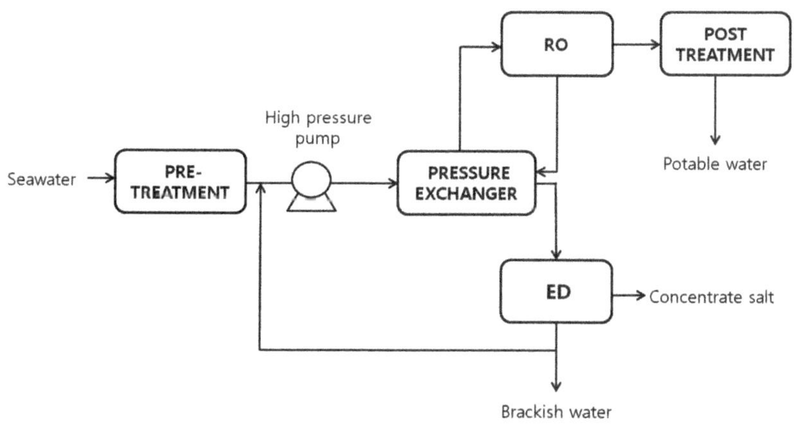

Figure 6.8 Integration of RO and ED for seawater desalination.

ED desalinates seawater to the concentration range of brackish water and produces a concentrated salt solution.

In addition, the possibility of producing drinking water from brackish groundwater was studied from brackish groundwater source using NF, RO, and ED processes [42]. NF reduces the concentrations of Ca^{2+} and Mg^{2+}, which are responsible for elevated hardness. Then, the integration of RO and ED is highly efficient in reducing the content of inorganic matters present in raw waters.

6.5.2 Process Intensification of the ED Desalination System

The process intensification consists of innovative equipment design and process development methods, resulting in cheaper sustainable technical solutions. The relevant improvements in manufacturing and processing decrease production costs, energy consumption, waste generation, and equipment size, and improve remote control, scale-up, and design and process flexibility [43]. Membrane desalination operations have well established applications in various industrial fields and remarkable progress can be expected.

Various ED desalination systems are available for a wide range of applications with the development of an innovative process such as continuous electrodeionization (CEDI), capacitive deionization (CDI), or reverse electrodialysis (RED). In many cases, the process intensification allows better performance in terms of product quality, plant compactness, environmental impact, and energy saving.

The process design of CEDI is very similar to that of desalting ED, which is widely used for the preparation of high quality deionized water, especially in the electrical industry and in analytical laboratories. The main difference is the stack construction, which affects the flow distribution of cations and anions in the cell. Generally, the diluate cell is filled with a mixed bed of ion exchange resins or in series of the cation and anion exchange resins [44, 45].

CDI is an electrochemical process that operates by adsorbing ions in the double layer formed at the electrodes by the application of a potential difference, sometimes referred to as electro-sorption. With increasing concerns about energy consumption and the environment, interest in CDI technology has increased significantly as a promising desalination process. CDI can be used to treat brackish water (800–10,000 mg/l) because of lower energy consumption compared with competing processes, ED and RO [46–48]. CDI can utilize with two trains of the ED system to desaline seawater. ED desalinates seawater to brackish water, followed by CDI to a TDS level of less than about 1,000 mg/L (Figure 6.9).

Among the intensified ED processes, RED is one of the possible processes for generating energy from the salt gradient between river and sea water [49, 50]. In the RED system, a number of cation and anion exchange membranes are stacked in an alternating pattern between a cathode and an anode, like a desalting ED cell arrangement. This electrical current and the potential difference over the electrodes can be used to generate electrical power when an external load or energy consumer is connected to the circuit. The development of RED should focus on system characteristics, such as optimization of the internal resistance, which is mainly determined by the width of the spacers [51–53].

Solar energy and RED for a sustainable production of desalinated water and electrical energy has been studied as an alternative hybrid concept. The concentration of the brine from the seawater desalination unit could be substantially increased by using solar energy while also producing additional condensate. Electricity then can be produced from the mixing energy of the highly concentrated brine and seawater by using RED. In this way the disadvantageous brine waste situation in seawater desalination can provide an opportunity regarding the production of a large amount of additional fresh water, a significant amount of electrical energy, and an answer to the brine disposal environmental problem [54, 55]

A low energy system and process for seawater desalination was considered in the integrated ED desalination system with ion exchange resin to produce partially desalinated water and a brine by-product, as an ion exchange softener. The ion exchange treats the partially desalinated water stream to remove or reduce the amount of scaling material to maintain deionization apparatus

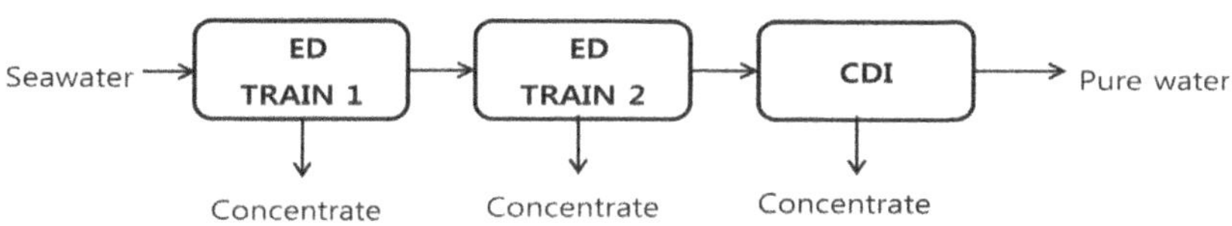

Figure 6.9 Integration of ED and CDI for seawater desalination.

efficiency and reduce energy consumption. ED desalinates seawater in a range of about 3,500 mg/L to about 5,500 mg/L, followed by ion exchange softening and final desalination to a TDS level of less than about 1,000 mg/L by a novel of CEDI. The power consumption to desalinate seawater to potable water (TDS 500 mg/L) was estimated to be very low as 1.4 kWh/m^3 using three trains of ED with softening and EDI devices to desalinate brackish and seawater in Figure 6.10 [56]. Such systems may be utilized to desalinate seawater as well as brackish water from estuaries, rivers and/or even groundwater.

6.5.3 Perspectives of ED Desalination

Often ion exchange membrane separation processes are in competition with other mass separation processes, whose application is determined by economic considerations. Since ED process provides higher quality products as environmentally friendly process, more comprehensive integration of ion exchange membrane processes with pressure-driven membrane separation processes can be found [12].

In most ED applications, high energy is consumed due to electrical resistance of ion transfer though membranes. Therefore, low resistant ion exchange membranes provide low operating costs and a novel process for their manufacture. Desalination of seawater requires thinner ion exchange membranes having low electrical resistance, high chemical and mechanical properties, and high permselectivity [57].

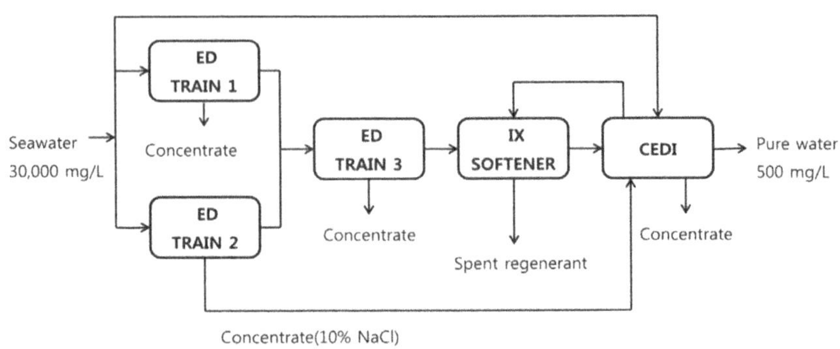

Figure 6.10 Low energy ED system and process for seawater desalination [56].

Even though ED has high water recovery rates and long operation life time and flexible operation compared to RO, the desalination of brackish water by ED has disadvantages because of microbes (viruses and bacteria) and corrosion problems in the surrounding of the desalination plant. Therefore, the process intensification should be concerned with the integration of ED process with different processes to utilize the effectiveness of the processes for the water desalination.

Notably, control and prevention of membrane fouling during ED operation is required to increase the effectiveness of the process, process performance and economical feasibility. More measures and methods including pretreatments can improve the ED process performance and contribute to enhancing and prolonging ED operation without deteriorating the product quality or increasing energy consumption [8, 12].

In total, the strategies to overcome the high energy consumption in ED process and to increase process performance are summarized as follows: (i) development of low resistance ion exchange membranes, (ii) optimization of stack design and operating process, and (iii) process intensification of ED with auxiliary desalination processes. More improvements and research are required for the economical ED applications in various areas for optimizing operation parameters and process integration.

6.6 Concluding Remarks

Desalination processes involve the separation of nearly salt-free fresh water from seawater or brackish water, where the salts are concentrated in the rejected brine stream. Membrane desalination processes are accepted as the most common and economical process among various desalination process. Electrodialysis (ED) is considered to be one of the promising membrane processes for desalination of seawater and brackish water. It has already been practiced successfully for decades and the technical and economical feasibility is well proven for salt production and desalination of brackish water, while membranes and processes are continuously developed for cost reduction in the desalination process.

A drawback of the ED process is the high energy consumption due to ohmic resistance of ion transfer through membranes. More improvements and research are required for the economical ED applications in various areas for optimizing operation parameters and process integration. To utilize the effectiveness of ion exchange membrane processes, the integration of ion exchange membrane is found in many areas consisting of membrane reactors using chemical and biochemical reactions, pressure-driven membrane processes and ion exchange.

Notably, membrane fouling during ED operation is the most significant problems to increase the effectiveness of the process. The ED desalination process performance and economic feasibility can increase through proper pretreatment, development of low resistance ion exchange membranes, optimization of stack design and operating conditions, integration with an auxiliary salt removal process and control of fouling and concentration polarization.

References

1. M.C. Anderson, A. L. Cudero and J. Palma, *Electrochimica. Acta* **37**, 3845 (2010).
2. R. Semiat and D. Hasson, Seawater and brackish-water desalination with membrane operations, in *Membrane Operations-Innovative Separations and Transformations*, E. Drioli and L. Giorno (Ed.), Weinheim, Wiley-VC (2009).
3. Mora Associates, Water desalination, http://www.moraassociates.com/publications/0712%20Water%20desalination.pdf (2011).
4. R.W. Baker, *Membrane Technology*, 2nd ed., John Wiley & Sons, Hoboken (2004).
5. H. Strathmann, Electrodialysis, in *Membrane Handbook*, W.S. Winston Ho and K.K. Sirkar (Ed.), New York, Van Nostrand Reinhold (1992).
6. M. Turek, *Desalination* **153**, 371 (2003).
7. H. Strathmann, *Desalination* **264**, 268 (2010).
8. H.J. Lee and S.H. Moon, Membrane integration processes in industrial applications, in *Advances in Membrane Science and Technology*, T. Xu (Ed.), New York, Nova Science Publishers (2009).
9. H. Strathmann, Ion-exchange membrane process in water treatment, in *Sustainable Water for the Future: Water Recycling versus Desalination*, I.C. Escobar and A.I. Schäfer (Ed.), Amsterdam, Elsevier (2010).
10. H. Strathmann, *Ion-Exchange Membrane Separation Processes*, Amsterdam, Elsevier (2004).
11. M.M. Nasefa and E.S.A. Hegazy, *Prog. Polym. Sci* **29**, 499 (2004).
12. T. Xu, *J. Membr. Sci* **263**, 1 (2005).

13. H. Strathmann, Fundamentals in electromembrane separation processes, in *Membrane Operations-Innovative Separations and Transformations*, E. Drioli and L. Giorno (Ed.), Weinheim, Wiley-VC (2009).
14. R.K. Nagarale, G.S. Gohil, and V.K. Shahi, *Adv. Colloid Interfac* **119**, 97 (2006).
15. H.J. Lee, J.H. Choi, J. Cho and S.H. Moon, *J. Membr. Sci* **203**, 115 (2002).
16. H.J. Lee and S.H. Moon, Electrochemical characterization of ion exchange membranes, in *Surface Electrical Phenomena in Membranes and Microchannels*, A. Szymczyk (Ed.), Kerala, Transworld Research Network (2008).
17. Y. Tanaka, *Ion Exchange Membranes-Fundamentals and Applications*, Amsterdam, Elsevier (2007).
18. H.J. Lee, H. Strathmann and S.H. Moon, *Desalination* **190**, 43 (2006).
19. H.J. Lee, F. Safert, H. Strathmann and S.H. Moon, *Desalination* **142**, 267 (2002)
20. J.H. Choi, H.J. Lee and S.H. Moon, *J. Colloid Interf. Sci* **238**, 188 (2001).
21. J.J. Shoeman and M.A. Thompson, Electrodialysis, in *Water Treatment Membrane Processes*, J. Mallevialle, P.E. Odendaal and M.R. Wiesner (Ed.), New York, McGraw-Hill (1996).
22. A.N. Rogers, Design and operation of desalting systems, in *Synthetic Membrane Process-Fundamentals and Water Applications*, G. Belfort (Ed.), New York, Academic Press (1984).
23. E.G. Lee, S.H. Moon, Y.K. Chang, I.K. Yoo, H.N. Chang, *J.Membr. Sci* **145**, 53 (1998).
24. B.V. der Bruggen and C. Vandecasteele, *Desalination* **143**, 207 (2002).
25. M. Wilf, *The Guidebook to Membrane Desalination Technology*, Awerbuch, Balaban Desalination Publications (2007).
26. T. Gullinkala, B. Digman, C. Gorey, R. Hausman and I.C. Escobar, Desalination: Reverse osmosis and membrane distillation, in *Sustainable Water for the Future: Water Recycling versus Desalination*, I.C. Escobar and A.I. Schäfer (Ed.),.Amsterdam, Elsevier (2010).
27. US DOI, *Desalting Handbook for Planner*, 3rd ed., (2003).
28. US AID, *Desalination Manual*, (1980).
29. P. Glueckstern and N. Arad, Economics of the application of membrane process Part I: Desalting brackish and seawater in *Synthetic Membrane Processes-Fundamental and Water Applications*, G. Belfort (Ed.), New York, Academic Press (1984).
30. H.J. Lee, S.H. Moon and S.P. Tsai, *Sep. Purif. Techmol* **27**, 89 (2002).
31. H.J. Lee, M.K. Hong, and S.H. Moon, *Desalination* **284**, 221 (2012).
32. J.C. Schippers and J. Verdouw, *Desalination* **32**, 137 (1980).
33. L. Song, *J. Membr. Sci* **139**, 183 (1998).
34. J. Veerman, M. Saakes, S.J. Metz and G.J. Harmsen, *J. Membr. Sci* **327**, 136 (2009).
35. H.J. Lee and S.H. Moon, *J. Colloid Interf. Sci* **270**, 406 (2004).

36. K. Wangnick, *Desalination Plants Inventory Report*, The International Desalting Association (2000).
37. A.A. Husseiny and H. L. Hamester, *Desalination* **39**, 171 (1981).
38. H. Schmoldt, H. Strathmann and J. Kaschemekat, *Desalination* **38**, 567 (1981).
39. A. Seigworth, R. Ludlum and E. Reahl, *Desalination* **102**, 81 (1995).
40. S. Thampy, G.R. Desale, V. K. Shahi, B. S. Makwana, and P.K. Ghosh, *Desalination* **282**, 104 (2011).
41. T.A. Vetter, E.M. Perdue, E. Ingall, J. F. Koprivnjak, and P.H. Pfromm, *Sep. Purif. Technol* **56**, 383 (2007).
42. K. Walha K, R.B. Amar, L. Firdaous, F. Quéméneur, and P. Jaouen, *Desalination* **207**, 95 (2007).
43. E. Drioli E and E. Fontananova, Integrated membrane processes, in *Membrane Operations-Innovative Separations and Transformations*, E. Drioli and L. Giorno (Ed.), Weinheim, Wiley-VC (2009).
44. A. Grabowski, G. Zhang, and H. Strathmann and G. Eigenberger, *J. Membr. Sci* **281**, 297 (2006).
45. A. Grabowski, G. Zhang, H. Strathmann and G. Eigenberger, *Sep. Purif. Technol* **60**, 86 (2008).
46. T.J. Welgemoed and C.F. Schutte, *Desalination* **18**, 327 (2005).
47. H. Li and L. Zou, *Desalination* **275**, 62 (2011).
48. C. J. Gabelich, P. Xu, and Y. Cohen, "Concentrate Treatment for Inland Desalting, in *Sustainable Water for the Future: Water Recycling versus Desalination*, I.C. Escobar and A.I. Schäfer (Ed.), Amsterdam, Elsevier (2010).
49. J.W. Post, J. Veerman, H.V.M. Hamelers, G.J.W. Euverink, S.J. Metz, K. Nymeijer and C.J.N. Buisman, *J. Membr. Sci* **288**, 218 (2007).
50. J. Veerman, J.W. Post, M. Saakes, S.J. Metz and G.J. Harmsen, *J. Membr. Sci*. **310**, 418 (2008).
51. D.A. Vermaas, M. Saakes and K. Nijmeijer, *J. Membr. Sci* ***385–386***, 234 (2011).
52. M. Turek and B. Bandura, *Desalination* **205**, 67 (2007).
53. P. Długołęcki, J. Dąbrowska, K. Nijmeijer, and M. Wessling, *J. Membr. Sci* **347**, 101 (2010).
54. E. Brauns, *Desalination and Water Treatment* **13**, 53 (2010).
55. M.A. Lichtwardt and D.M. Williams, Solar-powered direct current electrodialysis reversal system, US Patent 6042701 (2000).
56. G.C. Ganzi, L. Liang, F.C. Wilkins and S.H. Chua, Low energy system and method of desalting seawater, US Patent Application 20110180477 A1 (2011).
57. J.R. Lin and G.Y. Gu, Ion exchange membranes, US Patent Application 20110068002 A1 (2011).

7

Continuous Electrodeionization

Jonathan H. Wood, Joseph D. Gifford

Siemens Water Technologies, Lowell, MA, USA

Abstract

Continuous Electrodeionization is a hybrid of two well known processes, ion-exchange deionization and electrodialysis. It was developed to allow the production of deionized water without the use of the hazardous acid and caustic that is required to regenerate ion exchange resins. An electric field is used as the driving force to transfer ions from a feed stream through a selectively permeable ion-exchange membrane to a reject stream, while simultaneously splitting water to regenerate ion exchange resins and allow the removal of weakly ionized contaminants. This process was first commercialized in 1987 and is now used extensively worldwide in many industries, especially in the production of deionized water for pharmaceutical formulations, power generation and manufacture of microelectronics/semiconductor devices.

Keywords: Electrodeionization, EDI, Continuous Electrodeionization, CEDI, Filled-cell, Ultrapure water

7.1 Introduction

Electrodeionization (EDI) is a process that removes ionizable species from liquids using electrically active media and an electrical potential to effect ion transport. The electrically active media in EDI devices may function to alternately collect and discharge ionizable species, or to facilitate the transport of ions continuously by ionic or electronic substitution mechanisms. EDI devices may comprise media of permanent or temporary charge, and may be operated batchwise, intermittently, or continuously.

Jane Kucera (ed.) Desalination , (327–372) 2014 © Scrivener Publishing LLC

The *continuous electrodeionization* (CEDI) process, a subset of EDI, is distinguished from the EDI collection/discharge processes such as *electrochemical ion exchange* (EIX) or *capacitive deionization* (CapDI), in that CEDI performance is determined by the ionic transport properties of the active media, not the ionic capacity of the media. CEDI devices typically contain semi-permeable ion-exchange membranes and permanently charged media such as ion-exchange resin. The CEDI process is essentially a hybrid of two well-known separation processes - ion exchange deionization and electrodialysis, and is sometimes referred to as *filled-cell electrodialysis*.

The most common application of CEDI is as a polishing demineralization step downstream of reverse osmosis (RO) in the production of high purity water, as an alternative to conventional resin-based, chemically regenerated mixed-bed deionization (MBDI) equipment. CEDI systems eliminate the need for hazardous acid and caustic regeneration chemicals, instead using a small amount of electricity. A typical CEDI system will use approximately 0.3 kW-hr to deionize 1 m^3 of water from a feed conductivity of 50 microsiemen/cm to 0.1 μS/cm product conductivity. Since the CEDI concentrate (or reject) stream contains only the feed water contaminants at 5–20 times higher concentration, it can usually be discharged without treatment, or used for another process.

The CEDI process has a number of advantages over MBDI:

- Improvement in workplace health and safety conditions
- Reduction of operating labor
- Lowers cost of regulatory compliance
- Avoids issues associated with corrosive acid fumes
- Chemical storage and waste neutralization equipment is not required
- Continuous process eliminates need for duplexing
- Smaller system footprint

For the most part the elimination of regenerant chemicals is considered advantageous, but the chemicals do offer at least one benefit. In conventional demineralizers, acid and caustic is typically applied to the ion exchange resins at concentrations of 2%–8% by weight. At these concentrations the chemicals not only regenerate the resins but clean them as well. The electrochemical regeneration that occurs in a CEDI device does not provide the same level of resin cleaning. Therefore proper pretreatment is even more important with a CEDI device, in order to prevent fouling or scaling. This is one of the reasons that RO is normally employed upstream of a CEDI system. In general the feed water requirements for CEDI systems are stricter than for a chemically regenerated demineralizer.

7.2 Development History

CEDI theory and practice have been advanced by a large number of researchers throughout the world. It is believed that CEDI was first described in a publication by scientists at Argonne Labs in January 1955 as a method for removal of trace radioactive materials from water [1]. One of the earliest known patents describing a CEDI device and process was awarded in 1957 [2]. It is thought that the first pilot device incorporating mixed resins was developed by Permutit Company in the United Kingdom in the late 1950's for the Harwell Atomic Energy Authority, as described in a paper [3] and in more than one patent [4, 50]. One of the first detailed theoretical discussions of CEDI was written in December 1959 [5]. In April 1971, a Czechoslovakian researcher reported results of his experimental and theoretical work that advanced the theory of ionic transport within a CEDI device [6]. Layered bed devices were described in the patent literature in the early 1980's [7].

CEDI devices and systems were first fully commercialized in early 1987 [8] by a division of Millipore Corporation that is now part of Siemens AG. Since then, the theory and practice of CEDI has advanced worldwide, and commercial CEDI devices are now manufactured by a number of companies [9, 10, 11, 12]. There are now several thousand CEDI systems in commercial operation for the production of high purity water at capacities ranging from less than 0.1 to more than 250 m^3/h. This includes CEDI modules that have been in continuous operation for twenty four years [13].

7.3 Technology Overview

A typical CEDI device contains alternating selectively permeable anion exchange membranes (AEM) and cation exchange membranes (CEM). The spaces between the membranes are configured to create liquid flow compartments with inlets and outlets. A transverse DC electrical field is applied by an external power source using electrodes at the ends of the membranes and compartments, as shown in Figure 7.1.

When the compartments are subjected to an electric field, ions in the liquid are attracted to the oppositely charged electrodes. The result is that the compartments bounded by the anion membrane facing the anode and the cation membrane facing the cathode become depleted of ions and are thus called diluting compartments. The compartments bounded by the anion membrane facing the cathode and cation membrane facing the anode will then "trap" ions that have transferred in from the diluting compartments. Since the concentration of ions in these compartments increases relative to the feed, they are called concentrating compartments, and the water flowing through them is referred to as the concentrate stream (or sometimes, the reject stream).

In a CEDI device, the space within the ion depleting compartments (and in some cases in the ion concentrating compartments) is filled with electrically active media such as ion exchange resin. The ion-exchange resin enhances the transport of ions and can

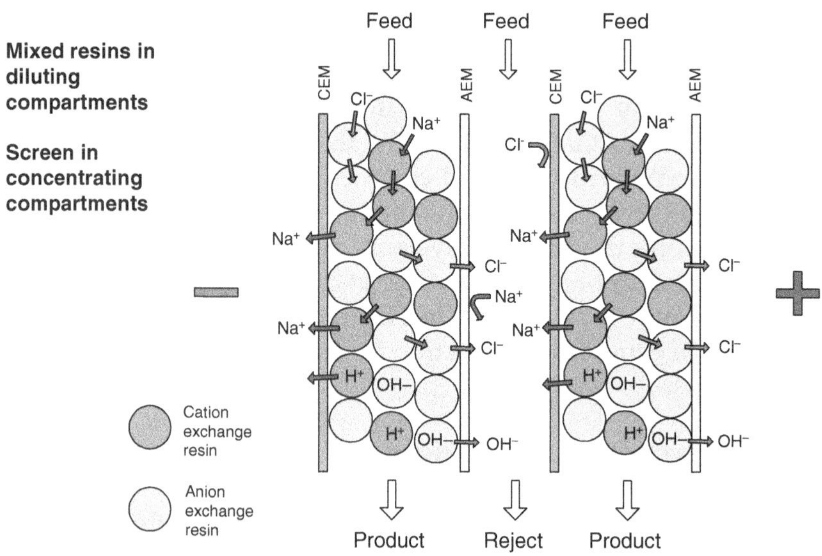

Figure 7.1 Ion transport and electrochemical regeneration in a CEDI cell

also participate as a substrate for electrochemical reactions, such as splitting of water into hydrogen (H^+) and hydroxyl (OH^-) ions. Different media configurations are possible, such as intimately mixed anion and cation exchange resins (mixed bed or MB) or separate sections of ion-exchange resin, each section substantially comprised of resins of the same polarity: e.g., either anion or cation resin (layered bed or LB and single bed or SB).

7.3.1 Mechanisms of Ion Removal

There are two distinct operating regimes for CEDI devices: enhanced transfer and electroregeneration [14]. In the **enhanced transfer** regime, the resins within the device remain in the salt forms. In low conductivity solutions the ion exchange resin is orders of magnitude more conductive than the solution, and acts as a medium for transport of ions across the compartments to the surface of the ion exchange membranes. This mode of ion removal is only applicable in devices that allow simultaneous removal of both anions and cations, in order to maintain electroneutrality.

The second operating regime for CEDI devices is known as the **electroregeneration** regime. This regime is characterized by the continuous regeneration of resins by electrically produced hydrogen and hydroxide ions. The dissociation of water preferentially occurs at bipolar interfaces in the ion-depleting compartment where localized conditions of low solute concentrations are most likely to occur [15]. The two primary types of interfaces in CEDI devices are resin/resin and resin/membrane. The optimum location for water splitting depends on the configuration of the resin filler. For mixed-bed devices water splitting at both types of interface can result in effective resin regeneration, while in layered bed devices water is dissociated primarily at the resin/membrane interface [16].

Regenerating the resins to their H^+ and OH^- forms allows CEDI devices to remove weakly ionized compounds such as carbonic and silicic acids, and to remove weakly ionized organic compounds. This mode of ion removal occurs in all CEDI devices that produce ultrapure water. Figure 7.1 is a representation of the process showing two diluting compartments, which illustrates the transport of ions and electrochemical regeneration of ion exchange resins in one type of CEDI cell.

7.4 CEDI Module Construction

7.4.1 Device Configurations

This description of module construction will first discuss the overall device and then the individual cell(s). Commercially available devices are produced in two main configurations: plate-and-frame or spiral wound. The plate-type devices are similar in concept to a plate-and-frame heat exchanger, with multiple fluid compartments sandwiched between a set of endplates (and electrodes) that are held in compression by bolts or threaded rods. The plates themselves can be either rectangular or disk-shaped [17]. The compartments alternate between diluting and concentrating, and are hydraulically in parallel but electrically in series. An exploded view of a typical plate-and-frame CEDI device is shown in Figure 7.2.

The spiral CEDI devices are analogous to a spiral wound membrane element, but with the membrane, resins, and spacers wound spirally around a center electrode rather than a permeate tube. Spiral wound devices must be installed inside a pressure vessel, while plate-and–frame devices incorporate some means of sealing on the individual fluid compartments, essentially making each a pressure vessel. A typical spiral device is shown in Figure 7.3.

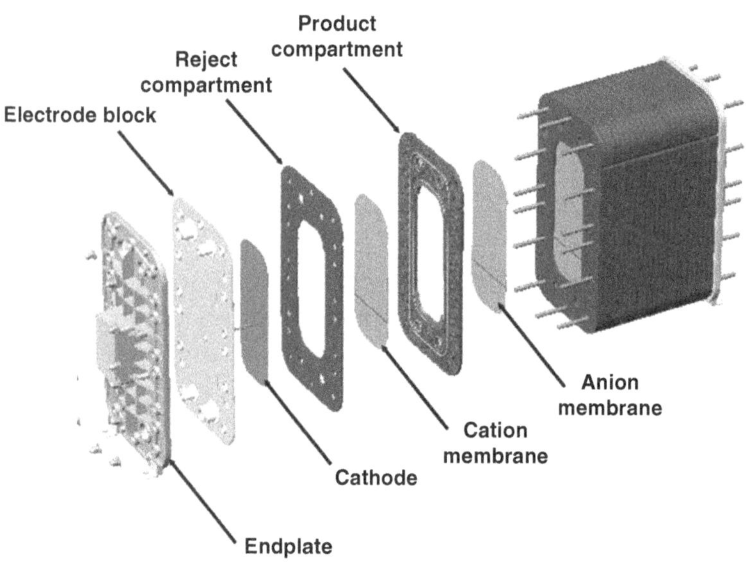

Figure 7.2 Plate-and-frame CEDI device

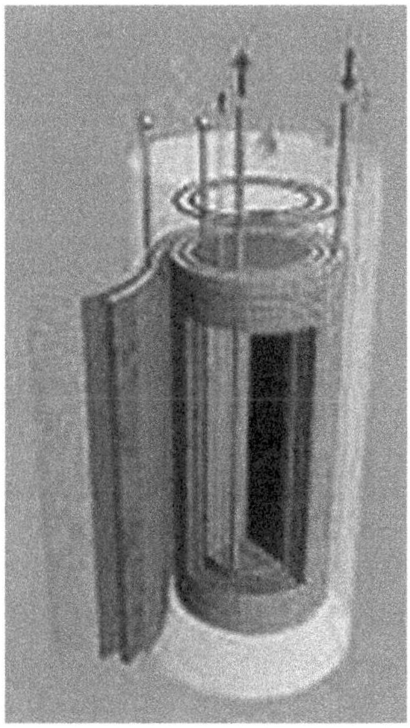

Figure 7.3 Spiral-wound CEDI device

The cells themselves can be classified as either thin cell or thick cell [18]. Thin-cell devices are those with a spacing of approximately 1.5–3.5 mm between the ion exchange membranes in the diluting compartments, while thick cell devices typically use intermembrane spacing of 8–10 mm. Both plate-and-frame and spiral wound configurations are suitable for either thin or thick cell construction. As will be shown below, thin-cell devices allow the use of intimately mixed anion and cation exchange resins in the product compartments, while thick cells work best with separate regions that contain primarily resins of the same polarity.

7.4.2 Resin Configurations

7.4.2.1 Mixed Bed Resin Filler (CEDI-MB) - Intermembrane Spacing

The first commercial CEDI devices used mixed-bed ion exchange resin as a conductive media in the diluting compartments. For devices using a mixed-bed resin filler, one of the most important design constraints is the distance between the ion exchange membranes [19]. In order for the resin to transport an ion to the membrane, there must be a continuous path of the appropriate type of ion exchange resin, i.e. cation resin for transfer of cations and anion resin for transfer of anions. For simple cubic packing and equal quantities of equal diameter anion and cation beads, the probability of a direct conductive path can be related to the number of resin beads between the membranes by Equation 7.1.

$$P = \frac{(N_b + 1)}{2^{N_b}} \tag{7.1}$$

This shows that the probability of a direct conductive path decreases as the intermembrane spacing increases. The effect of intermembrane spacing on salt removal in a CEDI-MB device has also been demonstrated experimentally, as shown in Table 7.1.

The experimental results show the optimum intermembrane spacing for this condition to be about 2.3 mm, which represents approximately three resin beads between the membranes. Because of this effect, thick-cell devices do not normally use a mixed-bed resin filler unless the feed water ionic loading is very low.

7.4.2.2 Mixed Bed Resin Filler (CEDI-MB) - Resin Packing

It has also been shown that the performance of a CEDI-MB device can be improved significantly by the use of uniform particle size ion exchange resins instead of conventional resins, which have a Gaussian distribution of bead sizes [20, 21]. The uniform beads allow a higher packing density, approaching a hexagonal close-packed structure. The effect of packing density on salt removal is illustrated by the data in Table 7.2.

7.4.2.3 Layered Bed Resin Filler (CEDI-LB)

In the late 1980s and early 1990s there was considerable activity in the development of layered bed (CEDI-LB) devices [7, 9]. In this

Table 7.1 Cell thickness vs. performance for a CEDI-MB device

Cell thickness, mm	Salt removal, %	Feed µS/cm	Product µS/cm	Velocity, cm/sec
1.0	99.8	600	1.2	0.86
2.3	99.9	600	0.6	0.86
4.7	94.3	600	34	0.86
7.2	71.7	600	170	0.86

Table 7.2 Resin particle size distribution vs. performance for a CEDI-MB device

Feed, µS/cm	Product Megohm-cm, Non-uniform beads	Product Megohm-cm, Uniform beads
145	0.4	0.7
87	0.8	1.5
65	1.5	4.2
41	3.4	10.5

configuration the media comprise separate, sometimes alternating layers (or in one variation, clusters) of ion-exchange resin, each layer containing mainly one type of resin: e.g., either anion or cation resin. Liquid to be deionized flows sequentially through the layers of resins.

For CEDI-LB devices there is essentially no "enhanced transfer" regime and less limitation on the intermembrane spacing. This is because transfer of only one type (polarity) ion is enhanced at any given time. In order to maintain electroneutrality, the ion that is transferred out is replaced by a co-ion resulting from splitting of water. This is illustrated in Figure 7.4. One of the main design constraints is the choice of ion exchange resin, which must catalyze the water splitting reaction at the resin/membrane interface. Resin selection must also ensure that the electrical resistance of the layers is similar, so that the DC current is fairly evenly distributed through the cell instead of preferentially passing through a single type of layer. It is likely that the use of uniform particle size resins will offer some benefit to the performance of thick-cell layered-bed devices, but that the difference will not be as dramatic as it is for a thin-cell mixed-bed.

One of the main advantages to the use of thicker cells is that it greatly reduces the amount of ion exchange membrane used to construct the device, which significantly reduces the assembly cost (both materials and labor). The tradeoff is that the performance for salt removal is lower than for thin cell devices, due to the higher flow per unit membrane area and greater distance that ions need to travel across the cell to reach the ion exchange membrane. The CEDI-LB module performance is more sensitive to increases in feed water concentration and to decreases in feed water temperature.

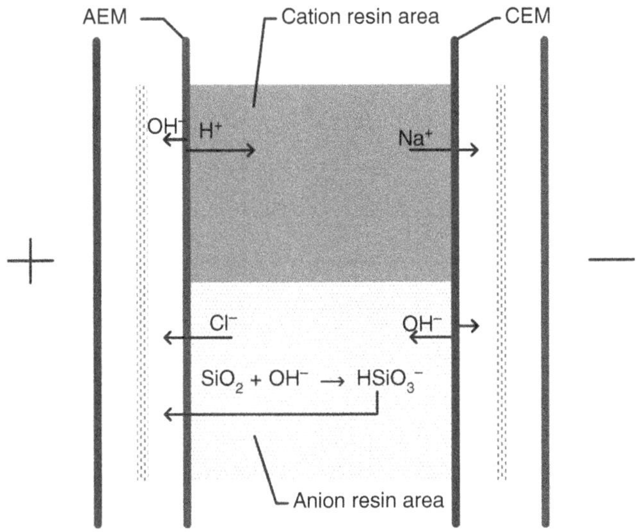

Figure 7.4 Removal mechanism in thick-cell, layered-bed CEDI cell

However, this is less important now than when CEDI was first commercialized, due to improvements in reverse osmosis and gas transfer membranes that have reduced the typical ionic load on the CEDI device. The performance of thick-cell CEDI devices is sufficient for their use in most ultrapure water applications, given proper system design.

The other significant advantage of thick-cell devices is that the thicker resin chambers are considerably stronger than thin spacers. They also offer more flexibility in the design of the intercompartment sealing, such as the use of grooves and O-ring seals. This allows construction of modules without external leaks and with higher pressure rating. Most of the commercial CEDI devices that are capable of operating continuously at a feed pressure of 7 bar are thick-cell type. Spiral-wound devices in a pressure vessel are generally limited to 4 bar or less.

7.4.2.4 Separate Bed Resin Filler (CEDI-SB)

Another electrodeionization device [22, 49] uses completely separate compartments for the cation and anion resins, and is somewhat analogous to a two-bed demineralizer. The cation exchange resin is placed in a compartment between a cation membrane and the anode, with the resin in direct contact with the electrode. The anion exchange resin is between an anion membrane and the cathode. The two ion exchange membranes create a concentrate compartment at the center of the cell. This configuration is shown in Figure 7.5.

Instead of splitting water at a resin/membrane or resin/resin interface, this process obtains the hydrogen (H^+) or hydroxyl (OH^-) ions needed to regenerate the resin from the electrode reactions – hydrogen ions being generated at the anode and hydroxyl ions at the cathode (electrode reactions are discussed below and shown in Equations 7.6 through 7.8).

In CEDI-SB devices the product water flows through the electrode compartments, picking up the O_2, H_2, and Cl_2 gas that is created by the electrode reactions (equations 7.6, 7.7, and 7.8, Section 7.6.4), which may require an additional gas removal process step. It is possible that the electrode reaction could produce enough chlorine to reduce the life of the ion exchange resin, depending upon the amount of chloride in the feed water.

It has been shown that the salt removal by CEDI-SB device with 10 mm intermembrane spacing, is not nearly as good as

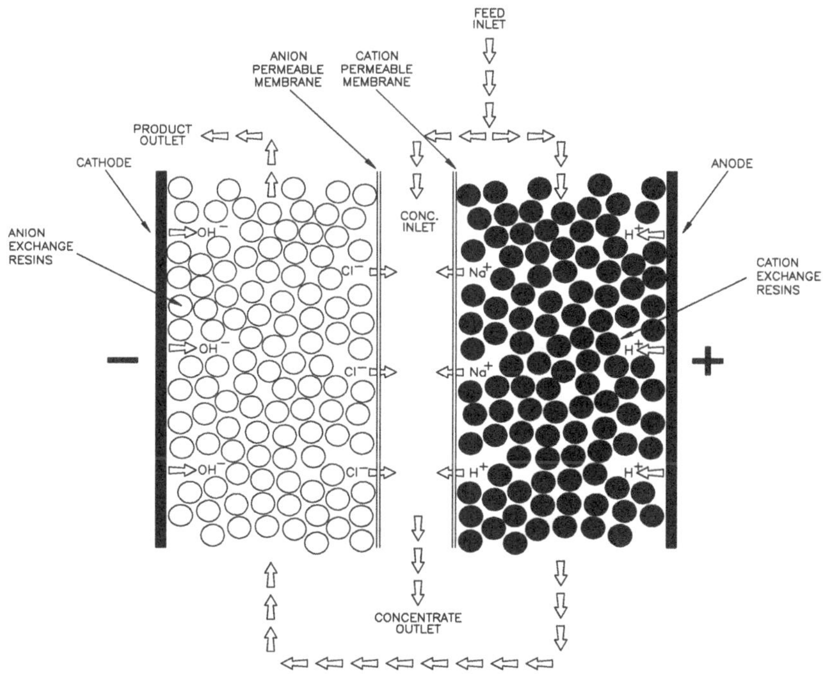

Figure 7.5 Removal mechanism in thick-cell, separate-bed CEDI cell

for a CEDI-MB device with 2.5 mm spacing [23]. But the main disadvantage of the CEDI-SB device is that it requires a set of electrodes for each cell. Since the electrodes are by far the most costly component of a CEDI device, this approach is only cost effective for low flow rate applications where a single cell is sufficient. There have been some attempts to produce a multi-cell device using bipolar ion exchange membranes, but these have not seen widespread use due to the high cost and short life of the bipolar membranes.

7.4.3 Flow Spacers

All commercial CEDI devices use ion exchange resin in the diluting compartments, and therefore require a component to contain the resin. This "resin spacer" consists of an inlet port, an inlet distributor, the resin compartment, an outlet distributor, and an outlet port. It is necessary to provide a means of sealing the ion exchange membrane against the spacer to form the sides of the resin compartment. Some designs will also include additional ports to allow slurrying the resin in and out of the cell. A typical dilute spacer for a plate-and-frame CEDI device is shown in Figure 7.6.

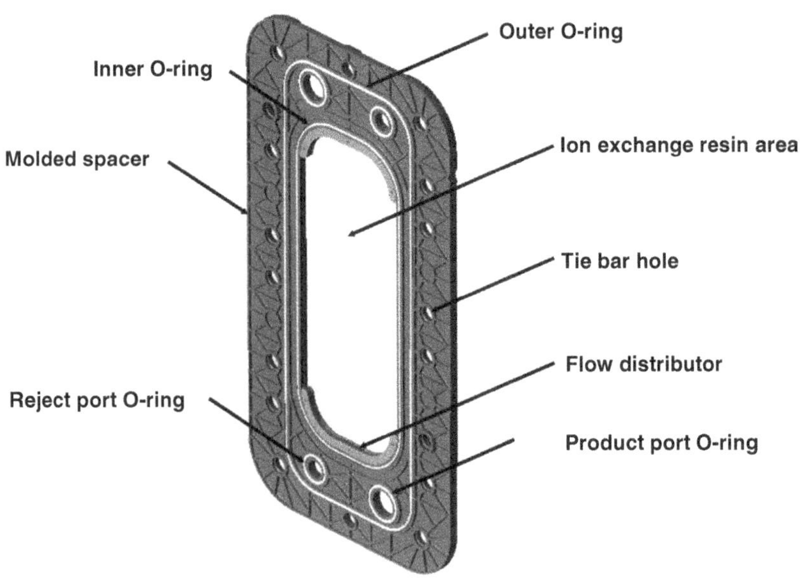

Figure 7.6 Dilute spacer from thick cell, layered bed CEDI module

All CEDI devices will also require flow compartments for the concentrate and electrode streams as well. The two types most commonly used are either a flow-through screen or a resin compartment. The flow-through screen is similar to a sheet-flow electrodialysis spacer. It generally consists of a woven plastic mesh screen (also like an RO feed spacer) that incorporates some sort of sealing mechanism, such as a rubber gasket impregnated in the perimeter of the screen.

The use of screen-type concentrate spacers is quite common in CEDI devices, as they are fairly inexpensive and relatively easy to fabricate. Their major disadvantage is that they are not conductive (there has been some work on making plastic screens conductive for electrodialysis applications, but nothing has yet been commercialized). Since the makeup water feeding the concentrate compartments is normally RO permeate (to avoid scaling and fouling), the concentrate stream is not very conductive, in spite of the ions transferring into the concentrate from the diluting compartments. For example, a CEDI system fed RO permeate with a conductivity of 5 μS/cm and operating at 90% water recovery would typically have a concentrate outlet conductivity of about 50 μS/cm. This is low enough to limit the amount of current that can be passed through the module (see discussion of resistance, below). Some manufacturers have recommended injection of salt into the concentrate to raise the conductivity to 300 μS/cm or more.

An alternative to the use of screen-type spacers for the concentrate and electrode compartments is to use a resin-filled compartment similar to the ones used for the diluting compartments. By employing a conductive filler the use of salt injection can be avoided. It has been found that injection of a salt solution into a resin-filled concentrate compartment has little effect on module resistance.

7.5 Electroactive Media Used in CEDI Devices

7.5.1 Ion Exchange Resin Selection

Ion exchange resins function much differently in CEDI devices than in a conventional demineralizer, or even than in a collection/discharge type EDI device. In CEDI, the ability of the resin filler to rapidly transport ions to the surface of the ion exchange membranes is much more important than the ion exchange capacity of the resin. The resins are therefore not optimized for capacity, but for other properties that influence transport, such as water retention and selectivity.

Membrane/resin combinations must also be carefully chosen to selectively catalyze the electrochemical splitting of water at various locations within the CEDI device, as mentioned previously. Considerable research has gone into optimization of resin fillers for CEDI devices, mostly by the manufacturers of the CEDI devices rather than the manufacturers of the ion exchange resins. The resulting knowledge is generally protected either through patents or as trade secrets.

7.5.2 Ion Exchange Membrane Selection

Ion exchange membranes are different from the many types of filtration membranes in that they are essentially impermeable to water. They combine the ability to act as a separation wall between two solutions (the diluting and concentrating streams) with the chemical and electrochemical properties of ion exchange resin beads [24] Ion exchange membranes are selectively permeable, as they will allow the passage of counter ions while excluding co-ions. When placed in a water solution and an electric field, a cation membrane will permit the passage of cations only, while an anion membrane will allow the passage of only anions. An in-depth discussion of the theory and properties of permselective membranes is available elsewhere [24, 25].

There are two main types of commercially available ion exchange membranes, heterogeneous and homogeneous. Homogeneous membranes consist of thin films of continuous ion exchange material, typically on a fabric support. These are essentially equivalent to an ion exchange resin bead, only in the form of a thin sheet. Heterogeneous membranes consist of small ion exchanger particles embedded in an inert binder, with or without any support.

Some of the more important properties of ion exchange membranes used in CEDI devices include the following:

Low water permeability
Low electrical resistance
High permselectivity
High strength
Resistance to contraction or expansion
Resistance to high and low pH

Ion exchange membranes that were developed for electrodialysis may not have sufficient mechanical strength and handling properties for use in assembly of CEDI devices, so most manufacturers have developed special ion exchange membranes that are optimized for their CEDI devices. Extruded heterogeneous membranes based on a polyolefin binder [26, 27] have become very popular for this application. They are relatively low in cost, offer flexibility in formulation, and have been shown to be fouling resistant [28].

7.6 DC Current and Voltage

7.6.1 Faraday's Law

In a continuous electrodeionization device the DC current is the driving force for the removal of ions, while the applied DC voltage is the means of obtaining the required current. Faraday's Law states that the electric charge required to liberate one gram-equivalent of a substance by electrolysis is 96,487 coulombs (a coulomb is the amount of electric charge that crosses a surface in one second when a steady current of one ampere is flowing across the surface). In both electrodialysis and electrodeionization, Faraday's Law is used to relate the transfer of salts through the membranes and the amount of current flowing through the membranes [29]. A common form of this relationship is given in Equation 7.2:

$$I_a = \frac{Q_d(C_{di} - C_{do})F}{\eta_c N_{cp}} \tag{7.2}$$

This shows that the amount of DC current required is directly proportional to the flow rate through the diluting compartments and the amount of ionic equivalents to be removed, and inversely proportional to the current efficiency.

7.6.2 Current Efficiency and E-factor

Current efficiency can be defined as the ratio of the theoretical minimum current predicted by Faraday's law (at 100% efficiency) to the actual current applied to the electrodes of the device, as shown in Equation 7.3:

$$\eta_c = \frac{I_t}{I_a} \tag{7.3}$$

In a CEDI device, current that does not cause the transfer of salt will cause water (HOH) to split into hydrogen (H^+) and hydroxyl (OH^-) ions, allowing electrochemical regeneration of the ion exchange resins within the device. For example in a module that is operating at 25% current efficiency and drawing 4 amps of DC current, 1 amp is causing the transfer of salt and 3 amps are causing water splitting that is unrelated to ion transfer. Cross-leakage, back diffusion and current leakage through manifolds can also cause some current loss, but in electrodeionization these are normally small compared to the water splitting.

In order to produce high purity water (over 1 megohm-cm resistivity) with a CEDI system, it is generally necessary to feed the system with low TDS (total dissolved solids) water such as RO permeate (normally less than 0.0005 equivalents/liter) and to operate at a current efficiency of less than 35%. For optimal removal of weakly ionized solutes such as silica and boron, current efficiencies as low as 5% are sometimes employed.

Some authors prefer to use the term E-factor [30]. This is defined as the ratio of the applied current to the theoretical current, and is therefore the reciprocal of the current efficiency:

$$EF = \frac{I_a}{I_t} \tag{7.4}$$

7.6.3 Ohm's Law and Module Resistance

Ohm's law states that the direct current flowing in an electric circuit is directly proportional to the voltage applied, and inversely proportional to the resistance of the element:

$$I = \frac{E}{R} \tag{7.5}$$

Most manufacturers of CEDI devices limit the applied voltage to 600 VDC, in order to avoid the need for the more expensive wiring construction that is required for higher voltages. Given such a voltage limitation, the electrical resistance of the module therefore controls how much current can be passed through the cells. Since the DC current determines how much water can be processed for a given product quality (or what the quality will be for a given flow rate), it is important to optimize the electrical resistance of the module.

The overall resistance of the CEDI module can be affected by the following:

Resistance of the anion-selective membranes
Resistance of the cation-selective membranes
Resistance of the ion exchange resins
Resistance of the concentrate stream
Resistance of the anolyte
Resistance of the catholyte
Feed water temperature
Ionic composition of the feed water

In addition to proper selection of resin and membranes, there are several methods that reduce the electrical resistance of the cell and therefore allow greater passage of DC current. The first technique used to accomplish this was to increase the water recovery and therefore the amount of salt in the concentrate compartments. This is generally done by incorporating a feed-and-bleed arrangement, using a pump to recirculate the concentrate stream and ensure adequate flow distribution, while decreasing the flow rate of the bleed that is sent to drain [31].

An alternative method of reducing the cell resistance is to inject a conductive salt such as NaCl into the feed to the concentrate compartments using a dosing pump. There are several possible drawbacks to this method. Increasing the TDS may prevent reclaiming the concentrate stream for other uses, and may increase the possibility of salt bridging and stray DC currents. If the concentrate is used to feed the electrode compartment, this can also lead to generation of chlorine gas at the anode.

A third method is to incorporate resin filler into the concentrate (and in some cases, electrode) compartments, which eliminates the need for injection of a conductive salt [18]. It has also been seen that the resin helps ions transfer away from the surface of the concentrate side of the ion exchange membrane. This reduces the ion concentration in the boundary layer, reducing the driving force for back-diffusion and improving salt removal.

7.6.4 Electrode Reactions and Material Selection

At the cathode, or negatively charged electrode, electrons are transferred from the external circuit to ions in the solution by the following reaction:

$$H_2O + e^- \rightarrow \tfrac{1}{2} H_2 + OH^- \tag{7.6}$$

Therefore an electrode that is stable in the presence of base and hydrogen is required. The most common cathode material for CEDI devices is stainless steel.

At the anode, or positively charged electrode, electrons are transferred from ions in solution to the external circuit by one or more of the following reactions:

$$\tfrac{1}{2} H_2O \rightarrow \tfrac{1}{4} O_2 + H^+ + e^- \tag{7.7}$$

$$Cl^- \rightarrow \tfrac{1}{2} Cl_2 + e^- \tag{7.8}$$

Commonly used anode materials include iridium-coated titanium and platinum-coated titanium.

Gases are evolved by the reactions at both the cathode and anode. These must be removed to prevent masking the surface of the electrode, which would result in a voltage drop and reduce the voltage applied to the cells. Removal of the gases is accomplished by maintaining a flow of water across the surface of the electrodes during operation. This requires the use of a flow compartment adjacent to the electrode. Such compartments could be either gasketed screen-type spacers or resin-filled compartments. Both downflow and upflow electrolyte streams have been shown effective at removing gas from CEDI electrode compartments.

Since copper wire is commonly used to conduct electric current to the electrodes, the junction of the copper wire and the non-copper electrode may be subject to corrosion, particularly if it is damp. It is best to have a projection of the electrode material that passes through the end plate of the module to an external connection that can be kept clean and dry [32].

7.7 System Design Considerations

When the CEDI process was first commercialized, equipment was available from the CEDI module manufacturers primarily in the form of complete systems including the module, power supply, piping, controls and instrumentation. Recently it has become more common for the device manufacturers to sell just the CEDI modules to OEMs (Original Equipment Manufacturers, sometimes called system integrators) (sometimes called system integrators) who then assemble the modules into complete systems. This section will describe the process controls required for construction of CEDI systems. The design of DC power supplies (rectifiers) is outside the scope of this chapter.

7.7.1 Required Process Control & Instrumentation

System designs vary slightly by CEDI module design and manufacturer. In general, all CEDI modules have at least two process streams, the dilute and concentrate, each with an inlet and outlet. In addition, modules may have electrode streams with a separate inlet and/or outlet. If a CEDI system has multiple modules in parallel on one skid, it is common practice to treat the modules as a single control block (controlled at the system level rather than at the module level), similar to RO membranes. Figure 7.7 shows a typical CEDI system piping and instrumentation diagram.

CEDI modules *must* be protected against overheating, which can occur due to operation with DC power on with little or no water flow. This can cause permanent damage to CEDI modules as well as create a safety hazard. Typical alarm set points are 50% of the lowest normal operating flow for both the product and reject streams with the alarms wired in series. This will shut down the power supplies if flow is lost on either stream.

Further protection from overheating due to loss of flow is accomplished by interlocking the CEDI power supply and the RO. This

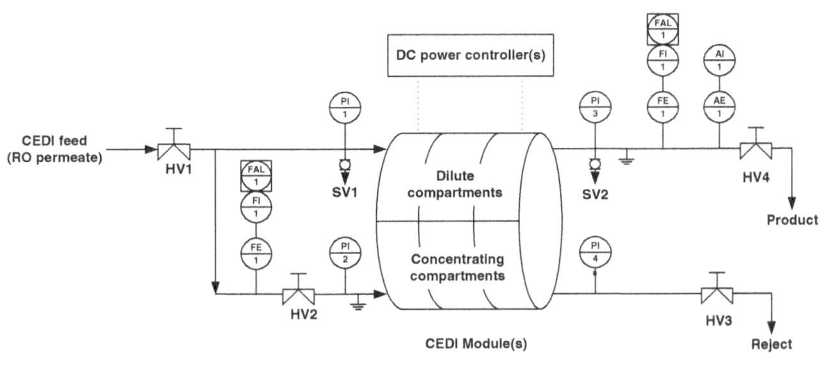

Figure 7.7 Typical CEDI system P&ID

should be set up so that when the RO is online and sending water to the CEDI system, power is sent to the DC power supply. When the RO is not sending water to the CEDI (offline or flushing), power is not sent to the DC power supply. If the CEDI system is fed from a pump, a similar arrangement should be made between the pump motor starter and the CEDI system.

The minimum requirements for CEDI system controls and instrumentation are shown in Table 7.3.

7.7.2 Optional Process Control & Instrumentation

There are other optional controls that may be used but are not critical to operation, as shown in Table 7.4. Optional alarms include one for the CEDI feed water quality and one for the CEDI product quality. An alarm on low feed water quality will protect the modules from damage due to poor quality feed water, which could contain hardness or other foulants. In addition, preventing high salinity feed water upsets, such as those observed on RO startup, will maintain the CEDI resins in a high state of regeneration and prevent changes in CEDI product quality. This is especially important when trying to maintain low product water concentrations of weakly ionized species such as silica. An alarm on low CEDI product quality will protect downstream equipment if the CEDI system is producing water that is out of specification. Automatic valves on the product stream can be set up to divert the CEDI product to drain on high conductivity, high silica, etc.

Table 7.3 Required Controls & Instrumentation

Process Stream	Controls and Instrumentation
Dilute (product)	Flow rate monitor and low flow alarm Inlet pressure, outlet pressure Inlet conductivity, outlet conductivity Inlet and outlet sample valves
Concentrate (reject)	Flow rate monitor and low flow alarm Inlet pressure, outlet pressure Inlet and outlet sample valves Throttling valve (inlet side)
Electrode (if separate stream)	Flow rate Inlet pressure, outlet pressure

Table 7.4 Optional Controls & Instrumentation

Process Stream	Controls
Dilute (product)	CIP Connections Individual module isolation valves Automatic valve for service (normally open) Automatic valve for product divert (normally closed) Feed quality alarm Product quality alarm
Concentrate (reject)	CIP Connections Individual module isolation valves Outlet conductivity

As discussed above, systems that use multiple CEDI modules typically do not include separate flow control and instrumentation for each module. In such cases it is important to ensure adequate sizing of the header pipes carrying water to and from the modules, to ensure fairly equal distribution of flow through all the modules. Good practice is to have the inlet and outlet connections at opposite ends of the feed manifolds.

7.8 Process Design Considerations

Since its commercial introduction in 1987, CEDI has gradually evolved into a polishing demineralization process that is usually employed downstream of a reverse osmosis system. There are several reasons for this: the CEDI devices are susceptible to hardness scaling, organic fouling, and physical plugging by particulates and colloids. In addition, the CEDI product water quality is somewhat dependent on the feed water quality. While some CEDI devices may be able to produce "two-bed quality" product water directly from a softened feed water, most process water applications now require "mixed-bed quality" water, which would not be produced by CEDI alone.

Using RO pretreatment ahead of the CEDI reduces the TDS to a level that allows the CEDI device to produce "mixed-bed" quality water. In addition the RO removes organics that could foul the ion exchange resins in the CEDI modules, and removes particulates that could clog the narrow flow channels in the resin compartments (spacers) or the resin bed itself.

7.8.1 Feed Water Requirements

CEDI feed water requirements can be divided into two categories. The first relates to the performance of the module. Feed water TDS (often measured as conductivity) and carbon dioxide must be limited so that the module can achieve the desired performance. Increasing the TDS or CO_2 in the feed water should not cause permanent damage to the module, but it may impact the product quality. Very cold feed water (1–5°C) will increase stack resistance and may reduce DC amperage and as a result the product water quality. The second category includes things that could damage the module, such as high temperature or pressure, foulants like organics and hardness, or strong oxidants.

A typical list of CEDI feed water requirements is shown in Table 7.5, below. This list includes ranges based on specifications from several different CEDI module manufacturers. Not all modules can accept the entire range, so it is important to determine and follow the guidelines specific to the device being used. Most of these requirements can be met by pretreatment with single or double-pass RO.

7.8.2 Hardness

By now the pretreatment of RO feed water is well understood. However, there are some issues regarding the use of RO/CEDI

Table 7.5 Typical CEDI Feed Water Requirements

Parameter	Units	Minimum	Maximum
Feed Conductivity	mS/cm	NA	25-50
Carbon Dioxide	mg/l as CO_2	NA	10-20
Temperature	°C (°F)	5-10 (41-50)	35-45 (95-113)
Pressure	psig (bar)	15-25 (1-1.7)	60-100 (4-6.9)
Free Chlorine	mg/l as Cl_2	NA	0.02-0.05
Fe, Mn, H_2S	mg/l	NA	0.01
Silica	mg/l as SiO_2	NA	0.5-1.0
Hardness	mg/l as $CaCO_3$	NA	0.5-1.0
TOC	mg/l as C	NA	0.5

systems that are sometimes overlooked, to the detriment of system performance and reliability. One such issue is that of hardness, since most CEDI modules can tolerate only about 1 mg/l total hardness as $CaCO_3$. This is often surprising to those familiar with saturation indices such as the Langelier Saturation Index (LSI), since in most cases the LSI of the bulk CEDI reject stream will be negative. However, the water splitting that is necessary for electrochemical regeneration of the ion exchange resin can also lead to localized pH shifts, creating regions (such as near the ion exchange membrane surface) where the scaling potential is greater than in the bulk solution.

Many systems will use ion exchange softening to protect both the RO and CEDI systems from scaling, but in some cases this is impractical. While an RO system alone may be able to produce *steady-state* permeate hardness of less than 1 mg/l, many process water systems operate in "start/stop" mode, only operating when the deionized water storage tank is calling for water. When an RO system starts up from a standby condition, the initial slug of RO permeate can be worse than the RO feed water (since the concentration gradient causes salts to continue to diffuse through the RO membrane after the permeate flow stops), and the first few minutes of RO permeate may not meet the CEDI system feed water specifications. This phenomenon is illustrated for permeate conductivity in Figure 7.8 [33] but the results are similar for most ionic constituents in the feed water. In this particular instance, it took about 2 minutes for the RO permeate to approach steady-state. Even though

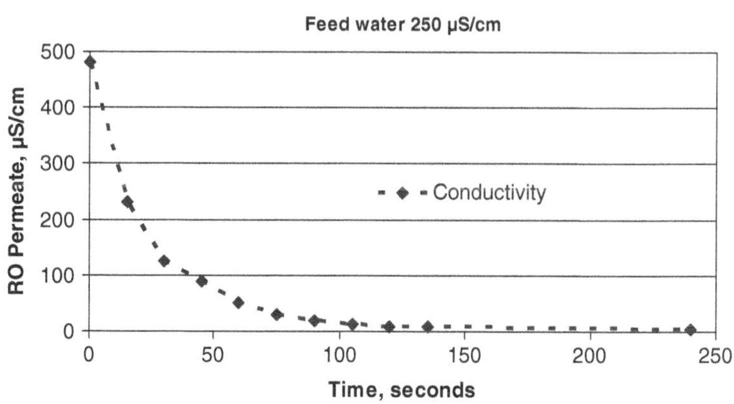

Figure 7.8 Flush up of RO system after standby

the volume is relatively small, it is important to flush this water to drain or upstream of the RO rather than send it to the CEDI system, especially since most RO/CEDI systems do not have a buffer tank between the RO and CEDI. An alternative approach is to flush the RO system with demineralized water before it shuts off.

In some cases the steady-state permeate hardness from an RO system will not meet the feed specification of the CEDI module. In this case, it is possible to install an ion exchange softener between the RO and CEDI. While this softener will still have to be sized to handle the flow of the RO permeate, it will be removing very low levels of hardness and regeneration frequency will be greatly reduced.

Whenever using a sodium cycle softener upstream of a CEDI system, it is important to thoroughly regenerate and flush the softener resin when first putting the system into service. This is because new cation exchange resin contains extractables that could foul the anion resin in the downstream CEDI system.

7.8.3 Carbon Dioxide

Another significant issue affecting CEDI operation is the presence of carbon dioxide (CO_2) in the feed water. Any CO_2 in the gaseous form will not be rejected by the RO membrane, and will impart an ionic "load" on the downstream CEDI system. For some CEDI devices, a concentration of 10 mg/l CO_2 in the CEDI feed may be enough to prevent the system from meeting the product water specifications. Therefore, RO/CEDI systems may employ a separate means of CO_2 removal. The most common methods are forced draft degasification, membrane degasification, or pH adjustment before the RO. In the latter case the pH is increased by the addition of a small amount of sodium hydroxide, which converts the CO_2 to sodium bicarbonate, which can be rejected by the RO:

$$CO_{2\,(aq)} + NaOH \rightarrow NaHCO_3 \tag{7.9}$$

While this method is very effective for the reduction of CO_2, it also increases the LSI of the RO feed and reject water. Therefore it may only be practical if the RO feed water is softened. It is also sometimes employed between passes of a two-pass (product-staged) RO system [34]

Since the reject stream from a CEDI system typically contains anywhere from one fifth to one half the salt concentration of the raw water, recycling the CEDI reject to the inlet of the RO is often desired. This may be possible, but it is necessary to consider that while the CEDI reject may be lower in salt, it typically has ten times the CO_2 of the RO feed water, as shown in Figure 7.9 [33]. In the absence of a degasification or pH adjustment step, recycling this water could result in as high as a three-fold increase in the CEDI feed CO_2 concentration, and could have a significant impact on the CEDI product water quality. In most cases this would be impractical without a CO_2 removal step as part of the process train.

7.8.4 Oxidants

Oxidants such as free and total chlorine, which are found in many raw waters, as well as ozone or hydrogen peroxide, which may be used in a plant for sanitization, can irreversibly damage the ion exchange resin and ion exchange membranes in a CEDI device. Oxidative attack can cause breakdown of the cross-linking in cation and anion resins as well as degradation of the functional exchange sites. This results in both poor salt removal performance as well as physical breakdown of the resin leading to increased pressure drop through the module.

Depending on the type of RO membrane used, residual oxidant in the raw water may have to be removed prior to the RO. For example, thin film composite polyamide RO membranes can tolerate only trace levels or chlorine (<0.05 mg/l), but cellulose acetate (CA) membranes can tolerate several mg/l. In either case, chlorine - and

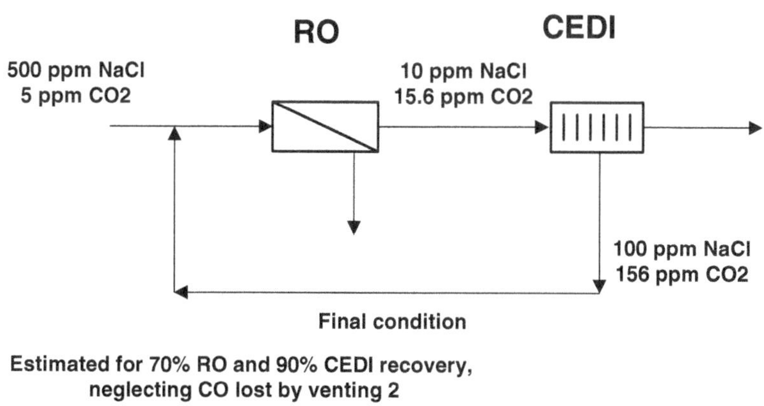

Figure 7.9 Salt and CO_2 concentrations in an RO/CEDI system

any other oxidants - must be removed to essentially non-detectable levels before the CEDI. Because of the relatively high flow per unit resin volume in a CEDI device and the high water content resins used to enhance ion transport, even low concentrations of oxidant can cause permanent damage in a relatively short amount of time.

Removal methods for chlorine vary depending on application. In smaller systems, activated carbon may be the method of choice given its reliable operation. If breakthrough does occur, it will be relatively gradual. Alternatively, chemical dechlorination by injection of a reducing agent (such as sodium sulfite or sodium bisulfite) can achieve complete reduction of chlorine, but failure will result in an immediate increase to the feed level. For this reason, it is common practice to use an oxidation reduction potential (ORP) or a residual bisulfite monitor as part of the control system for dosing of the reducing agent.

7.8.5 Temperature

Temperature can have either an adverse or beneficial effect on CEDI devices. As the feed water temperature decreases, reaction kinetics and diffusion rates slow and CEDI module electrical resistance increases, causing a decline in performance. Therefore, to maintain rated performance specifications, most CEDI manufacturers recommend a minimum temperature.

Warmer feed water temperatures will lower module resistance, increase diffusion rates and improve performance – within limits. As temperature increases, eventually some of the materials used to construct the CEDI module will start to degrade. Because of the applied DC voltage and need for high purity components, CEDI modules use plastic or elastomeric spacers. Most of these materials become weaker at higher temperature (> 35–45°C), making it more difficult to maintain integrity at higher temperature and pressure. At very high temperature (>90°C), the anion exchange resin can start to break down.

In most applications, water temperature is addressed during system design, by adding appropriate heating or cooling devices. However, occasionally RO and CEDI systems are used in recirculating loops. These loops can operate for long periods of time with little draw off. Under this condition, the water can heat up due to the recirculation pump. This can lead to unacceptably high water temperature and should be addressed with alarms and automatic cooling or draw off to maintain acceptable temperature.

Another cause of high temperature in a CEDI device could be the applied DC power. Under normal operation, the heat loss into the water due to the DC power is nearly imperceptible. However, if water flow through the module is reduced to a low enough level, this heating can cause the water temperature to exceed the manufacturer specification and cause permanent damage to components including the resins and membranes. For this reason, it is imperative that controls be designed to prevent the DC power from being on without water flow through the CEDI module.

7.8.6 Water Recovery

For most CEDI systems the water recovery is typically about 90%–95%. Water recovery is limited for several reasons:

- Limiting the concentrations of sparingly soluble salts helps to avoid precipitation (scaling) in the concentrate cells. The most common types of scale encountered in CEDI devices consist primarily of calcium or magnesium.
- As the recovery increases, the reject salt concentration also increases. This causes a large concentration gradient across the membrane, leading to the potential for back-diffusion and reduced product quality.
- In systems without concentrate recirculation, the flow through the concentrate compartments is directly related to the reject flow. As the reject flow is lowered, the concentrate velocity also reduces. Very low velocities can cause poor flow distribution, areas of high salt concentration in the boundary layer and gas buildup at the electrodes.

The concentration in the reject for any species removed by the CEDI can be calculated through a mass balance. If complete removal of salt by the CEDI device is assumed, a concentration factor can be calculated from the water recovery as follows:

$$CF = \frac{1}{(1-Y)} \tag{7.10}$$

Once the concentration factor is determined, it is multiplied times the feed concentration to estimate the reject concentration. This equation is useful in at least two instances. For design, it may be useful to know the concentration of all species in the concentrate to determine the feasibility of reusing the reject water or for determining scaling potential. For example, if the CEDI feed water contains 2 mg/l of sodium and the recovery is 90%, the reject will contain 20 mg/l. The concentration factor can also be useful in troubleshooting. Comparing the calculated and actual concentrate concentrations can help to determine if scaling has occurred in the past (actual >> calculated represents sloughage), or if scaling is presently occurring (calculated >> actual suggests accumulation in the module).

7.8.7 Recycling of CEDI Reject Stream

If high overall water recovery is desired but there is a limit on the minimum reject flow for the CEDI system, it may be feasible to reclaim some or all of the CEDI reject stream. If concentrate salt injection is not employed, the CEDI reject may be considerably purer than the raw water. In some applications, the CEDI reject water can be used for another application in the plant. In other applications, it is more desirable to recycle the CEDI reject water within the water system. This can be done by returning the water to the process prior to the RO system, which will prevent a buildup of salts in the CEDI feed water. However, in cases of high carbon dioxide, CO_2 buildup may prevent recycling without some type of CO_2 removal step (as discussed in Section 8.3). In addition, in some CEDI devices, the electrode stream is combined with the reject stream. If this is the case, this combined stream may contain gases that have to be vented. Also, this stream may contain chlorine, which must be removed prior to contact with TFC RO membranes and CEDI modules.

7.8.8 Total Organic Carbon

Most CEDI module manufacturers have set a specification of less than 0.5 mg/l of total organic carbon (TOC) in the feed water. This specification is empirical based on years of operational experience. It is very difficult to set an absolute value due to the fact that TOC is a non-specific analysis, and the actual organic species that make up this number vary significantly by location and also seasonally. For the most part, RO pretreatment will reduce TOC below the 0.5 mg/l level, and typically will remove the majority of the large organic molecules known to be fouling to ion exchange resin.

As for TOC removal by CEDI, there are normally no specific claims, again given that the actual makeup of the TOC varies so widely. In order for a CEDI device to remove TOC (by transport through the membrane, not adsorption) the organic molecule must be charged, and must be small enough to diffuse through the ion exchange membrane. It is the authors' experience that TOC removal by CEDI is typically in the range of 50 to 75%. In system designs where TOC removal is important, 185 nm UV treatment is placed upstream of the CEDI device [35]. The UV breaks down organics to carbon dioxide and charged organic fractions, which are then removed by the CEDI device to obtain high purity, low organic, water.

7.8.9 Electrode Gases

Gases are formed at the electrodes of a CEDI device, as shown in Equations 7.6, 7.7 and 7.8. The quantity of gas formed is easily calculated, since it is directly proportional to the amount of DC current passed through each set of electrodes. To simplify the calculation, ignore Equation 7.8 (the formation of chlorine), which only occurs if chloride is present at the electrode surface and varies depending on the amount of chloride available. Faraday's constant was defined previously as the amount of electricity associated with one mole of unit charges, or electrons. It has the value 96,487 ampere-second/equivalent.

From Equation 7.6, 96,487 amp-sec will produce ½ mole $H_{2(g)}$ & 1 mole OH^-

From Equation 7.7, 96,487 amp-sec will produce ¼ mole $O_{2(g)}$ & 1 mole H^+

Using the ideal gas law and a gas constant of 0.08206 (l-atm)/(mole-°K) the volume of a mole of gas is calculated for STP (298°K and 1 atm):

$$V = \frac{nRT}{P} = \frac{(1)(0.08206)(298)}{(1)} = 24 liters \qquad (7.11)$$

This allows us to determine the volume of gas produced per amp of current passed through the electrodes.
From Eq. 7.6:

$$\frac{1/2\,mole\,H_2}{98{,}487\,amp-\text{sec}}\,x\,\frac{60\text{ sec}}{\text{min}}\,x\,\frac{24\,liters}{mole}\,x\,\frac{1000\,ml}{liter}=\frac{7.46\;ml\,H_2}{amp-\text{min}} \quad (7.12)$$

From Eq. 7.7:

$$\frac{1/4\;mole\;O_2}{98{,}487\;amp-\text{sec}}\,x\,\frac{60\text{ sec}}{\text{min}}\,x\,\frac{24\;liters}{mole}\,x\,\frac{1000\;ml}{liter}=\frac{3.73\;ml\;O_2}{amp-\text{min}} \quad (7.13)$$

Therefore, approximately 11 ml/min of gas (7.5 ml/min H_2 + 3.7 ml/min O_2) are produced per amp of current, at standard conditions of 25°C and 1 atmosphere.

These gases are removed by flushing water over the surface of the electrodes during operation. In some CEDI devices, product water is passed over the electrodes. In this case a gas removal step may have to be employed prior to use of the product water. In other devices, a small stream of water is sent to the electrodes and then to drain or reuse in other plant applications. In still other devices, this water is combined with the reject water inside the module. In this case, the reject water may have to be vented before recycling.

Depending on the configuration of the electrode compartments and the feed water composition, some chlorine can be formed at the anode according to Equation 7.8. Values in the range of non-detectable to 8 mg/l Cl_2 have been observed [18]. This will also have to be taken into consideration when recycling or reusing the electrode and/or reject water.

People often express safety concerns regarding the electrode gases, since it is known that under certain conditions hydrogen can be explosive. However, the amount of gas that is produced by a CEDI module is so small that it does not present a safety hazard when the CEDI system is installed in an area with normal ventilation. Codes require that buildings have multiple air changes every day, one air change being a turnover of air equivalent to the building's internal volume. The number of air changes varies depending upon the building's use and the local codes, but a widely accepted value is half an air change per hour [36].

An example calculation will illustrate how little risk is presented by the CEDI electrode gases. Assume that a CDI system is installed in a 4m x 4m x 4m enclosed area that has the normal ventilation of half an air change per hour. This is equivalent to an air flow of 32 m^3/h. Assume the CEDI system is operating with a DC current of 10 amps, therefore producing hydrogen at the rate of 74.6 ml/min (equivalent to 0.0045 m^3/h). If all the hydrogen gas leaves the water and enters the room air, then the resulting concentration of hydrogen in the air would be about 0.014% (v/v), or about 141 mg/l. This is well below the explosive limit of a hydrogen/air mixture, which is 4.2% v/v at STP [37], and also well below the concentration at which asphyxiation would occur. It is good practice to perform this type of calculation as part of the CEDI system design process for each specific application. It may be worth noting that the related process of electrodialysis typically operates at much higher DC current density than CEDI, producing more hydrogen and often requiring active ventilation to disperse the hydrogen gas.

7.9 Operation and Maintenance

When compared to conventional demineralization, one of the advantages of the CEDI process is the reduction in labor required to operate and maintain the system. High-maintenance chemical handling equipment is replaced by a DC power supply, there is no resin separation/regeneration process to monitor, and the in-place electrochemical regeneration is much gentler and less likely to damage the ion exchange resin. Once the system has been properly set up, operation consists mainly of monitoring key performance parameters such as pressure drops and electrical resistance, much the same as you would monitor an RO system.

7.9.1 Estimation of Operating Current and Voltage

One of the most important steps in starting up a CEDI system is to determine the DC amperage and voltage required. The amperage can be calculated using Faraday's law as shown previously in Equation 7.2. This requires the product water flow rate, the concentration of ions in the feed water, the desired product water quality, and the current efficiency. Normally the value of the current efficiency will be based on the recommendation of the CEDI device manufacturer.

If the CEDI module electrical resistance is known, the required DC voltage can be determined from the calculated amperage by using Ohm's law. However, because of the many factors affecting module resistance, in most cases the resistance can only be determined by using the projection program provided by the CEDI device manufacturer.

7.9.2 Power Supply Operation

DC power supplies can be designed to operate either at constant voltage or at constant current. In constant voltage mode the circuitry of the power supply maintains the DC voltage at a fixed value, which means that the DC current will change in response to any change in module resistance. For example, if the water temperature decreases, the module electrical resistance will increase, and the DC current will go down. This would be expected to cause a drop in product water quality, requiring a manual adjustment to increase the applied voltage.

Since the product water quality of a CEDI system is directly related to the current, it has become popular to operate these systems in constant current mode. In this case the circuitry of the power supply maintains the DC amperage at a fixed value, which means that the DC voltage must change if the module resistance changes. For example, if the water temperature decreases, the power supply will automatically increase the DC voltage to keep the current at the set point. This mode of operation gives more consistent product water quality.

There have been some attempts to regulate the output of the DC power supply in order to maintain the CEDI system effluent conductivity at a given set point. These attempts have not been successful, due to the complex, non-linear relationship between DC voltage (or current) and CEDI product water quality. Therefore this practice is not recommended.

7.9.3 Power Consumption

Once the DC amps and volts are known, it is straightforward to calculate the power required to remove ions, as shown in Equation 7.14:

$$Rectifier\ Power = \frac{I_a \times E}{1000 \times \eta_R \times Q_p} = \frac{kwh}{m^3} \qquad (7.14)$$

This requires an assumption of the fractional efficiency of the DC power supply (rectifier), which typically is in the range of 0.85 to 0.95.

If the CEDI system also includes a concentrate recirculation pump, the power consumption of the pump can be estimated using Equation 7.15:

$$Pump\ Power = \frac{HP \times 0.7457}{Q_p} = \frac{kwh}{m^3} \tag{7.15}$$

7.9.4 Flows and Pressures

While the ion exchange membrane is for the most part impermeable to water, it is possible to have a small amount of water flow through the membrane, or across one of the seals separating the diluting compartments from the concentrating compartments. Such a flow of water between compartments is termed a cross-leak. Because of the difficulty of ensuring that all membranes and seals are absolutely perfect, most CEDI device manufacturers recommend that the dilute outlet pressure be maintained at a value slightly higher than the concentrate outlet pressure, typically 0.1–0.4 bar. This insures that if a cross-leak occurs on the product end of the module, that the leak is from the product compartment to the concentrate compartment and not vice-versa.

The majority of commercial CEDI devices operate co-flow, with the dilute, concentrate, and electrode streams all flowing in the same direction. The main reason for co-flow operation is to avoid an imbalance of the pressures between adjacent compartments, which could lead to cross-leaking or to physical stress on the ion exchange membrane. Figure 7.10 shows the pressure imbalance that can result from counter-flow operation.

One area where CEDI devices may differ is on the direction of flow: up or down. Most devices operate downflow, which helps maintain the integrity of the resin position in any filled cells. There are some devices that operate upflow, perhaps to assist in purging of gases from the electrode reactions. Both have been shown to be viable, but it is necessary to follow the manufacturer's recommendation for a specific module.

Initial startup of a CEDI system requires adjustment of the product, concentrate, and in some cases concentrate recycle and

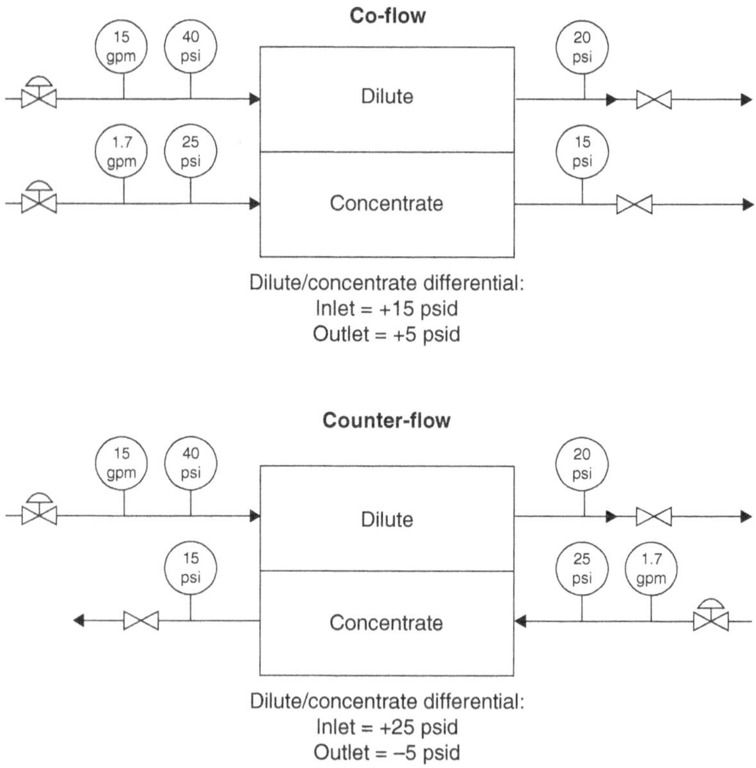

Figure 7.10 Co-flow and counter-flow pressure example

electrode flow rates. While setting the flow rates the pressures between compartments should be balanced.

In order to calculate the proper reject flow for a CEDI system, it is necessary to first know the maximum allowable recovery. With an RO system, this can usually be calculated using a scaling prediction index such as the Langelier Saturation Index. Unfortunately such indices are not very useful with CEDI devices, because the water splitting that occurs in the device can create significant localized risk of scale formation event though the bulk solution would not appear to be scale-forming. Therefore it is best to follow the recommendations of the CEDI device manufacturer regarding allowable recovery on a specific feed water. A typical recommendation would be to operate at 95% recovery if the CEDI feed hardness is less than 0.1 mg/l as $CaCO_3$, or to use 90% recovery for hardness of 0.1 – 1.0 mg/l.

Once the recovery is known, the reject flow is calculated using Equation 7.16:

$$Q_r = \frac{Q_p}{Y} - Q_p - Q_e \tag{7.16}$$

This calculation requires the desired product flow (design flow rate) as well as the electrode flow (which is usually a fixed flow per module, independent of overall system recovery). Some CEDI devices do not have a separate electrode stream (the electrode compartments are identical to the concentrate compartments), in which case the Q_e term is zero. In some other devices the electrode compartments are fed by a separate stream, but this stream is neglected when calculating reject flow from recovery.

7.9.5 Record Keeping

The most important part of operating an RO/CEDI system is to be diligent about recording the operating data on a regular basis. Should any problems arise, it can be very difficult to troubleshoot an RO/CEDI system without historical records. With such records, it may be possible to look at trends in operation over time and determine the most likely cause of a change in performance, and develop an appropriate strategy for remediation.

The most important items to record are shown in the sample log sheet provided below, in Table 7.6. Procedures for evaluation of CEDI module performance are available from standards organisations [46].

7.9.6 Cleaning and Sanitization

Some CEDI modules have remained in continuous service for over 20 years without any cleaning or sanitization, a testament to the effectiveness of RO as pretreatment to CEDI. But it is also possible to have pretreatment upsets which result in the exposure of CEDI modules to organic, inorganic and biological foulants that can reduce the quality or quantity of product water from the CEDI system. In such cases it is often possible to restore the performance of the CEDI system by performing a chemical

Table 7.6 Typical Operating Log Sheet: 2-Module CEDI System

	DATE		
	TIME OF DAY		
FEED	Temperature	°C	
WATER	Hardness	mg/l $CaCO_3$	
	Free chlorine	mg/l as Cl_2	
	Carbon dioxide	mg/l as CO_2	
	Conductivity	µS/cm	
	Silica	µg/l as SiO_2	
SYSTEM	Product conductivity	µS/cm	
DATA	Product silica	µg/l as SiO_2	
	Total product flow	m^3/h	
	Total reject flow	m^3/h	
	Feed (dilute inlet) pressure	bar	

	Product (dilute outlet) pressure	bar	
	Product pressure drop	bar	
	Concentrate inlet pressure	bar	
	Reject (outlet) pressure	bar	
	Reject pressure drop	bar	
MODULE	DC voltage	volts	
A	DC current	amps	
	Resistance (V/A)	ohms	
MODULE	DC voltage	volts	
B	DC current	amps	
	Resistance (V/A)	ohms	

cleaning. This would normally be done in response to one of the following conditions:

A decline in product water quality
A drop in flow rate or increase in pressure drop (dilute or concentrate)
An increase in electrical resistance.

Most CEDI devices can be chemically cleaned in much the same fashion as an RO system, by recirculating a chemical solution through the module at nominal service flow rates. A typical clean-in-place (CIP) apparatus is shown in Figure 7.11.

The manufacturer of the CEDI device must be consulted to confirm the compatibility of the CEDI module with a proposed cleaning solution, but Table 7.7 below gives cleaning solutions that are typically used.

The device manufacturer should also be consulted for specific cleaning instructions, as it is possible to damage a CEDI module by performing the cleaning incorrectly. In general a brine flush should be performed before an alkaline cleaning, in order to avoid the

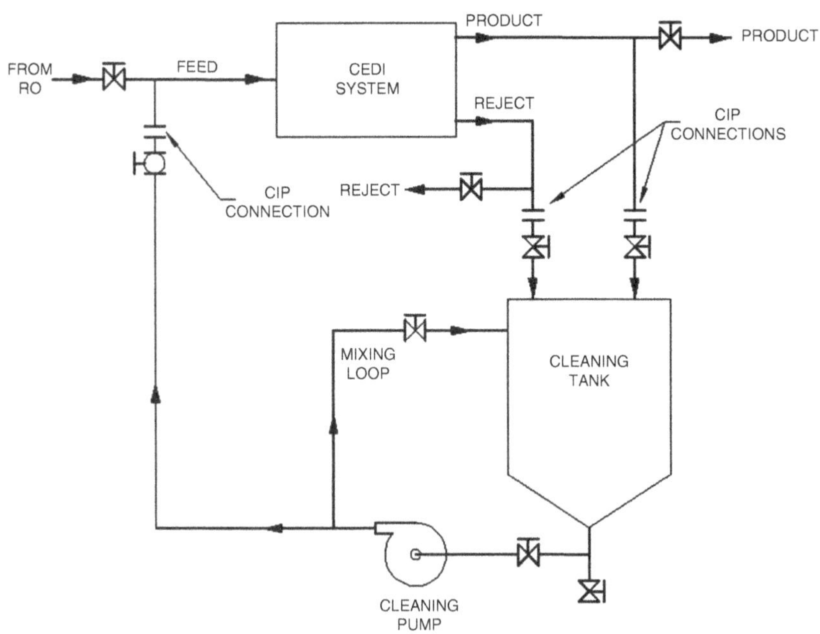

Figure 7.11 Typical CIP apparatus

Table 7.7 Typical Cleaning Solutions Used on CEDI Modules

Problem or foulant	Cleaning solution
Calcium carbonate scaling	2-4% hydrochloric acid
Biofouling or silica fouling	1-2% caustic (sodium hydroxide)
Organic fouling	5% salt / 1% caustic (mixture)
Routine sanitization	0.04% peracetic acid

formation of precipitates from hardness accumulated on the cation exchange resin. Recent findings from RO membrane cleaning suggest that high pH cleaning should be done before low pH cleaning, as the low pH can "harden" organic and biological foulants, making them more difficult to remove [39].

7.9.7 Preventive Maintenance

It is not normally recommended that CEDI systems be chemically cleaned or sanitized periodically as preventative maintenance. This would typically only be done as a response to a change in operation, as mentioned above.

The most important preventative maintenance items are the following:

- Test control interlocks
- Clean any salt deposits off the outside of the CEDI modules
- Check the torque on the tie bars nuts or endplate bolts, tighten as required.

In most case these tasks should be performed annually. However, it could be more or less frequent, depending on the CEDI device. Follow the module manufacturer's recommendations.

7.10 Applications

7.10.1 Pharmaceutical and Biotechnology

The first widespread adoption of CEDI was in the pharmaceutical industry, where it was shown that RO/CEDI systems were not only capable of removing particulate, organic and ionic contaminants but also provided very low levels of bacteria and bacterial endotoxins [47]. There was evidence that the electric field minimized the growth of bacteria in electrodialysis systems [39], which has led to system design incorporating product water recirculation so that the DC power remains on at all times. Improvement in construction allowed some CEDI devices to eliminate chemical sanitization by allowing sanitization with hot water at 65°C [40], and later modules were developed that could tolerate hundreds of sanitization cycles at 80°C [41]. Most new pharmaceutical and biotech process water systems will employ RO and CEDI. These are typically low flow rate systems, less than 10 m^3/h.

7.10.2 Steam Generation

The early nineties saw the first adoption of CEDI at power plants to produce DI water to feed a high pressure boiler [42, 48] or for NO_x control. With development and improvement of thick-cell CEDI devices, the cost of CEDI systems dropped significantly, making the technology more competitive with conventional deionization for higher flow rate applications. While boiler makeup water must meet strict limits for conductivity, sodium and silica, bacteria are not a concern, and systems are typically operated in start/stop mode. The higher flow rates also affect the design of the RO pretreatment equipment, introducing other factors that must be considered in the design of the overall RO/CEDI system [33]. The power industry has been somewhat cautious in implementing the use of membrane technologies for water treatment, but now seems to have widely accepted CEDI. It is believed that the world's largest CEDI system is one that provides boiler feed water for district heating [43].

7.10.3 Microelectronics/Semiconductor

The industry with the most demanding requirements regarding process water purity is likely semiconductor wafer fabrication, which desires DI water that is below detection limits for ionic, particulate and organic contaminants. Several related industry segments have similar requirements, such as manufacturing microelectronic devices, solar panels, or flat-panel displays. These water systems utilize extensive pretreatment, typically product-staged reverse osmosis and membrane degasification. CEDI is increasingly popular as an alternative to regenerable mixed-beds as a makeup deionizer, providing 16–18 megohm-cm to the DI water storage tank. While CEDI has been shown to be capable of providing water quality meeting or exceeding that of a regenerable mixed-bed deionizer [44], non-regenerable mixed-bed polishers are still more common in the final DI polishing loop.

7.10.4 System Sizing

Before designing a CEDI system for integration into a complete water treatment train, there are several pieces of information that must be known:

- Feed water composition and temperature
- Required product flow
- Required product quality and composition
- Allowable pressure loss

Once the above information has been gathered, projection programs provided by the various CEDI module manufacturers can be used to solve the design in an iterative fashion. Most of these programs allow for entering a feed water composition, i.e. RO permeate quality projection, and number of modules. Then the program estimates the product quality and composition as well as the pressure loss and power requirements. If any of these do not meet the process requirements, there are several options. First, the pretreatment can be modified to change the CEDI feed water composition. For example, softening can be implemented to reduce calcium or magnesium or a second pass of RO can be added to reduce the silica level. Some projection programs also allow the user to change the number of CEDI modules. More modules for a given product flow rate will lower the pressure loss and most likely improve the product quality. Another option that might be considered is to accept the effluent quality from the CEDI and add some type of polishing to meet process requirements.

7.11 Future Trends

The CEDI process presently occupies a relatively small niche in the field of water treatment, that of polishing RO permeate to produce ultrapure water. One possible area of development for the future would be to make the process tolerant of a broader range of feed water conditions, perhaps allowing relaxation of the feed water requirements. Such devices are described in the literature [45].

The past several years have seen a significant drop in the cost of CEDI modules, through such improvements in module design as the use of thick cells. But there is still considerable opportunity to further reduce the cost of CEDI systems by addressing the design of the skid, piping, power supplies and controls, which have not yet been optimized.

Nomenclature

C_{di}	Dilute inlet concentration, equivalents/liter
C_{do}	Dilute outlet concentration, equivalents/liter
E	DC voltage
EF	E-factor, as a fraction (dimensionless)
HP	Motor horsepower, concentrate recirculation pump
η_C	Current efficiency, as a fraction (dimensionless)
η_R	Rectifier efficiency, as a fraction (dimensionless)
F	Faraday's constant, 96,500 amp-seconds/equivalent
I	DC current, amps
I_a	Actual DC current applied to a CEDI module, amps
I_t	Theoretical DC current, from Faraday's Law at 100% current efficiency, amps
N_b	Number of beads (linearly, between ion exchange membranes)
N_{cp}	Number of cell pairs in the CEDI module
P	Probability, as a fraction (dimensionless)
Q_d	Dilute flow per CEDI module, liters/second
Q_e	Electrode flow per CEDI module, m^3/hour
Q_p	Product flow per CEDI module, m^3/hour
Q_r	Reject flow per CEDI module, m^3/hour
R	Electrical resistance, ohms
Y	Water recovery (conversion), as a fraction (dimensionless)

References

1. W.R. Walters, D.W. Weiser and L.J. Marek, Concentration of aqueous radioactive wastes *Ind. Eng Chem* 47, 61–67 (1955).
2. P. Kollsman, Method and Apparatus for treating ionic fluids by dialysis, U.S. Patent No. 2,815,320, December 3 (1957).

3. G.J. Gittens and R.E. Watts, *Some Experimental Studies of Electrodeionisation Through Resin Packed Beds*, United Kingdom Atomic Energy Authority Research Report (1964).
4. T.R.E. Kressman, Process for the removal of dissolved solids from liquids, U. S. Patent 2,923,674, February 2, (1960).
5. E. Glueckauf, Electro-deionisation Through a Packed Bed, British Chemical Engineering, 646–651 (1959).
6. Z. Mateka, Continuous production of high-purity water by electrodeionisation, *J. Appl. Chem. Biotechnol* 21, 117–120 (1971).
7. G. Kunz, Process and apparatus for treatment of fluids, particularly desalinization of aqueous solutions, U.S. Patent No. 4,636,296, January 13, (1987).
8. G.C. Ganzi, Y. Egozy, A.J. Giuffrida and A. Jha, High purity water by electro-deionization *Ultrapure Water J* 4(3), 43-50 (1987).
9. I.G. Towe, D.F. Tessier and M.P. Huehnergarrd, Modular apparatus for demineralization of liquids, U.S. Patent 6,193,869B1, February 27, (2001).
10. E.J. Parsi, K.J. Sims, I.D. Elyanow and T.A. Prato, Introducing and removing ion-exchange and other particulates from an electrodeionization stack, U. S. Patent 5,066,375, November 19 (1991).
11. P. Rychen, S. Alonso, H. Alt and D. Gensbittel, Apparatus for continuous electrochemical desalination of aqueous solutions, U. S. Patent 5,376,253, December 27, (1994).
12. B.M. Stewart and D. Darbouret, Advancements in the production of ultrapure water for ICP-MS analysis, American Laboratory, 36–38 (1998).
13. Wiklund, Lars, November 2012 personal communication regarding Ionpure CEDI system operating at the University of Stockholm Institute for Microelectronics in Sweden since April (1988).
14. G.C. Ganzi, Electrodeionization for high-purity water production AIChE Symposium Series, No. 261, Vol. 84, New Membrane Materials and Processes for Separation, Sirkar, K. K. and Lloyd, (Ed.), 73–83 (1988).
15. R. Simons, Electric field effects on proton transfer between ionisable groups and water in ion exchange membranes, *Electrochimica Acta* 29(2), 51–158 (1984).
16. G.C. Ganzi, A. Jha, F. DiMascio and J. Wood, The Theory and Practice of Continuous Electrodeionization, *Ultrapure Water J* **14(6)** (1997).
17. Liang, Apparatus for fluid purification and methods of manufacture and use thereof, U. S. Patent 7,572,359, August 11 (2009).
18. J. Gifford and D. Atnoor, An Innovative Approach to Continuous Electrodeionization Module and System Design for Power Applications, International Water Conference (2000).

19. A. Giuffrida, A. Jha. And G Ganzi, Electrodeionization Apparatus, U.S. Patent No. 4,632,745, December 30, (1986).
20. G.C. Ganzi, The Ionpure™ continuous deionization process: Effect of electrical current distribution on performance, Presented at the 80th annual AIChE meeting, Washington, D.C., November 28, (1988a).
21. Y. Oren, A Giuffrida and S. Ciaccio, Process for purifying water, U.S. Patent No. 5,154,809, October 13, (1992).
22. H. Neumeister, L Furst, R. Flucht and V.D. Nguyen, High Purity Water by Electrochemical Treatment, *Ultrapure Water J* 13(7), October (1996).
23. J.H. Wood, G.C. Ganzi and P. Springthorpe. *Continuous Electrodeionization: Module Design Considerations for the Production of High Purity Water* (2000). Ion Exchange at the Millennium; Proceedings of IEX 2000; Churchil College, Cambridge, England: Imperial College Press, Editor J.A. Greig
24. F. Helfferich, Ion-exchanger Membranes, in *Ion Exchange*, Dover Publications, 339–420 (1995).
25. R. Kunin, Permselective Membranes and Their Applications in Ion Exchange Resins, Robert Kreiger Publishing (1990).
26. A. Giuffrida, Heterogeneous Ion Exchange Materials Comprising Polyethylene of Linear Low Density or High Density High Molecular Weight", U.S. Patent 5,346,924, September 13, (1994).
27. J.M. Bernatowicz, M.J. Snow and R.J. O'Hare, Methods and Apparatus for the Formation of Heterogeneous Ion Exchange Membranes, U.S. Patent 6,503,957B1, January 7 (2003).
28. O. Kedem, L. Schechtmann, Y. Mirsky, G. Saveliev and N. Daltrophe, Low-polarisation electrodialysis membranes *Water Supply* 17(1), 305–314 (1999).
29. E.A. Mason and T.A. Kirkham, Design of Electrodialysis Equipment, Chemical Engineering Progress Symposium Series, 24, 55 (1959).
30. H. Barber, G. Towe and D.F. Tessier, EDI Operation for Removal of Weakly Ionized Impurities, International Water Conference (2000).
31. C. Griffin, Advancements in the Use of Continuous Deionization in the Production of High Purity Water, *Ultrapure Water J.* 8(8), 52–60 (1991).
32. P.A. Schweitzer, Electrodialysis and Dialysis in Handbook of Separation Techniques for Chemical Engineers, McGraw Hill, 454–465.
33. J. Wood and J. Gifford, Process and System Design for Reliable Operation of RO/CEDI Systems, Proceedings of the International Water Conference, October (2004)
34. G.A. Pittner, Reverse Osmosis System, U. S. Patent 4,574,049, March 4, (1986).
35. J.L. White, Electrodeionization and Ultraviolet Light Treatment Method for Purifying Water, U. S. Patent 5,116,509, May 26, (1992).

36. Gunderson, Roald, Breathing in Buildings: Indoor Air Pollution; The Problems and Solutions, http://www.mwt.net/~roald/indoorair.html (1999).
37. D. Bjerketvedt, J.R. Bakke and K. van Wingerden, Combustion Properties of Fuel-Air Mixtures, Chapter 4 in *Gas Explosion Handbook*, CMR, (1999).
38. C.A. Broden and C.D. Gilbert, Pushing the Limits: Improved Reverse Osmosis Membrane Cleaning Recommendations, International Water Conference (2007).
39. T. Sato, T. Tanaka and H. Ohya, Bactericidal effects of an electrodialysis system on E. Coli cells, *Bioelectrochem. Bioenerg* 21, 47–54 (1989).
40. J. Wood, J. Hirayama and S. Sato, Hot Water Sanitization of Continuous Electrodeionization Systems, *Pharm. Eng* November/December (2000).
41. J. Arba, Electrodeionization Device and Methods of Use, U.S. Patent 7,147,785 B2, December 12, (2006)
42. D. Auerswald, Regeneration Without Chemicals, *Ultrapure Water J.* April (1994).
43. C. Ho and J. Wood, Design, Construction and Operation of a 6,730 gpm RO/CEDI System for Con Edison's East River Repowering Project, Proceedings of the International Water Conference, October (2006).
44. A. Jha, L. Liang, J. Gifford, Advances in CEDI Module Construction and Performance, Proceedings of Semiconductor Pure Water and Chemicals Conference, February (2004).
45. C.J. Gallagher, F.C. Wilkins and G.C. Ganzi, Polarity Reversal and Double Reversal Electrodeionization Apparatus and Method, U. S. Patent 5,558,753, September 24 (1996).
46. ASTM Method D6807-02, "Standard Test Method for Operating Performance of Continuous Electrodeionization Systems on Reverse Osmosis Permeates from 2 to 100 μS/cm", Volume 11.02, 2009.
47. G. Ganzi and P. Parise, The Production of Pharmaceutical Grades of Water Using Continuous Deionization Post-Reverse Osmosis, *J. Parenter. Sci. Technol* 44(4), July-August (1990)
48. K. Leyton, M. Pacek and C. Fournier, Upgrading Boiler Makeup Feed Water Quality While Reducing Production Costs at the Dave Johnston Plant, International Water Conference, November (1994).
49. S. Thate, A comparison of Different EDI Concepts Used for the Production of High Purity Water, *Ultrapure Water J* October (1999).
50. F.L. Tye, U. K. Patent 815,154 (1959).

8

Membrane Distillation: Now and Future

Xing Yang*,[1,2], Anthony G. Fane[1,2], Rong Wang[1,2]

[1]*School of Civil and Environmental Engineering, Nanyang Technological University, Singapore 639798*
[2]*Singapore Membrane Technology Centre, Nanyang Technological University, Singapore 639798*

Abstract
This chapter provides a comprehensive introduction to the state-of-the-art of Membrane Distillation, including basic theoretical principles, system design specifications, technical advancements, limitations, and future prospects for water desalination. Focus will be on the historical perspectives of MD, process engineering aspects including membrane characteristics, module design, applications and cost evaluation. The technical requirements together with the benefits and limitations will also be addressed in terms of economic consideration and process engineering. It is expected to provide future prospects to implement the MD process industrialization as a promising desalination technology.

Keywords: membrane distillation, desalination, mass and heat transfer, renewable energy, process engineering, water production cost (*WPC*)

8.1 Introduction

In the past decades, membrane technology has made an enormous progress and membrane separation processes have become competitive to the conventional approaches for a wide range of applications, particularly in desalination and water reclamation. Due to the increasingly deteriorating situation in water resources and quality, over one billion people cannot access safe drinking water all over

**Corresponding author:* yang0243@e.ntu.edu.sg

Jane Kucera (ed.) Desalination , (373–424) 2014 © Scrivener Publishing LLC

the world [1]. Therefore, many attempts have been made to obtain drinking water from seawater, including conventional distillation, membrane-based separation processes such as reverse osmosis (RO), nano-filtration (NF). In recent years, a non-isothermal membrane process, namely membrane distillation (MD), is emerging as a promising alternative to conventional distillation and RO, as it may be potentially cost effective by utilizing low-grade waste heat or renewable green energy sources [2].

Membrane distillation (MD) is a process for water treatment that is driven by a temperature gradient across a microporous hydrophobic membrane between a hot feed solution and a cold permeate. It involves simultaneous mass- and heat-transfer phenomena through the membrane: the evaporation of the water molecules at the hot interface, the transport of water vapor across the porous partition (the membrane) and the condensation of water vapor at the cold interface [3]. MD is a promising technique for water desalination because of several advantages: low sensitivity to salt concentration and theoretically 100% salt rejection; feasibility to utilize low-grade heat and renewable energy (e.g., industrial waste heat, solar power or geothermal energy); low vulnerability to membrane fouling, low equipment cost and good performance under mild operating conditions as compared to conventional, multi-stage distillation or pressurized process like RO; the versatility of applying different MD configurations for various applications based on specific requirements [4]. Four main configurations including the direct contact membrane distillation (DCMD), vacuum membrane distillation (VMD), air gap membrane distillation (AGMD) and sweeping gas membrane distillation (SGMD) will be discussed in a later section.

Granted with attractive benefits to acquire high quality water, MD has been a targeted subject and studied extensively by numerous experimentalists and theoreticians for over forty years worldwide. Although its principles and benefits have been proved by lab- or pilot-scale studies, MD has only gained little acceptance and yet to be fully implemented in industry. Potential challenges impeding its application include the following: developing appropriate MD membranes to avoid pore wetting; increasing the permeation rates; mitigating flow maldistribution and/or poor hydrodynamics and severe temperature polarization (TP) that compromise module performance; and assessing the energy consumption and water production cost with reliable standards . However, attempts have been made to gradually overcome the above challenges and prove

MD as a potential alternative technique. Also, new strategies have been explored to enhance the thermal efficiency by heat recovery and achieve more separation purposes via systematic integration, such as solar-powered MD (SPMD), integrated reverse osmosis membrane distillation (ROMD), etc.

Hence, this chapter will provide a comprehensive introduction to the state-of-the-art of MD process, including basic theoretical principles, configurational variations, and technical advancement and limitations, and future prospects in terms of desalination. Focus will be given to the historical perspectives of MD, process engineering aspects including membrane characteristics, module design, applications and cost evaluation. Technical requirements together with the benefits and limitations will also be addressed in terms of economic consideration and process engineering. It is expected to provide future prospects to implement the MD process industrialization as a promising desalination technique.

8.2 MD Concepts and Historic Development

8.2.1 MD Concepts and Configurations

Membrane-based separation processes have found numerous applications in various fields such as water, energy, chemical, petro-chemical and pharmaceutical industries. This growth has been primarily due to two developments: firstly, the ability to produce high permeability and essentially defect-free membranes on a large scale; secondly, the ability to assemble these membranes into compact, efficient and economical membrane modules with much higher contact surface area [5, 6].

As a thermal process, MD presents very straightforward working principles: on one side an aqueous feed solution is heated and brought to the surface of a hydrophobic and porous membrane, which acts as a physical interface between hot and cold streams. The hydrophobic property of the membrane helps to prevent the water penetrating into the membrane pores, while only allowing the water vapor to pass through. This helps to achieve a vapor-liquid equilibrium at each pore entrance.

A variety of approaches arises when it comes to the vapor or distillate collection on the other side of the membrane, i.e., different methods can be adopted to impose a driving force (vapor pressure

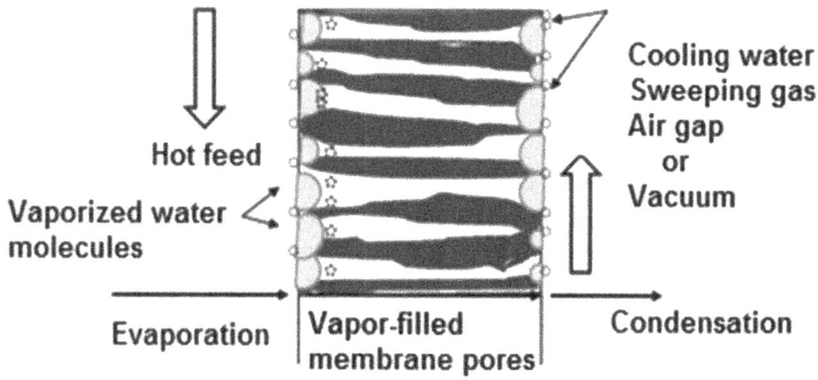

Figure 8.1 Basic working principles of MD (redrawn from [7])

difference) across the membrane matrix to produce flux. Fig. 8.3 gives the schematics of four commonly studied MD configurations, namely the direct contact membrane distillation (DCMD), air gap membrane distillation (AGMD), sweep gas membrane distillation (SGMD), and vacuum membrane distillation (VMD), which are schematically shown in Fig. 8.2(a)-(d), respectively.

The original concept of membrane distillation was initiated as direct contact membrane distillation (DCMD) [9], where both hot and cold aqueous streams are in direct contact with the membrane surfaces. The operation is simple and requires the least equipment. The vapor pressure gradient caused by temperature difference across the membrane drives the water vapor from the hot feed surface through the pores to the cold permeate side and condenses. The mechanism is given in Fig. 8.2(a).

As shown in Fig. 8.2(b), in AGMD, the water vapor permeates through the membrane pores to the other side and diffuses through an air gap to a cold surface where it condenses. It introduces a heat insulating air gap between the membrane and the distillate water, so the conductive heat loss across the membrane is greatly reduced. However, the presence of air gap increases the mass/heat-transfer resistance, thus results in lower permeate flux [10].

SGMD is similar to AGMD, both having permeate collected externally. The sweep gas in SGMD carries the vapor which is cooled down outside of the chamber (Fig. 8.2(c)). The concept of VMD is to reduce the absolute pressure in the downstream side of the membrane by employing a vacuum, which establishes the

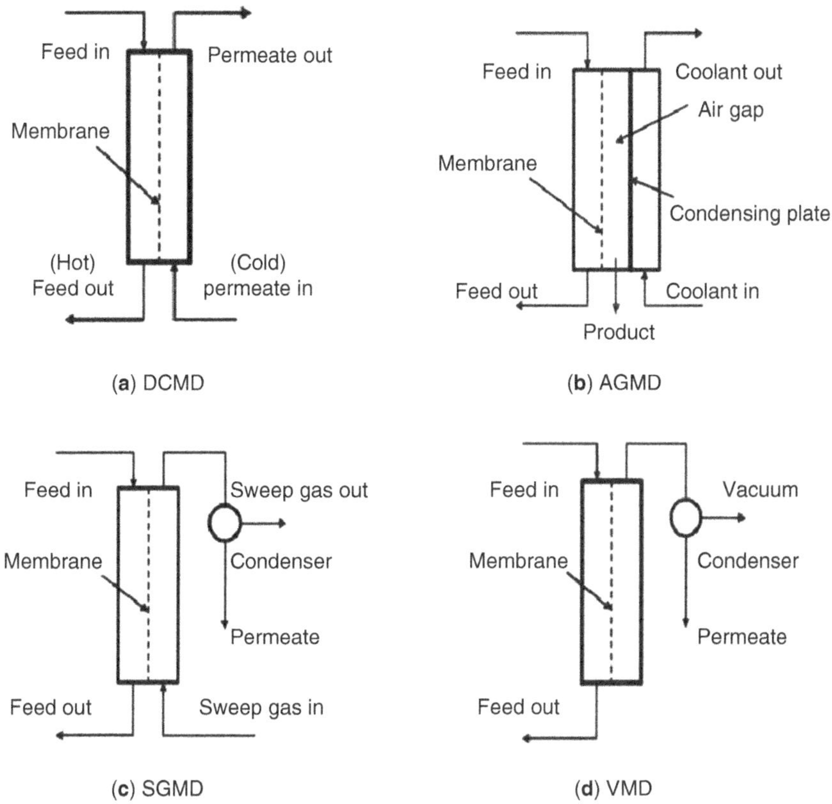

Figure 8.2 Configurational variations of MD [8]

vapor pressure difference, as shown in Fig. 8.2(d). VMD overcomes the problem of air resistance to transfer.

Generally, the direct contact membrane distillation (DCMD) is the most studied and simplest mode among the various MD processes because no external condenser is required as compared to vacuum membrane distillation (VMD) and sweep gas membrane distillation (SGMD) [4], which are the least studied configurations due to the extra system complexity and equipment cost. However, regardless of the slight distinction on the flux collection for different configurations, the membrane itself plays an insignificant role in selecting different species, except under extreme circumstances where the membrane pores are smaller than the mean free path of the diffusing molecules and hence acts as a screener. Fortunately, most nonvolatile species, such as ions, colloids, macromolecules, etc, are theoretically 100% rejected in MD [11].

8.2.2 Historic Development

As early as in 1963, Bodell [9], for the first time, proposed the concept of membrane distillation and filed a patent. However, the first related paper was published by Findley few year later in 1967, regarding the basic theories of MD operation and experimental evidence for the most desirable membrane characteristics needed for MD [12], i.e., high hydrophobilicity surface. Later that year, another patent was filed on the MD process designs using the single, multi-staged units and even spiral wound module [13]. More interestingly, some of very recently studied MD membrane concepts were already discussed in this patent, such as composite hydrophobic-hydrophilic membrane. In 1968, Bodell again proposed alternative configurations namely SGMD and VMD for improving MD efficiency in desalting from various water sources [14]. To have a deeper understanding of MD, Findley [15] studied the heat and mass transfer and suggested based on the preliminary MD modeling work "if low cost, high temperature, long-life membranes with desirable characteristics can be obtained, this method could become an economic method of evaporation, as well as an important possibility in the conversion of water". Therefore, the interest towards MD has been vastly lost several years later due to the low productivity (compared to RO) and unknown energy requirement. Even though several important concepts were proposed through the investigations of enhancement strategies for thermal efficiency by partially recovering the latent heat from evaporation [16, 17].

In the late 1980s, with the rapid development of membrane fabrication technologies, the spotlight of MD was brought back around by enthusiastic academic community. By then the membrane permeability/productivity could achieve 100-fold higher than that of those membranes Findley used in the 1960s. Several industrial companies such as Gore and Associates (USA) [18], Swedish Development Co. [19, 20] and Enka AG (Germany) [11, 21, 22] have tried commercialize their MD units. However, the enthusiasm for MD died out again after several commercial attempts due to the discouraging outcomes. Fortunately, since 1990s the progress on module design and understanding on the temperature and concentration phenomena have helped to revive the confidence of MD researchers and again make it an attractive and competitive alternative for water treatment [23–25]. The growth rate of the research interest on MD from the 1960s till 1990s has been reported in prior reviews [8]. As the heat continues to keep the attention with the fact that this process embodies many fundamental engineering concepts, particularly for academic research, the importance and popularity of the MD technology as a promising separation method can be seen through the increasing number of publications and citations since 2000, as shown in Fig. 8.3, about 70% of which are on desalination. Not only larger scale laboratorial studies but also commercial MD units have been successfully demonstrated by membrane companies such as Memsys® [26, 27] and Memstill® [28]. It is evident that extensive research has focused on novel MD membrane development and fabrication, optimization of operating parameters, module design and energy analysis and cost estimation. Moreover, the advancement on the MD technology also brings a new future for the utilization of low-grade waste heat and/or alternative renewable green energy sources such as geothermal and solar power.

Overall, the MD process has been proposed and investigated for over forty years, and the principles and benefits have been proved by numerous lab-scale studies. Despite the great efforts made, MD is still under evaluation and yet to fulfill all the expectations from industry, due to the potential disadvantages.

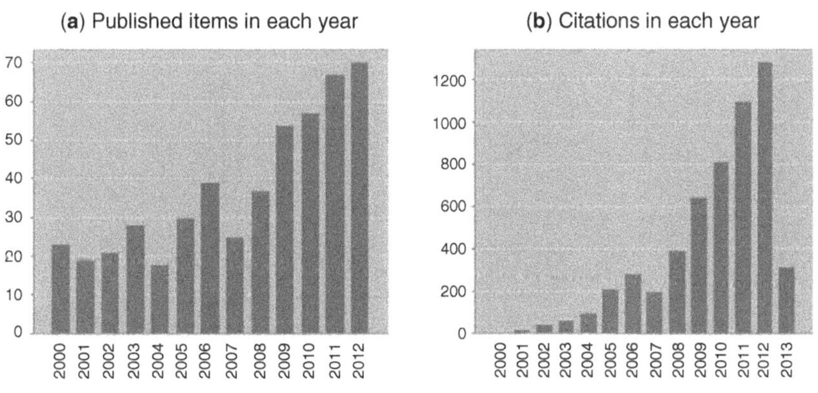

Figure 8.3 A surge of interest in MD from 2000 to 2012 (data obtained from Web of Science)

8.3 MD Transport Mechanisms

MD is a unique membrane process with a combination of mass and heat transfer, a full understanding on each aspect and their coupling effects is essential. Fig. 8.4 shows the combination of mass and heat transfer leading to two important phenomena in MD involving the treatment of aqueous solutions containing more than one component: concentration polarization (CP) and temperature polarization (TP). The CP refers to the build-up of the retained species at the membrane surface (here, salt is particularly studied as the main solute), and it exists in all membrane processes. TP refers to the thermal profiles developed near the membrane surface in the bulks. Both the CP and TP lead to smaller temperature differences than that of the bulk streams.

8.3.1 Mass Transfer in MD

The mass transport of the volatile species takes place in three steps: (i) from the bulk feed to the membrane wall; (ii) through the

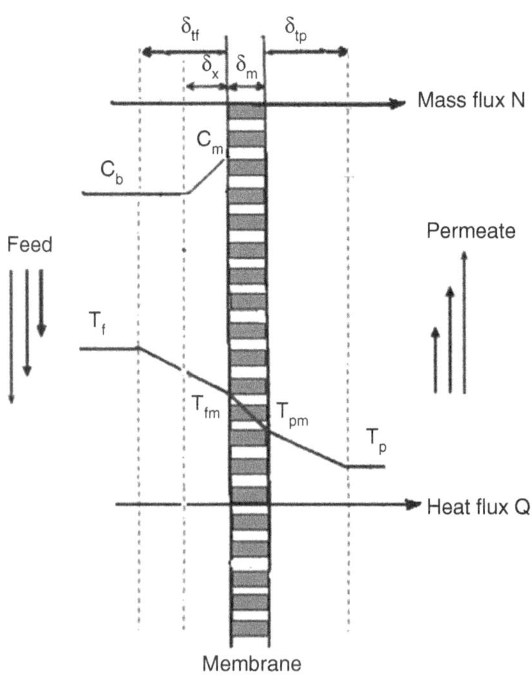

Figure 8.4 Temperature- and concentration-polarization effects in MD (redrawn from [29])

membrane pores in gas phase; (iii) from the membrane wall at the permeate side to the bulk permeate. The individual resistances are described as $1/k_f$, $1/k_m$ and $1/k_p$ for the feed, membrane and permeate sides, respectively. Both of the mass-transfer resistances in the feed and permeate are mainly controlled by the molecular diffusion through the liquid boundary layers adjacent to the membrane surfaces. They are dependent on the flow conditions (velocity, etc), fluid properties (viscosity, temperature, etc) and channel dimensions, and commonly determined using empirical models [30].

The overall resistance of mass transfer $1/K$ can be expressed using the resistance-in-series model:

$$\frac{1}{K} = \frac{1}{k_f} + \frac{1}{k_m} + \frac{1}{k_p} \tag{8.1}$$

8.3.1.1 *Mass Transfer Through the Feed Boundary Layer (CP Effect)*

In MD desalination, the component of interest is the solvent itself (water, considered as volatile) in the feed solution. Therefore, no discussions on the transport of nonvolatile solutes (e.g., salts), which is assumed to be 100% rejected, will be addressed in this chapter. However, due to the existence of liquid boundary layers, the chance of solute accumulation on the membrane surface will still result in additional mass-transfer resistance and lead to CP and wetting problems, which is detrimental to MD performance. As depicted in Fig. 8.4, the CPC is written as:

$$CPC = \frac{C_m}{C_b} \tag{8.2}$$

where C_b and C_m are the concentrations of nonvolatile solute in the bulk feed and on the membrane surface, respectively. Although it is well recognized that the effect of CP on flux decline is not as significant as pressure-driven processes until saturation, care must be taken to avoid scaling and membrane pore wetting [8].

It is noted that the membrane selectivity might be affected if there is other volatile impurities from the feed solution, other than nonvolatile salt solutes. In this case, more considerations will be given to the improvement of membrane surface properties, or pretreatment requirements.

8.3.1.2 Mass Transfer Through Membrane Pores

In an MD desalination process, where highly purified distillate is produced, the mass transfer across the membrane is driven by the transmembrane vapor pressure gradient, ΔP, caused by temperature difference $\left(T_{fm} - T_{pm}\right)$. Therefore, the general form of vapor flux N is defined based on Darcy's law:

$$N = K\left(T_{fm} - T_{pm}\right) = K\frac{\Delta P}{\delta_m} \tag{8.3}$$

From the above equation, a correlation between mass and heat transfer is clearly indicated. Determined by Antoine equation [31]:

$$P = \exp(23.20 - \frac{3816.44}{T - 46.13}) \tag{8.4}$$

Where T is the operating temperature and P is the vapor pressure. It is clear that the MD flux N exhibits an exponetial increase with the increase of operating temperature. As defined in Eq (8.1), K is the overall mass-transfer coefficient of MD which characterizes the permeablity of the porous membrane and is commonly determined empirically via a function of pore geometries (pore size, porosity, tortuosity, etc) and operating temperature.

However, the empirical equations used in the calculations are widely based on the dusty gas model (DGM) [32] applied across porous media. It is described as a combination of viscous flow (momentum transfer), Knudsen (molecule--pore wall collision) and molecular diffusion (molecule-molecule collision), and surface diffusion involving molecule – membrane interaction (usually ignored). To describe the individual resistance k_i from each transport mechanism in DGM, an electronic analog can be used, as shown in Fig. 8.5a.

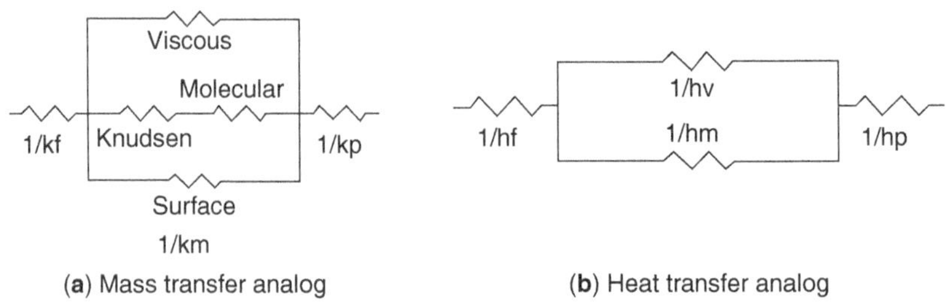

Figure 8.5 Mass- and heat-transfer analogs in MD (redrawn from [4])

Often, more than one mechanism may occur through the membrane matrix due to the characteristic pore size distribution of MD membrane. Since the surface diffusion term is usually ignored, a complete expression of the DGM in MD is given by the following three equations [32]:

$$\frac{N_i^D}{K_0 v_{M,i}} + \frac{p_j N_i^D - p_i N_j^D}{K_1 P D_{ij}} = \frac{-\nabla p_i}{RT} \tag{8.5}$$

$$N_i^V = \frac{-p_i B_0}{RT\eta} \nabla P \tag{8.6}$$

$$N_i = N_i^V + N_i^D \tag{8.7}$$

Where N_i^D is the diffusive flux, N_i^V is the viscous flux and total flux is N_i. P is the total vapor pressure and p_i is the partial pressure of component i, ∇x is the derivative of x in all directions. R is the gas constant, T is the operating temperature, $v_{M,i}$ is the mean molecular velocity of component i, μ is the fluid viscosity. $D_{i,j}$ is the binary/ordinary diffusion coefficient for components i and j. K_0, K_1 and B_0 are the membrane structural parameters for Knudsen diffusion, molecular and viscous flow, respectively. These constants are again determined experimentally based on the membrane structural values such as pore radius r, tortuosity τ and porosity ε (uniform cylindrical pores in the membrane are commonly assumed) [33–38].

Based on the analog in Fig. 8.5a, the arrangement of respective transport resistances could be different based on the portion of individual contributions affected by the membrane surface properties, matrix structure and feeding components. For instance, although the surface diffusion term is usually ignored with water as the main component due to the extremely low ratio of surface area to pore area, the molecule-membrane interaction can be dominant for MD process employing composite hydrophobic-hydrophilic membranes, or membranes with exceptionally small pores (< 0.02 μm) [4]. Hence, the diffusion mechanisms through the boundary layers and membrane matrix could be significantly different whenever, there is strong affinity between the compounds in the bulk and the membrane surface. However, there is no literature available on this aspect [39].

Also, since in most cases the VMD system employs membranes with much smaller pore size, which is smaller than the mean free path of the molecules, the molecule-molecule collision (Molecular diffusion) has negligible contribution to the mass-transfer resistance; while the molecule-pore wall (Knudsen flow) is dominant. Thus, the vapor flux across the membrane in VMD is Knudsen diffusion limited:

$$N_i = \frac{K_0}{RT}\left\{\frac{8RT}{\pi M}\right\}^{1/2}\frac{\Delta p_i}{\delta_m} \tag{8.8}$$

where the transmembrane pressure gradient $-\nabla p_i$ is replaced by a specific value $\nabla p_i/\delta_m$. K_0 can be measured with gas permeation of a non-condensable gas such as nitrogen. Therefore, VMD has been extensively used for the removal of volatile organic carbons or inert gas from water [32, 40, 41].

On the other hand, the molecular diffusion limited DGM can be applied in DCMD and AGMD applications due to the dominant resistance from the stagnant air trapped within the membrane pores or the air gap of AGMD. So the dissolved air in the membrane pores acts a stagnant film, even for membranes with pores as small as 0.2μm. The DGM only gives the diffusion of water vapor through a stagnant film equation:

$$N_i = \frac{D_{12}^0}{T^b}\frac{T^{b-1}}{R\delta}\frac{\Delta p_i}{|p_2|_{\ln}} \tag{8.9}$$

Experimental results on AGMD confirmed that the vapor flux N is proportional to the air gap thickness δ [42]. Also, a significant increase on the membrane permeability after a complete deaeration of both streams in DCMD was observed [43, 44].

In reality, due to the uniformity of membrane pores, the actual application of DGM in MD is more likely on transitional models. For example, the Knudsen-molecular diffusion mechanism for membranes with pores filled with air (pore size < 0.5 μm), when the frequency of molecule-pore and molecule-molecule collisions take place simultaneously [4]; empirical Knudsen-viscous transition DGM models are applied to degassed DCMD systems and VMD with relatively large pores (mean pore size $\langle r \rangle \approx$ mean free path λ) [4, 36].

8.3.2 Heat Transfer in MD

With rapid development on membrane fabrication technology, the heat transfer in all MD configurations is playing a more dominant role as the controlling steps has gradually shifted away from the mass transfer limitation. To realize phase change and perform separation in MD, a certain amount of heat must be provided externally (from the feed solution) for the latent heat of evaporation. Along with the water vapor, this amount of heat is transferred across the liquid films and through the membrane matrix until condenses on the cold surface of the permeate side. Similar to the mass transfer, the heat-transfer resistance in MD can also be described via an analog, as shown in Fig. 8.5b, consisting three consecutive steps: heat flux q transferred from the bulk feed solution to the membrane surface, where the water turns into vapor phase; heat transferred across the membrane matrix, where two aspects are considered: latent heat carried by vapor (q_{MD}) and conductive heat loss through the membrane material (q_{HL}), which is considered as heat loss; finally, heat is released on cold membrane surface due to condensation. The corresponding heat transfer resistances for the feed, membrane and the permeate sides are $1/h_f$, $1/h_m$ (where $h_m = h_{HL} + h_{MD}$) and $1/h_p$, respectively, as depicted in the electric analog in Fig 8.5b. Hence, the overall resistance is given as below:

$$\frac{1}{H} = \frac{1}{h_f} + \frac{1}{h_{HL} + h_{MD}} + \frac{1}{h_p} = \frac{1}{h_f} + \frac{1}{h_{HL} + N\Delta H_T/\Delta T_m} + \frac{1}{h_p} \quad (8.10)$$

where the N is the vapor flux; ΔH_T is the molar heat of evaporation; ΔT_m is the membrane difference between two membrane surfaces; $N\Delta H_T/\Delta T_m$ is the heat transfer coefficient of evaporation, h_{MD} and the coefficient of conduction through the membrane is h_{HL}. The total heat transferred through the membrane (including evaporation heat and conduction loss) q (W·m^{-2}), is written as follows:

$$q = q_{HL} + q_{MD} = h_{HL}(T_{fm} - T_{pm}) + N\Delta H_T \quad (8.11)$$

$$q = h_f(T_f - T_{fm}) = h_p(T_{pm} - T_p) \quad (8.12)$$

where T_{fm}, T_{pm}, T_f and T_p are membrane wall temperatures and fluid bulk temperatures at the feed and permeate side, respectively, as illustrated in Fig. 8.4. Through the above Eqs. (8.10)–(8.12), the membrane wall temperatures, mass- and heat-transfer coefficient can be iterated and solved using empirical equations [4, 45] or obtained through computational fluid dynamics (CFD) modeling [46, 47].

8.3.2.1 *Heat Transfer on the Feed Side (TP Effect)*

Similar to any other thermal process, a larger temperature difference will lead to a faster heat flow rate. The thermal boundary layer formed at each side of the membrane surface was reported to impose up to 80% reduction in the driving force due to the TP effect [4, 48–51]. Thus, the temperature-polarization coefficient (*TPC*) was defined as the fraction of transmembrane temperature difference to the bulk temperature difference [23]:

$$TPC = \frac{T_{fm} - T_{pm}}{T_f - T_p} \tag{8.13}$$

All related temperatures are schematically shown in Fig. 4. It is noted that the *TPC* calculation has a slight variation in VMD [4]. Nevertheless, an effective reduction on the thickness of both boundary layers, which are functions of fluid properties and hydrodynamic conditions, is favorable for mitigating the TP effect in any MD configurations. Enhanced heat transfer can be achieved by improving the design of flow passage, membrane arrangement and fiber configuration or applying turbulence promotes like spacers/baffles [52]. Operating at a higher circulating flowrate is always seen as an option to achieve better mixing conditions and minimizing the TP effect. However, sometimes it will not be economically feasible under optimum conditions, and it is also necessary to ensure a low pressure drop along the fiber length in order to avoid membrane pore wetting, as the pressure drop along the fiber is a function of the second power of the fluid velocity [8].

8.3.2.2 *Heat Transfer Across the Membrane—Conduction and Evaporation*

In MD, the heat transferred by conduction through both the membrane matrix and the gas-filled pores is considered heat loss and should be minimized in order to decrease the TP effect and increase the process efficiency. The heat flux transferred by conduction through the membrane can be expressed by [23]:

$$q_{HL} = h_{HL}(T_{fm} - T_{pm}) \tag{8.14}$$

where h_{HL} $(= \lambda_m/\delta_m)$ is the equivalent heat-transfer coefficient for conduction through the membrane. It is obvious that a larger membrane wall thickness δ_m is favorable for reducing the conduction heat loss but disadvantageous for the mass-transfer resistance, as suggested in Eq. (8.3). Therefore, a compromise has to be made for determining an optimum membrane thickness. The λ_m is the average thermal conductivity of the membrane, which takes into account both thermal conductivity of the membrane matrix and thermal conductivity of the gas-filled pores.

$$\lambda_m = \varepsilon\lambda_s + (1-\varepsilon)\lambda_g \tag{8.15}$$

where ε is the membrane porosity, λ_s and λ_g are the thermal conductivities ($W{\cdot}m^{-1}{\cdot}K^{-1}$) of the membrane material and the gas in the pores, respectively. It must be pointed out the thermal conductivity of air/gases is an order of magnitude lower than that of the membrane material. Thus the heat loss by conduction can be minimized by using membranes with high porosity. On the other hand, MD relies on phase change to perform the required separation process. Therefore, the amount of heat contributing to the evaporation step is considered as the efficient heat. The efficiency of the process will be maximized if the TP effect, the internal heat loss by conduction and the external heat loss to the environment are reduced.

8.4 Strategic Development for an Enhanced MD System

8.4.1 MD Membranes

In the past decades, most MD research activities have been focused on the theoretical modeling and process/system design. However, it is well-recognized the availability of qualified and cheap membranes is one of the major barriers impeding the MD industrialization. To achieve an efficient MD process, the membrane should satisfy the following requirements: firstly, contains at least one hydrophobic and porous layer with appropriate pore sizes and a narrow pore size distribution to prevent membrane pore wetting; secondly, has a hydrophobic selective layer with a high porosity, which is usually corresponding to a high vapor flux; thirdly, exhibits an optimum membrane thickness, which can compromise the conduction heat loss and mass-transfer resistance, as suggested in Eq. (8.3); lastly, has excellent thermal and chemical stability to maintain a stable long-term performance with high salt rejection. In brief, membrane pores should be filled with water vapor or non-condensable gases and must not be wetted by the liquid. To avoid pore wetting, the membrane material has to be hydrophobic with a contact angle as high as possible and the membrane should have a relatively small maximum pore size, highest possible porosity, appropriate wall thickness and tortuosity. Normally, the hydrophobic micro-porous membranes such as those made from polytetrafluoroethylene (PTFE), polyvinylidene (PVDF), polyethylene (PE) or polypropylene (PP) can fulfill the basic requirement of hydrophobicity. However, since most of these currently available membranes are fabricated for other processes such as microfiltration (MF), they suffered from the presence of some large pore sizes when applied to MD processes. The pore sizes are 'nominal' mean sizes and there will be a distribution including larger pores. The presence of larger pores is a possible reason causing membrane wetting even though the membranes are highly hydrophobic [53]. To have a better sense of the currently-used commercial membranes in MD, a summary is provided in Table 8.1. It is evident that there is a lack of commercial hollow fiber or capillary membranes used in MD. Although some of these membranes exhibit extremely high permeability, such as 3M membranes [54], the long term performance and system cost is to be evaluated.

Considering the characteristic wetting phenomenon and low permeability problem in current MD applications, there is a need for specialized MD membranes as has been proposed by many researchers [4, 55–64]. Hence, the quest for qualified membrane materials and the optimization of fabrication parameters using homogenous hydrophobic materials/dope solutions to achieve desired structures is of great interest [65–69]. Characteristic hydrophobic porous membranes can be prepared using various formation mechanisms / techniques such as stretching, sintering, phase inversion or thermal phase separation (TIPS). It is mainly based on the specific material properties such as the solubility/compatibility with solvents, thermal and mechanical properties, and fabrication cost and process complexity [70]. Major progress has been made in the past 6 years on the fabrication and modification of polymeric MD membranes and hence a significant increase in reliability for the MD process is anticipated [71]. Nevertheless, the development of MD membranes is often constrained by the conventional spinning methods and the limitation of suitable material properties due to its unique process characteristics. Thus, novel techniques of making composite and mix matrix membranes rise and greatly extend the possibility of producing more permeable MD membranes, including surface modified (coating, grafting and plasma polymerization, etc) and electro-spun nanofiberous membranes [39, 71, 72]. A summary of the state-of-the-art of lab-scale MD membrane development is listed in Table 8.2. Relatively, the flat sheet membranes are better studied due to their simplicity. With great versatility and flexibility, composite membranes are of great interest in satisfying MD needs. In brief, there are two the main strategies to modify the hydrophobic membrane surface and minimize membrane wetting: one is a hydrophobic coating or grafting which helps to minimize the pores size while maintaining similar porosity and higher contact angle [29, 73–76]; the other is a dense hydrophilic coating which can protect the effective membrane pores from wetting [61, 64, 77–81].

8.4.2 MD Module Design

Despite many attractive characteristics and numerous results from lab-scale studies, MD has not yet been fully implemented in industry [8, 120]. One of the major challenges impeding its application is the mitigation of flow maldistribution and/or poor hydrodynamics and subsequent severe temperature polarization (TP) that compromise module performance [4]. In recent years a surge of studies have focused on membrane development [118, 121–124] and energy analysis [94, 120, 125–127]. However, there has not been a comprehensive investigation of MD module design [39]. Yet the potential benefits of improved module design indicate that a more intensive effort is needed in this area.

Thus far most of the work on hydrodynamic improvement in MD studies has focused on flat sheet membrane modules including net-type spacers [128–132], which is illustrated in Fig. 8.6. Although the flat sheet membranes have small membrane areas and thus are limited to laboratory units, which exhibit great simplicity and ease for making small-scale multi-stack plate-and-frame [133] or spiral wound modules [18]. For example, Swedish company XZero AB [133] has commercialized a small-scale AGMD module for portable drinking water production but it was found that the poor hydrodynamics has resulted in low permeability and high pressure

Table 8.1 Summary of currently-used commercial membranes for MD (information extracted and modified from [39, 71])

Membrane type	Manufacturer	Materials	Nominal pore size (μm)	Porosity %	Reported permeability N (10^{-3} kg/m^2 s)	References
Flat sheet membranes						
TF 200	Gelman	PTFE/PP (PTFE supported by PP layer, with an active layer of 5–10nm)	0.20	80	18.69	DCMD, distilled water as feed, T_f=80.1°C, T_p=20.1°C [82]
					5.2	SGMD, distilled water as feed. T_f=70°C, T_p=20 °C (air) [83]
TF 450			0.45	80	6.10	SGMD, distilled water as feed, T_f=70°C, T_p=20°C (air) [71]
TF 1000			1.0	80	--	--
Taflen			0.8	50	--	--
Metricel		PP	0.1	55	--	--
TS22	Osmonics Corp.	PTFE/PP (PTFE supported by PP layer, with an active layer of 5–10nm)	0.22	70	21.67	DCMD, 0.6 g/L NaCl feed, T_f=60°C, T_p=20 °C [84]
TS45			0.45	70	22.22	DCMD, 0.6 g/L NaCl feed, T_f=60°C, T_p=20 °C [84]
TS1.0			1.0	70	--	--
PP22		PP	0.22,70	70	--	--

GVHP	Millipore	PVDF	0.22	75	13.52	DCMD, distilled water as feed, T_f =90.7 °C, T_p=19.7 °C [82]
					1.81	VMD, distilled water as feed, T_f =25 °C, P_p=1666.5 Pa [85]
HVHP (Durapore)			0.45	75	10.80	DCMD, distilled water as feed, T_f =70 °C, T_p =20 °C [53]
GVSP			0.22	80	--	
FGLP		PTFE/PE (PTFE supported by PE layer, with an active layer of 5–10nm)	0.20	70	8.56	DCMD, distilled water as feed, T_f =57.2 °C, T_p =20 °C[86]
FHLP			0.5	85	--	--
Gore	Gore	PTFE	0.2-0.45	89-90	--	[87]
Gore		PTFE/PP (PTFE supported by PP layer, with an active layer of 5–10nm)	0.2	44-78	--	[88]
Sartorious	Sartorious	PTFE	0.2	70	14.00	DCMD, distilled water as feed, T_f =70 °C, T_p =20 °C [53]
Enka		PP	0.1–0.2	75	--	--
Celgard 2500	Hoechst Celanese Co.	PP	0.05,	45	--	--
Celgard 2400			0.02	38	--	--

(Continued)

Table. 8.1 (*Cont.*)

Membrane type	Manufacturer	Materials	Nominal pore size (μm)	Porosity %	Reported permeability N (10^{-3} kg/m^2 s)	References
Flat sheet membranes						
Membrana	Membrana, Germany	PP	0.2	--	15.61	Cross flow VMD, distilled water as feed, T_f =59.2°C; P_p=103Pa [89]
Vladipore	Vladipore	--	0.25	70	--	[90]
3MA	3M Corporation	PP	0.29	66	102.6	VMD, distilled water as feed, T_f =74 °C; P_p=3·103 Pa[54]
3MB	3M Corporation	PP	0.40	76	117.0	VMD, distilled water as feed, T_f =74 °C; P_p=3·103 Pa [54]
3MC	3M Corporation	PP	0.51	79	160.2	VMD, distilled water as feed, T_f =74 °C; P_p=3·103 Pa [54]
3MD	3M Corporation	PP	0.58	80	27	DCMD, distilled water as feed, T_f =70 °C; T_p=20 °C [91]
3ME	3M Corporation	PP	0.73	85	40.5	DCMD, distilled water as feed, T_f =80 C; T_p=20 °C [91]
Teknokrama	Teknokrama	PTFE	0.5	80	--	[92]
G-4.0-6-7	GoreTex Sep GmbH	PTFE	0.20	80	--	--

Hollow fiber/ capillary membranes						
Membrane type	**Manufacturer**	**Materials**	**Nominal pore size (μm)**	**Porosity %**	**Reported permeability N (10^{-3} kg/ m² s)**	**References**
Accurel® S6/2	AkzoNobel	PP	0.2	70	8.56–6.36	DCMD, T_f =80°C, T_p=20°C, distilled water as feed [93]
					7.9	VMD, T_f =65°C, P_p=3.5-103 Pa [93]
					5.1	SGMD, T_f =70°C; T_p=20°C (air) [94]
MD080CO2N	Enka Microdyn	PP	0.2	70	0.83	DCMD, T_f =55°C, T_p=15°C, 35 g/L NaCl [95]
MD020TP2N			0.2	70	0.97	DCMD, T_f =70°C, T_p=15°C, 35 g/L NaCl [95]
Accurel®	Enka A.G.	PP	0.43	70	--	--
			0.5	66	--	[96]
			0.6	74	--	
Celgard X-20	Hoechst Celanese Co.	PP	0.03	35	--	--
Capillary membranes	Membrana GmbH	PP	0.2	75	15.83	VMD, dye solution, T_f =60°C, P_p=103 Pa [97]
EHF270FA-16	Mitsubishi	PE	0.1	70	--	[98]
UPE test fiber	Millipore		--	--	--	[98]
PTFE	Sumitomo Electric	POREFLON	0.8	62	--	[99]
PTFE	Gore-tex	TA001	2	50	--	[100]

Table 8.2 Lab-scale MD membrane fabrication

Membrane type		Materials	Preparation methods	References
Flat sheet	Single layer	PVDF	Phase inversion	[57, 85, 101]
		PTFE	Extrusion sintering or etching	[70]
		PP	Thermally induced phase separation (TIPS)	[70]
	Composite	Hydrophobic coated PES	plasma polymerization	[102]
		Copolymer PVDF-TFE	Phase inversion	[103]
		Copolymer PVDF-HFP	Phase inversion	[104]
		Modified CA	radiation polystyrene grafting	[105]
		Modified CN	Plasma polymerization	[105]
		SMM/PEI	Surface modifying macromolecules (SMMs)	[61, 64, 77-81]
		SMM/PS		
		SMM/PES		
		PFS/anodiscs	Surface modification	[63]
		PVA/PEG/PVDF	Hydrophilic modification via polymer blending and crosslinking	[106]
		Modified nanospiked glass	differential chemical etching	[107]
	Nanofiberous	Carbon Nantube Bucky Paper (CNT-BP)	chemical vapor deposition	[108]
		PVDF nano-fibers	Electro-spinning	[109–111]

		PTFE	Extrusion, sintering or etching	[70]
		PVDF 1#	Dry/jet wet spinning	[55]
		PVDF #2	Dry/jet wet phase inversion	[65]
Hollow fibers/	Single layer	PVDF #3	Wet phase inversion	[112, 113]
		Microvoid-free PVDF	Co-extrusion	[66, 112, 113]
		PP	Melt extruded/cold-stretched	[59, 114]
		PE	Melt extruded/cold-stretched.	[59, 67, 115]
	Composite	PVDF/PTFE	dry/jet wet phase inversion	[114]
		PVDF-HFP	dry/wet spinning	[67, 73, 74, 115, 116]
		Hydrophobic coating +PVDF	Surface modification (fluoro-compound solution and plasma polymerization)	[29]
		Hydrophobic coating + PP	Plasma polymerization using silicone fluoropolymer	[73–76]
		Hydrophobic coating +PES	plasma polymerization	[102]
		PVDF/Cloisite clay (mixed matrix)	Dry/jet wet phase inversion	[116, 117]
		Chitosan coated PVDF	Surface modification	[118]
		Hydrophilic coated PTFE	plasma polymerization	[119]
		Dual layer PVDF	co-extrusion dry/jet wet spinning	[117]
		Modified inorganic (Ceramic/TiO_2)	Surface grafting	[75, 76]

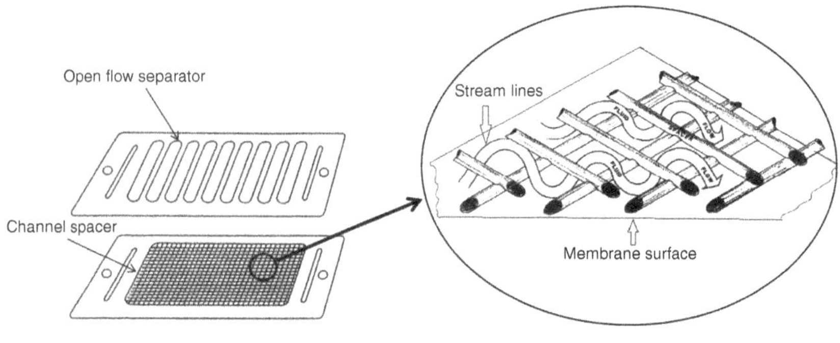

Figure 8.6 Net-type spacer design in flat-sheet membrane module [135]

drop. Another AGMD module was manufactured by Scarab [134]. However, the unit price of these modules is far more expensive to compete with any of commercial RO units. In recent years, there are two popular module concepts emerging, namely Memstill® [28] and Memsys® [27], respectively. The former was initiated and commercialized by a research organization TNO in Netherlands, as shown in Fig. 8.7a. The Memstill® module is a plate-and-frame configuration housing a continuum of evaporation stages in an almost ideal countercurrent flow process, which allows a high recovery of evaporation heat. Similarly, the latter is claimed to be a vacuum-multi-effect-membrane-distillation (V-MEMD) module, as shown in Fig. 8.7b, with alternating membrane frames and foil frames for evaporation and condensation, the vapor generated in the steam raiser under vacuum condition exchanges heat with the first-stage feed, which circulates to the next stage to generate vapor. The distillate is produced in each stage and in the condenser for the last stage vapor.

Nevertheless, in industry hollow fiber-based membrane modules are preferable due to their larger membrane area per unit volume and reduced vulnerability to TP [3]. The first commercial MD module was manufactured by Enka AG using their PP membrane [11, 21, 22]. However, it has been often reported that poorly-designed hollow fiber modules resulting in reduced productivity, increased energy consumption and shortened membrane lifespan [136], there are limited studies on improving fluid dynamics and designing hollow fiber modules for MD applications in the open literature [8, 22, 52, 137–139]. It is well-recognized that by incorporating

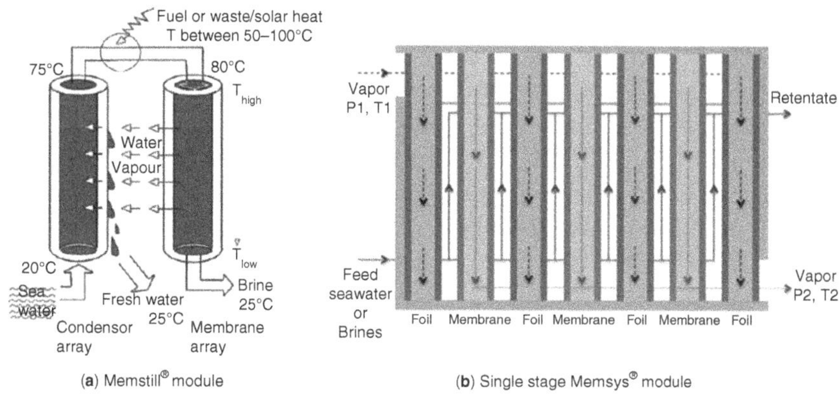

Figure 8.7 Working principles of (a) Memstill [28] and (b) Memsys [27] MD modules

proper flow alteration aids or modifying fiber geometries to create secondary flows or eddies (such as novel fiber configurations or turbulence promoters, e.g. spacers or baffles), the permeation flux can be enhanced and TP can be mitigated. An early exploration on hollow fiber module design by Schneider *et al.* [22] in 1988 investigated the effects of module size and modified fiber geometries on the transmembrane flux of the DCMD process. It showed that larger modules could achieve uniform flow more easily than smaller ones and certain capillary arrangements (twisted and braided geometries) could lead to much higher fluxes than those with straight woven fabric designs. In 2008 Teoh et al. [139] studied different hollow fiber configurations in the DCMD process and found that the introduction of baffles could increase the feed-side heat-transfer coefficients leading to 20%–28% flux enhancement. In addition, they also explored the concepts of wavy geometries (twisted and braided) of hollow fibers that were able to achieve as high as a 36% flux enhancement compared to the unaltered conventional modules. Recently, Yang *et al.* [124] described strategies to improve PVDF-based module performance in the DCMD process that included the investigations on the module size, packing density and critical fiber length combined with heat-transfer analysis. Although the existence of simultaneous concentration polarization (CP) and TP will lead to the reduction of mass- and heat-transfer driving force, it is well-established that the effect of CP in the DCMD system is negligible in comparison to that of TP [50, 140]. Therefore, the quantification of the TP effect, which is used to assess the thermal efficiency, is essential for the implementation of an MD system [3, 121, 140]. A recent module design work has addressed the associated conductive heat loss or the mitigation of TP by altering the fiber geometries or introducing turbulence aids, as shown in Fig. 8.8 [52]. To confirm the module performance in this MD module study, the fluid dynamics of the modules can be investigated using low-field bench-top nuclear magnetic resonance imaging (MRI), in terms of the cross-section imaging, diffusion and propagator experiments, and velocity maps in the shell-side flow [141–143]. Furthermore, the mass- and heat-transfer fundamentals in MD modules can be studied and optimized using numerical aids such as Computational fluid dynamics (CFD) modeling [46, 144].

Overall, a well-performed MD module should present high mass- and heat-transfer rates with low CP and TP effects as well as less scaling/fouling to maintain a high permeation rate. However, in reality a lack of adequate modules of industrial scale is still impeding the MD commercialization. Most of the currently used pilot-scale hollow fiber or spiral wound modules were originally design for other membrane processes such as membrane contactors or RO. Although it is well-recognized that MD plants can be scaled up simply by upsizing the module capacity, it is necessary to identify what are the important characteristics for a qualified industrial MD module: a high packing density corresponding with high surface area, in a cross-flow or transversal-flow operating mode, even fiber arrangements to avoid flow maldistribution, bypassing and dead

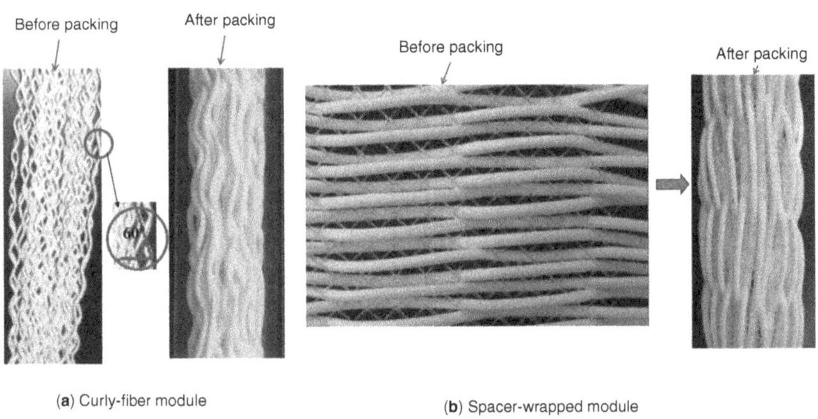

Figure 8.8 Novel module design and fabrication: (a) curly-fiber module; (b) spacer-knitted module [52]

zones (for hollow fiber/ capillary or tubular membranes), proper spacer insertion or fiber geometry modification to improve surface renewal and reduce TP/CP or potential fouling, and plastic housing with high resistance to temperature and chemicals should be used to avoid corrosion. Moreover, the module should allow easy drying (in case of membrane pore wetting), regular manual inspection and membrane replacement.

8.4.3 MD Process Parameters

As mentioned previously, the MD process performance can not only be enhanced by fabricating specialized membranes and designing adequate modules, but also through fluid management. Therefore, to facilitate the heat/ mass transfer and improve the permeation flux, an optimization of operating parameters is also essential in designing an MD system, such as the feed/permeate inlet temperatures, flow rates, feed concentration, etc.

Driven by vapor-pressure difference (temperature gradient), the feed and permeate temperatures show exponential influence to the vapor generation based on the Antoine equation [31]. Hence, a higher permeation flux will be obtained with an increased feed inlet temperature T_f (in a range of 40–90°C) in all MD configurations, which is unfortunately corresponding to a higher overall thermal consumption and more severe temperature-polarization effect (TP, Eq. (8.13)) [8, 132, 145, 146]. Also, a decrease on the species selectivity over water molecules was observed when treating brines containing volatile organic compounds [54, 147, 148]. Nevertheless, the more significant increase in vapor flux would contribute to a higher thermal efficiency, which will be further discussed in Section 5. In brief, to reduce the thermal boundary layer and mitigate TP phenomenon, a homogeneous temperature distribution at the feed side should be created by improving the flow distribution, as indicated in Fig. 8.4. On the contrary, an increase on the permeate temperature (in a range of 15–25°C) decreases the flux, because of the resulting smaller vapor-pressure difference across the membrane [4, 8, 94, 149]. However, changes on the permeate temperature (*e.g.*, temperature of the cooling plates) have negligible effect in AGMD and SGMD, where the heat-transfer resistance at the permeate side dominates [150–153].

Although it is stated that MD is much less sensitive to the salt concentration compared to pressure-driven processes, the water permeation flux still slightly decreases with increasing feed concentration (non-volatile solutes) in all MD configurations [4, 84, 91, 150]. There are two main reasons: a decrease on the vapor pressure due to the presence of solutes and concentration-polarization (CP) phenomenon. Nevertheless, the effect of feed concentration is not significant unless saturation is reached, where potential scaling or foulant deposition occurs [48–50, 148]. In this case, the addition of antiscalant as well as an improvement on the hydrodynamics of the membrane surface would be helpful.

It has been widely studied that the permeation rate will be enhanced via optimizing the flow conditions, *i.e.*, the feed and permeate flow rates. It is noted that the permeate flow condition can only be investigated in DCMD and SGMD. The effect of the feed flow rate is significant in DCMD, VMD and AGMD applications but negligible in SGMD [83, 94, 154]. Generally, a higher recirculating velocity (i.e., higher mixing intensity) up to turbulent condition can help to reduce the thickness of the liquid boundary layers adjacent to the membrane surface and maximize the driving force between the feed and permeate sides [94, 150, 155], which is favorable for the mitigation of concentration and temperature polarizations. However, increased pressure loss along the module length at high flow velocities might inevitably result in liquid penetration into the membrane pores (wetting problem) and subsequently distillate contamination. Also, increased pumping energy is a concern when employing a higher flow velocity. Hence, a compromise among the permeation flux, hydrostatic pressure loss and pumping energy should be made.

The above-mentioned variables are applicable to all MD configurations. Other parameters, including the air-gap distance in AGMD, selection of sweeping gas in SGMD or vacuum condition in VMD, are comprehensively reviewed [8]. Nevertheless, most of the prior studies focused on the individual effect by keeping other variables constant. It is well-recognized that there is an interplay between various parameters and hence an optimization should be carried out to achieve an enhanced MD operation [39].

8.5 Energy and Cost Evaluation in MD

Although with milder operating conditions and substantially less chemical addition involved, desalination using membrane separation is still considered as a capital and energy intensive approach to access clean water, especially for treating certain feed with higher salt concentration [156, 157]. In general, the energy and cost analysis in traditional distillation (e.g., multistage flush (MSF)) and membrane desalination process such as RO are well studied [156]. However, only a handful of literature is available for the energy analysis and cost evaluation in MD [71, 158]. Moreover, most prior studies only included very brief analyses on the MD energy consumption and cost evaluation, some of which are even different and conflicting results [158]. For instance, the unit water production cost could range from \$0.26/m^3 to \$130/m^3; while total energy consumption could be from 1 to 9000 kWh/m^3, depending on the types and size of the MD systems, operating conditions, energy sources and recovery approaches, cost estimation procedures, etc [39, 158]. Although it is clear that energy consumption and cost are becoming one of the main focuses for MD research [8, 159, 160], most of the prior results were obtained from laboratorial or small scale pilot plant studies [28, 160–162]. One of the main reasons is that MD is not yet fully industrialized. This has led to in unstable/fluctuating capital investment cost (e.g., plant size, capacity, membranes, modules and utility devices) as well as lacking of cost-related information such as pretreatment, optimized operating parameters, energy sources, long-term performance and fouling control. Also, the variation of four different configurations causes difference in capital and operating cost as well as permeation flux. More importantly, the cost of highly concentrated brine discharge, which is the top concern of the ecosystem, is yet to be acquired and taken into account [39]. Overall, thus far there are no widely agreed standards for MD energy estimation and cost calculations.

8.5.1 Thermal Efficiency and Cost Evaluation

In general, the energy consumption in MD systems includes the necessary thermal energy for heating the feed solution and cooling the permeate stream, and possible condensation step in the condenser (for VMD, SGMD, AGMD) as well as the electricity needed for the pumps and auxiliary devices. In most lab-scale or pilot plant studies, the MD energy consumption are evaluated via three thermally related metrics namely the thermal efficiency (η_h), gain output ratio (GOR) and water production cost (WPC). Firstly, as the total heat flux transferred through the membrane is split into two parts: the latent heat of evaporation (the effective portion that generates a certain amount of vapor) and the conduction across the membrane matrix (considered as heat loss), the MD thermal efficiency η_T is defined as the fraction of the evaporation heat:

$$\eta_T = \frac{NA_0 \Delta H_T}{q_m} = \frac{q_{MD}}{q_{MD} + q_{HL}} \tag{8.17}$$

Fundamentally, the reduction of inherent conductive heat loss should be minimized in the core MD system. Prior investigations [163] on the factors affecting the membrane conduction showed that it is mainly determined by the intrinsic MD coefficient C and operating temperatures. In general, the η_T increases with increasing membrane temperature, membrane thickness and porosity.

Some researchers used a similar term namely energy efficiency $\eta_E = \frac{A_0 \times N \Delta H_T}{E_{in}}$, which has further taken into account the overall energy input of an MD system E_{in}, including both thermal energy E_t and electrical consumption E_e. Interestingly, the η_E has exactly the same definition as our second metric GOR, which characterizes the ability of a particular MD plant on allocating energy for creating the actual driving force of the system [23], is expressed as:

$$GOR = \eta_E = \varepsilon_E H_R = \varepsilon_E \frac{q_{HE}}{q_m} \tag{8.18}$$

where H_R is the heat recovery factor, indicating the maximum amount of heat recoverable with certain heat transferred across the membrane. In a word, the GOR is a dimensionless parameter which reflects how well the energy input is utilized for water production in a system. Many attempts have been made to increase the GOR by incorporating heat recovery devices, improved module designs, effective insulation to the environment, optimized piping system and multi-staged operation, etc [25, 161, 164]. However, a trade-off relationship is found between the H_R and permeation rate N and it affects the *GOR* [164–166]. *i.e.*, high H_R could be achieved by designing a system with large membrane area, low flow velocity and more recovery stages; while the N decreased due to severe TP and CP effects. A wide dispersion on the *GOR* values from 0.3 to 8.1 is found in available literature because of different system settings. Therefore, compromise has to be made between the design of operating parameters and heat recovery units.

Thirdly, the *WPC* is a cost estimation based on the water production and total investment, the expression is given as:

$$WPC = \frac{C_{total}}{fM365} \tag{8.18}$$

Where f is plant availability, assumed as 90%; M is daily capacity kg/day and C_{total} is the capital cost (CC) of the MD plant \$/year (e.g., assuming an amortization life of 20 years with regular membrane replacement), which can be divided into three parts such as direct capital cost (*DCC*), indirect capital cost (*ICC*) and annual operating cost (*AOC*), including all necessary cost elements listed in Table. 8.3 [158]. An estimated percentage/contribution of each part to the overall cost is indicated. In general, the major cost comes from the process equipment purchase (mainly on membrane modules); while the installation and building is assumed as 25% of the equipment cost. Both *ICC* and *AOC* are roughly estimated as 10% of the *DCC*. Although a distinct dispersion was spotted on the inconsistent *WPC* evaluation for MD units of different specifications [158], more attention has been paid to the detailed cost study in recent years [95, 158, 165, 167, 168]. Sensitivity analyses on the effects of operating parameters, membrane properties as well as utility price were carried out. A more comprehensive evaluation package, which includes pilot plant data and formulas together with necessary but minimum assumptions, is available for economic calculations [95, 158].

8.5.2 Current Status of MD Cost and Energy Resources

The inherent conduction heat loss and high demand on thermal energy still seems to present a substantial barrier in implementing MD industrialization [120]. Nevertheless, it was reported that the *WPC* of MD desalination has decreased significantly in the past 50 years. For example, 25 years ago, a small solar-powered MD pilot plant with a production capacity of 500 kg/day had a WPC of \$10–\$15/m^3 [160]; while the modeling results showed that a conceptual DCMD plant with a capacity of 24,000 m^3/day performed a minimum *WPC* of \$1.23/m^3 using 55°C /25°C feed/permeate temperatures without heat recovery [95]. The same system with heat recovery was \$1.17/m^3 at even higher operating temperatures of 60°C /30°C feed/permeate. The *WPC* continued to drop to

Table 8.3 Cost elements involved in MD *WPC* calculations

Cost elements	**Estimated cost**
Direct capital cost (*DCC*)	
- MD Process equipment (membrane modules, heat exchangers, pumps)	>50% of total capital cost Membranes \$36-116/m2
- Auxiliary Equipment (open intakes, pipe lines, valves, transmission piping, storage tanks, generators and transformers, electric wiring, brine disposal line).	N.A
- Systems pre-treatment - Systems post-treatment.	N.A
- Installation and building (workshop, control room, laboratory, water analysis).	25% of equipment cost
- Renewable energy conversion and storage	N.A
- Land cost	N.A
Indirect capital cost (ICC), 10% of DCC	
- Administration and legal fees - Insurance	15% of DCC
Construction overhead	15% of DCC
- Labor burden - Field supervision	Variable, depends on the plant size
- Temporary facilities - Small tools and miscellaneous	5%–15% of DCC
- Contingency	10% of DCC
Annual operating cost (*AOC*)	
- Grid energy - Labor - Membrane replacement - Operation and maintenance (O&M) - Brine disposal - Consumables and chemicals for pretreatment and post-treatment: (Antiscalent, cleaning agents, etc.). - Amortization or fixed charges.	10% of total capital cost

\$0.64/m^3 if low grade waste heat was used [95]. An on-site study of the commercialized AGMD technology by Memstill® claimed to achieve a very competitive *WPC* of \$0.26/m^3, which is even lower than that of RO, when they operated MD using cheap industrial waste heat [28, 169]. However, a larger scale Memstill® system should be tested to confirm the potential desalination cost achieved by MD. The water cost was between \$0.56 and 0.97/m^3 when treating the concentrated NF and RO brines in an integrated MD system [167]. The encouraging results and continuous effort have built great confidence for a further reduction with more rigorous system design and cheaper energy sources, which is one of the main directions for MD application development in the coming years [158].

As mentioned previously, alternative low grade energy resources can be used to minimize energy cost and hence the *WPC* [39, 170]. Other than conventional low grade waste heat available from industrial processes (nuclear/diesel plants) [156], renewable energy such as salt gradient solar ponds (SGSP), solar and geothermal energy are widely considered as promising alternatives [171–173]. Some of these energy supplies can be integrated into the MD system for either direct or indirect (converted to electricity) heating to produce drinking water for special needs, particularly in arid rural areas short of electricity or regions with solar/geothermal abundant [39]. In the past decades until 1990s, either electrical power or "imaginary" low grade waste heat was employed to drive the evaporation process in most MD experiments and pilot plants. Therefore, assumptions have to be made on waste heat supplies and recovery capacity when conducting economic evaluation. Enormous efforts have been made to improve the thermal efficiency and overall MD performance, both experimentally and theoretically [39, 174]. For example, to maintain constant operating temperatures, an innovative integration of MD module and heat exchanger (MDHX) as one unit was proposed to facilitate independent heating/cooling for respective feed/permeate streams and heat recovery [175]. In recent years, with an increasing concern on the environmental sustainability, there has been a surge of interest on the development of clean renewable energy sources for desalination purpose. In 1991, the concept of solar-powered MD (SPMD) process was initiated, incorporated with photovotaic (PV) panel and heat recovery exchanger [161]. Major progress on hybrid solar MD systems was made on solar-powered AGMD units in various rural regions

(Germany, Jordan, Egypt, Morocco, Spain and Jordan) supported by the SMADAS project since 2002 [168, 176–180]. Simulation results showed that the MD process efficiency could be improved by optimizing the flow velocities, membrane area, PV panel area and heat exchanger capacity [181, 182]. In order to make a more compact SPMD system, some researchers introduced the integration of DCMD membrane modules and solar absorber [183]. However, confirmation on its long-term performance and the feasibility to integrate with other MD configurations are to be explored. Also, it is noted that in these prior SPMD studies only flat-plate solar collectors (SC) were employed. In fact, other thermal collectors such as parabolic solar concentrators or spherical collectors incorporated heat storage tank could be beneficial for performance improvement. Other than that, experimental and simulation work on the SGSP assisted MD, which has been proved to greatly enhance the thermal energy storage and utilization, was carried out in respective DCMD, AGMD and VMD configurations [170, 184, 185]. Among all potential energy resources used in MD, the geothermal was least explored and is less competitive in terms of current cost [162]. Hence, more intensive effort is needed to improve the process efficiency and decrease the *WPC*.

In a word, the discovery of new and cheap energy resources will enable more and more developing or under-developed countries to access safe desalted water. However, further interdisciplinary investigations should be conducted for larger scale hybrid MD systems with actual energy supplies. Also, comprehensive and standardized cost studies on these MD processes are needed.

8.6 Innovations on MD Application Development

In the past decades, the total cost and increased environmental concerns had limited the widespread adoption of desalination technologies. Nevertheless, for a state-of-the-art membrane process like MD is expected to see a massive increase in production of water as well as reduction in cost in the next 10–20 years. Especially for family and small community systems in remote locations or developing countries, solar or geothermal hybrid MD plants may find a demanding position, particularly in inland semi-arid areas with access to saline lakes and aquifers. As discussed previously in Section 4.4.2, with the development of new hybrid membrane systems, there is an increasing potential to achieve higher water quality and efficiency of the overall installation [186, 187]. The possibility of utilizing any industrial waste heat and renewable energy enables MD to cooperate with other processes (membrane or non-membrane based) and makes it a more versatile separation technique for desalination [187–189].

To increase the water recovery ratio and alleviate the environmental impact due to brine discharge, MD can be integrated with other pressure-driven membrane processes to further enhance the brine concentration. For example, Combined with cheaper pretreatment processes such as UF or NF, which can remove large particles or hardness and subsequently prolong the lifespan of MD membranes from a longer lifespan, the new hybrid MD system was proved to be cost effective [186]. A series of integrations of MD with NF, RO as well UF were found to greatly improve the product quality for drinking purpose. Drioli et al [189] proposed a combined RO and MD process to treat seawater based on the unique characteristic of MD with little sensitivity to salt concentration. It was reported that this hybrid process produced twice as much water as RO does (typical recovery of 39.1%) at the same cost, but with a much higher distillate quality; while saved 5% cost compared to a stand-alone MD system with the same water recovery of 85.6%. With such combination, the zero discharge concept could be implemented even saturation of salt solution is reached. However, in treating real brines from NF or RO plants the scaling problem was found to be unavoidable but could be mended by proper acid cleaning [190]. In these hybrid systems, all four MD configurations could be feasible based on the feed properties and local needs.

There are other hybrid MD processes benefiting from the uniqueness of MD to achieve desired separation purposes but lacking of details. For instance, the combination of MD and crystallization (MDC), which can not only help reduce brine discharge substantially down to zero, but also recover valuable solutes [191–193]. In fact, it was found that the profit of recovered solutes could cover the energy cost in an MD process [39]. However, further investigations are needed on this subject, such as enhancement strategies to achieve a higher process efficiency and avoid crystal scaling when approaching the super-saturation point [194]; comprehensive energy analysis is lacking [195]. To support this research thrust, computational simulations such as Aspen Plus will be invaluable. Another emerging hybrid MD concept is a combination of MD and membrane bioreactor (MBR), namely MDBR [196, 197]. Compared to the traditional activated sludge process (ASP) and ROMBR for water treatment, the MDBR has a high retention time that is required for an effective removal of refractory organics, especially with the potential of the low grade energy utilization. Nevertheless, the principal challenge of MDBR is the development of specialized membranes that can minimize fouling and other process inefficiencies. In recent years, with a resurgence of interest in forward osmosis (FO), which is also called direct osmosis (DO) that exhibits spontaneous water transfer through a semi-permeable membrane under transmembrane osmotic pressure gradient provided by the feed and draw solution, an interesting hybrid MD concept is proposed, namely FOMD. In such system, the diluted draw solution (i.e., concentrated salt solutions) from FO or DO can be regenerated in an MD unit to maintain a constant osmotic driving force [198, 199]. Despite the attractive benefits, intensive effort is still needed to develop specialized membranes, overcome internal and external concentration polarization effects (ICP and ECP) in FO, select proper draw solutions in terms of water recovery, improve the regeneration efficiency and energy utility.

In a word, more inspirations to achieve more challenging desalination demands with an enhanced MD system are to be explored. As mentioned previously in Section 4.4.2, the integration of energy supply with MD can still be further investigated. Other than adding the heat source into the feed stream, the overall driving force can also be enhanced by reducing the temperature of the permeate with cooling tower [200]. Also, although desalination has become the targeted application for MD, other traditional processes such as fermentation, distillation (multi-effect distillation, MED, or multi-stage flash, MSF) or photocatalysis can be integrated into MD to achieve desired separation purposes, even for solutions containing volatile compounds and/or organisms. In addition, innovative approaches can be developed for a simultaneous improvement on the productivity and system compactness, *e.g.*, designing an enhanced vacuum MD or applying deaeration in the feed and/or permeate streams in DCMD [43, 84].

8.7 Concluding Remarks and Future Prospects

With intensive research work dedicated in the past 50 years, tremendous progress has been made on the MD development academically. Fortunately, commercialized modular products that employed MD concepts are emerging for desalination needs, such as the vacuum multi-effect MD (V-MEMD) from Memsys and multi-stage DCMD Memstill® from TNO. Nevertheless, MD is still considered as under evaluation and yet to be fully implemented in an industrial scale, due to the potential technical challenges and unconvincing/inconsistent test results from engineering perspectives.

However, it MD will certainly become one of the most promising desalination techniques with the following aspects intensified in the future:

1. Membranes properties play an essential role in achieving high permeation rate and preventing pore wetting as well as fouling/scaling in MD. Unfortunately, effort is still needed to commercialize specialized MD membranes with competitive cost, especially hollow fibers. Therefore, the exploration of new materials and development of specialized MD membranes deserve more emphasis.
2. Although new thoughts have been given to module design in the past five years, most of MD evaluation results were obtained using non-standardized modules fabricated in lab-scale. Unfortunately, inconsistent flux data from module performance tests caused by non-standardized module specifications and irregular fiber arrangement (potential cause of flow maldistribution) was observed. Also, more attention is to be paid for module design for less studied MD configurations such as VMD, AGMD and SGMD. A rigorous technique developed for MD modularity is very important for implementing industrialization, particularly on hollow fiber or spiral wound modules.
3. Relatively, DCMD is the most studied configuration due to its technical simplicity, followed by VMD that usually exhibits much higher permeation flux, the other two configurations are rarely investigated. However, they still present attractive features from an engineering perspective.
4. Although the effects of operating conditions have been relatively well-studied, it is important to identify the interactions among parameters and optimize the MD system to achieve enhanced process efficiency and

reduced energy consumption. This can be done by using process control techniques such as design of experiments (DOE), statistical process control (SPC), etc.

5. A distinct dispersion on the system evaluation, including the gain-output-ratio (*GOR*), specific energy consumption and economic analysis, were reported depending on the type and size of the MD units, feed solutions, energy sources and cost, heat recovery facilities and cost study approaches. Also, there is insufficient evidence of large-scale MD operations. These remain as uncertainties in MD application development that failed to gain industry-wide recognition as an economically competitive desalination technique. Therefore, rigorous MD systems should be designed and developed.
6. The exploration on potential low grade energy resources has contributed tremendously to the progress of the MD development in the past decade. The integration of energy supply and MD configuration brings great added benefits in reducing the operation cost and advancing towards commercialization. There is a growing interest on collocating the MD units and thermal desalination plants for efficient water recovery in the Middle East. Nevertheless, a further reduction on the overall cost by mitigating conduction heat loss and intensifying energy recovery is worth exploring.
7. Initial attempts were made on hybrid MD systems (e.g., RO, UF/MF/NF) leading to an enhancement on the process efficiency by improving the product quality, increasing the overall water recovery and significantly reducing the disposal of RO or oil and gas-field brines, which have caused serious environmental problems. Continuous effort should be launched to further improve the efficiency and compactness of existing hybrid MD units and confirm their good performance in large scale. Moreover, innovative integrations, including MD with one or more MD configurations and MD with one or more traditional processes, are worth exploring for achieving more challenging separation needs with complex compositions.

Overall, with attractive advantages and interesting engineering characteristics, MD has great potential to play an important role in desalination. With continuous effort in academia and industry, the MD desalination capacity is projected to grow with a decrease in energy consumption and water production cost. These reductions are attributed to the continual technological improvements in the MD application development, including highly permeable membranes, efficient modules and energy recovery approaches, innovative hybrid integration, utility of cheap clean energy, multistage or cascade modular arrangements and brine management technologies, etc.

References

1. W. Jury and H. Vaux, The emerging global water crisis: Managing scarcity and conflict between water users, *Adv. Agronomy* **95**, 1–76 (2007).
2. A. G. Fane, Membranes for water production and wastewater reuse, *Desalination* **106(1)**, 1–9 (1996).
3. R.W. Schofield, A.G. Fane and C. J. Fell, Heat and mass transfer in membrane distillation, *J. Membr. Sci* **33**, 299–313 (1987).
4. K. W. Lawson and D. R. Lloyd, Membrane distillation, *J. Membr. Sci* **124(1)**, 1–25 (1997).
5. M. C. Porter, *Handbook of Industrial Membrane Technology*, William Andrew Inc, (1990).
6. V. Abetz, T. Brinkmann, M. Dijkstra, K. Ebert, D. Fritsch, K. Ohlrogge, D. Paul, K. V. Peinemann, S. Pereira-Nunes, N. Scharnagl and M. Schossig, Developments in membrane research: from material via process design to industrial application, *Adv. Eng. Mater* **8(5)**, 328–358 (2006).
7. X. Yang, *Membrane Distillation: Module Design and Modeling* (Ph.D dissertation), Nanyang Technological University, (2012)
8. M. S. El-Bourawi, Z. Ding, R. Ma and M. Khayet, A framework for better understanding membrane distillation separation process, *J. Membr. Sci* **285(1–2)**, 4–29 (2006).
9. B. R. Bodell, Silicone rubber vapor diffusion in saline water distillation, U.S Patent 285 032 (1963)
10. J. Phattaranawik, R. Jiraratananon, A. G. Fane and C. Halim, Mass flux enhancement using spacer filled channels in direct contact membrane distillation, *J. Membr. Sci* **187(1–2)**, 193–201 (2001).
11. K.Schneider and T. J. V. Gassel, Membrane distillation, *Chem. Eng. Technol* **56(7)**, 514–521 (1984).

12. M. E. Findley, Vaporization through porous membranes, *Ind. Eng. Chem. Res* **6,** 226–230 (1967).
13. P. K. Weyl, Recovery of demineralized water from saline waters,U.S Patent 3,340,186, (1967).
14. B. R. Bodell, Distillation of saline water using silicone rubber membrane, US Patent, Serial No 3,361,645 (1968),
15. M. E. Findley, V. V. Tanna, Y. B. Rao and C. L. Yeh, Mass and heat transfer relations in evaporation through porous membranes, *AIChE J* **15,** 483–489 (1969).
16. F. A. Rodgers, Stacked microporous vapor permeable membrane distillation system, US Patent Serial No. 3,650,905 (1972).
17. F. A. Rodgers, Compact multiple effect still having stacked impervious and previous membranes, US Patent, Serial No. Re.27,982 (1974).
18. D. W. Gore, Gore-Tex membrane distillation, Proc. of the Tenth Ann. Convention of the Water Supply Improvement Assoc., Honolulu, HI, USA, (1982).
19. L. Carlsson, The new generation in sea water desalination: SU membrane distillation system, *Desalination*, **45,** 221–222 (1983).
20. S. I. Andersson, N. Kjellander and B. Rodesjo, Design and field tests of a new membrane distillation desalination process, *Desalination*, **56,** 345–354 (1985).
21. T. J. V. G. A. K. Schneider, An Energy-Efficient Membrane Distillation Process, in *Membranes and Membrane Processes* E. Drioli and M. Nagaki (Ed.), 343-348, Plenum Press., New York, (1986).
22. K. Schneider, W. Holz and R. Wollbeck, Membranes and modules for transmembrane distillation, *J. Membr. Sci* **39,** 25–42 (1988).
23. R. W. Schofield, A. G. Fane and C. J. Fell, Heat and mass transfer in membrane distillation, *J. Membr. Sci* **33,** 299–313 (1987).
24. S. Kimura, S.I. Nakao and S. I. Shimatani, Transport phenomena in membrane distillation, *J. Membr. Sci* **33(3)**, 285–298 (1987).
25. R. W. Schofield, A. G. Fane and C. J. D. Fell, The efficient use of energy in membrane distillation, *Desalination*, **64,** 231–243 (1987).
26. W. Heinzl, Membrane distillation device, patent WO 2010127819, (2010).
27. K. Zhao, W. Heinzl, M. Wenzel, S. Büttner, F. Bollen, G. Lange, S. Heinzl and N. Sarda, Experimental study of the memsys vacuum-multi-effect-membrane-distillation (V-MEMD) module, *Desalination*, (In Press) (2013)
28. J. H. Hanemaaijer, J. van Medevoort, A. E. Jansen, C. Dotremont, E. van Sonsbeek, T. Yuan and L. De Ryck, Memstill membrane distillation - a future desalination technology, *Desalination*, **199(1–3),** 175–176 (2006).
29. X. Yang, R. Wang, L. Shi, A. G. Fane and M. Debowski, Performance improvement of PVDF hollow fiber-based membrane distillation process, *J. Membr. Sci* **369(1–2),** 437–447 (2011).

30. V. Gekas and B. Hallstrom, Mass transfer in the membrane concentration polarization layer under turbulent cross flow. I. Critical literature review and adaptation of existing Sherwood correlations to membrane operations, *J. Membr. Sci* **30,** 153–170 (1987).
31. R. C. Reid, J. M. Prausnitz and T. K. Sherwood, *The Properties of Gases and Liquids,* 3rd edn., McGraw-Hill, New York, (1977).
32. S. Bandini, C. Gostoli and G. C. Sarti, Separation efficiency in vacuum membrane distillation, *J. Membr. Sci* **73,** 217–229 (1992).
33. E. A. Mason and A. P. Malinauskas, *Gas Transport in Porous Media: The Dusty-Gas Model,* Elsevier, Amsterdam, (1983)
34. C. M. Guijt, I. G. Racz, T. Reith and A. B. d. Haan, Determination of membrane properties for use in the modelling of a membrane distillation module, *Desalination,* **132,** 255–261 (2000).
35. R. W. Schofield, A. G. Fane and C. J. Fell, Gas and vapour transport through microporous membranes. I. Knudsen-Poiseuille transition, J. Membr. Sci **53(1),** 159–171 (1990).
36. K. W. Lawson, M. S. Hall and D. R. Lloyd, Compaction of microporous membranes used in membrane distillation. I. Effects on gas permeability, *J. Membr. Sci* **101,** 99–108 (1995).
37. U. Beuscher and C. H. Gooding, Characterisation of the porous support layer of composite gas permeation membranes, *J. Membr. Sci* **132,** 213–227 (1997).
38. J. W. Veldsink, G. F. Versteeg and W. P. M. v. Swaaij, An experimental study of diffusion and convection of multicomponent gases through catalitic and noncatalytic membranes, *J. Membr. Sci* **92,** 275–291 (1994).
39. M. Khayet and T. Matsuura, *Membrane Distillation: Principles and Applications,* Elsevier, Amsterdam, The Netherlands, (2011)
40. G. C. Sarti, C. Gostoli and S. Bandini, Extraction of organic components from aqueous streams by vacuum membrane distillation, *J. Membr. Sci* **80,** 21–33 (1993).
41. S. Bandini, G. C. Sarti and C. Gostoli, *Vacuum membrane distillation: Pervaporation through porous hydrophobic membranes* (1988), Proc. Third Int. Conf. Pervaporation processes in the Chemical Industry, Nancy, France, Bakish materials, 117–126.
42. S. Kimura and S. Nakao, Transport phenomena in membrane distillation, *J. Membr. Sci* **33,** 285–298 (1987).
43. R. W. Schofield, A. G. Fane, C. J. D. Fell and R. Macoun, Factors affecting flux in membrane distillation, *Desalination* **77,** 279–294 (1990).
44. R. W. Schofield, A. G. Fane and C. J. D. Fell, Gas and vapor transport through microporous membranes. II. Membrane distillation, *J. Membr. Sci* **53,** 173–185 (1990).
45. S.P. Agashichev and A. V. Sivakov, Modeling and calculation of temperatureconcentration polarization in the membrane distillation process (MD), *Desalination* **93,** 245–258 (1993).

46. X. Yang, H. Yu, R. Wang and A. G. Fane, Analysis of the effect of turbulence promoters in hollow fiber membrane distillation modules by computational fluid dynamic (CFD) simulations, *Journal of Membrane Science* **415** (2012) 758–769
47. X. Yang, H. Yu, R. Wang and A. G. Fane, Optimization of microstructured hollow fiber design for membrane distillation applications using CFD modeling, *Journal of Membrane Science*, **421** (2012) 258–270
48. H. Yu, X. Yang, R. Wang and A. G. Fane, Analysis of heat and mass transfer by CFD for performance enhancement in direct contact membrane distillation, *J. Membr. Sci* **405–406,** 38–47 (2012).
49. M. Shakaib, S. M. F. Hasani, I. Ahmed and R. M. Yunus, A CFD study on the effect of spacer orientation on temperature polarization in membrane distillation modules, *Desalination* **284,** 332–340 (2012).
50. M. Khayet, M. P. Godino and J. I. Mengual, Study of asymmetric polarization in direct contact membrane distillation, *Sep. Purif. Technol*, **39(1),** 125–147 (2004).
51. L. Martinez-Diez and M. I. Vazquez-Gonzalez, Effect of polarization on mass transport through hydrophobic porous membranes, *Ind. Eng. Chem. Res* **37** 4128–4135 (1998).
52. L. Martinez-Diez and M. I. Vazquez-Gonzalez, Temperature and concentration polarization in membrane distillation of aqueous salt solutions, *J. Membr. Sci* **156,** 265–273 (1999).
53. C. Zhu and G. Liu, Modeling of ultrasonic enhancement on membrane distillation, *J. Membr. Sci* **176(1),** 31–41 (2000).
54. X. Yang, R. Wang and A. G. Fane, Novel designs for improving the performance of hollow fiber membrane distillation modules, *J. Membr. Sci* **384(1–2),** 52–62 (2011).
55. J. Phattaranawik, R. Jiraratananon and A. G. Fane, Effect of pore size distribution and air flux on mass transport in direct contact membrane distillation, *J. Membr. Sci* **215,** 75–85 (2003).
56. K. W. Lawson and D. R. Lloyd, Membrane distillation. I. Module design and performance evaluation using vacuum membrane distillation, *J. Membr. Sci* **120(1),** 111–121 (1996).
57. Y. Fujii, S. Kigoshi, H. Iwatani and M. Aoyama, Selectivity and characteristics of direct contact membrane distillation type experiment. I. Permeability and selectivity through dried hydrophobic fine porous membranes, *J. Membr. Sci.* **72(1),** 53–72 (1992).
58. O. d. Zárate, J. M. Pena and L.Mengual, Characterization of membrane distillation membranes prepared by phase inversion, *Desalination* **100(1),** 139–148 (1995).
59. M. Tomaszewska, Preparation and properties of flat-sheet membranes from poly(vinylidene fluoride) for membrane distillation, *Desalination* **104(1),** 1–11 (1996).

60. M. Khayet and T. Matsuura, Application of surface modifying macromolecules for the preparation of membranes for membrane distillation, *Desalination* **158(1–3),** 51–56 (2003).
61. J. Li, Z. Xu, Z. Liu, W. Yuan, H. Xiang, S. Wang and Y. Xu, Microporous polypropylene and polyethylene hollow fiber membranes. Part 3. Experimental studies on membrane distillation for desalination, *Desalination,* **155(2),** 153–156, (2003).
62. C. Feng, B. Shi, G. Li and Y. Wu, Preparation and properties of microporous membrane from poly(vinylidene fluoride-co-tetrafluoroethylene) (F2.4) for membrane distillation, *J. Membr. Sci* **237(1–2),** 15–24 (2004).
63. D. E. Suk, T. Matsuura, H. B. Park and Y. M. Lee, Synthesis of a new type of surface modifying macromolecules (nSMM) and characterization and testing of nSMM blended membranes for membrane distillation, *J. Membr. Sci* **277(1–2),** 177–185 (2006).
64. S. Bonyadi and T. S. Chung, Highly porous and macrovoid-free PVDF hollow fiber membranes for membrane distillation by a solvent-dope solution co-extrusion approach, *J. Membr. Sci* **331(1–2),** 66–74 (2009).
65. Z. D. Hendren, J. Brant and M. R. Wiesner, Surface modification of nanostructured ceramic membranes for direct contact membrane distillation, *J. Membr. Sci* **331(1–2),** 1–10 (2009).
66. M. Qtaishat, M. Khayet and T. Matsuura, Guidelines for preparation of higher flux hydrophobic/hydrophilic composite membranes for membrane distillation, *J. Membr. Sci* **329(1–2)**, 193–200 (2009).
67. K. Y. Wang, T. S. Chung and M. Gryta, Hydrophobic PVDF hollow fiber membranes with narrow pore size distribution and ultra-skin for the fresh water production through membrane distillation, *Chem Eng Sci*, **63**, 2587–2594 (2008).
68. S. Bonyadi and T.-S. Chung, Highly porous and macrovoid-free PVDF hollow fiber membranes for membrane distillation by a solvent-dope solution co-extrusion approach, J. Membr. Sci **331(1–2),** 66–74 (2009).
69. M. C. García-Payo, M. Essalhi and M. Khayet, Effects of PVDF-HFP concentration on membrane distillation performance and structural morphology of hollow fiber membranes, J. Membr. Sci **347,** 209–219 (2010).
70. P. Wang, M. M. Teoh and T.-S. Chung, Morphological architecture of dual-layer hollow fiber for membrane distillation with higher desalination performance, *Water Res* **45(17),** 5489–5500 (2011).
71. M. Khayet, C. Cojocaru, M. Essalhi, M. C. García-Payo, P. Arribas and L. García-Fernández, Hollow fiber spinning experimental design and analysis of defects for fabrication of optimized membranes for membrane distillation, *Desalination,* **287,** 146–158 (2011).

72. M. Mulder, *Basic Principles of Membrane Technology*, springer, (1996).
73. M. Khayet, Membranes and theoretical modeling of membrane distillation: A review, *Adv. Colloid Interface Sci* **164(1–2),** 56–88 (2011).
74. L.M. Camacho, L. Dumée, J. Zhang, J.-de. Li, M. Duke, J. Gomez and S. Gray, Advances in membrane distillation for water desalination and purification applications, *Water* **5,** 94–196 (2013).
75. B. Li and K.K. Sirkar, Novel membrane and device for direct contact membrane distillation-based desalination process, *Ind. Eng. Chem. Res 43(17),* 5300–5309 (2004).
76. B. Li and K. K. Sirkar, Novel membrane and device for vacuum membrane distillation-based desalination process, *J. Membr. Sci* **257(1–2),** 60–75 (2005).
77. A. Larbot , L. Gazagnes, S. Krajewski, M. Bukowska and W. Kujawski, Water desalination using ceramic membrane distillation, *Desalination* **168,** 367–372 (2004).
78. S. Cerneaux, I. Stuzynska, W.M. Kujawski, M. Persin and A. Larbot, Comparison of various membrane distillation methods for desalination using hydrophobic ceramic membranes, *J. Membr. Sci* **337,** 55–60 (2009).
79. M. Khayet, T. Matsuura and J.I. Mengual, Porous hydrophobic/hydrophilic composite membranes: estimation of the hydrophobic layer thickness, *J. Membr. Sci* **266,** 68–79 (2005).
80. M. Qtaishat, D. Rana, M. Khayet and T. Matsuura, Preparation and characterization of novel hydrophobic/hydrophilic polyetherimide composite membranes for desalination by direct contact membrane distillation, *J. Membr. Sci* **327(1–2),** 264–273 (2009).
81. M. Qtaishat, D. Rana, T. Matsuura and K. M, Effect of surface modifying macromolecules stoichiometric ratio on composite hydrophobic/hydrophilic membranes characteristics and performance in direct contact membrane distillation, AIChE J, 55 (2009) 3145-3151
82. M. Qtaishat, M. Khayet and T. Matsuura, Novel porous composite hydrophobic/hydrophilic polysulfone membranes for desalination by direct contact membrane distillation, *J. Membr. Sci* **341,** 139–148 (2009).
83. M. Khayet, D. Suk, R. Narbaitz, J. Santerre and T. Matsuura, Study on surface modification by surface-modifying macromolecules and its applications in membrane-separation processes, *J. Appl. Polymer. Sci* **89,** 2902–2916 (2003).
84. K. M, I. AO and M. T, Monte carlo simulation and experimental heat and mass transfer in direct contact membrane distillation, *Int. J. Heat Mass Transfer* **53,** 1249–1259 (2010).
85. M. Khayet, P. Godino and J. I. Mengual, Theory and experiments on sweeping gas membrane distillation, J. Membr. Sci **165** 261–272 (2000).

86. T. Y. Cath, V. D. Adams and A. E. Childress, Experimental study of desalination using direct contact membrane distillation: a new approach to flux enhancement, J. Membr. Sci **228,** 5–16 (2004).
87. M. Khayet, K. Khulbe and T. Matsuura, Characterization of membranes for membrane distillation by atomic forcemicroscopy and estimation of theirwater vapor transfer coefficients in vacuum membrane distillation process, *J. Membr. Sci* **238,** 199–211 (2004).
88. M. Qtaishat, T. Matsuura, B. Kruczek and M. Khayet, Heat and mass transfer analysis in direct contact membrane distillation, *Desalination* **219(1–3),** 272–292 (2008).
89. M.A. Izquierdo-Gil, M.C. García-Payo and C. Fernández-Pineda, Air gap membrane distillation of sucrose aqueous solutions, *J. Membr. Sci* **155,** 291–307 (1999).
90. M. Gryta, M.D. Osmotic and other membrane distillation variants, *J. Membr. Sci* **246,** 145–156 (2005).
91. A. Criscuoli, M.C. Carnevale and E. Drioli, Evaluation of energy requirements in membrane distillation., *Chem. Eng. Process* **47,** 1098–1105 (2008).
92. P.P Zolotarev, V.V. Ugrosov, I.B. Volkina and V.N. Nikulin, Treatment of waste water for removing heavy metals by membrane distillation, *J. Hazard. Mater* **37,** 77–82 (1994).
93. K. W. Lawson and D. R. Lloyd, Membrane distillation. II. Direct contact MD, *J. Membr. Sci* **120(1),** 123–133 (1996).
94. M.I. Vázquez-Gónzález and L. Martínez, Nonisothermal water transport through hydrophobic membranes in a stirred cell, *Sep. Sci. Technol* **29,** 1957–1966 (1994).
95. M. Gryta, Influence of polypropylene membrane surface porosity on the performance of membrane distillation process, *J. Membr. Sci.***287,** 67–78 (2007).
96. M. Khayet, M. P. Godino and J.I. Mengual, Possibility of nuclear desalination through various membrane distillation configurations: a comparative study: International Journal of Nuclear Desalination, Fuel and Energy Abstracts **44(6)** 372–372 (2003).
97. S. Al-Obaidani, E. Curcio, F. Macedonio, G. Di Profio, H. Al-Hinai and E. Drioli, Potential of membrane distillation in seawater desalination: Thermal efficiency, sensitivity study and cost estimation, *J. Membr. Sci* **323(1),** 85–98 (2008).
98. M. Tomaszewska, M. Gryta and A.W. Morawski, Study on the concentration of acids by membrane distillation, *J. Membr. Sci* **102,** 113–122 (1995).
99. A. Criscuoli, J. Zhong, A. Figoli, M.C. Carnevale, R. Huang and E. Drioli, Treatment of dye solutions by vacuum membrane distillation, *Water Res* **42,** 5031–5037 (2008).

100. C.M. Guijt, G.W. Meindersma, T. Reith and A.B. de Haan, Method for experimental determination of the gas transport properties of highly porous fibre membranes: a first step before predictive modeling of a membrane distillation process, *Desalination* **147** 127–132 (2002).
101. C.H. Lee and W. Hong, Effect of operating variables on the flux and selectivity in sweep gas membrane distillation for dilute aqueous isopropanol, *J. Membr. Sci* **188,** 79–86 (2001).
102. R.L. Calibo, M. Matsumura, J. Takahashi and H. Kataoka, Ethanol stripping by pervaporation using porous PTFE membrane, *J. Ferment. Technol.***65,** 665–674 (1987).
103. M. Khayet and T. Matsuura, Preparation and characterization of polyvinylidene fluoride membranes for membrane distillation, *Ind. Eng. Chem. Res* **40(24),** 5710–5718 (2001).
104. X. Wei, B. Zhao, X.-M. Li, Z. Wang, B.-Q. He, T. He and B. Jiang, CF4 plasma surface modification of asymmetric hydrophilic polyethersulfone membranes for direct contact membrane distillation, J. Membr. Sci **407–408,** 164–175 (2012).
105. C. Feng, B. Shi, G. Li and Y. Wu, Preliminary research on microporous membrane from F2.4 for membrane distillation, *Sep. Purif. Technol* **39** 221–228 (2004).
106. M. Khayet, C. Cojocaru and M. García-Payo, Experimental design and optimization of asymmetric flat-sheet membranes prepared for direct contact membrane distillation, *J. Membr. Sci* **351,** 234–245 (2010).
107. Y. Wu, Y. Kong, X. Lin, W. Liu and J. Xu, Surface-modified hydrophilic membranes in membrane distillation, *J. Membr. Sci* **72(2),** 189–196 (1992).
108. P. Peng, A. Fane and X. Li, Desalination by membrane distillation adopting a hydrophilic membrane, *Desalination* **173,** 45–54 (2005).
109. Z. Ma, Y. Hong, L. Ma and M. Su, Superhydrophobic membranes with ordered arrays of nanospiked microchannels for water desalination, *Langmuir. Lett* **25,** 5446–5450 (2009).
110. L. Dumée, K. Sears, J. Schütz, N. Finn, C. Huynh and S. Hawkins, Characterization and evaluation of carbon nanotube Bucky-Paper membranes for direct contact membrane distillation, *J. Membr. Sci* **351,** 36–43 (2010).
111. C. Feng, K. Khulbe, T. Matsuura, R. Gopal, S. Kaur and S. Ramakrishna, Production of drinking water from saline water by air-gap membrane distillation using polyvinylidene fluoride nanofiber membrane, *J. Membr. Sci* **311**, 1–6 (2008).
112. R.Gopal, *Influence of Electrospun Nanofibrous Architecture in Separation Technology* (Ph.D dissertation), National University of Singapore (2010)
113. J.A. Prince, G. Singh, D. Rana, T. Matsuura, V. Anbharasi and T.S. Shanmugasundaram, Preparation and characterization of highly

hydrophobic poly(vinylidenefluoride) - Clay nanocomposite nanofiber membranes (PVDF - clay NNMs) for desalination using directc ontact membrane distillation, *J. Membr. Sci* **397–398,** 80–86 (2012).

114. B. Wu, X. Tan, K. Li and T. WK, Removal of 1,1,1-trichloroethane from water using a poly(vinylidene fluoride) hollow fiber membrane module: vacuum membrane distillation operation, *Sep. Purif. Technol* **52,** 301–309 (2006).
115. B. Wu, K. Li and T. WK, Preparation and characterization of poly(vinylidene fluoride) hollow fiber membranes for vacuum membrane distillation, *J. Appl. Polymer. Sci* **106,** 1482–1495 (2007).
116. M.M. Teoh and T.S. Chung, Membrane distillation with hydrophobic macrovoid-free PVDF-PTFE hollow fiber membranes, *Sep. Purif. Technol* **66(2),** 229–236 (2009).
117. M.C. García-Payo, M. Essalhi and K. M, Preparation and characterization of PVDF/HFP copolymer hollow fiber membranes for membrane distillation, *Desalination* **245**, 469–473 (2009).
118. K.Y. Wang, S.W. Foo and T.S. Chung, Mixed matrix PVDF hollow fiber membranes with Nanoscale pores for desalination through Direct Contact Membrane Distillation, *Ind. Eng. Chem. Res* **48(9),** 4474–4483 (2009).
119. S. Bonyadi and T.S. Chung, Flux enhancement in membrane distillation by fabrication of dual layer hydrophilic-hydrophobic hollow fiber membranes, *J. Membr. Sci* **306(1–2)**, 134–146 (2007).
120. A. Chanachai, K. Meksup and R. Jiraratananon, Coating of hydrophobic hollow fiber PVDF membrane with chitosan for protection against wetting and flavor loss in osmotic distillation process, *Sep. Purif. Technol* **72(2),** 217–224 (2010).
121. C.-L. Lai, R.-M. Liou, S.-H. Chen, G.-W. Huang and K.-R. Lee, Preparation and characterization of plasma-modified PTFE membrane and its application in direct contact membrane distillation, *Desalination* **267(2–3),** 184–192 (2011).
122. H. Susanto, Towards practical implementations of membrane distillation, *Chem. Eng. Process* **50,** 139–150 (2011).
123. M. Khayet, Membranes and theoretical modeling of membrane distillation: a review, *Adv. Colloid Interface Sci* In Press (2011)
124. M. Su, M. M. Teoh, K. Y. Wang, J. Su and T.-S. Chung, Effect of inner-layer thermal conductivity on flux enhancement of dual-layer hollow fiber membranes in direct contact membrane distillation, *J. Membr. Sci* **364(1–2),** 278–289 (2010).
125. J. Zhang, J.-D. Li, M. Duke, Z. Xie and S. Gray, Performance of asymmetric hollow fibre membranes in membrane distillation under various configurations and vacuum enhancement, *J. Membr. Sci* **362(1–2),** 517–528 (2010).

126. M. Gander, B. Jefferson and S. Judd, Aerobic MBRs for domestic wastewater treatment: a review with cost considerations, *Sep. Purif. Technol* **18(2),** 119–130 (2000).
127. A. Zhu, P. D. Christofides and Y. Cohen, Minimization of Energy Consumption for a Two-Pass Membrane Desalination: Effect of Energy Recovery, Membrane Rejection and Retentate Recycling, *J. Membr. Sci* **339(1–2),** 26–137 (2009).
128. C. Cabassud and D. Wirth, Membrane distillation for water desalination: How to chose an appropriate membrane?, *Desalination* **157(1–3),** 307–314 (2003).
129. L. Martínez-Díez, M. I. Vázquez-González and F. J. Florido-Díaz, Study of membrane distillation using channel spacers, *J. Membr. Sci* **144(1–2),** 45–56 (1998).
130. M. N. Chernyshov, G. W. Meindersma and A. B. d. Haan, Comparison of spacers for temperature polarization reduction in air gap membrane distillation, *Desalination* **183,** 363 (2005).
131. L. Martínez-Díez and J. M. Rodríguez-Maroto, Characterization of membrane distillation modules and analysis of mass flux enhancement by channel spacers, *J. Membr. Sci* **274,** 123–137 (2006).
132. J. Phattaranawik and R. Jiraratananon, Direct contact membrane distillation: effect of mass transfer on heat transfer, *J. Membr. Sci* **188,** 137–143 (2001).
133. J. Phattaranawik, R. Jiraratananon and A. G. Fane, Heat transport and membrane distillation coefficients in direct contact membrane distillation, *J. Membr. Sci* **212**, 177–193 (2003).
134. http://www.xzero.se,
135. http://www.scarab.se,
136. L. Martínez-Díez and J. M. Rodríguez-Maroto, Characterization of membrane distillation modules and analysis of mass flux enhancement by channel spacers, *J. Membr. Sci* **274,** 123–137 (2006).
137. X. Yang, R. Wang, A. G. Fane, Chuyang Y.Tang and I. G. Wenten, Membrane module design and dynamic shear-induced techniques to enhance liquid separation by hollow fiber modules: a review, *Sep. Purif. Technol*, submitted (2010).
138. X. Yang, R. Wang, A. G. Fane, Chuyang Y.Tang and I. G. Wenten, Membrane module design and dynamic shear-induced techniques to enhance liquid separation by hollow fiber modules: a review, *Desalination and Water Treatment*, **51** (16–18) (2013) 3604–3627.
139. L. Cheng, P. Wu and J. Chen, Modeling and optimization of hollow fiber DCMD module for desalination, *J. Membr. Sci* **318**, 154–166 (2008).
140. M. M. Teoh, S. Bonyadi and T.-S. Chung, Investigation of different hollow fiber module designs for flux enhancement in the membrane distillation process, *J. Membr. Sci* **311(1–2),** 371–379 (2008).

141. M. Khayet, M. P. Godino and J. I. Mengual, Study of Asymmetric Polarization in Direct Contact Membrane Distillation, *Sep. Sci. Technol* **39(1),** 125–147 (2005).
142. B. J. Pangrle, E. G. Walsh, S. C. Moore and D. DiBiasio, Investigation of fluid flow patterns in a hollow fiber module using magnetic resonance velocity imaging, *Biotechnol. Tech* **3** (1989)
143. S. Yao, M. Costello, A. G. Fane and J. M. Pope, Non-invasive observation of flow profiles and polarisation layers in hollow fibre membrane filtration modules using NMR micro-imaging, *J. Membr. Sci* **99(3),** 207–216 (1995).
144. X. Yang, E. O. Fridjonsson, M. L. Johns, R. Wang and A. G. Fane, A non-invasive study of flow dynamics in membrane distillation hollow fiber modules using low-field nuclear magnetic resonance imaging (MRI), *Journal of Membrane Science, In Press (2013), ISSN 0376-7388.*
145. A. Cipollina, A. D. Miceli, J. Koschikowski, G. Micale and L. Rizzuti, CFD simulation of a membrane distillation module channel, *Desalination Water Treat* **6,** 177–183 (2009).
146. M. Khayet, M. P. Godino and J. I. Mengual, Thermal boundary layers in sweeping gas membrane distillation processes, *AIChE J* **48(7),** 1488–1497 (2002).
147. F. Lagana, G. Barbieri and E. Drioli, Direct contact membrane distillation: modelling and concentration experiments, *J. Membr. Sci* **166,** 1–11 (2000).
148. S. Bandani, A. Saavedra and G.C. Sarti, Vacuum membrane distillation: experiments and modeling, *AIChE J* **43,** 398–408 (1997).
149. M. Tomaszewska, M. Gryta and A. W. Morawski, The influence of salt in solution on hydrochloric acid recovery by membrane distillation, *Sep. Purif. Technol* **14,** 183–188 (1998).
150. J. I. Mengual and L. Peña, Membrane distillation, *J. Colloid Int. Sci* **1,** 17–29 (1997).
151. F. A. Banat and J. Simandl, Theoretical and experimental study in membrane distillation, *Desalination* **95,** 39–52 (1994).
152. F.A. Banat and J. Simandl, Desalination by membrane distillation: a parametric study, *Sep. Purif. Technol* **33(2),** 201–226 (1998).
153. C. A. Rivier, M. C. Garcia-Payo, I. W. Marison and U. V. Stockar, Separation of binary mixtures by thermostatic sweeping gas membrane distillation. I. Theory and simulations, *J. Membr. Sci* **201,** 1–16 (2002).
154. M. C. Garcia-Payo, C. A. Rivier, I. W. Marison and U. V. Stockar, Separation of binary mixtures by thermostatic sweeping gas membrane distillation. II. Experimental results with aqueous formic acid solutions, *J. Membr. Sci* **198,** 197–210 (2002).
155. M. Khayet, M. P. Godino and J. I. Mengual, Nature of flow on sweeping gas membrane distillation, *J. Membr. Sci* **170,** 243–255 (2000).

156. S. Bouguccha, R. Chouilkh and M. Dhahbi, Numerical study of the coupled heat and mass transfer in membrane distillation, *Desalination* **152**, 245–252 (2000).
157. M. Elimelech and W. A. Phillip, The Future of Seawater Desalination: Energy, Technology, and the Environment, *Sci* **333**, 712–717 (2011).
158. M. A. Shannon, P. W. Bohn, M. Elimelech, J. G. Georgiadis, B. J. Marinas and A. M. Mayes, Science and technology for water purification in the coming decades, *Nat* **452,** 301–310 (2008).
159. M. Khayet, Solar desalination by membrane distillation: Dispersion in energy consumption analysis and water production costs (a review), *Desalination* **308,** 89–101 (2013).
160. A. Alkhudhiri, N. Darwish and N. Hilal, Membrane distillation: A comprehensive review, *Desalination* **287,** 2–18 (2012).
161. A. G. Fane, R. W. Schofield and C. J. D. Fell, The efficient use of energy in membrane distillation, *Desalination* **64**, 231–243 (1987).
162. P. A. Hogan, Sudjito, A. G. Fane and G. L. Morrison, Desalination by solar heated membrane distillation, *Desalination* **81**, 81–90 (1991).
163. S. Bouguecha, B. Hamrouni and M. Dhahbi, Small scale desalination pilots powered by renewable energy sources: case studies, *Desalination* **183,** 151–165 (2005).
164. H. Yu, X. Yang, R. Wang and A. G. Fane, Numerical simulation of heat and mass transfer in direct membrane distillation in a hollow fiber module with laminar flow, *J. Membr. Sci* **384(1–2),** 107–116 (2011).
165. J. Gilron, L. Song and K. K. Sirkar, Design for cascade of crossflow direct contact membrane distillation, *Ind. Eng. Chem. Res* **46(8),** 2324–2334 (2007).
166. G. Zuo, R. Wang, R. Field and A.G. Fane, Energy efficiency evaluation and economic analyses of direct contact membrane distillation system using Aspen Plus, *Desalination* **283**, 237–244 (2011).
167. H. Lee, F. He, L. Song, J. Gilron and K. K. Sirkar, Desalination with a cascade of cross-flow hollow fiber membrane distillation devices integrated with a heat exchanger, *AICHE J* **57**, 1780–1795 (2011).
168. F. Macedonio, E. Curcio and E. Drioli, Integrated membrane systems for seawater desalination: energetic and exergetic analysis, economic evaluation, experimental study, *Desalination* **203**, 260–276 (2007).
169. F. Banat and N. Jwaied, Economic evaluation of desalination by small-scale autonomous solar-powered membrane distillation units, *Desalination* **220,** 566–573 (2008).
170. G. W. Meindersma, C. M. Guijt and A. B. d. Haan, Desalination and water recycling by air gap membrane distillation, *Desalination* **187,** 291–301 (2006).
171. J.Walton, H. Lu, C. Turner, S. Solis and H. Hein, *Solar and waste heat desalination by membrane distillation*, Desalination and water purification research and development program report No. 81, University of Texas, El Paso, USA, (2004)

172. A. Hepbasli and Z. Alsuhaibani, A key review on present status and future directions of solar energy studies and applications in Saudi Arabia, Renewable and Sustainable Energy Reviews, **15**, 5021–5050 (2011).
173. H. A, Desalination using renewable energy sources, *Desalination* **97,** 339–352 (1994).
174. R. Sarbatly and C.-K. Chiam, Evaluation of geothermal energy in desalination by vacuum membrane distillation, *Appl. Energy* in press (2013)
175. R. Sarbatly and C.-K. Chiam, Evaluation of geothermal energy in desalination by vacuum membrane distillation, *Appl. Energy,* **112**(2013), 737–746.
176. A. Hausmann, P. Sanciolo, T. Vasiljevic, M. Weeks and M. Duke, Integration of membrane distillation into heat paths of industrial processes, *Chem. Eng. J* **211–212,** 378–387 (2012).
177. J. Koschikowski, M. Wieghaus and M. Rommel, Solar Thermal-driven desalination plants based onmembrane distillation, *Desalination* **156,** 295–304 (2003).
178. F. Banat, N. Jwaied, M. Rommel, J. Koschikowski and M. Weighaus, Desalination by a "compact SMADES"autonomous solar-powered membrane distillation unit, *Desalination* **217**, 29–37 (2007).
179. F. Banat, N. Jwaied, M. Rommel, J. Koschikowski and M. Weighaus, Performance evaluation of the "large SMADES" autonomous desalination solar-driven membrane distillation plant in Aqaba, Jordan, *Desalination* **217**, 17–28 (2007).
180. H. E. S. Fath, S. M. Elsherbiny, A. A. Hassan, M. Rommel, M. Wieghaus, J. Koschikowski and M. Vatansever, PV and thermally driven small-scale, stand-alone solar desalination systems with very low maintenance needs, *Desalination* **225**, 58–69 (2008).
181. J. Koschikowski, M. Wieghaus, M. Rommel, V. S. Ortin, B. P. Suarez and J. R. B. Rodríguez, Experimental investigations on solar driven stand-alone membrane distillation systems for remote areas, *Desalination* **248**, 125–131 (2009).
182. Z. Ding, L. Liu, M. S. El-Bourawi and R. Ma, Analysis of a solar-powered membrane distillation system, *Desalination* **172**, 27–40 (2005).
183. H. Chang, G. B. Wang, Y. H. Chen, C. C. Li and C. L. Chang, Modeling and optimization of a solar driven membrane distillation, *Renewable Energy* **35**, 2714–2722 (2010).
184. C. D. H. T.C. Chen, Immediate assisted solar direct contact membrane distillation in saline water desalination, *J. Membr. Sci* **358,** 122–130 (2010).
185. J. Mericq, S. Laborie and C. Cabassud, Evaluation of Systems coupling vacuum membrane distillation and solar energy for seawater desalination, *Chem. Eng. J* **166,** 596–606 (2011).
186. F. Suárez, S. W. Tyler and A. E. Childress, A theoretical study of a direct contact membrane distillation system coupled to a salt-gradient solar pond for terminal lakes reclamation, *Water Res* **44**, 4601–4615 (2010).

187. R. Ray, R. W. Wytcherley, D. Newbold, S. Mccray, D. Friesen and D. Brose, Synergistic, membrane-based hybrid separation systems, *J. Membr. Sci* **62**, 347–369 (1991).
188. D. E. Suk and T. Matsuura, Membrane-based hybrid processes: a review, *Sep. Sci. Tech* **41**, 595–626 (2006).
189. F. Lipnizki and R. W. Field, Pervaporation-based hybrid processes in treating phenolic wastewater: technical aspects and cost engineering, *Sep. Sci. Tech* **36**, 3311–3335 (2001).
190. F. L. E. Drioli, A. Criscuoli, G. Barbieri, Integrated membrane operations in desalination process, *Desalination* **122**, 141–145 (1991).
191. K. Karakulski, M. Gryta and A. Morawski, Membrane processes for portable water quality improvement, *Desalination* **145**, 315–319 (2002).
192. C. M. Tun, A. G. Fane, J. T. Matheickal and R. Sheikholeslami, Membrane distillation crystallization of concentrated salts—flux and crystal formation, *J. Membr. Sci* **257(1–2)**, 144–155 (2005).
193. X. Ji, E. Curcio, S. Al Obaidani, G. Di Profio, E. Fontananova and E. Drioli, Membrane distillation-crystallization of seawater reverse osmosis brines, *Sep. Purif. Technol* **71(1)**, 76–82 (2010).
194. R. J. M. Creusen, J. van Medevoort, C. P. M. Roelands and J. A. D. v. R. v. Duivenbode, Brine Treatment by a Membrane Distillation-crystallization (MDC) Process, *Procedia Eng* **44(0),** 1756–1759 (2012).
195. G. Chen, X. Yang, R. Wang and A. G. Fane, Performance enhancement and scaling control with gas bubbling in direct contact membrane distillation, *Desalination* DOI:10.1016/j.desal.2012.07.018 (2012)
196. G. Guan, R. Wang, F. Wicaksana, X. Yang and A. G. Fane, Analysis of membrane distillation crystallization system for high salinity brine treatment with zero discharge using Aspen flowsheet simulation, *Ind. Chem. Eng. Res* **51**, 13405–13413 (2012).
197. T.-H. Khaing, J. Li, Y. Li, N. Wai and F.-s. Wong, Feasibility study on petrochemical wastewater treatment and reuse using a novel submerged membrane distillation bioreactor, *Sep. Purif. Technol* **74(1)**, 138–143 (2010).
198. J. Phattaranawik, A. G. Fane, A. C. S. Pasquier and W. Bing, A novel membrane bioreactor based on membrane distillation, *Desalination* **223(1–3)**, 386–395 (2008).
199. K. Y. Wang, M. M. Teoh, A. Nugroho and T.-S. Chung, Integrated forward osmosis–membrane distillation (FO–MD) hybrid system for the concentration of protein solutions, *Chem. Eng. Sci* **66(11),** 2421–2430 (2011).
200. T. Y. Cath, V. D. Adams and A. E. Childress, Experimental study of desalination using direct contact membrane distillation: a new approach to flux enhancement, *J. Membr. Sci* **228(1)**, 5–16 (2004).
201. J. Wang, B. Fan, Z. Luan, D. Qu, X. Peng and D. Hou, Integration of direct contact membrane distillation and recirculating cooling water system for pure water production, *J. Cleaner Prod* **16**, 1847–1855 (2008).

SECTION IV

NON-TRADITIONAL DESALINATION PROCESSES

9

Humidification Dehumidification Desalination

G. Prakash Narayan and John H. Lienhard V

Rohsenhow Kendall Heat Transfer Laboratory, Massachusetts Institute of Technology, Cambridge, MA 02139-4307 USA.

Abstract

Humidification-dehumidification (HDH) desalination involves vaporizing water from a saline liquid stream into a carrier gas stream and then condensing the vapor to form purified water. This chapter describes various forms of the HDH cycle, with analysis of the energy consumption of various realizations of the process. Bubble column dehumidifiers are described in detail.

Keywords: Humidification-dehumidification desalination, Carrier gas extraction, Bubble column dehumidifier, Thermodynamic balancing, Mass injection and extraction, Effectiveness, Gained-output-ratio, Enthalpy pinch

9.1 Introduction

More than a billion people lack access to safe drinking water worldwide [1]. A large majority of these people live in low income communities. The United Nations acknowledges this fact in its millennium development goals [2] by highlighting the critical need for impoverished and developing regions of the world to achieve self-sustenance in potable water supply. Figure 9.1 further

Jane Kucera (ed.) Desalination, (427–472) 2014 © Scrivener Publishing LLC

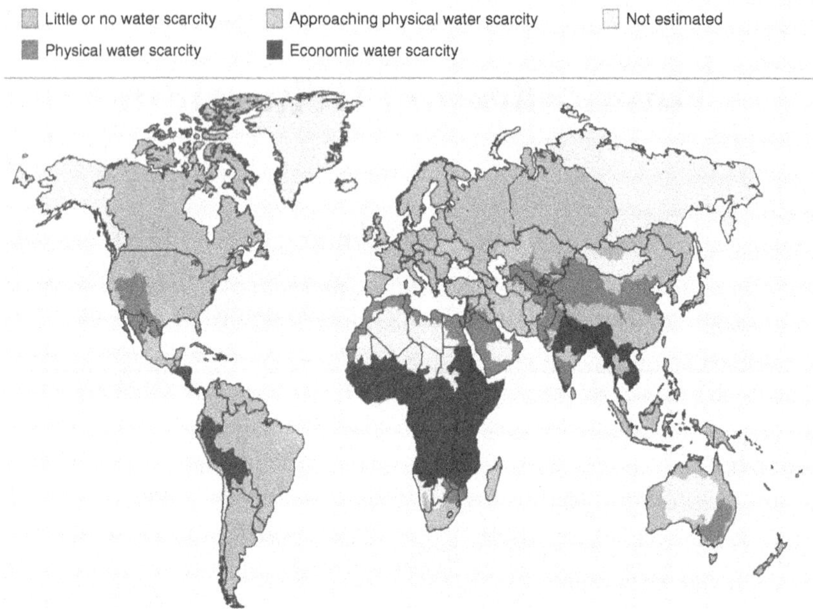

Figure 9.1 World map showing areas of physical and economic water scarcity [4].

illustrates how intense water scarcity exists mainly in the developing parts of the world[1]. For example, in India alone, 200,000 villages (and several peri-urban communities) lack access to safe potable water [3]. There is a clear need to help create a sustainable solution to the rural water problem in order to solve the global water crisis.

Most of the villages lacking safe drinking water are small communities with a population between 1,000 and 10,000 people. Thus, the minimum water needs (for drinking and cooking) for each one of those communities is between 10 and 100 cubic meter of pure water per day (at a consumption rate of 10 liters per person

---

[1] Often, lack of a potable water supply to the general population (or water scarcity) is misunderstood as absence of freshwater in a community. This is only one of the forms of water scarcity known as physical water scarcity. There is also scarcity in areas where there is plenty of rainfall and/or freshwater. This is primarily because of the lack of infrastructure to purify and transport the fresh water from aquifers or water bodies like lakes and rivers to the people who need it. This is termed as economic water scarcity.

per day). Systems that produce such amounts of pure water are relatively small-scale compared to conventional water treatment systems (for example, most existing state-of-the-art desalination systems are of the order of 100,000 to 1 million cubic meters per day [5]).

Any potential small-scale solution to the problem needs to be both implementable and scalable. For the solution to be implementable, it has to be cost effective and resource-frugal. Currently, the price of safe drinking water (in the rare case when it is available) in these low income communities is very high relative to the cost of centrally distributed municipal drinking water in nearby "developed" regions. For example, in some parts of rural India the cost of water is up to $10/m^3 which is roughly 40 times the cost of municipal drinking water available a few miles away in a nearby city [6]. Furthermore, in many villages, resources including skilled labor, a continuous energy supply, and raw materials are not readily available. The solution should, hence, be implementable within these constraints too.

An implementable solution is truly worthwhile only if it is scalable and can reach a large number of people (say, a million or more). For such scalability, the solution should be able to handle an array of contaminants in the water to be treated. In India alone, the contaminants range from high fluoride content to bacterial contamination to water being very brackish. Sixty-six million people have been reported to be consuming water with elevated levels of fluoride in India [7]. Most of these people live in the states of Rajasthan and Gujarat where fluoride contents reach up to 11 mg/L. Some districts in Assam Orissa have very high iron content in water (1 to 10 mg/L - red water) and some in Rajasthan, Uttar Pradesh and Bihar have yellow water (>1 mg/L of iron) [8]. Certain places in Haryana, Gujarat, and Andhra Pradesh were also found to have dangerously high levels of mercury. The problems associated with high levels of arsenic in ground water (in West Bengal) are well documented [9]. At least 300,000 people are affected by drinking water with arsenic above the permissible limit of 0.05 mg/L in this region. In parts of coastal Tamil Nadu, because of seawater intrusion, there is the problem of high salinity in ground water supplies (as high as 10,000 ppm in some cases) [10]. These problems are almost exclusively limited to rural and peri-urban communities. In all, almost one in three of the 600,000

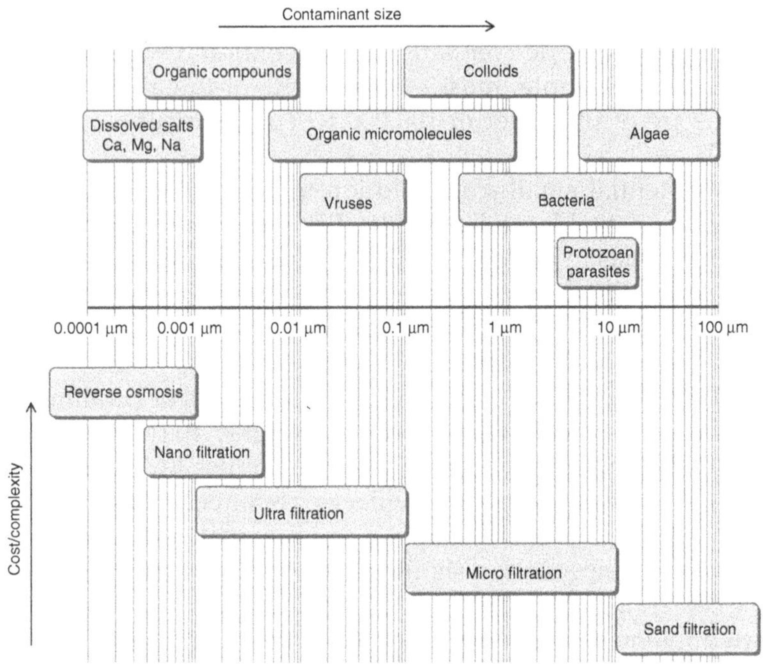

Figure 9.2 Illustration of various membrane technologies and the various contaminants they remove (Figure by MIT OpenCourseWare; data source: A. Twort et al. [11]).

Indian villages face problems of brackish or contaminated water and scarcity of fresh water. The Indian example is typical in that most developing and under-developed nations face similar water problems.

Desalination technologies can remove all contaminants including almost all dissolved ions, micro organisms and so on. For example, as illustrated in Fig. 9. 2, reverse osmosis removes even the smallest contaminants (albeit at a higher cost and complexity compared to other water treatment techniques shown in figure). Furthermore, thermal desalination technologies (MSF, MED, HDH and so on) are commonly known to remove all contaminants (producing what is in principle de-ionized water). The challenges in implementing these technologies are, however, to make them low cost (< $5/m^3$), to provide them at a community-scale (10–100 m^3/day), and, relatively maintenance-free (or at least maintable by non-technical laborers). Humidification dehumidification desalination offers a small-scale desalination technology which can meet these challenges.

9.1.1 Humidification Dehumidification (HDH) Desalination

Nature uses air as a carrier gas to desalinate seawater by means of the rain cycle. In the rain cycle, seawater gets heated (by solar irradiation) and evaporates into the air above to humidify it. Then the humidified air rises and forms clouds. Eventually, the clouds 'dehumidify' as rain and that which falls over land can be collected for human consumption. The man-made version of this cycle is called the humidification-dehumidification desalination (HDH) cycle. The simplest form of the HDH cycle is illustrated in Figure 9.3. The cycle consists of three subsystems: (a) an air and/or a brine heater (only a brine heater is shown in the figure), which can use various sources of energy like solar, thermal, geothermal or combinations of these; (b) the humidifier or evaporator; and (c) the dehumidifier or condenser.

The HDH cycle has received some attention in recent years and many researchers have investigated the intricacies of this technology. It should be noted here that the predecessor technology of the HDH cycle is the simple solar still. Several researchers [12–14] have reviewed the numerous works on the solar still. It is important to understand the demerits of the solar still concept.

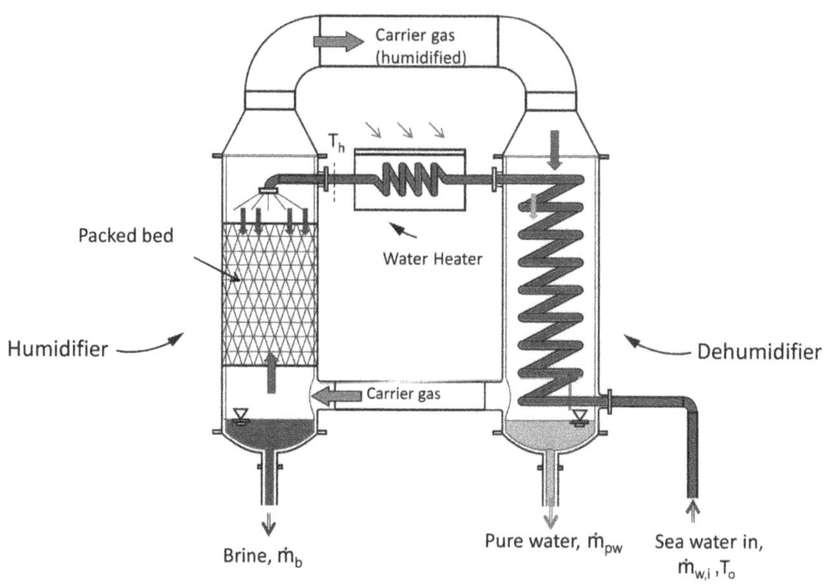

Figure 9.3 Simplest embodiment of HDH process [17].

The most prohibitive drawback of a solar still is its low efficiency (Gained-output-ratio, or GOR[2], less than 0.5) which is primarily the result of the immediate loss of the latent heat of condensation through the glass cover of the still. Some designs recover and reuse the heat of condensation, increasing the efficiency of the still. These designs (called multi-effect stills) achieve some increase in the efficiency of the still, but the overall performance is still relatively low. The main drawback of the solar still is that the various functional processes (solar absorption, evaporation, condensation, and heat recovery) all occur within a single component. By separating these functions into distinct components, thermal inefficiencies may be reduced and overall performance improved. This separation of functions is the essential characteristic of the HDH system. For example, the recovery of the latent heat of condensation, in the HDH process, is effected in a separate heat exchanger (the dehumidifier) wherein the seawater, for example, can be preheated. The module for heat input (like a solar collector) can be optimized almost independently of the humidification or dehumidification component. The HDH process, thus, promises higher productivity due to the separation of the basic processes.

HDH systems are ideal for application in small-scale systems. They have no parts which require extensive capital cost and maintenance like membranes or high temperature steam lines. There is also no barrier to applying HDH for varied and difficult feedwater qualities.

HDH systems may be classified under three broad categories. One is based on the form of energy used such as solar, thermal, geothermal, or hybrid systems. This classification brings out a promising merit of the HDH concept: the promise of water production by use of low grade energy, especially from sources of industrial waste heat or from renewable resources like solar energy or biomass.

The second classification of HDH processes is based on the cycle configuration (Figure 9.4). As the name suggests, a closed-water open-air (CWOA) cycle is one in which ambient air is taken into the humidifier where it is heated and humidified and sent to the dehumidifier where it is partially dehumidified and let out in an open cycle as opposed to a closed air cycle wherein the air is circulated in

---

[2] See Sec. 9.1.2 for definition of GOR.

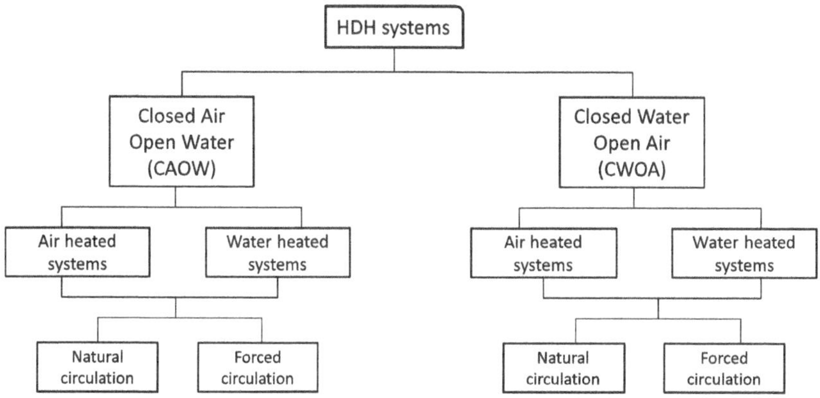

Figure 9.4 Classification of HDH systems based on cycle configurations [15].

a closed loop between the humidifier and the dehumidifier. In this cycle, the brine is recirculated until a desirable recovery is attained. The air in these systems can be circulated by either natural convection or mechanical blowers and feedwater is typically circulated by a pump. It is of pivotal importance to understand the relative technical advantages of each of these cycles and choose the configuration that is best in terms of efficiency and cost of water production.

The third classification of the HDH systems is based on the type of heating used - water or air heating systems. The performance of the system depends greatly on which fluid is heated.

9.1.2 Review of Systems in Literature

As a first step for understanding different works in literature the following performance parameters are defined.

1. Gained-Output-Ratio (GOR): is the ratio of the latent heat of evaporation of the water produced to the net heat input to the cycle.

$$\mathrm{GOR} \equiv \frac{\dot{m}_{pw} \cdot h_{fg}}{\dot{Q}_{in}} \tag{9.1}$$

This parameter is, essentially, the effectiveness of water production, which is an index of the amount of the heat recovery affected in the system. This is the primary performance parameter of interest in HDH (and in thermal desalination, in general) and is very similar to the performance ratio (PR) defined for MED and MSF systems. For steam-driven desalination systems (like in most state-of-the-art MSF and MED systems), PR is approximately equal to GOR:

$$\text{GOR} = \frac{\dot{m}_{pw} \cdot h_{fg}}{\dot{m}_s \cdot \Delta h_s} \tag{9.2}$$

$$\approx \underbrace{\frac{\dot{m}_{pw}}{\dot{m}_s}}_{\text{PR}} \tag{9.3}$$

It is worthwhile to note that GOR is also defined as the ratio of the latent heat (h_{fg}) to the specific thermal energy consumption. The latent heat in the equations above is calculated at the average partial pressure of water vapor (in the moist air mixture) in the dehumidifier.

2. Recovery ratio (RR): is the ratio of the amount of water produced per kg of feed. This parameter is also called the extraction efficiency [16]. This is, generally, found to be around 5% for the HDH system in single pass and can be increased to higher values (up to 90% depending on feed salinity) by brine recirculation.

$$\text{RR} \equiv \frac{\dot{m}_{pw}}{\dot{m}_w} \tag{9.4}$$

The literature has been reviewed in detail previously by the present authors [15] and the review was also updated recently [17]. Based on this review, we can benchmark the key performance metrics of existing HDH systems: (1) the cost of water production; (2) the heat and mass transfer rates in the dehumidifier; and (3) the system energy efficiency (GOR).

The total cost of water production in HDH systems is mostly a sum of the energy cost (captured by the GOR of the system) and the capital cost. A large fraction of the capital investment in typical HDH systems is the dehumidifier cost[3]. This is driven by the low heat and mass transfer rates common in such devices. It has been reported that the 'equivalent' heat transfer coefficient in the dehumidifier is between 1 and 100 W/m²K [18, 19]. This is two orders of magnitude lower than for pure vapor condensers.

[3] The HDH system has relatively minimal maintenance requirements.

Using the data given in various papers, GOR for the reported systems was calculated. It was found that the maximum GOR among existing HDH systems was about 3. Figure 9.5 illustrates the GOR of a few of the studies. The GOR varied between 1.2 to 3. These values of GOR translate into energy consumption rates from 215 kWh_{th}/m^3 to 550 kWh_{th}/m^3. The low value of GOR achieved by Ben Bacha et al. [20] was because they did not recover the latent heat of condensation. Instead, they used separate cooling water from a well to dehumidify the air. Lack of a systematic understanding of the thermal design of HDH systems, which can help to optimize performance, is the reason behind such inefficient designs. The higher value of GOR achieved by Müller-Hölst et al. [21] was because of higher heat recovery. These results tell us the importance of maximizing heat recovery in minimizing the energy consumption and the operating and capital cost of HDH systems. It is also to be noted that the GOR fluctuated between 3 to 4.5 in Müller-Hölst's system because of the inability of that system to independently control the air flow under the natural convection design that was applied. It is, therefore, desirable to develop forced convection based systems which have a sustainable peak performance.

Based on a simple thermodynamic calculation, the GOR of a thermodynamically reversible HDH system can be evaluated to be 122.5 for typical boundary conditions [22]. When compared to a GOR of 3 for existing systems, the reversible GOR of 122.5 shows that there is significant potential for improvement to existing HDH

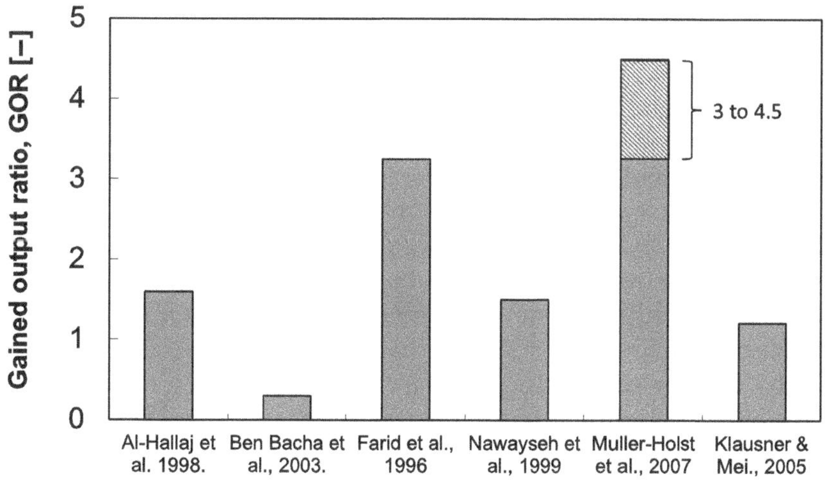

Figure 9.5 Performance of the older HDH systems in the literature [17].

systems in terms of reducing thermodynamic losses. This observation gives ample motivation to study the thermal design of these systems in detail.

A few studies in literature actually report the overall cost of water production in a HDH system [21, 23, 24]. This cost is found to be about \$30 per cubic meter of water produced, which is very high. More recent work, which we describe in subsequent sections, has reduced the cost to affordable levels (< \$5 per cubic meter).

9.2 Thermal Design

When finite time thermodynamics is used to optimize the energy efficiency of thermal systems, the optimal design is one which produces the minimum entropy within the constraints of the problem (such as fixed size or cost). In this section, we apply this well-established principle to the thermal design of combined heat and mass exchange devices (dehumidifiers, and humidifiers) for improving the energy efficiency of HDH desalination systems.

The theoretical framework for design of heat and mass exchange (HME) devices for implementation in the HDH system has been developed in a series of recent papers [22, 25–31]. The linchpin in this theoretical work is the definition of a novel parameter known as the 'modified heat capacity rate ratio' (HCR). A brief summary of the definition of this parameter and its significance to thermal design of HME devices and the HDH system is given below.

Modified heat capacity rate ratio

In the limit of infinite heat transfer area, the entropy generation rate in a regular heat exchanger will be entirely due to what is known as thermal imbalance. This is associated with conditions fort which the heat capacity rate of the streams exchanging heat are not equal [32]. In other words, a heat exchanger (with constant heat capacity for the fluid streams) is said to be thermally 'balanced' at a heat capacity rate ratio of one. This concept of thermodynamic balancing, very well known for heat exchangers, was previously extended by the present authors to HME devices [25].

In order to define a thermally 'balanced' state in HME devices, a modified heat capacity rate ratio (HCR) for combined heat and mass exchangers was defined by analogy to heat exchangers as the ratio of the maximum change in the total enthalpy rate of the cold stream to that of the hot stream.

$$\text{HCR} = \left(\frac{\Delta \dot{H}_{\max,c}}{\Delta \dot{H}_{\max,h}} \right) \tag{9.5}$$

The maximum changes are defined by identifying the ideal states that either stream can reach at the outlet of the device. For example, the ideal state that a cold stream can reach at the outlet will be to match the inlet temperature of the hot stream and that a hot stream can reach at the outlet will be to match the inlet temperature of the cold stream. The physics behind this definition is explained in detail in a prior publication [25].

9.2.1 Design Models

HME devices can be studied under the constraint of a fixed performance (with size varying to maintain this performance under varying inlet conditions) or as a fixed piece of hardware (with varying performance under varying inlet conditions). The former is known as an on-design analysis and the latter is known as an off-design analysis. Here we review an on-design model developed in previous work [26, 28] - the energy effectiveness model.

Effectiveness model

An energy-based effectiveness, analogous to the effectiveness defined for heat exchangers, is given as:

$$\varepsilon = \frac{\Delta \dot{H}}{\Delta \dot{H}_{\max}} \tag{9.6}$$

This definition is based on the maximum change in total enthalpy rate that can bc achieved in an adiabatic heat and mass exchanger. It is defined as the ratio of change in total enthalpy rate ($\Delta \dot{H}$) to the maximum possible change in total enthalpy rate ($\Delta \dot{H}_{max}$). The maximum possible change in total enthalpy rate can be of either the cold or the hot stream, depending on the heat capacity rate of the two streams. The stream with the minimum heat capacity rate dictates the thermodynamic maximum amount of heat transfer that can be attained between the fluid streams. This concept is explained in detail in a previous publication [26]. Thus,

$$\Delta \dot{H}_{max} = \min(\Delta \dot{H}_{max,c}, \Delta \dot{H}_{max,h}) \tag{9.7}$$

9.2.2 Analysis of Existing Embodiments of the HDH System

From the literature review, it has been found that no study has carried out a detailed thermodynamic analysis in order to optimize the system performance of existing HDH cycles for either the water and air heated designs. In this chapter, the thermodynamic performance of these HDH cycles is analyzed by way of a theoretical cycle analysis. Control-volume based models for the humidifier and the dehumidifier are used to perform this analysis. The governing equations for the control-volume based models are presented in detail in previous publications [22, 26].

In performing the analysis, the following approximations have been made:

- The processes operate at steady-state conditions.
- There is no heat loss from the humidifier, the dehumidifier, or the heater to the ambient.
- Pumping and blower power is negligible compared to the energy input in the heater.
- Kinetic and potential energy terms are neglected in the energy balance.
- The water condensed in the dehumidifier is assumed to leave at a temperature which is the average of the humid air temperatures at inlet and outlet of the dehumidifier.
- It was previously shown that the use of pure water properties instead of seawater properties does not significantly affect the performance of the HDH cycle at optimized mass flow rate ratios [33]. Hence, only pure water properties are used in the present discussion.

9.2.2.1 *Water Heated HDH Cycle*

One of the most commonly studied HDH cycles is the closed-air open-water water-heated (CAOW) cycle (see Figure 9.6). A comprehensive study of parameters which affect the performance of this cycle has not been reported in literature. Such a study will help to understand the ways by which the performance of this basic cycle

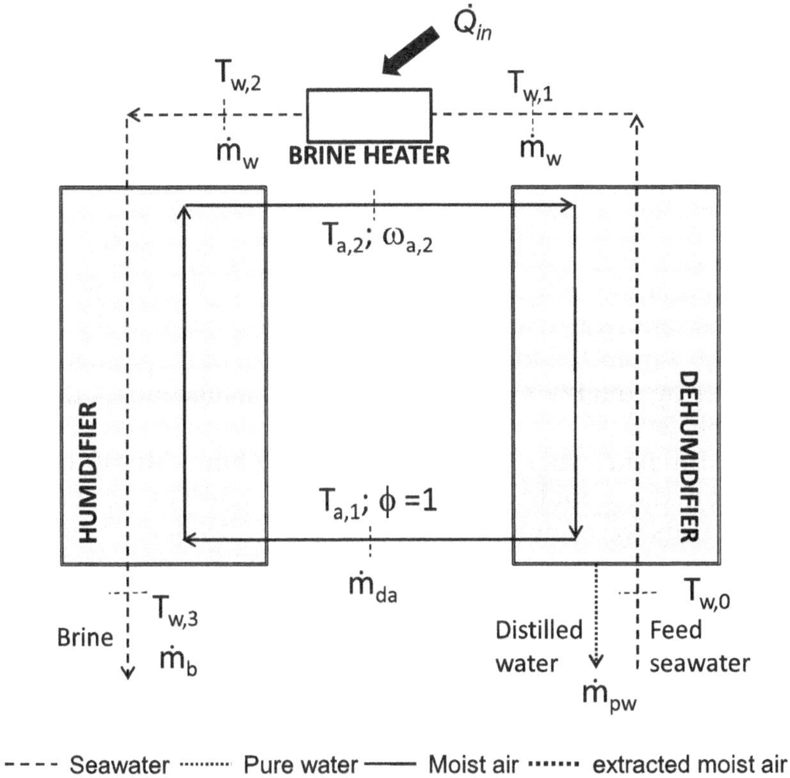

Figure 9.6 Schematic diagram of a water-heated closed-air open-water HDH cycle [22].

can be improved and hence, is reported below. The parameters studied include top and bottom temperatures of the cycle, mass flow rate of the air and water streams, the humidifier and dehumidifier effectivenesses and the operating pressure. The performance of the cycles depends on the mass flow rate ratio (ratio of mass flow rate of seawater at the inlet of the humidifier to the mass flow rate of dry air through the humidifier), rather than on individual mass flow rates. Hence, the mass flow rate ratio is treated as a single variable. This variation with mass flow rate ratio has been noted by many investigators [33, 35–37].

Effect of relative humidity of the air entering and exiting the humidfier ($\varphi_{a,1}$, $\varphi_{a,2}$)

The humidifier and dehumidifier can readily be designed such that the relative humidity of air at their exit is one. Hence, the exit air from these components is usually considered to be saturated when analyzing these cycles. However, the exit relative humidity is indicative of the performance of the humidifier and the dehumidifier; and hence, understanding how the variation of these parameters changes the performance of the system is important.

Figure 9.7 illustrates the effect that relative humidity of air at the humidifier inlet and exit can have on the performance of the cycle (GOR). For this particular case, the top ($T_{w,2}$) and bottom temperatures ($T_{w,0}$) were fixed at 80°C and 35°C respectively. Humidifier and dehumidifier effectivenesses ($\varepsilon_h, \varepsilon_d$) were fixed at 90%. Mass flow rate ratio was fixed at 5. It can be observed that for a variation of $\varphi_{a,2}$ from 100 to 70% the performance of the system (GOR) decreases by roughly 3%, and for the same change in $\varphi_{a,2}$ the effect is roughly 34%.

This difference suggests that the relative humidity of the air at the inlet of the humidifier has a much larger effect on performance. These trends were found to be consistent for all values of mass flow rate ratios, temperatures and component effectivenesses. This, in turn, suggests that the dehumidifier performance will have a larger impact on the cycle performance. This issue is further investigated in the following paragraphs.

Effect of component effectiveness (ε_h, ε_d)

Figure 9.8 and 9.9 illustrate the variation of performance of the cycle at various values of component effectivenesses. In Fig. 9.8, the top

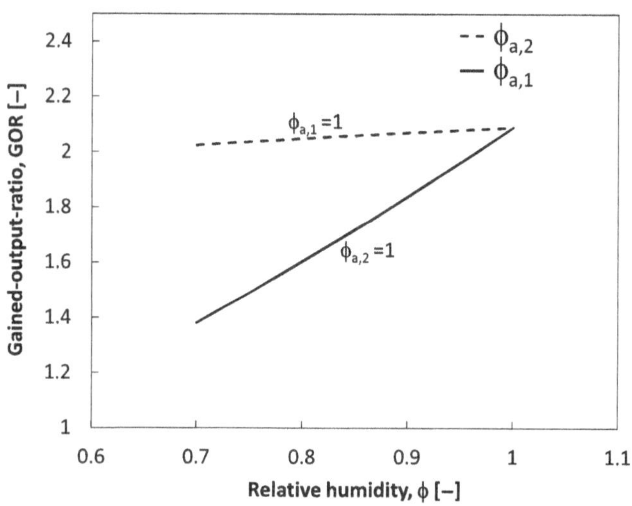

Figure 9.7 Effect of relative humidity on performance of the WH-CAOW HDH cycle [22].

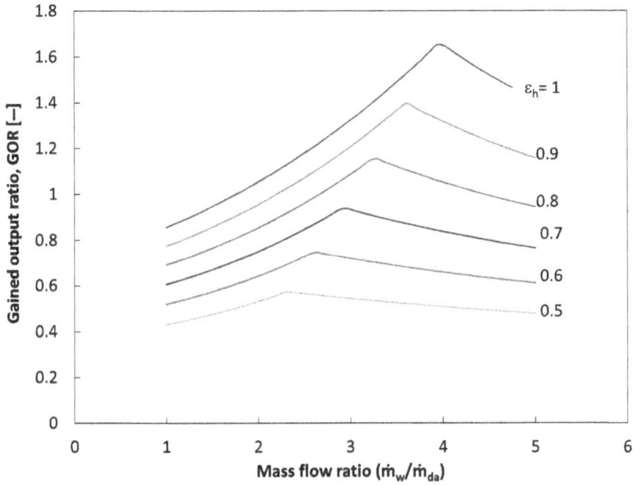

Figure 9.8 Effect of component effectiveness of humidifier on performance of the WH-CAOW HDH cycle [22].

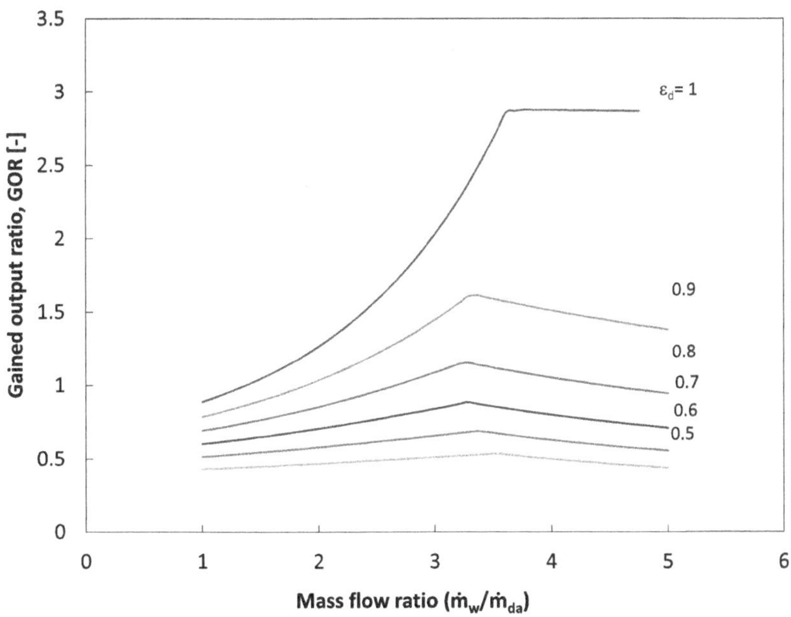

Figure 9.9 Effect of component effectiveness of dehumidifier on performance of the WH-CAOW HDH cycle [22].

temperature is fixed at 80°C, the bottom temperature is fixed at 30°C and the dehumidifier effectiveness is fixed at 80%. The mass flow rate ratio was varied from 1 to 6. It is important to observe that there exists an optimal value of mass flow rate ratio at which the GOR peaks. It can also be observed that the increase in performance is fairly linear with increasing humidifier effectiveness, ε_h. In Fig. 9.9, the top temperature is fixed at 80°C, the bottom temperature is fixed at 30°C and the humidifier effectiveness is fixed at 80%. The cycle performance changes more dramatically for higher values of dehumidifier effectiveness. These trends are consistent for various values of top and bottom temperatures. Hence, a higher dehumidifier effectiveness is more valuable than a higher humidifier effectiveness for the performance (GOR) of the cycle.

In the previous discussion, we have observed that the dehumidifier exit air relative humidity ($\varphi_{a,1}$) is more important than the humidifier exit air relative humidity ($\varphi_{a,2}$). Hence, based on these results, we can say that for a water heated cycle the performance of the dehumidifier is more important than the performance of the humidifier.

Effect of top temperature ($T_{w,2}$)

Figure 9.10 illustrates the effect of the top temperature on the cycle performance (GOR). For this particular case, the bottom temperature ($T_{w,0}$) was fixed at 35°C and humidifier and dehumidifier

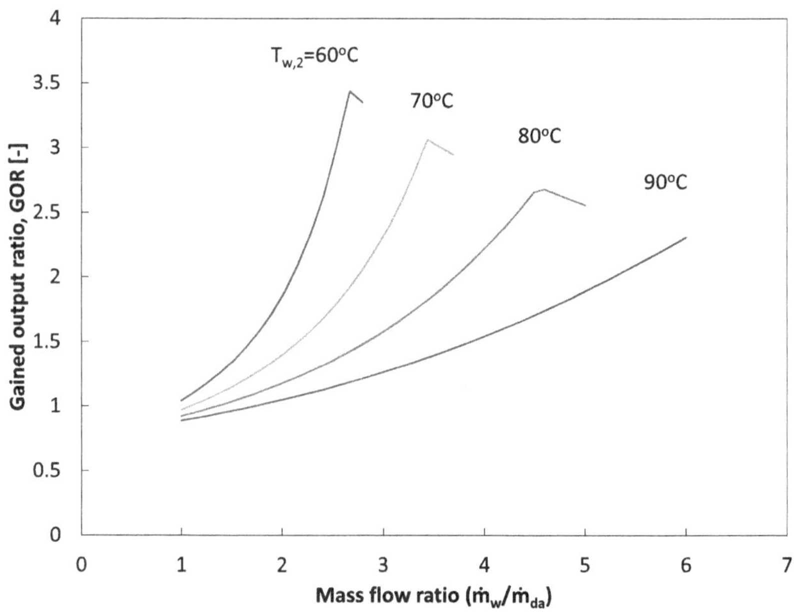

Figure 9.10 Effect of top brine temperature on performance of the WH-CAOW HDH cycle [22].

effectivenesses were fixed at 92%. Top temperature ($T_{w,2}$) was varied from 50°C to 90°C. The optimal value of mass flow rate ratio increases with an increase in top temperature. Depending on the humidifier and dehumidifier effectiveness itself this trend changes. At lower component effectivenesses, the top temperature has no or little effect on the cycle performance. This result is counter-intuitive. However, it can be explained using the modified heat capacity rate ratio.

The modified heat capacity rate ratio (HCR) as the ratio of maximum possible enthalpy change in the cold stream to the maximum possible enthalpy change in the hot stream. It was found that the entropy generation in a heat and mass exchange device is minimized (for a given effectiveness and inlet conditions) when HCR=1 ('balanced' condition). We are going to use this understanding here to explain the trends obtained at various top temperatures.

Figure 9.11 shows the variation of GOR with the heat capacity rate ratio of the dehumidifier (HCR_d). It can be seen that GOR reaches a maximum at $HCR_d = 1$. The maximum occurs at a balanced condition for the dehumidifier which, as we have shown in the preceeding paragraphs is the more important component. Further, it

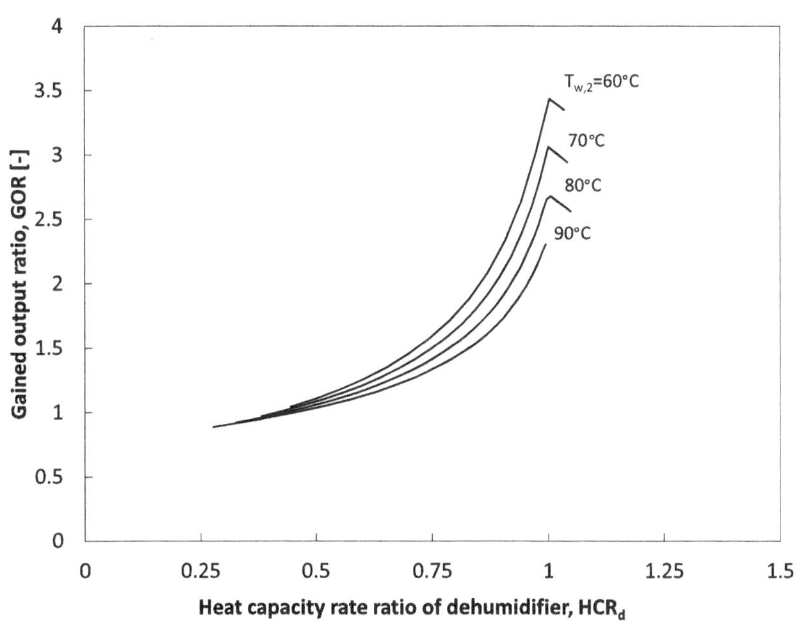

Figure 9.11 HCR of dehumidifer versus GOR at various top brine temperatures [22].

can be noticed from Fig. 9.12 that the degree of balancing of the humidifier at the optimum GOR condition reduces (HCR_h moves farther away from 1) as the top temperature increases. Hence, the irreversibility of the humidifier (and the total irreversibility of the system) increases with increase in top temperature. A system with higher total irreversibility has a lower GOR [33]. This explains the decrease in GOR with an increase in top temperature.

Also, as the top temperature increases, the dehumidifier is balanced at higher mass flow ratio and hence the optimum value of GOR occurs at higher mass flow ratios.

Effect of bottom temperature ($T_{w,0}$)

The bottom temperature of the cycle ($T_{w,0}$) is fixed by the feedwater temperature at the location where the water is drawn. Figure 9.13 illustrates a case with top temperature of 80°C and component effectivenesses of 92%. A higher bottom temperature of the cycle results in a higher value of GOR as illustrated in the figure. This result can again be understood by plotting HCR of humidifier and dehumidifier versus the GOR of the system (Figs. 9.14 and 9.15).

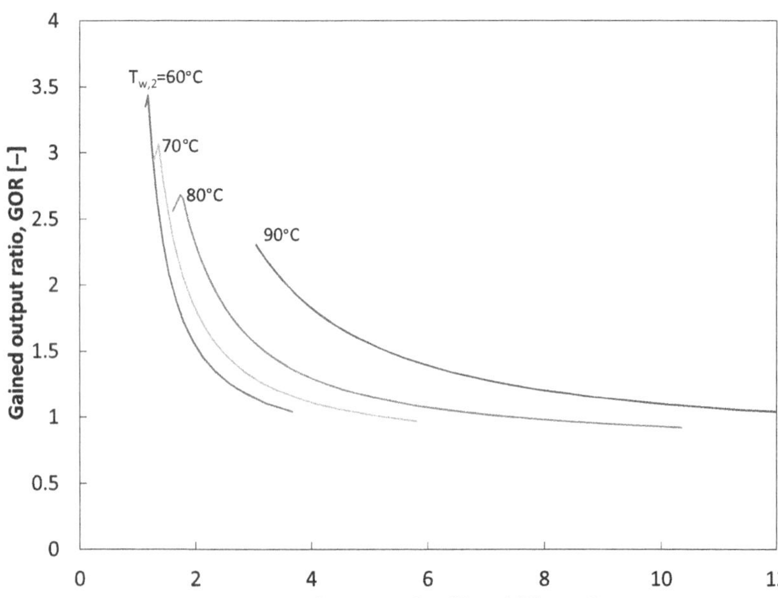

Figure 9.12 HCR of humidifier versus GOR at various top brine temperatures [22].

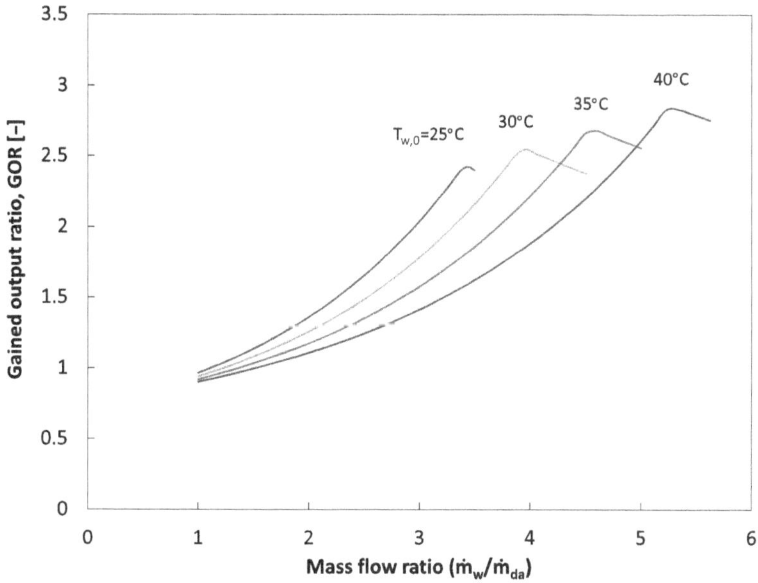

Figure 9.13 Effect of feedwater temperature on performance of the WH-CAOW HDH cycle [22].

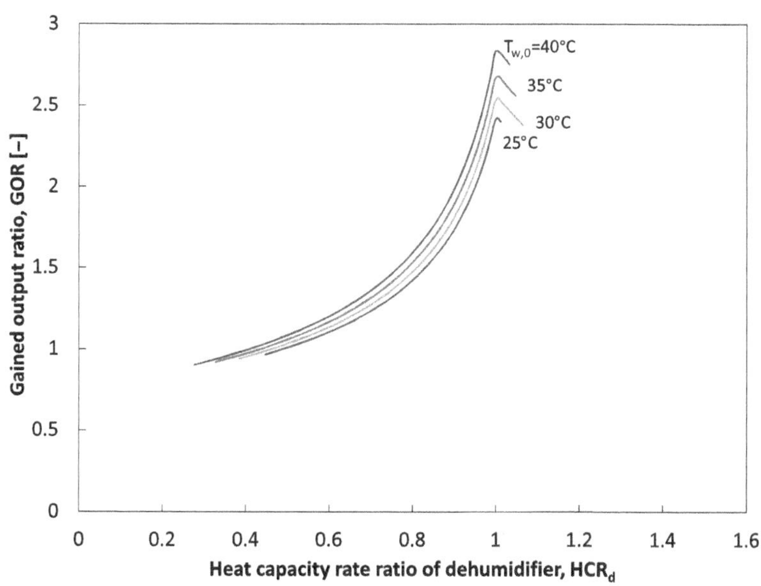

Figure 9.14 HCR of dehumidifer versus GOR at various feedwater temperatures [22].

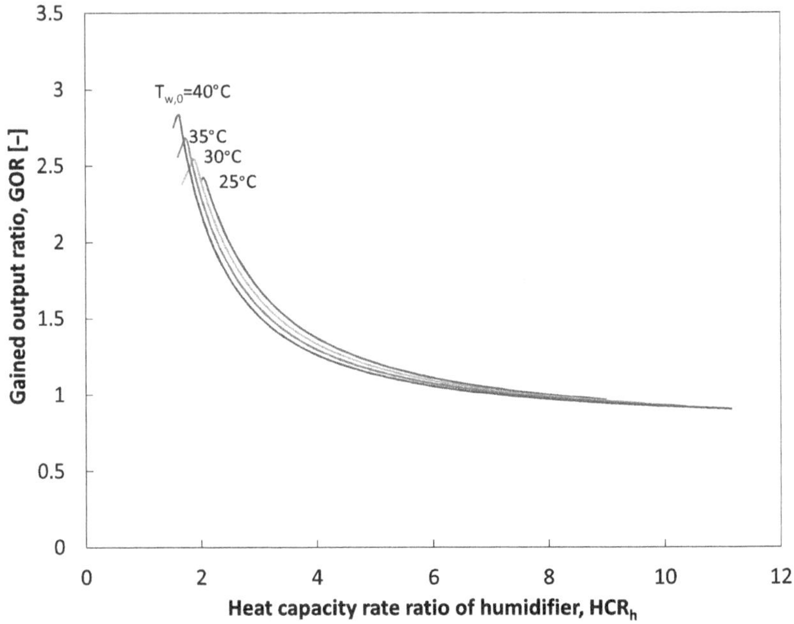

Figure 9.15 HCR of humidifier versus GOR at various feedwater temperatures.

The degree of balancing of the humidifier at the optimum condition for GOR decreases with a decrease in bottom temperature. Hence, the irreversibilities in the humidifier (and the total irreversibility of the system) increase with decreasing bottom temperature, and the GOR declines.

From the discussions in this subsection we have observed that the performance of the cycle (GOR) has a functional dependence as follows:

$$\text{GOR} = f(\text{HCR}_h, \text{HCR}_d, \varepsilon_h, \varepsilon_d, T_{w,2}, T_{w,0}, \varphi_{a,2}, \varphi_{a,1}) \qquad (9.8)$$

The numerically computed values of GOR reported in this section for the CAOW water-heated cycle are within 20% of the experimental value obtained by Nawayseh [38] for the same boundary conditions. Further experimental validation is presented later in this section.

9.2.2.2 *Single and Multi-Stage Air Heated Cycle*

A simple [23, 24, 39, 40] air-heated cycle is one in which air is heated, humidified, and dehumidified. Current simulations have found that the GOR for this cycle is very low (GOR<1; only slightly better than a solar still). It is important to understand the reasons for this poor performance. The air in this cycle is heated and immediately sent to a humidifier where it is saturated. The air also gets cooled during the humidification process since it is at a higher temperature than the water stream. Thus, heat is lost to the water stream in the humidifier. In the water-heated cycle, the air stream is heated in the humidifier. This further facilitates heat recovery in the dehumidifier, which is absent in an air heated system. Hence, the performance is much lower in an air-heated system.

To improve the performance of air-heated systems, Chafik [23, 41] proposed a multi-stage cycle. The air in this cycle is heated and sent to a humidifier where it is saturated. It is then further heated and humidified again. The idea behind this scheme was to increase the exit humidity of the air so that water production can be increased. Chafik was able to increase the exit humidity from 4.5% (by weight) for a single stage system to 9.3% for a 4 stage system. We reproduce this result for the same cycle under similar operating conditions. However, we also observe that the GOR of the cycle rises by only 9% (Fig. 9.16). This is because the increased water production comes at the cost of increased energy input. This, in turn, is because the multi-staging does not improve the heat recovery in the humidification process. Chafik reported

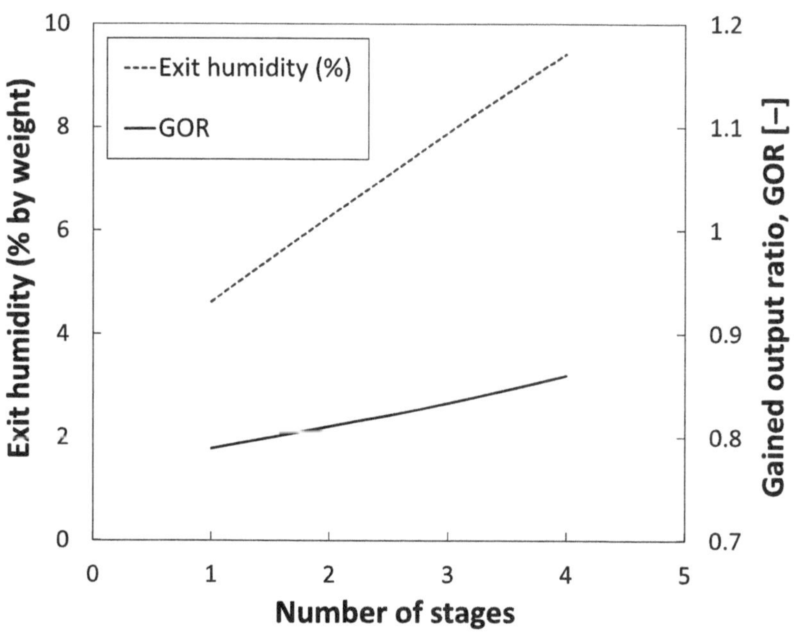

Figure 9.16 Effect of number of stages on performance of air heated CAOW HDH system [22].

very high cost of water production of 28.65 Euro/m^3 caused in part by the low energy efficiency of the system.

9.2.2.3 *Summary*

In this section, we have examined the effect of a number of parameters and configurations on the performance of basic HDH cycles. The following significant conclusions are reached:

1. The performance of a basic water-heated cycle depends on: (a) the water-to-air mass flow rate ratio; (b) the humidifier and dehumidifier effectivenesses; (c) top and bottom temperatures; and (d) relative humidity of air at the exit of the humidifier and the dehumidifier.
2. At a specific value of the water-to-air mass flow rate ratio, the performance of the system is optimal. This optimal point is characterized by a thermodynamically balanced condition in the dehumidifier. The balanced condition, as explained previously, occurs at a modified heat capacity rate ratio of 1. This finding is important, as it is also fundamental to design algorithms for HDH systems with mass extraction and injections.
3. As shown in the table below, previously studied multistage and single-stage air heated cycles have low energy efficiency compared to the water heated HDH cycle.
4. The performance of existing HDH systems is only about 1/60th of the thermodynamically reversible GOR. This shows the extent of the thermodynamic losses in these systems. Much of the remainder of this chapter is dedicated to improving the thermal design of the HDH cycle so as to reduce the thermodynamic irreversiblity.

Table 9.1 Comparison of GOR of HDH cycles under representative boundary conditions.

CYCLE	GOR
Single stage air heated cycle	0.78
Four-stage air heated cycle	0.85
Water heated cycle	2.5
Reversible cycle	122.5

9.2.3 Systems with Mass Extraction and Injection

Studies have been conducted on the effect of entropy generation on the thermal design of the HDH system [29, 33, 42, 43], and it has been found that reducing the total entropy generated (per unit amount of water distilled) improves the energy efficiency (measured in terms of the gained-output-ratio or GOR). It has also been reported that incorporating mass extractions and injections to vary the water-to-air mass flow rate ratio in the combined heat and mass transfer devices (like the humidifier and the dehumidifier) can potentially help in reducing entropy production in those devices [25]. A comprehensive method of thermodynamic analysis is available for the design of mass extractions and injections in the HDH system. This approach (discussed in the subsequent sections) draws upon the fundamental observation that there is a single value of water-to-air mass flow rate ratio (for any given boundary conditions and component effectivenesses) at which the system performs optimally [22].

A schematic diagram of a representative the HDH system with mass extractions and injections is shown in Fig. 9.17. The system shown here is a water-heated, closed-air, open-water system with three air extractions from the humidifier into the dehumidifier. States a to d are used to represent various states of the seawater stream and states e and f represent that of moist air before and after dehumidification. Several other embodiments of the system are possible based on the various classifications of HDH listed earlier in this chapter.

Enthalpy pinch model

McGovern et al. [44] proposed that it is advantageous to normalize enthalpy rates by the amount of dry air flowing through the system for easy representation of the thermodynamic processes in enthalpy versus temperature diagrams. We use this concept here and derive the following equation from Eq. (9.6) by dividing the numerator and the denominator by the mass flow rate of dry air ($\dot{m}_{da}$):

$$\varepsilon = \frac{\Delta h^*}{\Delta h^*_{\max}} \tag{9.9}$$

$$= \frac{\Delta h^*}{\Delta h^* + \Psi_{TD}} \tag{9.10}$$

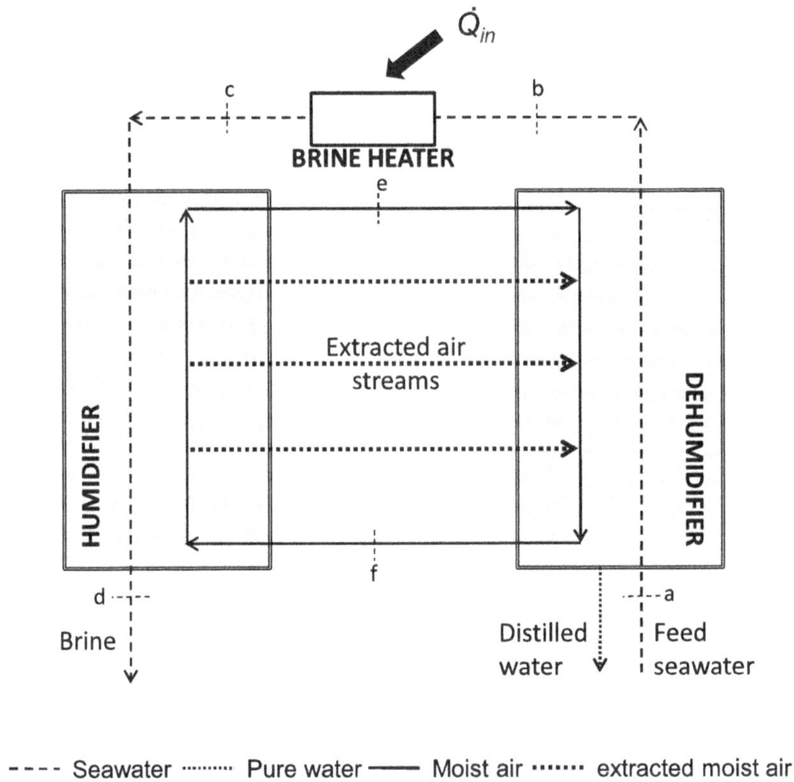

Figure 9.17 Schematic diagram of a water-heated, closed-air, open-water humidification-dehumidification desalination system with mass extraction and injection of the moist air stream [28].

Ψ_{TD} is the loss in enthalpy rates at terminal locations because of having a "finite-sized" HME device, and it is defined as follows:

$$\Psi_{TD} = \min\left(\frac{\Delta \dot{H}_{\max,c}}{\dot{m}_{da}} - \Delta h^*, \frac{\Delta \dot{H}_{\max,h}}{\dot{m}_{da}} - \Delta h^*\right) \quad (9.11)$$

$$= \min(\Psi_c, \Psi_h) \quad (9.12)$$

In the case of a heat exchanger, Ψ_{TD} will be analogous to the minimum terminal stream-to-stream temperature difference (TTD). TTD is seldom used to define performance of a heat exchanger in thermodynamic analyses; the temperature pinch is the commonly used parameter. The difference is that pinch is the minimum stream-to-stream temperature difference at any point in the heat exchanger and not just at the terminal locations. Like temperature pinch, Ψ can be defined as the minimum loss in enthalpy rate due to a finite device size at any point in the HME device and not just at the terminal locations. Thus, the general definition of Ψ will be as follows:

$$\Psi = \min_{local}(\Delta h^{*}_{max} - \Delta h^{*}) \tag{9.13}$$

Hence, based on the arguments presented in this section, we can say that Ψ for an HME device is analogous to temperature pinch for a heat exchanger, and it can be called the 'enthalpy pinch'. We recommend that, because of the presence of the concentration difference as the driving force for mass transfer in HME devices, a temperature pinch or a terminal temperature difference should not be used when defining the performance of the device. Further details about the enthalpy pinch and its significance in thermal design of HME devices are given in detail in Reference [28]. Balancing of HDH cycles has been studied in further detail in References [30], [31], and [45].

9.2.3.1 *System Balancing Algorithms*

In a previous publication, we used the concepts of thermodynamic balancing developed for HME devices and applied them to the HDH system design [28]. Detailed algorithms for systems with zero, single, and infinite extractions were developed. Temperature-enthalpy diagrams were used to model the systems. Figure 9.18 illustrates temperature versus enthalpy of a system with a single extraction and injection. In the illustrated case, the air was extracted from the humidifier at the state 'ex' and injected in a corresponding location in the dehumidifier with the same state 'ex' to avoid generating entropy during the process of injection. This criteria for extraction is applied for all the cases reported in this paper since it helps us study the effect of thermodynamic balancing, independently, by separating out the effects of a temperature and/or a concentration mismatch between the injected stream and the fluid stream passing through the HME device (which when present can make it hard to quantify the reduction in entropy generated due to balancing alone).

The effect of the number of extractions (at various enthalpy pinches) on the performance of the HDH system was studied using the developed algorithms and is shown in Fig. 9.19. Several important observations can be made from this chart.

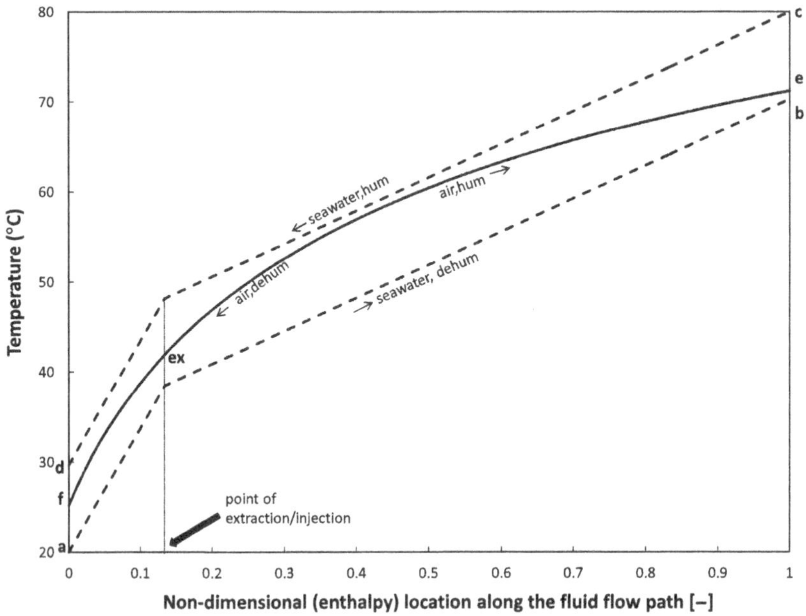

Figure 9.18 Temperature profile representing the HDH system with a single extraction. Boundary conditions: $T_a = 20^{\circ}$ C; $T_c = 80^{\circ}$ C; $\Psi_{deh} = \Psi_{hum} = 20$ kJ/kg dry air [28].

First, it may be observed that thermodynamic balancing is effective in HDH cycles only when the humidifier and the dehumidifier have an enthalpy pinch less than about 27 kJ/kg dry air. For various boundary conditions, it has been found that above the aforementioned value of enthalpy pinch the difference in performance (GOR) with that of a system without any extractions or injections is small (less than 20%). Further, at very low values of the enthalpy pinch ($\Psi \leq 7$ kJ/kg dry air) in the humidifier and the dehumidifier, the limiting case of continuous balancing with infinite number of extractions and injections was found to give much better results than that with a single extraction and injection. For a top brine temperature of 80°C, a feed water temperature of 20° C and 'infinitely' large humidifier and dehumidifier ($\Psi_{hum} = \Psi_{deh} = 0$ kJ/kg dry air), the GOR was found to be 8.2 for a single extraction (compared to a GOR of 109.7 for a similar system with an infinite number of extractions, i.e. continuous extraction). At higher values of enthalpy pinch ($7 < \Psi \leq 15$), a single extraction reduced the entropy generation of the total system roughly by a similar amount as continuous extractions. At even higher values of enthalpy pinch ($15 < \Psi \leq 27$), a single extraction outperforms continuous extractions.

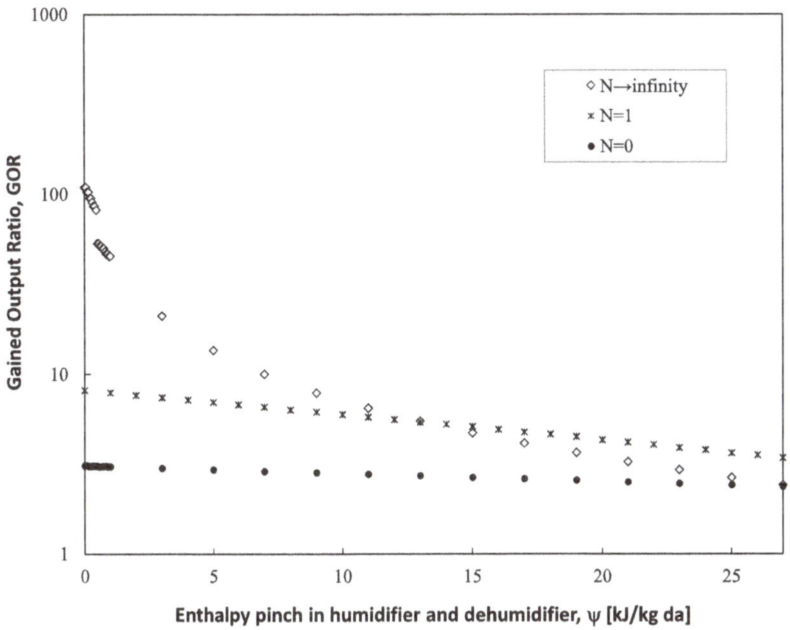

Figure 9.19 Effect of number of extractions (for thermodynamic balancing) on the performance of the HDH system with finite and infinite size HME devices. Boundary conditions: $T_a = 20^\circ$ C; $sal = 35$ g/kg; $T_c = 80^\circ$ C; $HCR_{deh} = 1$ [28].

9.2.4 Experimental Realization of HDH with Extraction

A pilot scale HDH unit producing up to 700 liters of pure water per day has been built [29]. The unit was fully instrumented and both component and system experiments were carried out.

9.2.4.1 HME Balancing

As described previously, theoretical considerations show that a modified heat capacity rate ratio (HCR) of 1 will lead to minimum entropy generation in a fixed effectiveness or fixed hardware device [25, 26, 29] and that the condition should be established to optimize the thermal performance of the HDH cycle [22]. In this section, this important conclusion is investigated experimentally.

Figure 9.20 illustrates that there exists a particular mass flow rate at which non-dimensional entropy generated in the device is minimized (fixed inlet air condition and fixed inlet water temperature). At different values of these boundary conditions, the same result was found to be true. The minimum that is observed also

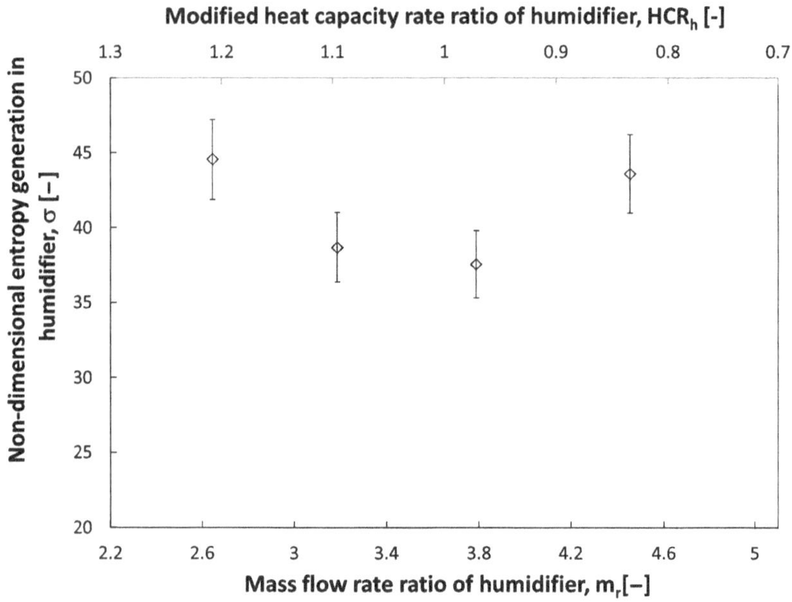

Figure 9.20 Effect of mass flow rate ratio on non-dimensional entropy generation in the humidifier. Boundary conditions: $T_e = 32^\circ C$; $T_c = 60°C$; $T_{wb,e} = 20^\circ C$; $P = 101.3$ kPa; $V_h = 0.27$ m^3[29].

corresponds to the case closest to an HCR of 1. This is consistent with the theoretical observation that irreversiblity is minimized at HCR of unity [25, 27, 29].

9.2.4.2 System Balancing

Figure 9.21 illustrates the effect of mass flow rate extracted on the increase in performance of the HDH system. The increase in performance of the HDH system is calculated as the ratio of the GOR with extraction to that without extraction. In cases with and without extraction the top brine temperature, the feedwater temperature, the water flow rate and total air flow entering the humidifier from the dehumidifier (measured at state 'f' shown in Fig. 9.3) are held fixed. In the zero extraction case, this aforementioned ratio of GORs is 1 and increases with better balancing. The amount of air extracted is also normalized against total air flow.

It may be observed that the performance is optimal at a particular amount of extraction. In this particular case, where the top temperature is 90°C and the feed temperature is 25°C, the optimum amount of extraction is around 33%. The GOR is enhanced by up

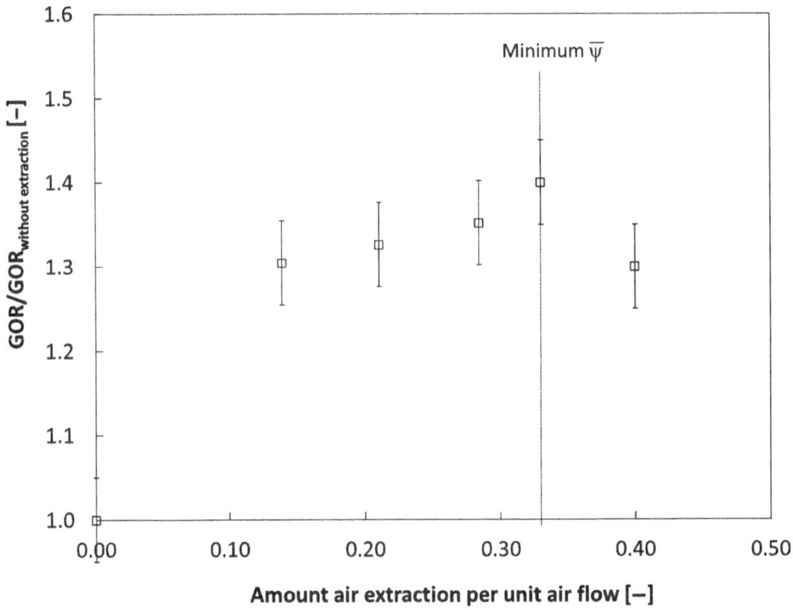

Figure 9.21 Effect of mass flow rate of air extracted on the performance of the HDH system. Boundary conditions: $T_a = 25^\circ$ C; $T_c = 90^\circ$ C; $N = 1$; $V_h = 0.27$ m^3; $A_d = 8$ m^2[29].

to 40%. The trends are similar at different boundary conditions and the maximum enhancement in GOR with a single extraction of air was found to be about 55%.

As would be expected, the maximum performance corresponded to the minimum average of local enthalpy pinches in the dehumidifier ($\overline{\Psi}_{local,d}$). This is consistent with the principal purpose of thermodynamic balancing: to drive the process at minimum driving force and correspondingly smaller entropy generation at a fixed system size.

9.2.4.3 *Summary of HDH Characteristics Related to Extraction*

1. HCR=1 (i.e. the point at when the maximum change in enthalpy rates of either stream exchanging energy is equal) represents a thermally balanced state for a simultaneous heat and mass exchange device.
2. For a water heated CAOW HDH system without mass extractions, the HCR_d=1 represents the state at which the GOR is maximized.

3. HDH systems without mass extractions need to be operated at as high a top brine temperature as is possible in order to ensure a high GOR.
4. Mass extractions from the humidifier to the dehumidifier increase the GOR of the water heated CAOW HDH system by up to 55%.
5. It is found that thermodynamic balancing is effective in HDH only when the HME devices have an appropriately low enthalpy pinch ($\Psi < 27$ kJ/kg dry air).
6. The optimum extraction mass flow rate corresponds to the case in which a minimum average local enthalpy pinch is achieved in the device.

9.3 Bubble Column Dehumidification

When a non-condensable gas is present, the thermal resistance to condensation of vapor on a cold surface is much higher than in a pure vapor environment. This is, primarily, caused by the diffusion resistance to transport of vapor through the mixture of non-condensable gas and vapor. Several researchers have previously studied and reported this effect [46–54]. When even a few mole percent of non-condensable gas are present in the condensing fluid, the deterioration in the heat transfer rates can be up to an order of magnitude [56–60]. From experimental reports in literature, it can be observed that the amount of deterioration in heat transfer is a very strong (almost quadratic) function of the mole fraction of non-condensable gas present in the condensing vapor.

In HDH systems, a large percentage of air (60–90% by mass) is present by default in the condensing stream. As a consequence, the heat exchanger used for condensation of water out of an air-vapor mixture (otherwise known as adehumidifier) has very low heat and mass transfer rates (an 'equivalent' heat transfer coefficient as low as 1 W/m^2K in some cases [19, 61–63]). This leads to very high heat transfer area requirments in the dehumidifier (up to 30 m^2 for a 1 m^2/day system). In this section, we describe how to achieve a substantial improvement in the heat transfer rate by condensing the vapor-gas mixture in a column of cold liquid, rather than on a cold surface, by using a bubble column heat and mass exchanger.

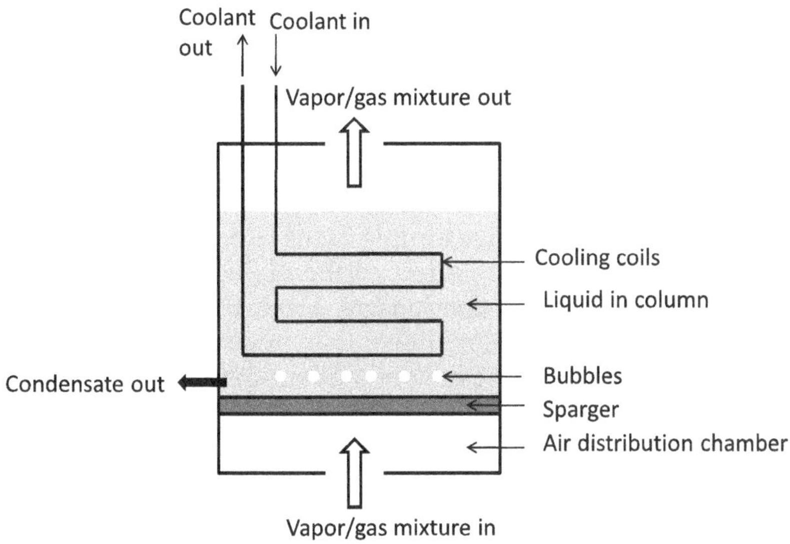

Figure 9.22 Schematic diagram of the bubble column dehumidifier [66].

In this device, moist air is sparged through a porous plate (or any other type of sparger [64]) to form bubbles in a pool of cold liquid. The upward motion of the air bubbles causes a wake to be formed underneath the bubble which entrains liquid from the pool, setting up a strong circulation current in the liquid pool [65]. Heat and mass are transferred from the air bubble to the liquid in the pool in a direct contact transport process. At steady state, the liquid, in turn, losses the energy it has gained to the coolant circulating through a coil placed in the pool for the purpose of holding the liquid pool at a steady temperature. The system is illustrated in Fig. 9.22.

9.3.1 Modeling and Experimental Validation

A thermal resistance models for the condensation of water from an air-vapor mixture in a bubble column heat exchanger were introduced in Reference [66] and have been revised and refined in References [67-69]. The primary temperatures in the resistance network are: (1) the average local temperature of the air-vapor mixture in the bubbles (T_{air}), (2) the average temperature of the liquid in the pool (T_{column}), [64] and (3) the average local temperature of the coolant inside the coil ($T_{coolant}$). Between T_{air} and T_{column} there is direct contact heat and mass transfer. The liquid pool is well-mixed by the bubbles, and may be considered to hold a constant temperature. The local heat transfer from the pool to the coolant can be represented by heat transfer coefficients inside and outside the coil, and the temperature change of the coolant can be modeled as a single-stream heat exchanger. The heat transfer between the moist air stream may be modeled similarly.

Figure 9.23 illustrates an example of the experimental and modeling results [66]. A strong effect of the mole fraction is seen, as is also the case in steam condensers. From the experiments, we observe that the effect is more linear than quadratic (in the studied range). Hence, the presence of non-condensable gas is affecting the heat transfer to a much lesser degree than in the film condensation situations of a standard dehumidifier. This demonstrates the superiority of the bubble column dehumidifier technology [70]. This observation is further discussed in Sec. 9.3.2. Figure 9.23 also illustrates that the model predicts the effect of inlet mole fraction very accurately.

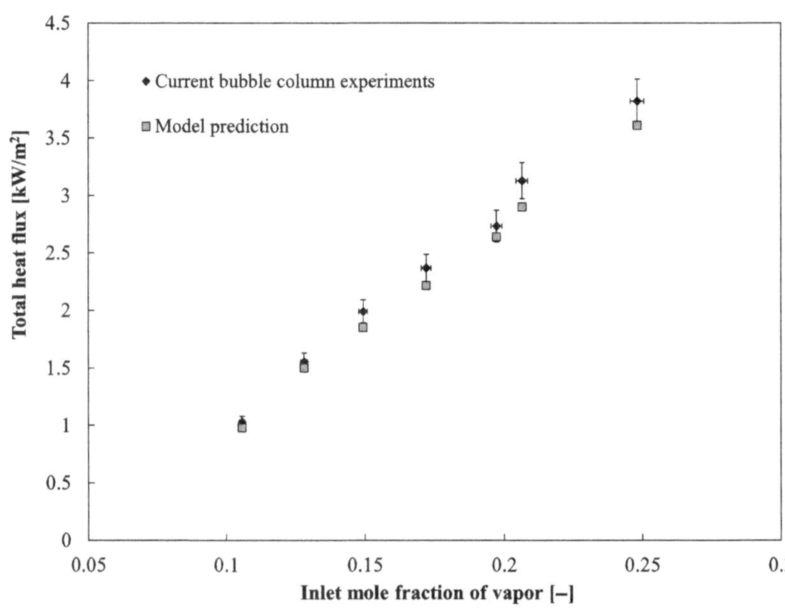

Figure 9.23 Effect of inlet mole fraction of the vapor on the total heat flux in the bubble column measured and evaluated at $Vg = 3.8$ cm/s; $D_b = 4$ mm; $H = 254$ mm [66].

9.3.2 Prototype

In an HDH system, the isothermal nature of the column liquid in the bubble column dehumidifier reduces the temperature to which seawater can be preheated to (in the coils). This limits the energy effectiveness of the device [26]. A low effectiveness in the dehumidifier, reduces the HDH system performance significantly [22]. In this section, we detail an innovation which increases the energy effectiveness of these devices.

A schematic diagram of a multi-stage bubble column is shown in Fig. 9.24. In this device, the moist air is sparged successively from the bottom-most (first) stage to the top-most (last) stage through pools of liquid in each stage. The coolant enters the coil in the last stage and passes through the coil in each stage and leaves from the first stage. Thus, the moist air and the coolant are counter-flowing from stage to stage the condensate is collected directly from the column liquid in each stage.

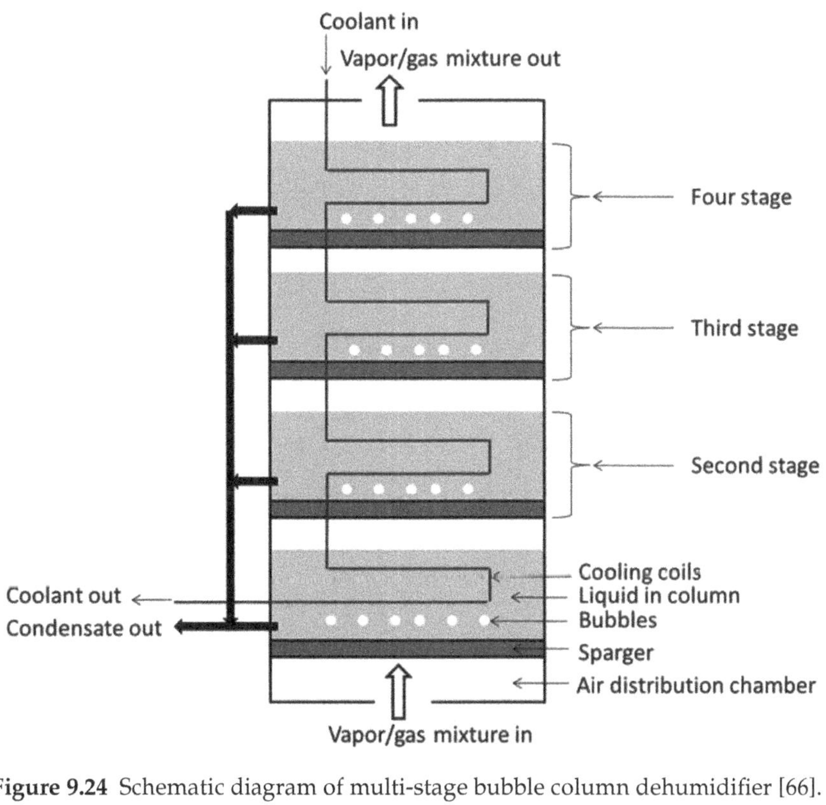

Figure 9.24 Schematic diagram of multi-stage bubble column dehumidifier [66].

Figure 9.25 illustrates the temperature profiles in a single stage and multi-stage bubble column. In both cases, the moist air comes in fully saturated at a temperature of 353 K and leaves dehumidified at 310 K. In the process, the pool of liquid in the bubble column gets

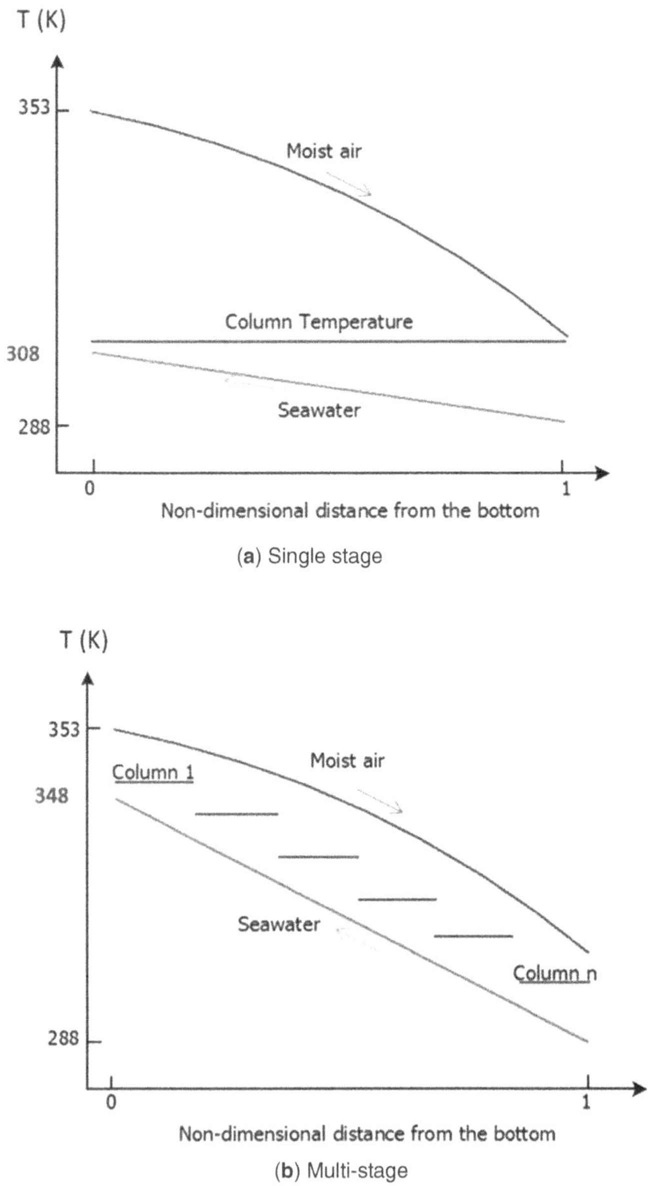

Figure 9.25 An illustration of the temperature profile in the bubble columns for (a) single stage and (b) multi-stage [66].

heated and in turn preheats the seawater going through the coil. In the single stage case, the coolant gets preheated to a temperature of 308 K only (limited by the air exit temperature of 310 K). This corresponds to a very low effectiveness of 30%. In the case of multi-stage bubble columns, the column liquid in each stage is at a different temperature limited by the temperature of the air passing through the respective stage. Hence, the outlet coolant temperature is only limited by the exit temperature of the air from the first stage. In this example, we see that the coolant reaches 348 K (40 K higher than in the single stage case). This corresponds to an increase in effectiveness from 30% in the single stage device to 92% in the multi-stage device.

Figure 9.26 illustrates the increase in effectiveness of the device with multistaging. The data presented here is for an air inlet temperature of 65°C, inlet relative humidity of 100%, a water inlet temperature of 25°C and a water-to-air mass flow rate ratio of 2.45. It can be seen that the energy effectiveness of the device is increased from around 54% for a single stage to about 90% for the three stage

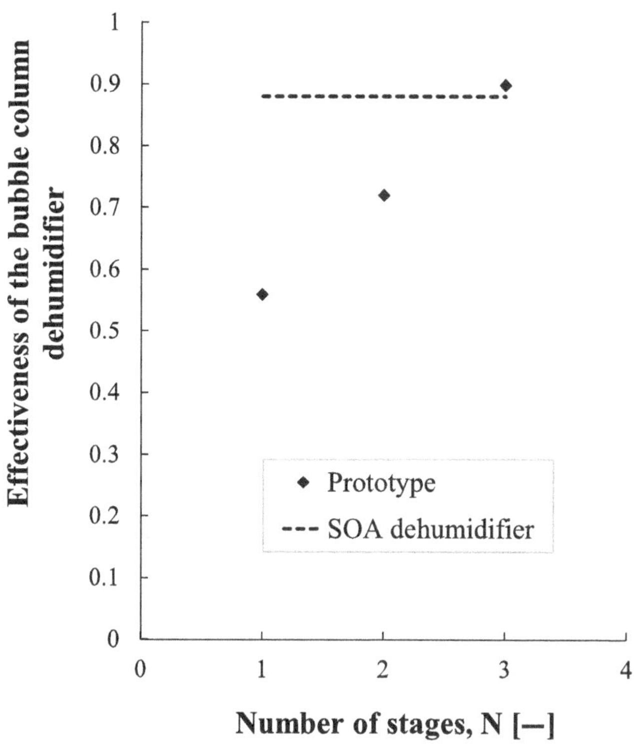

Figure 9.26 Effect of multistaging the bubble column on energy effectiveness of the device.

device. Further, owing to the higher superficial velocity (because of smaller column diameter), the heat fluxes were much higher (up to 25 kW/m^2) than for film-condensation dehumidifiers. Also, the total gas side pressure drop of this device was modest at 800 Pa.

9.3.3 Comparison to Existing Devices

A state-of-the-art dehumidifier (which operates in the film condensation regime) procured from Fischer LLC was found to yield a maximum heat flux of 1.8 kW/m^2 (as per the design specification) compared to a maximum of 25 kW/m^2 obtained in the bubble column dehumidifier, demonstrating the superior performance of the novel device. This comparison was carried out at the same inlet conditions for the vapor-air mixture and the coolant streams. Also, the streamwise temperature differences were similar in both cases. Further, the energy effectiveness of a three-stage bubble column dehumidifier was found to be similar to the conventional dehumidifier mentioned here.

By way of this innovation, the heat transfer area requirement is reduced to a fraction of that in existing HDH systems and is brought close to that for pure vapor systems (such as MED systems). This trend is illustrated in Fig. 9.27.

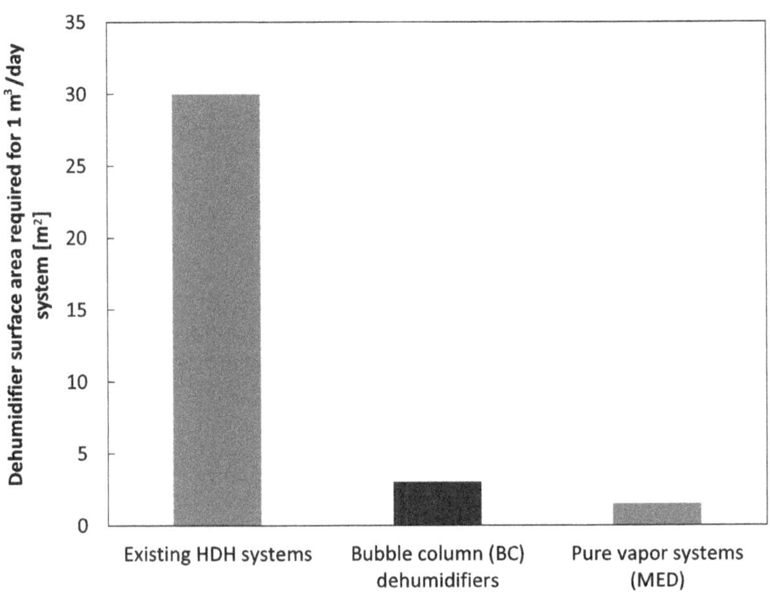

Figure 9.27 Dehumidifier area requirement for bubble columns compared to existing technology.

9.3.4 Summary of Bubble Column Dehumidification in HDH Systems

1. Bubble column dehumidifiers have an order of magnitude better performance than existing state-of-the-art dehumidifiers operating in the film condensation regime.
2. In order to minimize pressure drop, the liquid height can be kept to a minimum such that the coil is entirely submerged in the liquid. This is possible because the height been shown to have no effect on the performance of the device for column heights down to 4 cm [68], and is likely to have minimal effect so long as the pool depth is somewhat greater than the bubble diameter (≈ 4–6 mm).
3. A three-stage bubble column with a manageable air side pressure drop of < 1 kPa, an high effectiveness of 92%, and a very high heat rate of 25 kW/m^2 can be constructed at a fraction of the cost of a regular dehumidifier operating in the film condensation regime.

Implementation of the novel dehumidifier described for application in HDH systems will reduce the capital cost of the system leading to a reduced cost of water production. The volume is reduced to $1/18^{th}$ of the regular dehumidifier.

9.4 Cost of Water Production

The cost of water production is calculated by a standard method used in the desalination industry [71, 72].

Figure 9.28 illustrates a three-dimensional model of a trailer-mounted, single air extraction HDH system operating under sub-atmospheric pressure with a 12 foot (3.6 m) tall packed bed humidifier and a four-stage bubble column dehumidifier. This unit is designed to produce 10 m^3 per day. We have calculated the cost of this unit as an example of the cost of the state-of-the-art in HDH systems.

The total thermal energy consumed by this system is 156 kWh_{th} per cubic meter of water produced and the electrical energy consumption is 1.2 kWh per cubic meter of water produced. The thermal energy is provided from compressed natural gas tanks on the trailer at an assumed cost of $4 per 1000 cubic feet (this is the current average price in India [73]). The electrical energy is provided using a diesel generator at the rate of $0.20 per kWh. The total energy cost per m^3 of water produced is $2.17.

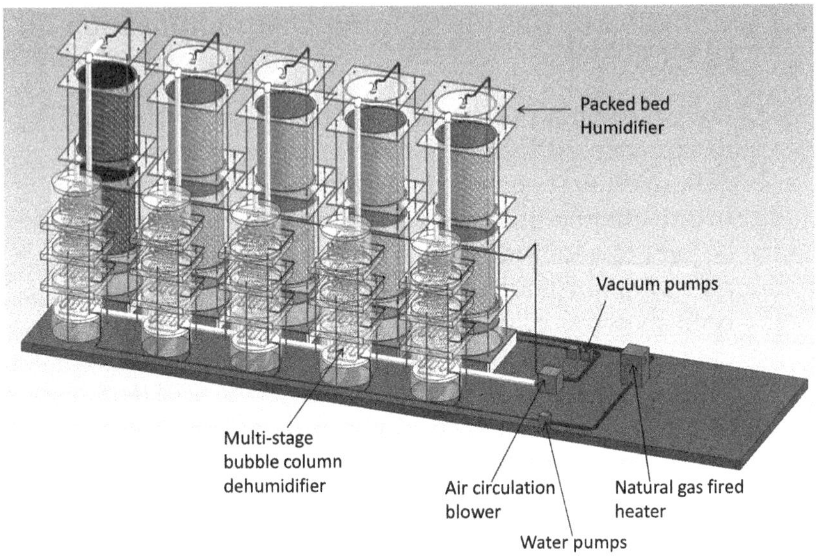

Figure 9.28 Three-dimensional model of a trailer-mounted, sub-atmospheric-pressure, natural-gas-fired, single air extraction HDH system with a 12 foot tall packed bed humidifier and a four-stage bubble column dehumidifier [73].

The capital cost is the sum of the various costs listed in Table 9.2. These costs are obtained from different manufacturers. The parts for the humidifier and dehumidifier are to be obtained from these manufacturers and the assembly and fabrication is to be done by sub-contractors.

The total capital cost is amortized over a life of 20 years. The amortization factor (CAF) is calculated as follows.

$$\text{CAF} = \frac{I}{1-(1+I)^{-20}} \tag{9.14}$$

where, I is the interest rate (taken to be 6% in this calculation). Annual amortization is the product of CAF and CAPEX. This comes to $4,266.38. In the desalination industry, fixed O & M is often considered to be a maximum of 5% of the CAPEX [74]. The total annual levelized cost is the sum of the annual amortization and annual fixed cost divided by the total amount of water produced in a year. This comes to $2.30 per cubic meter of water produced. Further, we also assume that the plant is available for only 90% of the time, which is reasonable for thermal desalination systems [74].

Table 9.2 Various components of capital expenditure (CAPEX) for a 10 m^3 per day HDH system.

Component	Cost [$]
Vacuum pumps	1,250.00
Column containers	2,500.00
Pumps	1,600.00
Blowers	2,000.00
Dehumidifier	8,000.00
Humidifier	8,000.00
NG combustor	1,500.00
Generator	1,000.00
Sub contractor costs	15,000.00
Assembly	2,585.00
Controls	5,500.00
Total	48,935.00

Hence, the total cost of water is $4.91 per cubic meter of water produced[4], which is significantly lower than the costs reported for previous HDH systems. Thus, thermal balancing and bubble column dehumidifiers provide substantial improvements to the HDH system which may make them afforable for small scale applications. These applications may include drinking water production in remote settings or remediation of water produced during oil and gas extraction.

Acknowledgments

The authors would like to thank the King Fahd University of Petroleum and Minerals for funding the research reported in this paper through the Center for Clean Water and Clean Energy at MIT and KFUPM (project # R4-CW-08).

[4] This is based on conservative upper limit estimates for CAPEX and is bound to be lower when the advantages of economies of scale are realized.

Nomenclature

Acronyms

CAF	Captial Amortization Factor
GOR	Gained Output Ratio
HCR	Heat Capacity rate Ratio
HDH	Humidification Dehumidification
HE	Heat Exchanger
HME	Heat and Mass Exchanger
RR	Recovery Ratio

Symbols

c_p	specific heat capacity at constant pressure (J/kg.K)
$\dot{H}$	total enthalpy rate (W)
h	specific enthalpy (J/kg)
h^*	specific enthalpy (J/kg dry air)
h_{fg}	specific enthalpy of vaporization (J/kg)
m_r	water-to-air mass flow rate ratio (-)
$\dot{m}$	mass flow rate (kg/s)
N	number of extraction (-)
P	absolute pressure (Pa)
$\dot{Q}$	heat transfer rate (W)
s	specific entropy (J/kg.K)
sal	feed water salinity (g/kg)
$\dot{S}_{gen}$	entropy generation rate (W/K)
T	temperature (°C)

Greek

Δ	difference or change
ε	energy based effectiveness (-)
Ψ	enthalpy pinch (kJ/kg dry air)
Ψ_D	terminal enthalpy pinch (kJ/kg dry air)
η_{tvc}	reversible entrainment efficiency for a TVC (-)

η_e	isentropic efficiency for an expander (-)
φ	relative humidity (-)
ω	absolute humidity (kg water vapor per kg dry air)

Subscripts

a	humid air
c	cold stream
deh	dehumidifier
d	dry air
h	hot stream
hum	humidifier
HE	heat exchanger
in	entering
int	water-vapor interface
max	maximum
local	defined locally
out	leaving
pw	pure water
rev	reversible
w	seawater

Thermodynamic states

a	Seawater entering the dehumidifier
b	Preheated seawater leaving the dehumidifier
c	Seawater entering the humidifier from the brine heater
d	Brine reject leaving the humidifier
e	Moist air entering the dehumidifier
ex	Moist air state at which mass extraction and injection is carried out in single extraction cases
f	Relatively dry air entering the humidifier
g	Air at an arbitary intermediate location in the dehumidifier
i	Seawater at an arbitary intermediate location in the dehumidifier

References

1 Water for people, water for life, executive summary, *Tech. rep.*, The United Nations World Water Development Report (2003).
2 The millennium development goals report, *Tech. rep.*, United Nations, New York, (2008).
3 A. Hammond, The next 4 billion: Market size and business strategy at the base of the pyramid, *Tech. rep.*, Published by World Resource Institute (2007).
4 C. K. Prahalad and A. Hammond, Serving the world's poor profitably, *Har. Bus. Rev.* 4–11 (2002).
5 J. Fawell, K. Bailey, J. Chilton, E. Dahi, L. Fewtrell, and Y. Magara, Fluoride in drinking-water, in *WHO Drinking-water Quality Series, IWA Publishing, London* (2006).
6 B. Bhuyan, A study on arsenic and iron contamination of groundwater in three development blocks of lakhimpur district, assam, india, *Rep. Opin.* **2 (6)**, 82–87 (2010).
7 Appello, Arsenic in ground water a global problem, in *Proceedings of Seminar Utrecht, Netherlands National Committee of IAH*, J. P. Heedreik (2006).
8 A. Annapoorania, A. Murugesan, Ramu, and N. G. Renganathan, Groundwater quality assessment in coastal regions of chennai city, tamil nadu, india – case study, in *Proceedings of India Water Week 2012 - Water, Energy and Food Security*, 10–14, New Delhi, April (2012).
9 H. Strathmann, Ion-Exchange Membrane Separation Processes. *Elsevier*, New York (2004).
10 F. F. Alshare, Investment opportunities in the desalination sector of the kingdom of saudi arabia resulting from privatization and restructuring, in Proceedings of Saudi Water and Power Forum, Jeddah (2008).
11 A. Twort, D. Rathnayaka, and M. Brandt, *Water Supply.* IWA Publishing (2000).
12 G. N. Tiwari, H. N. Singh, and R. Tripathi, Present status of solar distillation, *Solar Energy*, **75(5)**, 367–373 (2003).
13 H. E. S. Fath, Solar distillation: a promising alternative for water provision with free energy, simple technology and a clean environment, *Desalination* **116**, 45–56 (1998).
14 T. A. Lawand, Systems for solar distillation, *Tech. rep.*, Brace Research Institute., Report No. R 115, Sep. (1975).
15 G. P. Narayan, M. H. Sharqawy, E. K. Summers, J. H. Lienhard V, S. M. Zubair, and M. A. Antar, The potential of solar-driven humidification-dehumidification desalination for small-scale decentralized water production, *Renewable and Sustainable Energy Rev.* **14**, 1187–1201 (2010).
16 Y. Li, J. F. Klausner, and R. Mei, Performance characteristics of the diffusion driven desalination process, *Desalination* **196** 188–196 (2006).

17 G. P. Narayan, *Thermal Design of Humidification Dehumidification Systems for Affordable, Small-scale Desalination*. PhD thesis, Massachusetts Institute of Technology (2012).
18 G. Al-Enezi, H. Ettouney, and N. Fawzi, Low temperature humidification dehumidification desalination process, *Energy Conversion Manag.* **47**, 470–484 (2006).
19 B. M. Hamieh, J. R. Beckman, and M. D. Ybarra, Brackish and seawater desalination using a 20 sq. ft. dewvaporation tower, *Desalination* **140**, 217–226 (2001).
20 H. B. Bacha, T. Damak, and M. Bouzguenda, Experimental validation of the distillation module of a desalination station using the smcec principle, *Renewable Energy* **28**, 2335–2354 (2003).
21 H. Müller-Holst, *Solar thermal desalination using the Multiple Effect Humidification (MEH) method in Solar Desalination for the 21st Century*, ch. 15, pp. 215–225. NATO Security through Science Series, Springer, Dordrecht (2007).
22 G. P. Narayan, M. H. Sharqawy, J. H. Lienhard V, and S. M. Zubair, Thermodynamic analysis of humidification dehumidification desalination cycles, *Desalination and Water Treatment* **16**, 339–353 (2010).
23 E. Chafik, A new type of seawater desalination plants using solar energy, *Desalination* **156**, 333–348 (2003).
24 I. Houcine, M. B. Amara, A. Guizani, and M. Maâlej, Pilot plant testing of a new solar desalination process by a multiple-effect-humidification technique, *Desalination* **196**, 105–124 (2006).
25 G. P. Narayan, J. H. Lienhard V, and S. M. Zubair, Entropy generation minimization of combined heat and mass transfer devices, *Int. J. Thermal Sci* **49**, 2057–2066 (2010).
26 G. P. Narayan, K. H. Mistry, M. H. Sharqawy, S. M. Zubair, and J. H. Lienhard V, Energy effectiveness of simultaneous heat and mass exchange devices, *Frontiers in Heat and Mass Transfer* **1**, 1–13 (2010).
27 G. P. Thiel and J. H. Lienhard V, Entropy generation in condensation in the presence of high concentrations of noncondensable gases, *Int. J. Heat Mass Transfer* **55**, 5133–5147, May (2012).
28 G. P. Narayan, K. M. Chehayeb, R. K. McGovern, G. P. Thiel, S.M. Zubair, and J. H. Lienhard V, Thermodynamic balancing of the humidification dehumidification desalination system by mass extraction and injection. *Int. J. Heat Mass Transfer* **57(2)**, 756 770 (2013).
29 G. P. Narayan, M. S. St. John, S. M. Zubair, and J. H. Lienhard V, "Thermal design of the humidification dehumidification desalination system: an experimental investigation," *Int. J. Heat Mass Transfer* **58**, 740–748 (2013).
30 J. A. Miller and J. H. Lienhard V, "Impact of Extraction on a Humidification-Dehumidification Desalination System," *Desalination* **313**, 87–96 (2013).

31 G. P. Thiel, J. A. Miller, S. M. Zubair, and J. H. Lienhard V, "Effect of mass extractions and injections on the performance of a fixed-size humidification-dehumidification desalination system," *Desalination* **314**, 50–58 (2013).
32. A. Bejan, *Entropy generation minimization: the method of thermodynamic optimization of finite size systems and finite time processes.* Boca Raton, FL: CRC Press (1996).
33 K. H. Mistry, J. H. Lienhard V, and S. M. Zubair, "Effect of Entropy Generation on the Performance of Humidification-Dehumidification Desalination Cycles," *Int. J. Thermal Sci.* **49(9)**, 1837–1847 (2010).
34 S. A. Klein, Engineering Equation Solver, Academic Professional, Version 8, 08 (2009).
35 S. Al-Hallaj, M. M. Farid, and A. R. Tamimi, Solar desalination with humidification-dehumidification cycle: performance of the unit, *Desalination* **120**, 273–280 (1998).
36 S. Soufari, M. Zamen, and M. Amidpour, "Performance optimization of the humidification-dehumidification desalination process using mathematical programming," *Desalination* **237**, 305–317 (2009).
37 M. M. Farid and A. W. Al-Hajaj, "Solar desalination with humidification-dehumidification cycle," *Desalination*, **106**, 427–429 (1996).
38 N. K. Nawayseh, M. M. Farid, S. Al-Hajaj, and A. R.Tamimi, Solar desalination based on humidification process-i evaluating the heat and mass transfer coefficients, *Energy Conversion Manage.* **40**, 1423–1449 (1999).
39 M. Ben Amara, I. Houcine, A. Guizani, and M. Maalej, Experimental study of a multiple-effect humidification solar desalination technique, *Desalination* **170**, 209–221 (2004).
40 C. Yamali and I. Solmus, A solar desalination system using humidification–dehumidification process: experimental study and comparison with the theoretical results, *Desalination* **220**, 538–551 (2008).
41 E. Chafik, Design of plants for solar desalination using the multi-stage heating/humidifying technique, *Desalination* **168**, 55–71 (2004).
42 K. H. Mistry, A. Mitsos, and J. H. Lienhard V, Optimal operating conditions and configurations for humidification-dehumidification desalination cycles, *Int. J. Thermal Sci.* **50**, 779–789 (2011).
43 K. H. Mistry, R. K. McGovern, G. P. Thiel, E. K. Summers, S. M. Zubair, and J.H. Lienhard V, "Entropy generation analysis of desalination technologies," *Entropy* **13(10)**, 1829–1864 (2011).
44 R. K. McGovern, G. P. Thiel, G. P. Narayan, S. M. Zubair, and J. H. Lienhard V. Performance Limits of Single and Dual Stage Humidification Dehumidification Desalination Systems, *Applied Energy* **102**, 1081–1090, Feb. 2013
45 K. M. Chehayeb, G. P. Narayan, S. M. Zubair, and J. H. Lienhard V, "Use of multiple extractions and injections to thermodynamically balance the humidification dehumidification desalination system," *Int. J. Heat Mass Transfer*, accepted, (2013).

46 A. P. Colburn and O. A. Hougen, Design of cooler condensers for mixtures of vapors with noncondensing gases, *Ind. Eng. Chem.* **26**, 1178–1182 (1934).
47 W. Nusselt, Die oberflachenkondensation des wasserdampfes, *Zeitschrift des VereinsDeutscherIngenieure* **60**, 541–546 (1916).
48 E. M. Sparrow and E. R. G. Eckert, Effects of superheated vapor and noncondensable gases on laminar film condensation, *AIChE J.* **7**, 473–477 (1961).
49 E. M. Sparrow, W. J. Minkowycz, and M. Saddy, Forced convection condensation in the presence of noncondensables and interfacial resistance, *Int. J. Heat Mass Transfer* **10**, 1829–1845 (1967).
50 W. J. Minkowycz and E. M. Sparrow, Condensation heat transfer in the presence of noncondensables: interfacial resistance, superheating, variable properties, and diffusion, *Int. J. Heat Mass Transfer* **9**, 1125–1144 (1966).
51 V. E. Denny, A. F. Mills, and V. J. Jusionis, Laminar film condensation from a steam-air mixture undergoing forced flow down a vertical surface, *J. Heat Transfer* **93**, 297–304 (1971).
52 V. E. Denny and V. J. Jusionis, Effects of noncondensable gas and forced flow on laminar film condensation, *Int. J. Heat Mass Transfer* **15**, 315–326 (1972).
53 C. Y. Wang and C. J. Tu, Effects of non-condensable gas on laminar film condensation in a vertical tube, *Int. J. Heat Mass Transfer* **31**, 2339–2345 (1988).
54 T. Kageyama, P. F. Peterson, and V. E. Schrock, Diffusion layer modeling for condensation in vertical tubes with noncondensable gases, *Nucl. Eng. Des.* **141**, 289–302 (1993).
55 H. A. Hasanein, *Steam condensation in the presence of noncondensable gases under forced convection conditions.* PhD thesis, Massachusetts Institute of Technology (1994).
56 S. Z. Kuhn, *Investigation of heat transfer from condensing steam-gas mixture and turbulent films flowing downward inside a vertical tubes.* PhD thesis, University of California at Berkeley (1995).
57 N. K. Maheshwari, P. K. Vijayan, and D. Saha, Effects of non-condensable gases on condensation heat transfer, in *Proceedings of 4th RCM on the IAEA CRP on Natural Circulation Phenomena* (2007).
58 H. A. Hasanein, M. S. Kazimi, and M. W. Golay, Forced convection in-tube steam condensation in the presence of noncondensable gases, *Int. J. Heat Mass Transfer* **39**, 2625–2639 (1996).
59 M. Siddique, M. W. Golay, and M. S. Kazimi, Local heat transfer coefficients for forced convection condensation of steam in a vertical tube in the presence of air, *Two-Phase Flow and Heat Transfer (3rd ed.)*, ASME, New York, **197**, 386–402 (1992).
60 V. D. Rao, V. M. Krishna, K. V. Sharma, and P. M. Rao, Convective condensation of vapor in the presence of a non-condensable gas of

high concentration in laminar flow in a vertical pipe, *Int. J. Heat Mass Transfer* **51**, 6090–6101 (2008).

61 B. M. Hamieh and J. R. Beckman, Seawater desalination using dewvaporation technique: experimental and enhancement work with economic analysis, *Desalination* **195**, 14–25 (2006).

62 M. A. Sievers and J. H. Lienhard V, "Design of flat-plate dehumidifiers for humidification dehumidification desalination systems," *Heat Transfer Engrg.* **34(7)**, 543–561 (2013).

63 M. A. Sievers and J. H. Lienhard V, "Design of Plate-Fin Tube Dehumidifiers for Humidification-Dehumidification Desalination Systems," *Heat Transfer Engnrg. accepted* (2014).

64 A. V. Kulkarni and J. B. Joshi, Design and selection of sparger for bubble column reactor. part i: Performance of different spargers, *Chem. Eng. Res. Des.* **89**, 1972–1985 (2011).

65 J. B. Joshi and M. M. Sharma, A circulation cell model for bubble columns, *Chem. Eng. Res. Des.* **57a**, 244–251 (1979).

66 G. P. Narayan, M. H. Sharqawy, S. Lam, and J. H. Lienhard V, Bubble columns for condensation at high concentrations of non-condensable gas: heat transfer model and experiments, *AIChE J.* **59(5)**: 1780–1790, May 2013.

67 E. W. Tow and J. H. Lienhard V, "Analytical Modeling of a Bubble Column Dehumidifier," Proc. ASME 2013 Summer Heat Transfer Conf. HT2013-17763, Minneapolis MN (2013).

68 E. W. Tow and J. H. Lienhard V, "Heat flux and effectiveness in bubble column dehumidifiers for HDH desalination," Proc. IDA World Congress on Desalination and Water Reuse, Tianjin, China (2013).

69 E. W. Tow and J. H. Lienhard V, "Experiments and Modeling of Bubble Column Dehumidifier Performance," *Int. J. Thermal Sci.*, to appear (2014).

70 G. P. Narayan, G. P. Thiel, R. K. McGovern, M. H. Sharaqawy, and J. H. Lienhard V. Vapor mixture condenser. US Patent 8523985 B2.

71 C. Sommariva, *Economics and Financing*. Balaban Desalination Publications (2010).

72 A. K. Plappally and J. H. Lienhard V, Cost of water supply, treatment, end-use, and reclamation, in *Proceedings of European Desalination Society Conference, Barcelona, April 22–26* (2012).

73 J. H. Huang, Design of a mobile community level water treatment system based on humidication dehumidication desalination. SB Thesis, Massachusetts Institute of Technology, June (2012).

74 US Energy Information Administration. URL: http://www.eia.gov/countries/cab.cfm?fips=IN. 2012.

10

Freezing-Melting Desalination Process

Mohammad Shafiur Rahman and Mohamed Al-Khusaibi

Sultan Qaboos University, Oman

Abstract

This chapter introduces the basic principle and advantages of freezing-melting process followed by its applications in desalination. It introduces different types of freezing-melting process by describing its components and operations as well as their advantages and disadvantages. Finally the future prospects, challenges and other potential hybrid freezing-melting processes for desalination are discussed.

Keywords: Freeze-concentration, sea water desalination, nucleation, crystallization, melting unit, refrigerants, wash column, vacuum freezing, solar desalination

10.1 Introduction

Desalination refers to a water treatment process that separates water from salt solution and its use has grown steadily since the 1960s. Freezing-melting (FM) process is essentially capable of removing water by freezing it out from saline solution as ice crystals. A wide variety of FM systems are currently discussed in the literature. The main advantages of the FM process are the requirement of low energy and low temperature operation compared to the thermal desalination. The reduction in energy costs results as the latent heat of fusion of ice is only one-seventh the latent heat of vaporization. Other advantages are less scaling or fouling and corrosion problems, ability to use inexpensive plastics or low cost

Jane Kucera (ed.) Desalination, (473–502) 2014 © Scrivener Publishing LLC

material, and absence of pretreatment. It is clear from the literature review that the above two points (low energy and low temperature) are the main drive to look at the FM process. However the technology is still lagging behind from the commercial success point, especially in the area of desalination.

The choice of a technology is usually based on the product quality, operating economy, energy cost, initial investment, and complexity of the process. The main factors affecting the use of FM process is the capital cost and complexity of the process. This is clearly evident from the wide varieties of alternatives available in the state-of-the art. The technology was successful only when the above mentioned two factors were compensated by other advantages. First of all we need to identify which industry is now using FM process successfully and to explore the possible reasons of success. Some sectors of food and chemical industries have used the technology successfully. In food industry it has been utilized commercially for the concentration of citrus fruit juices, for vinegar, coffee, and tea extracts, sugar syrup, maple syrups, milk, and whey concentration, and for concentration of beer, wines and aroma extracts. In this case only indirect FM process is used to avoid refrigerant contamination. The success was mainly due to its ability of producing high quality products as compared to the available thermal technology in the market. In addition food is high value product as compared to water. In the case of chemical industry it is mainly adopted when there are no other alternatives. In this situation the principal attraction of applying FM process is its capability for concentrating heat-sensitive mixtures without damaging them, and separating hazardous and flammable chemicals, and azeotropic fluid. It would be difficult to utilize the above advantages to progress the FM for desalination. In addition, misconceptions or negative attitude also affected the progress of FM process. In the case of desalination, various types of technologies are available. The practical difficulties of FM process limiting its commercial success are complexity of the process, and high capital costs. The failure is due to many folds, such as technical, political, and financial. The pilot studies in many countries indicated that hybrid techniques of combining FM process and other desalination methods have high potential for future development. In this case solar assisted systems may have high potential.

10.2 Background or History of Freezing Melting Process

The Danish physician Thomas Bartholinus (1616–1680) was apparently the first to report that water obtained by the melting of ice formed in seawater was fresh. Almost at the same time, Robert Boyle (1627–1691) reported the same observation, foreseeing the phenomenon as a source of fresh water, and Jesuit Athanasios Kircher (1602–1680) discussed the reason why ice formed in the sea is fresh [1]. At the end of the eighteenth century, the Italian scientist Anton Maria Lorgna (1735–1796) described a method to purify seawater and impure water by freezing and then melting of ice. In 1786 Lorgna published his first paper on water desalination by freezing and he was wondering why nobody had previously applied it in an artificial process initiating what nature does so well and easily in the cold seas (i.e. blocks of fresh water ice from seawater). He also identified that a single freezing of seawater produced an ice block having salinity, although much less than that of seawater. This justified the needs of multistage freezing-melting (FM) process. Freezing in large bodies of water occurs in nature on the surface of oceans, lakes, and bays [1]. This natural system has been put to advantage in Siberia in order to freeze salt water and store the ice, which was then melted in summer, thus providing a complete fresh water source for village communities [2]. Recently this natural process was used in north Chile. The climate of dry and transparent atmosphere causes strong temperature drops due to evaporation and radiation heat losses. Water evaporation rates of 5 mm per night as well as heat losses of 150 W/m^2 due to radiation make it possible to freeze salt water. Fournier et al. [3] placed salt water in pans which was oriented towards the open sky and observed ice formation approximately 9 liters of ice per m^2 of pan surface per day. This process is simple and low cost. They used the produced melted fresh water in greenhouse crop cultivation in the dessert area. This FM was more practical form after the development of refrigerating machines. The interest in the process for obtaining fresh water from seawater by freezing was revived in the late 1930s, and an experimental desalting plant (i.e. indirect freezing process) had been operated for some years near Rome by the Istituto Superore di Sanita. FM process was first used commercially in the 1950s. Research in the 1960s and 1970s for desalination, petroleum, chemical and food processing applications provided many technical innovations [4].

10.3 Principles of Freezing-Melting Process

First we could analyze the natural process of water freezing to ice in seas and rejecting salts at the interface. The combined heat transfer associated with heat removal by the environment, and latent heat release at the water-ice interface results in natural convection flows of water. Freezing seawater releases fluid at the water-ice interface, which is denser that the ambient water. The resulting solute buoyancy force therefore acts downward in addition to the thermal buoyancy force. Flow visualization revealed that the flow was downward, below the freezing surface; and convection heat transfer was strongly affected by solute rejection upon freezing [5].

In a FM process, first the solution is partially frozen, the ice crystals are physically separated from residual solution (i.e. concentrated solution), and the ice is melted to form the product water. The FM process is accomplished in two major stages: ice crystallization (Stage I), and separation and melting (Stage II) (Figure 10.1). In stage I, nucleation occurs at a suitable supercooling temperature. The nuclei in solution grow to become large ice crystals in a crystallization unit. In stage II, the crystals are separated from the concentrate by a separator (mechanical) and then melted to produce pure water. In some cases, a precooling step is used on the feed, which reduces heat load in the freezer.

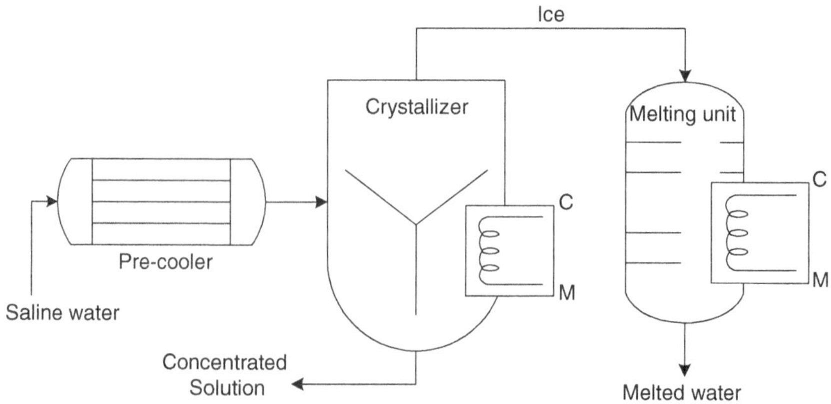

Figure 10.1 Freezing-Melting (FM) process showing pre-cooler, crystallizer, and melting unit.

The advantages of FM process are: (i) low energy requirement as compared to that of distillation processes (i.e. latent heat of fusion of ice is only one-seventh the latent heat of vaporization of water), (ii) it has low operating temperature, which minimizes scaling and corrosion problems, ability to use inexpensive or low cost materials, causes less thermal damage to the components in solution, and ability to have high surface area and heat transfer coefficient by direct contact between brine and refrigerant, (iii) requires no or little pretreatment, and (iv) insensitive to fouling as it affects the membrane and thermal desalination [6].

The main disadvantages of FM process as compared with evaporation and reverse osmosis are higher capital and operating costs during the ice separation. Other disadvantages are: (i) retention of undesirable flavors and aromas (initially present in the feed saline water) that may come into the produced fresh water [7], (ii) washing steps, which uses a certain amount of fresh water, are necessary for ice crystals, (iii) compressor represent an expensive method of furnishing the energy requirements of the system although energy requirements for freezing is much lower than boiling [8], (iv) complexity of the unit operations in the freezing unit (especially ice crystallization and growth), wash-separation column, and melting unit, and limited knowledge availability in designing the process and determining its efficiency, (v) the trapping of salts solution in the ice during crystallization, and (vi) high quality energy is required for crystallization as compared to low quality energy used in many evaporation processes.

10.4 Major Types of Freezing-Melting Process

The major types of FM process are thoroughly reviewed by [6, 9]. The classification is mainly based on the methods of freezing process. These are: (i) direct contact FM, (ii) indirect contact FM, (iii) vacuum FM, (iv) eutectic separation FM, and (v) block FM.

10.5 Direct Contact Freezing

10.5.1 Ice Nucleation

Direct-contact freezing uses crystallization by intimate mixing between the refrigerant and the product to be frozen. A typical

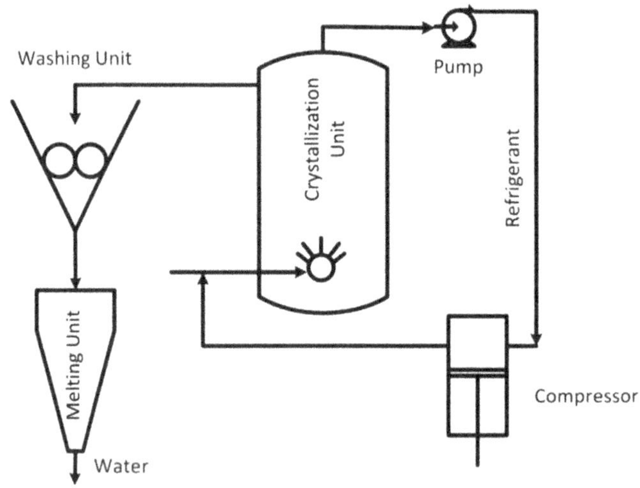

Figure 10.2 Direct contact FM showing crystallization, washing and melting units.

direct FM system is composed of ice nucleation and crystallization, ice crystal separator, ice washing, and melting units (Figure 10.2) [10]. The main advantages of this method are a high production rate per unit volume at a low driving force, power consumption is small, absence of moving parts, and the unit is compact and efficient [11]. The successful design of a direct-contact freeze desalination plant significantly depends upon the availability of a suitable refrigerant [12, 13].

10.5.1.1 Ice-Crystallization Unit

The refrigerant in the liquid form and maintained under pressure is expanded through a nozzle into the product liquid, where it vaporizes at low pressure. This vaporization provides a refrigeration effect and causes the formation of ice and solutes crystals with in the product. Orcutt and Hale [14] used mathematical models to study the operational-design economics of a freezing process and predict the best operating conditions. Optimization computations showed that the economics of process operation depend largely on the temperature maintained in the freezer and the overall difference in the refrigerant and equilibrium freezing temperature. An analysis of the linearized freezer dynamic equations showed the freezer to be stable and did not indicate regions of difficult control. The cost of the washing unit and melting unit was influenced by the operation of the freezer, which determines the value of the crystal size. The freezer operating costs depend on the brine temperature, which influenced both the crystal size and refrigerant vaporization rate. Barduhn [12] emphasized the following points in designing the crystallization unit: (i) adequate dispersion of the liquid refrigerant into the solution (paramount importance), (ii) use of normal butane since it is the cheapest and probably does not hydrate, (iii) avoid smaller crystals and poor wash ability although with a short residence time, (iv) ability to handle short contact time, and (v) include several methods or modes of agitation.

There are certain thermodynamic, chemical, physical, and economical requirements, which the refrigerant must meet to be suitable for use in the process. These are: (i) the refrigerant should have a normal boiling point of –4°C or less and have a vapor pressure below 2.8×10^5 Pa at room temperature; (ii) refrigerant must be nontoxic, preferably nonflammable and chemically stable in solution, (iii) the fluid should be virtually immiscible with water and possesses such molecular size factors so that it does not form a hydrate under the freezing conditions employed in the process, and (iv) the refrigerant should be cheap and readily available from commercial suppliers [15].

The refrigerants that can be used are butane, carbon dioxide, nitrous oxide, Freon–114 and Freon–318. The details of the thermodynamic properties of normal butane at refrigeration temperatures are compiled by Kurnik and Barduhn [16]. Freon–114 and Freon–318 are better choice based on the above factors, however these materials are relatively expensive when compared to other refrigerants such as butane. The appropriate choice of a suitable refrigerant for the process is important.

The cooling process could be used by linking other endothermic processes. In many cases, gas companies import huge quantities of liquefied natural gas (LNG), which is vaporized from low temperature to the ambient temperature at the terminals, and then transported through gas pipelines. In this case, seawater could be used as a source of heat for vaporization of LNG. The cost would be substantially reduced if seawater could be cooled from the ambient to near freezing [17, 18].

Simpson et al. [19] studied the evaporation process of refrigerant (i.e. butane) by describing the dynamics of the bubble motion in a more meaningful way, and recorded this more readily with the visual evidence of the bubbles' motion. It was found that the rate of evaporation of butane droplets increased rapidly with a diameter ratio compared with the initial one, up to a critical value, and then gradually with the 1/6 power, implying that evaporation was controlled by heat transfer through the transient liquid butane film on the inside surface of the bubble. The quality of the ice crystals depends on the circulation of immiscible refrigerant [20]. Some of the best possible ways of mixing include [12]: (i) using fine spray nozzles to introduce the liquid refrigerant under the brine, (ii) pumping vapor from the vapor space through spurge system that reintroduces it under the brine, (iii) pumping the entire liquid content of the freezer rapidly around a closed path, and (iv) using conventional mechanical agitators. The later method is most difficult to scale up and, furthermore, multiple mixers appear to complicate the design and increase costs in large plats. The combination of (i) and (ii) could be a viable option. The refrigerant (i.e. butane) can introduced at the bottom thru stripping from a draft tube gave good vertical movement to the slurry and 25% ice suspensions could be handled. This small flow of vapor (containing a negligible amount of incondensable gas and little super heat) is still effective in causing circulation even when the pressure in the bottom half of the crystallizer is above the vapor pressure of butane. It was found quite unnecessary to use a mechanical agitation and unit operates satisfactorily without any additional agitation, and a comparatively small flow of butane vapor, purged near the bottom of the crystallizer, greatly improves the mixing and ensures reliable operation under all circumstances [12, 21, 22].

Other option is the use of spray freezer. In this approach, refrigerants and salt water are sprayed into a low pressure space and slurry forms virtually instantaneously and the ice production rates in a spray freezer can be 10–30 times than those in stirred tanks. The particle size averages only 40 μm, and the temperature gradient between refrigerant and slurry is very large, 18°C [23]. In the case of RC–318, the ice crystal size was a strong function of salinity and a marked size maximum occurred at about 0.5% sodium chloride. Similar phenomenon is also seen in single crystal growth rates from many aqueous solutions [24].

Another option is to use exchange crystallization unit. It is a horizontal vessel, operating at atmospheric pressure, consisting of three distinct sections [25]: (i) an ice-brine and hydro-carbon disengaging section, (ii) an agitated or contacting section, and (ii) a brine and hydrocarbon disengaging section. The brine entering the agitated section is broken up into small droplets by turbine agitators and counter currently contacted with a partially solidified stream of normal straight chain hydrocarbons. The melted hydrocarbon, which contains entrained brine, flows into the disengaging section where separation is affected by gravity and electrostatic coalescence. The electrical coalesce is a horizontal unit consisting of three vertically stacked grids.

Applying ultrasound to crystallizing systems offers a significant potential for modifying and improving the process. The most important mechanism by which ultrasonic can influence crystallization is ultrasonic cavitation, which is particularly effective for inducing nucleation. Using ultrasound to generate nuclei in a relatively reproducible way offers a well-defined starting point for the crystallization process, and allows the focus to be on controlling the crystal growth for the remainder of the residence time in the crystallizing-unit. This approach can successfully manipulate crystal size-distribution and hence to modify solid/liquid separation behavior, washing, and fresh water purity [26]. Otero *et al.* [27] identified that pressure-shift nucleation could be a potential tool for FM process. It presents four important advantages over the conventional crystallization. First, temperature in the high-pressure vessel can be set considerably higher temperature in order to save energy. Second, ice crystals are instantaneously produced in the whole volume of the solution instead only at the surface close to the heat exchanger. Therefore, no complex scraping devices are needed. Third, the final concentration is achieved in the whole sample just after expansion and there is no need to recirculate the product to progressively increase the amount of ice formed to reach the desired concentration. Fourth, after release of pressure, the ice crystals formed are round without pockets and indentations, thus low entrapment of solutes.

The ice nucleation unit produces small ice crystals, which are transferred into the crystallization unit and grown by ripening (i.e. larger in size) at the expense of smaller ones. In the crystallization unit, the formation and growth of ice crystals should be controlled in such a way that a uniform distribution of large ice crystals suitable for separation is formed. The optimum size distribution for most separators is a mono-disperse distribution (narrow range of sizes) with a large mean size. This facilitates the washing step and reduces the amount of product carryover into the separated ice stream. The key technology in this system is how to grow ice crystals in the crystallization unit large enough to facilitate the separation of the ice crystals and solutes [28]. Super-cooling and secondary nucleation were identified to be major factors preventing ice crystals from growing large [29, 30, 31]. Very rapid super-cooling can create a large number of smaller new crystals, therefore lower rates of super-cooling are desirable to prevent excessive nucleation.

Lower nucleation rates are required to produce reasonably large ice crystals at an acceptable residence time [32]. Thijssen [33, 35] suggested increasing the agitation rates, within certain limits, to lower the nucleation rates, since high mixing rates may promote smaller size due to mechanical damage. Garabedian and Strickland-Constable [36] found that fluid shear does not produce crystal breeding, and collision of a single crystal in pure water produces high rates of nucleation. Polycrystals may be formed by agglomeration or growing together of fine crystals, but little is understood about agglomeration. The nucleation rate at low stirring rates is determined primarily by cooling rate, while with intensive stirring it depends primarily on the hydraulic factor [37]. At least two factors, namely diffusion and surface adsorption, could be controlling the rate in a continuous crystallizing unit [38]. Depending on the liquid depth and temperature conditions in the freezer, several liquid refrigerant zones may exist [39]. Refrigerant at depths sufficient to suppress vaporization is said to be in the *inactive zone*. Vaporizing refrigerant is in the *active zone*, and if liquid refrigerant accumulates on the surface of the slurry, it is said to form an *excess zone*. Nucleation occurs mainly in the active zone, while crystal growth proceeds throughout the entire brine. The depth of the active zone can be calculated from the relationship between the refrigerant vapor pressure and temperature. It is good to prevent the formation of an excess zone, which generally interferes with freezer control [14]. Polymers could suppress the secondary nucleation of ice crystals, thus larger crystal size could be achieved. The suppression depends on the types of polymer and concentration, and also is related to the increase in viscosity [31].

10.5.1.2 Hydrate Formation

From the cost and stability, hydrocarbons with four carbon atoms have been recommended. When n-butane is used, the operation is to be carried out carefully because it is at a vapor pressure lower than the atmospheric pressure, while i-butane has a considerably higher vapor pressure than atmospheric pressure and can form solid hydrate in contact with aqueous phase under specific conditions. Formation of hydrates results in the elimination of ice crystals formed. The mixture with less than 73.8% i-butane cannot form a hydrate at –0.7°C with 1.6% sodium chloride aqueous solution, and thus limiting ratio increases with increasing temperature [40].

The refrigerant-brine interactions affect the process. The brine carried with refrigerant adversely affects compressor performance, thus separation devices need to be used. The dissolved refrigerant is rather easily removed by vacuum stripping of the effluent stream. Excessive loss of refrigerant should be avoided. These losses could occur by entrainment, solubilize in brine, and hydrolysis. Hydrolysis is an irreversible process, and undesirable levels of soluble compound could build up in the product water, for example soluble fluoride in the case of Freon. Solubility and hydrolysis decreased with the increase of salinity [41–43].

10.5.2 Ice Separation Unit

The ice crystals formed in the crystallization unit are collected and transferred to the ice crystal separator and washed with water to remove the brine or solution from the ice crystal surface. From a separation point of view, the formation of a few large ice crystals is desirable. Separation devices can be classified as: presses, gravity drainage, centrifuges, filter, and wash columns [44, 33, 34, 35, 40]. Filtration showed less effective for crystal separation, and difficult to use for washing the crystals. In many cases, screens or filter showed a history of freezing up (brine freezing in the openings of the weave). Columns are commonly used for separation and the important factors controlling the separation efficiency of the columns were the axial diffusion of impurity and the mass transfer between the adhering and free liquids around the crystal phase, types of the screw conveyer used [45].

10.5.3 Wash Columns

Effective washing of ice is one of the most difficult unit operations. The purpose of wash columns is to separate concentrated liquid and other crystalline components and impurities from the crystal surface and to produce pure crystals. Wash column could be flooded column and the drained column [32]. There are two types of wash columns: pressurized and gravity. In the pressurized wash column, the crystals raise to the top and hydraulic pressure forces a wash liquid, derived from the melted pure crystals, to flow down. As the wash liquid flows down the column, it removes impurities from the surface of the crystals. At the interface between washed and unwashed crystals, called the wash front, the wash liquid comes in contact with colder crystals and crystallizers on them. In this way, the wash liquid does not mix with the concentrated liquid.

The gravity wash column is simple in design but larger than the pressurized wash column. Its greater height creates the pressure needed to compact the ice bed. It works in much the same way as the pressurized column but at lower pressures. An ice pack is still formed and moved hydraulically up the column. To overcome the difficulties associated with surface tension forces, clean fresh water can serve as a displacing liquid. The separation by drainage can greatly improve when in addition to draining the brine from the bottom of ice crystals. In this option one could add pure water at the top of the batch and lets it filter through the interstices of the ice bed to displace the brine.

In another option, the slurry of brine and ice crystals is introduced into the bottom of a vertical column from which the brine drains through screens at the bottom. The ice crystals move upward by virtue of their own buoyancy force forming a porous ice plug at the top of the column where wash water is added from the top. As the ice plug moves upward through the layer of wash water the brine is displaced from the interstices of the porous ice plug and the salt free ice crystals are harvested at the top of the column and transferred to the melting-unit. The rise of the ice crystals in the column is, however, rather slow, being a limiting velocity for a particle moving through a fluid by buoyancy (or gravity) forces alone. This limiting velocity determines an upper limit to the production rate for such a gravity wash column, a value which is much too low for economic consideration [46]. In order to enhance the process moving brine could be used as driving force, and the discharge screens are located in the vertical walls of the column, about midway between the top and bottom [47].

The performance of wash columns depends on the crystal size and shape, and also on the viscosity of the mother liquid. Uniformity of crystal size and shape is important 1.to avoid the wash water follows least resistance with evenly distributed through the crystals.

The design criteria for economic processing condition are also proposed in the literature [14, 48].

Centrifuge has also been used to separate ice crystals from the concentrated liquid. Filtering centrifuge utilizes the difference in specific gravity between the ice crystal and liquid concentrate to separate ice from liquid, with the liquid being forced through the filter basket by centrifugal force. Generally water rinsing is required to assure that there is no salt carryover.

Ice particles from brine can also be separated by a filter. Vacuum filters have been used for separating ice slurry made by direct freezing and vacuum systems [49, 50].

10.5.4 Melting Unit

Two types of melting unit are used: direct-contact melting and indirect-contact melting. Energy recovery is one of the important aspects of this process. The heat pump options could be used in the system, when input saline water could be pre-cooled by melting of ice.

10.6 Direct Contact Eutectic Freezing

Barduhn [51] devised the process of eutectic freezing and Pangborn [51] tested the idea. In eutectic FM process, salts separate as solids and fresh water as ice from brine. In the progress of freezing, ice formation concentrates and at its eutectic point salts precipitate simultaneously. The ice and salt crystals nucleate and grow independently and are easily separated since the ice floats and the salt sinks. The eutectic FM may be one of the meaningful to brine disposal [52]. In addition, different byproducts could also be produced from salts. The essential feature is that at –20°C both ice and hydrated sodium chloride crystals precipitate (Barduhn and Manudhane, 1979). In normal FM process, the freezer operates at about –5°C, whereas eutectic FM process freezers operates –20 to –25°C. Thus, eutectic FM needs more energy. Operation below –25°C other ions may interfere the process. The system could include a stirred-tank crystallizing unit, hydro-cyclone separator and a floating wash column [52]. The salts separated from brine could be filtered and then dried [53]. Two stage eutectic FM is more economical [12, 53, 52]. A combination of eutectic FM and distillation or membrane could be feasible since product stream (i.e. ice) contains less salt (around 3–6 times less) [54].

10.7 Indirect-Contact FM Process

In indirect-contact FM process, the energy removal must be passed through the walls of some form of heat exchanger [55, 34]. The dendrite type ice growth need to be explained based on heat-flow and molecular growth kinetics [56, 57]. Indirect-contact FM process can be subdivided into: (i) internally cooled and (ii) externally cooled.

10.7.1 Internally Cooled

In this process the brine is cooled internally to the solution and crystals are form on the solid surface. The internally cooled can be classified as: (i) static layer growth and (ii) dynamic layer growth, and (iii) suspension crystallization.

10.7.1.1 Static Layer Growth System

In this system, the liquid from which the crystal mass is grown is stagnant (i.e. no stirring). It is relatively reliable and requires very simple equipment without moving parts (i.e. easier to handle) or no need for a solid-liquid separation device. The residence time in this process is large because the mass transfer is only promoted by free convection. High purification efficiency can only be obtained with relatively low rates of ice growth ($<10^{-7}$ m/s). Fast crystallization produce impure ice crystals. Two mechanisms of ice crystal growth are observed: (i) ice crystals grow larger by the usual growth, governed by heat or mass transfer resistance, and (ii) ice crystals agglomerated and the agglomerated fused into a very large ice crystal (1–3 mm in diameter) due to long residence time [58]. Large equipment volumes are required due to the batch-wise operation and slow crystallization rate [59, 60]. The crystal-solution interface per unit equipment volume can be increased by using plate-type contact surface.

10.7.1.2 Layer Crystallization on Rotating Drum

In this process, ice forms in thin layers on the heat exchange surface and after a suitable period of time for the ice layer to build up, the ice is removed from the surface. A rotating drum immersed in the fluid to be concentrated and refrigerant is circulated within the drum. Ice formed on the suface is then scraped free as the drum rotates past a knife [10].

Progressive Crystallization Unit: Progressive FM process utilizes the concentration phenomenon of a solute at the ice-solution interface moving from one end of a vessel to the other end [61, 62]. It is characterized by producing a single ice crystal and easy to separate from concentrated solution. Distribution coefficient (solute ratio in the ice and solution) is defined to determine the purity (i.e. efficiency). The high efficiency depends on: (i) the formation of smooth solid-liquid interface, (ii) lower moving speed of the freezing front, (iii) higher agitation inside the solution (i.e. ice-solution interface), (iv) application of supersonic radiation, (v) increased dissolved air concentration, (vi) avoidance of super-cooling, (vii) enzymes, (viii) type of geometry, and (ix) low solute concentration [61, 63, 28, 64, 55, 65, 66]. The ice growth rate depends on the ionic salts, surfactants, antifreeze proteins, and water soluble polymers [28, 54]. The growth-rate enhancements (up to a factor of 5) and retardations (up to a factor of 3) could be possible in the presence of additives [67]. The ice growth magnitude depends on the nature of additive, its concentration, and degree of super-cooling.

10.7.1.3 Dynamic Layer Growth (Falling Film Type)

In this type, solution flows down over the wall of the heat exchanger (well-mixed). Crystals are formed on the wall surface under the falling film. Shear caused by solution flow at the crystal-solution interface increases the mass transfer coefficient and promotes the transport of impurities from the interface to bulk. This process is easy to scale-up because of its modular design. Local equilibrium at the interface exists. Solutes entrapment occurred depending of the rate of ice crystal growth due to physical entrapment and adsorption on the crystal surface [68, 69]. Sanchez *et al.* [70] studied the pilot scale falling film FM process and observed that purity increased with the decrease of solution concentration.

10.7.1.4 Dynamic Layer Growth (Circular Tube Type)

In this process, ice is formed from a solution flowing through a tube cooled from outside [71–74]. The thickness of the ice layer first increases rapidly, reaches a maximum, and then decreases. This phenomenon corresponds to the growth and subsequent melting of dendrites. Super-cooling can be achieved in a liquid before it solidifies when forced to flow inside circular tubes. It depends on the local tube wall temperature, tube diameter, Reynolds number, and dimensional constant [75]. Ice adhering and retained in the interstices by capillary forces cases high salt retained in the ice [76]. Rane and Padiya [9] developed tube-tube heat exchanger heat pump FM process for energy efficiency and avoiding the ice water separation step. Mandri et al. [77] proposed a seating steps in melting of ice in order to produce more pure ice. The network formed by the grain boundaries should ease the draining out of the impurities under the influence of the gravity. In the case of the purer ice layer, the formation of cracks or bridges between the pockets is required to allow in depth sweating.

10.7.1.5 Suspension Growth

In suspension growth, the product to be concentrated is agitated in a vessel cooled by heat transfer through the walls of the jacketed vessel and it results suspension of ice crystals, which are then separated. Independent control of ice nucleation and crystal growth is very difficult [10, 34]. This vessel may be either a scraped surface heat exchanger on simply a jacketed kettle vessel. Nucleation occurs mainly in the heat exchanger, usually scraped-surface, while most of the growth occurs in the main vessel.

10.7.2 Externally cooled

Externally cooled crystallizers employ a heat transfer device external to the main crystallization vessel. In this process, a heat exchanger produces supercool liquid and feed into the main vessel. Both nucleation and subsequent crystal growth occur in the main vessel [78]. Conditions of the heat exchanger must be closely controlled to avoid nucleation where it is not wanted and the primary aim to avoid heterogeneous crystallization within the main vessel. The inside wall of the heat exchanger has to be highly polished or coated with a hydrophobic plastic to minimize changes of minimal nucleation. In the vessel ice crystals with 1 mm diameter could be produced by applying the process of ripening with long residence time [79, 80]. A method of producing large ice crystals, which uses the Oswald ripening effect, was developed and is now widely used in the industry. Shirai et al. and Kobayashi [57, 81] proposed another strategy to make large ice crystals by agglomerating the small ice crystals produced.

10.8 Vacuum Process

Vacuum-freezing and vapor-compression system can be used for FM process [82]. In this option, a high vacuum is employed to vaporize a portion of water, which then provides the refrigeration effect by lowering the temperature of the product and causing ice crystallization to occur. The pressure in the vessel is maintained to the triple point (610 Pa). In this process compressor must handle a very large volume of low-density water vapor due to very low vapor pressure of water. Whereas when a relatively volatile refrigerant such as butane is used, the freezer pressure is raised to approximately the atmospheric pressure and the volume of vapor to be compressed is greatly reduced. In addition, compressor technology for butane close to atmospheric pressure is much better developed than for water vapor compressor from 400 Pa. In the case of vacuum-freezing, the complexity and expense of refrigerant recovery, make-up, and explosion protection could be avoided. However, it needs more efficient designing of melting unit for removal of non-condensable gas in the system [8, 12]. The washed ice is melted by direct-contact condensation of water vapor in the melting-condensing unit. Based on the method by which the vapors are removed, these may be further classified as: (i) vacuum-freeze and vapor-compression system, (ii) absorption-freeze and vapor-compression system, (iii) vacuum-freeze and ejector-absorption system. Vacuum freezing with high-pressure ice-melting process was introduced by [83], and improvements were proposed by [84].

10.8.1 Vapor-Compression System

There are two components: (i) vapor removal unit to keep the slurry at or below its triple point, and (ii) a freezing/evaporation unit to keep ice particles suspended with a fluid slurry/vapor interface. In the vacuum freezing vapor compression process, a large multiple compressor is used to compress vapor from the freezer to the melting unit. For a plant having a capacity of 227 m3/day, the compressor was more than 3 m diameter and needed a fairly high moment of inertia for starting. For larger FM plants of 4000 m3/day and above, it is difficult to find a practical compressor [85, 86] identified the following module for increasing the compressor efficiency: multi-compressor modules, direct contact and evaporation feed pre-coolers, and multistage heat removal compressors of flexible blade type. Burton and Lloyd [87] examined the design considerations of the primary and secondary compressors considering safety and environment.

10.8.2 Vapor-Absorption

In an absorption vacuum-freezing process, water evaporates from freezing solution and condenses on a cold salt solution. In this process, water vapor is compressed by a combination of steam ejector and absorber loop with primary energy source being thermal rather than mechanical.

10.8.3 Multiple-Phase Transformation

When a sub-triple point vapor is cooled at a constant pressure, it condenses as solid solvent as ice and this operation is called de-sublimation [88]. In the cases of concentrated solution, this method is competitive to reverse osmosis and distillation or evaporation process.

10.9 Block FM Process

Block FM was proved to be an effective method for syrup concentration [89]. In this process solution is first frozen by an indirect contact cooling by low temperature air. The temperature was lowered until reaching the values of –10 or –19°C. After freezing, two thawing modes could be used to separate the whole frozen block into two phases as concentrated and ice fractions. The first thawing mode was a passive defrosting during which the frozen solution can be left at a temperature of 22°C, and second thawing mode was a microwave-assisted defrosting. In both cases, 50% of the initial frozen solution can be thawed and separated as unfrozen ice block and concentrated solution. Multistage operation could be used to generate more pure ice. Aider and de Halleux [89] hypothesized that during freezing ice crystals migrated toward the cold surface to form bigger ice crystals and then subsequent melting (i.e. thawing) produces water with low concentration of solutes. However other mechanisms may also involve, which need to be explored.

10.10 Applications

FM process was found to be mostly applied when there was no alternative available. Desalination trial using FM process is mainly limited to the direct contact refrigeration system due to its processing efficiency and economics. However it has a number of drawbacks, such as residue of refrigerant in the water, formation of hydrates, and complexity of separating refrigerants from the ice. Technologically all the FM methods used in the food industry could be used for desalination. All the methods could go up to 100 ppm level with multiple effects. The only point needs to be considered is the economic analysis of the FM process for desalination since water is a low value product compared to foods. In many cases food product, such as dairy items, functional extracts may have extremely high price where it is easy to justify more processing cost when the methods give high quality. However, detailed economic analysis needs to be done for desalination before reaching a conclusion. In food industry the FM was successful mainly due to its ability of producing high quality products as compared to other available technology in the market. It would be expected to face more challenge for applying FM process in the case of desalination since numbers of viable existing technology are available at present although it offers energy savings. Johnson [90] identified the main points that need to be considered for commercial potential of the FM process: simplicity (as compared to other freezing processes and no refrigerant contamination, containment or removal problems), a totally closed cycle refrigeration system (more integrated and compact), and all element would be at the state-of-the-art. In addition reduction of capital cost and use of hybrid techniques should also be considered.

The choice of a technology is usually based on the product quality, operating economy, energy cost, initial investment, and complexity of the process [91]. At present no commercial FM plant is available for desalination of sea water [92]. Since the process of FM is almost a century old, this question may fairly be asked: why is it, then that FM is not today in wider use. The reasons of this state of affairs are manifold, such as technical, political, and financial [25]. The applications of FM process in other industries have been explored in order to know how they feel about this technology and their learning curves. Three applications seem to be winning favor of FM process are: treating hazardous wastes, concentrating fruit juices, and purifying organic chemicals. The main reasons of these successes are due to the development of more efficient and high capacity process, and high purity or quality products. For examples, boosting the appeal of the technology today is the use of new direct-contact refrigeration cycles that are 50% more efficient than traditional ones, new commercially available crystallizing-units and crystal-washers that enable production of ultra-pure chemicals (99.99%) and continuous processes that permit throughputs as high as 75 million kg/yr (about five times higher than previously) [93, 4].

Based on the success in the other industry, especially food and chemical process, it is evident that the principal attraction of applying FM is its capability for concentrating heat-sensitive mixtures without damaging them, and separating hazardous and flammable chemicals, and azeotropic fluid. This technology is mainly adopted when there are no other alternatives. This may be the main cause why it has not been used widely in the desalination industry, where numbers of alternative technology exist.

Traditionally, the FM process has been limited by high capital costs- two to three times those of distillation or evaporation systems and production hang-ups caused by a greater degree of mechanical complexity [93]. We need to justify the capital costs of FM process if we are going to replace it for thermal processes [4]. In case of chemical processes, FM system manufacturers see this as a relatively hard sell, so they are going for markets where either the existing technologies can't fill the need or the conventional processes can't do it in one step. The practical difficulties and developmental stage reduce its wide commercial success [94].

FM is one of the most complex processes for desalination. It involves six steps plus the difficulty of handling liquid/solid slurry at its freezing point [25]. FM for desalination is an old standby, but today new process designs may open up applications. Continued areas of development for FM systems include improved crystal growth, more efficient refrigeration, better design methods, and better heat recovery. FM must also demonstrate that it can match the level of concentration available with other technologies [4]. Several other drawbacks encumber the industrial use of FM systems. These include limited capacity, relatively high production costs, and limited maximum concentration of saline water (one of the major drawbacks) [4]. In many countries pilot FM processes, such as the large solar-powered water desalination plant of 210m^3/day in Saudi Arabia, are built by applying hybrid techniques [4].

There has always been natural reluctance to accept a fundamentally new technique, especially when acceptable results are being obtained from an old and proven technique [95]. There should be a need to change the attitude of the industry towards the potential technology application for desalination. The support of FM process from industry is very low compared to the RO or MSF processes. This has created an atmosphere conducive to easy explanations and sweeping statements.

Persisted misconceptions or myths proved damaging to the commercial development of freezing as a competitive desalination process [96]. Weigandt and Von Berg [96] ten of these myths with respect to their engineering validity and a comparison is made between FM and other important desalination processes. This could reduce further damage of the FM technology since all represent incomplete, fuzzy or totally incorrect thinking. In 1952 the Organization for European Economic Cooperation convened a working party and promoted a negative impact on the development of sea-water desalting [96]. This team, with delegates from Belgium, Denmark, England, France, Germany, The Netherlands, and Sweden, and observers from Australia, South Africa, United States, and the Rockfeller Foundation, made its report a year later. For atmospheric distillation and the FM method it recommended that no further work be considered; distillation because there was already sufficient knowledge of the process, and FM because the method appeared to be impossibly difficult and economically impractical. The engineering problems that needed solving to develop an FM process are more in number that for RO or MSF distillation and far more complex than the later. Fortunately there are a determined group of dedicated people in the universities, in industry, and in the US Office of Saline Water who realized that there would someday be a FM process for which the concepts would be sound and the engineering goals achievable [96]. This is evident from the literature that huge numbers of alternatives in the FM process components are being developed.

10.11 Future Challenges

There is a high potential of combining the FM process with other desalination techniques. This hybrid approach could provide a synergy to the desalination process. One of the most promising one is the combination of reverse osmosis and freezing melting process. A zero-discharge direct-contact freezing/solar evaporator desalination complex is proposed as an efficient system to reduce the environmental impact of concentrated rejected brine from seawater desalination plants. The proposed method produced fresh water, Na, Mg and K salts and bromide [97, 78] identified different possible desalination techniques in association with wind, tidal, and solar energy sources. An economic analysis showed that FM might be competitive with solar distillation in suitable locations. Ice collection methods, washing by natural drainage and the coupling of FM with solar distillation should be further studied [53, 85] presented a design analysis and economic evaluation of solar assisted vacuum freezing ejector absorption desalination (VFEA) plant with a capacity of 1 mgd and located in Abu Dhabi. The parameters that affect the design and plant costs are: seawater salinity and temperature, and solar collector outlet temperature. The collector outlet temperature was set at 90° or 120°C using flat plate and evacuated tube collectors. The absorber loop of the VFEA system uses a sodium hydroxide solution with concentration ranging from 0.5 (dilute stream) to 0.6% (concentrated stream). The capital cost of the system increases with increasing seawater salinity and temperature, whereas cost decreases with higher collector outlet temperature (120°C). The thermal load on the concentrator increased with the seawater salinity, whereas it dropped substantially with the collector outlet temperature of 120°C as compared to 90°C. Life cycle savings in fuel costs of the solar-assisted VFEA plant were also estimated using a set of economic ground rules with the objective of specifying the optimum collector area, which yields the maximum life cycle savings. It was observed that the optimum area increases with increasing seawater salinity. Abdul-Fattah [98] evaluated the alternative solar power systems compatible with freezing process considering the special case of Saudi Arabia. He pointed that FM can be a viable water system since freezing units of small scale are proven technology. Four designs of solar freezing are considered to select the most promising option. The decision is made on the basis of fuzzy set analysis of the criteria surrounding the choice. Taking the case of Saudi Arabia as an example, photovoltaic driven indirect freezing seems to be the most promising technology. The dual vapor absorption freezing using thermal collectors is second in ranking.

Combined wind or tidal power-desalination systems include vapor compression, reverse osmosis, electro-dialysis and freezing melting techniques. Solar energy in direct thermal form or through its conversion to electric power has the potential of usages with almost all desalination technologies. Photovoltaic solar power can be used with FM process [78].

Schwartzberg [99] suggested that the combination of reverse osmosis and a cheaper FM could provide economical alternative for concentrating liquids. Disposal of the rejected concentration brine from reverse osmosis plants may cause serious environmental impacts. Different ways for disposal handling are adopted, including pumping into lined evaporation ponds, injection into underground rock formation, or spreading on unusable arid land. All of these are short time solutions due to the large amount of rejected brine to be disposed of [100]. The amount of rejected brine from reverse osmosis plants could be minimized by a further desalination of the rejected brine. The high concentration of the RO rejected brine limits the choice of the second stage desalination unit. The energy efficiency of FM makes it a promising choice since the process is independent of fouling, and low corrosion due to the operation at low temperature. Madani and Aly [101] conducted economical and energy comparisons between the combined system and separate RO and direct FM units of 200 m^3/h. The combined system can reduce the energy consumption by about 13% and 17% compared to separate RO and direct FM plants, respectively. The combined system can reduce the rejected brine by over 90% of that of separate RO plant at the same water production.

The use of electric field and ultra sound aided process could be used in the freezing-unit in order to enhance the performance. However, all these addition will make the process more complicated.

Acknowledgment

This chapter is updated based on the report of a project funded by the Middle East Desalination Research Center (MEDRC), Muscat, Sultanate of Oman (Project No. 98-BS–032A).

References

1. G. Nebbia, G.N. Menozzi, Early experiments on water desalination by freezing. *Desalination* **5**, 49–54 (1968).
2. C.I. Gellen, *Importance of Water Desalination by Natural Freezing for Agricultural Purposes*. Geographical Institute of the Russian Academy of Sciences, 5 (1962).

3. J. Fournier, J.L. Grange, S. Vergara, Water desalination by natural freezing. *Desalination* **15**, 167–175 (1974).
4. J. Rosen, Freeze concentration beats the heat. *Mech. Eng* **112(12)**, 46–50 (1990).
5. R.A. Brewster, B. Gebhart, The effects of supercooling and freezing on natural convection in seawater. *Int. J. Heat Mass Transfer* **37(4)**, 543–552 (1994).
6. M.S. Rahman, M. Ahmed, X.D. Chen, Freezing-Melting Process in Liquid Food Concentration. In: *Handbook of Food Preservation*. CRC Press, Boca Raton, FL. p. 667–690 (2007).
7. R.J. Braddock, J.E. Marcy, Quality of freeze concentrated orange juice. *J. Food Sci* **52(1)**, 159–162 (1987).
8. P.L.T. Brian, Potential advantages and development problem in water desalination by freezing. *Chem. Eng* 191–197 (1971).
9. M.V. Rane and Y. S. Padiya, Heat pump operated freeze concentration system with tubular heat exchanger for seawater desalination. *Desalination* **15**, 184–191 (2011).
10. R.W. Hartel, Evaporation and freeze concentration. In: *Handbook of Food Engineering*. D.R. Heldman, D. Lund, M Dekker (Ed.), p. 341–392, New York. (1992).
11. W. Gibson, D. Emmermann, G. Grossman, R. Johnson, A. Modica, Spray freezer and pressurized counter washer for freeze desalination. *Desalination* **14**, 249–262 (1974).
12. A.J. Barduhn, The state of the crystallization processes for desalting saline water. *Desalination*. **5**, 173–184 (1968a).
13. A.J. Barduhn, The freezing process for water conversion in the United States. In: *Selected Papers on Desalination and Ocean Technology*. S.N. Levine (Ed.), P, 414, Dover, New York. (1968b)
14. J.C. Orcutt and F.J. Hale, The secondary refrigerant freezing process: a modeling study. *Desalination* **7**, 201–227 (1970).
15. G.L. Stepakoff, A.P. Modica, The hydrolysis of halocarbon refrigerants in freeze desalination processes. *Desalination* **12**, 85–105 (1973).
16. R.T. Kurnik, A.J. Barduhn, Thermodynamic properties of normal butane at refrigeration temperature. *Desalination* **26**, 211–283 (1978).
17. A. Antonelli, Desalinated water production at LNG-terminals. *Desalination* **45**, 383–390 (1983).
18. P.J. Schroeder, Freezing processes-the standard of the future-an update. *Desalination* **33**, 299–310 (1980).
19. H.C. Simpson, G.C. Beggs, M. Nazir, Evaporation of butane drops in brine. *Desalination* **15**, 11–23 (1974).
20. H.F. Wiegandt, A. Madani, P. Harriott, Ice crystallization developments for the butane direct-contact process. *Desalination* **67**, 107–126 (1987).
21. M. Landau, A. Martindale, Assessment of crystallization designs for a butane freeze desalination process. *Desalination* **3**, 318- (1967).

22. W.H. Denton, M.J.S. Smith, J.T. Klaschka, R. Forgan, H.R. Diffey, C.H. Rumary and R.W. Dawson, Experimental studies on washing and melting ice crystals in the immiscible refrigerant freezing process. *Desalination* **14**, 263–290 (1974).
23. D.S. Hubbard, J.P. Leinroth, H.F. Wiegandt, Spray freezing of brine with an immiscible refrigerant. Report to OSW on Control No. 1141, Cornell University (1968).
24. C.A. Johnson, S.J. Moore, N.D. Wagman, D.J. Sandell, Freezing process using R-C318. OSW Progress Report No. 256 (1967).
25. W.E. Johnson, State-of-the-art of freezing processes, their potential and future. *Desalination* **19**, 349–358 (1976).
26. I.J. MaCausland, P.W. Cains, P.D. Martin, Use the power of sonocrystallization for improved properties. *Chem. Eng. Pro* **97(7)**, 56–61 (2001).
27. L. Otero, P. Sanz, B. Guignon, P.D. Sanz, Pressure-shift nucleation: A potential tool for freeze concentration of fluid foods. *Innov. Food Sci. Emerg. Technol* **13**, 86–99 (2012).
28. L. Liu, O. Miyawaki, K. Nakamura, Progressive freeze-concentration of model liquid food. *Food Sci. Technol. Int. Tokyo* **3(4)**, 348–352 (1997).
29. R.W. Hartel, M.S. Chung, Contact nucleation of ice in fluid dairy products. *J. Food Eng* **18**, 281–296 (1993).
30. A.M. Omran, C.J. King, Kinetics of ice crystallization in sugar solutions and fruit juices. *AIChE J* **20**, 795–803 (1974).
31. Y. Shirai, K. Nakanishi, R. Matsuno, T. Kamikubo, Mechanism of ice growth in a batch crystallizer with an external cooler for freeze concentration. *Agricu. Biol. Chem* **51**, 2359–2366 (1985).
32. S.S. Deshpande, M. Cheryan, S.K. Sathe, D.K. Salunkhe, Freeze concentration of fruit juices. CRC Crit. Rev. Food Sci Nutr **20(3)**, 173–248 (1984).
33. H.A.C. Thijssen, Apparatus for separation and treatment of solid particles from a liquid suspension, US Patent 3872009 (1975a)
34. H.A.C. Thijssen, Current developments in the freeze concentration of liquid foods. In: Freeze Drying and Advanced Food Technology. Goldblith, S. A., Rey, L., Rothmayr, W. W. eds. Academic Press, New York (1975b)
35. H.A.C. Thijssen, Freeze concentration of food liquids. *Food Manufacture* **44(7)**, 49–54 (1969).
36. H. Garabedian, R.F. Strickland-Constable, Collision breeding of ice crystals. *J. Cryst. Growth* **22**, 188–192 (1974)
37. L.N. Matusevich, N.P. Blinova, Isohydric crystallization at different solution cooling rates. Journal of Applied Chemistry of the USSR. **37(10)**, 2310–2314 (1964).
38. S.H. Bransom, Factors in the design of continuous crystallisers. *Br. Chem. Eng* **5,** 838–844 (1960).
39. D.M. Lupu, The dynamics of sea water desalting through zone-freezing. *Desalination* **31**, 415 (1979).

40. I. Tuchida, Hydrate formation in i-butane, b-butane or a butane isomer water system at a given temperature. *Desalination* **3**, 373–377 (1967).
41. R. Battino and H.L. Clever, The solubility of gases in liquids. *Chem. Rev* **59**, 395 (1966).
42. S. Benson, *The Foundations of Chemical Kinetics*. McGraw-Hill, New York (1960).
43. S. Colten, F.-S.Lin, T.-C.Tsao, S.A. Stern and A.J. Barduhn, Hydrolysis losses in the hydrate desalination process, rate measurement and economic analysis. *Desalination* **11**, 31 (1972).
44. J. Shwartz and R.F. Probstein, Experimental study of slurry separators for use in desalination. *Desalination* **6**, 239–266 (1969).
45. G.T. Arulampalam, C. Bates, L.F. Khaw, L. McGrath, Investigation of purification mechanism in a column crystallizer operating under total reflux and continuous conditions. *Desalination* **36**, 87–97 (1981).
46. C.M. Bosworth, S.S. Carfagno, D.J. Sandell, Development of a Direct-Freezing Continuous Wash-Separation Process for Saline Water Conversion. Office of Saline Water Research Development Progress Report No. 23.(1959).
47. W.J. Hahn, R.C. Burns, R.S. Fullerton and D.J. Sandell, OSW R&D Progress Report N. 113, US Department of Interior, Washington DC (1964).
48. J. Shwartz, R.F. Probstein, An analysis of counter washers for freeze-distillation desalination. *Desalination* **4**, 5–29 (1968).
49. M.F. Dallmer, L.C. Dickey, E.R. Radewonuk, Horizontal cross flow filtration and rinsing of ice from saline slurries – has a slurry inlet, a filter plate with two slotted sections, a heater, and two pumps to induce flow through the slots, US Patent 5816057 (1998).
50. D.L. McKay, Crystal separation and purification, United States Patent 3,487,652 (1970).
51. J.B. Pangborn, Observations of the eutectic freezing of salt solution. Report to A. J. Barduhn, Department of Chemical Engineering, Syracuse University (1963).
52. G.L. Stepakoff, D. Siegelman, R. Johnson, W. Gibson, Development of a eutectic freezing process for brine disposal. *Desalination* **14**, 25–38 (1974).
53. P.J. Schroeder, A.S. Chan, A.R. Khan, Freezing process- the standard of the future. *Desalination* **21**, 125–136 (1977).
54. P.C. Wankat, Desalination by natural freezing. *Desalination* **13**, 147–157 (1973).
55. H.M. Curran, Water desalination by indirect freezing. *Desalination* **7**, 273–284 (1970).
56. H.R. Pruppacher, Interpretation of experimentally determined growth rates of ice crystals in supercooled water, *Chem. Phys* **47(5)**, 1807–1813 (1967a).

57. H.R. Pruppacher, Some relations between the structure of the ice-solution interface and the free growth rate of ice crystals in supercooled aqueous solutions. *J. Colloid Interface Sci* **25**, 285–294 (1967b).
58. Y. Shirai, T. Sugimoto, M. Hashimoto, K. Nakanishi and R Matsuno, Mechanism of ice growth in a batch crystallizer with an external cooler for freeze concentration. *Agricultural Biol Chem* **51**, 2359–2366 (1987).
59. M. Hassence, MSC offers purity advantages. *Asia-Pac Chem* **10**, 30–32 (1993).
60. S. Rittner, R. Steiner, Die schmelzkristalization von organischen stiffen and ihre grobtechiche anwendung. *Chemie Ingenieur Technik* **57(2)**, 91–102 (1985).
61. S.K. Bae, O. Miyawaki. and S. Arai, Control of freezing front structure and its effect on the concentration-efficiency in the progressive freeze-concentration. *Cryobiol. Cryotechnol.* **40**, 29–32 (1994).
62. T.H. Gouw, Normal freezing. In: *Progress in Separation and Purification.* Vol. 1. E.S. Perry,(Ed.) Wiley, New York. Pp. 57–82 (1968).
63. O. Miyawaki, L. Liu, and K. Nakamura. Effective partitioning constant of solute between ice and liquid phases in progressive freeze-concentration. *J. Food Sci.* **63(5)**, 756–758 (1998).
64. A. Matsuda and K. Kawasaki, Concentration and separation of impurities in liquid by freezing with supersonic radiation. *J. Chem. Eng. Japan* **30(5)**, 825–830 1997.
65. O. Miyawaki, L. Liu, Y. Shirai, S. Sakashita and S. Kagitani, Tubular ice system for scale-up of progressive freeze-concentration. *J. Food Eng* **69**; 107–113 (2005).
66. F.A. Ramos, J.L. Delgado, E. Bautista, A.L. Morales, and C. Duque, Changes in volatiles with the application of progressive freeze-concentration to Andes berry (*Rubus glaucus benth*). *J. Food Eng.* **69**, 291–297 (2005)
67. A.S. Michaels, P.L.T. Brian and P.R. Sperry, Impurity effects on the basal plane solidification kinetics of supercooled water. *J. Appl. Phys* **37(13)**, 4649–4661 (1966).
68. J.C. Baker,J.W. Cahn, Solute trapping by rapid solidification. *Acta Metallurgica.* **17**, 575–578 (1969).
69. L. Bayindirli, M. Ozilgen, S. Ungan, Mathematical analysis of freeze concentration of apple juice. *J Food Eng.* **19**, 95–107 (1993).
70. J. Sanchez, E. Hernandez, J.M. Auleda and M. Raventos, Freeze concentration of whey in a falling-film based pilot plant: Process and characterization. *J. Food Eng* **103**, 147–155 (2011).
71. C.A. Depew and R.C. Zenter, Laminar flow heat transfer and pressure drop with freezing at the wall. International *J. Heat Mass Transfer* **12**, 1710–1714 (1969)
72. G.S.H. Lock and R.H. Nyren, Analysis of fully developed ice formation in a convectively cooled circular tube. *Int. J. Heat Mass Transfer* **14**, 825–834 (1971)

73. M.N. Ozisik and J.C. Mulligan, Transient freezing of liquids in forced flow circular tubes. *J. Heat Transfer* **91**, 385–389 (1969)
74. R.D. Zerkle and J.E. Sunderland, The effect of liquid solidification in a tube upon laminar-flow heat transfer and pressure drop. *J. Heat Transfer* **90**, 183–189 (1968).
75. A.P.S. Arora, J.R. Howell, An investigation of the freezing of supercooled liquid in forced turbulent flow inside circular tubes. *Int. J. Heat Mass Transfer* **16,** 2077–2085 (1973).
76. E.F. Janzow and B.T. Chao, Salt entrainment in ice crystallized from brine. *Desalination* **12**, 141–161 (1973).
77. Y. Mandri, A. Rich, D. Mangin, S. Abderafi, C. Bebon, N. Semlali, J. Klein,T. Bounahmidi, A. Bouhaouss, Parametric study of the sweating step in the seawater desalination process by indirect freezing. *Desalination* **269**, 142–147 (2011).
78. A. Hanafi, Desalination using renewable energy sources. *Desalination* **97**, 339–352 (1994).
79. N.J.J. Huige and H.A.C. Thijssen, Production of large crystals by continuous ripening in a stirred tank. *J. Cryst. Growth* **13/14**, 483–487 (1972).
80. N.S. Tavare, Simulation of ostwald ripening in a reactive batch crystallizer. *AIChE J* **33**, 152–156 (1987).
81. T. Kobayashi, Freeze concentrations and their application. *Reitou* **66**, 1008–1018 (1991).
82. L.C. Dickey, Evaporation of water from agitated freezing slurries at low pressure. *Desalination* **104**, 155–163 (1996).
83. C.Y. Cheng, S.W. Cheng, US Patent 3690116, issued September 12 (1972).
84. C.Y. Cheng, S.W. Chang and W.C. Cheng, US Patent 4235382, issued December 1 (1980).
85. A. El-Nashar, Solar desalination using the vacuum freezing ejector absorption (VFEA) process. *Desalination* **49**, 293–319 (1984).
86. M. Pachter and A. Barak, The vacuum freezing vapor compression (Zarchin) process present status and future trends. *Desalination* **2**, 358–367 (1967).
87. W.R. Burton, A.I. Lloyd, Design features of secondary refrigerant freezing plants. *Desalination* **14**, 151–161 (1974)
88. C.Y. Cheng, W.C. Cheng, M.D. Yang, The vacuum freezing multiple phase transformation process. *Desalination* **67**, 139–153 (1987).
89. M. Aider and D. de Halleux, Passive and microwave-assisted thawing in maple sap cryoconcentration technology. *J. Food Eng* **85**, 65–72 (2008).
90. W.E. Johnson, Indirect freezing. *Desalination* **31**, 417–425 (1979).
91. J.G. Moore and W.E. Hesler, Evaporation of heat sensitive materials. *Chem. Eng. Prog* **59(2)**, 87–92 (1963).

92. K. Wangnick, 2003. 2004 IDA worldwide desalting plants inventory. Report No. 18, Wangnick Consulting, Germany.
93. J. Chowdhury, CPI warm up to freeze concentration. *Chem. Eng* **95(6)**, 24 (1988).
94. W. Rice, D.S.C. Chau, Freeze desalination using hydraulic refrigerant compressors. *Desalination* **109**, 157–164 (1977).
95. J.G. Muller, Freeze concentration of liquids: theory, practice and economics. *Freeze Technol* **21**, 49–60 (1967).
96. H.F. Wiegandt, R.L. von Berg, Myths about freeze desalination. *Desalination* **33**, 287–297 (1980).
97. A.A. Madani, Zero-discharge direct-contact freezing/solar evaporator desalination complex. *Desalination* **85**, 179–195 (1992).
98. A.F. Abdul-Fattah, Use of low grade in driving small freezing units for desalination. *Desalination* **61**, 169–183 (1987).
99. H.G. Schwartzberg, Energy requirement for liquid food concentration. *Food Technol* **31(3)**, 67–76 (1977).
100. I. Al-Mutaz, Water resources development in Riyadh, Saudi Arabia. *Desalination* **64**, 193–202 (1987).
101. A.A. Madani, S.E. Aly, A combined RO/freezing system to reduce inland rejected brine. *Desalination* **75**, 241–258 (1989).

11

Desalination by Ion Exchange

Bill Bornak

Recirculation Technologies LLC, Warminster, PA, USA

Abstract
Ion exchange is not used in direct desalination; limitations in resin capacity preclude this in favor of membrane-based options. Ion exchange resins, however, play a key role in desalination systems by way of pre-softening the RO feed and in the selective removal of contaminants leaking past the membrane step. One of the most commercially successful post-treatment resin applications is the selective removal of boron from RO permeate. Fixed bed geometry is used in most large-scale B-removal systems, and the advent of Uniform Particle Size resin products allows for smaller vessels for a given flowrate. Even the resulting packed bed geometry is still used in a batch/cycling sequence. Continuous geometries have been proposed in patents and the technical literature, but have seen limited commercial success.

Keywords: Ion exchange, resins, uniform particle size beads, packed beds, boron removal, softening, strong acid cation exchange resin, weak acid cation exchange resin, carousel geometry, post-treatment for permeate, powdered resins, continuous ion exchange, Desal ™ Process [Desal is a registered trademark of the Dow Chemical Company]

Ion Exchange in Desalination

Welcome to what is possibly the shortest chapter in this book. The short length is not due to a lack of industry on the part of the writer, but stems directly from the nature of ion exchange: a finite resin

Jane Kucera (ed.) Desalination, (503–520) 2014 © Scrivener Publishing LLC

capacity used, for the most part, in a batch operation. That's why "ion exchange" and "desalination" are not often used in the same sentence, let alone together for a full chapter.

11.1 Introduction

Ion exchange predates commercial RO by at least two decades, but as RO moved from lab bench to commercial practice, three advantages were touted.

1. Firstly, in the earlier years of RO it was always important (for us resin folks) to calculate the Total Dissolved Solids (TDS) at which ion exchange demineralization became uneconomical compared to the new, competing membrane processes. Initially ion exchange prevailed to a very high TDS, easily over 1500 ppm. This was due to the high cost of early membrane elements. As the cost of membranes came down, the cross-over point also dropped, clearly favoring membranes.
2. Secondly, the virtually continuous operation of a membrane system was recognized as a strong advantage over the service/regeneration batch cycling of the resin systems.
3. And, thirdly, membrane systems were advertised as "chemical free." Every pro-RO technical or trade paper from that era included a bullet point emphasizing the intrinsic dangers associated with handling acid and caustic in a demineralizer water plant, although all the regenerant handling systems were safely hard-piped.

We all know, of course, that membrane systems do require periodic cleaning and most of the protocols use the same "dangerous" chemicals decried in the advertising, and the solutions are made down in the *open tank* of a Clean In Place (CIP) skid. Although this neutralizes the third claim, the first two advantages of membranes are still valid. And there seems nothing on the horizon of ion exchange technology which is likely to change that. On high TDS waters ion exchange is simply not cost-effective against membrane-based processes. That said, however, we also state there is a place for resins at the desalination station.

11.2 Early Ion Exchange Desalination Processes

From a historical standpoint, the highpoint of the proposed use of ion exchange in desalination was Dr. Robert Kunin's Desal ™ process, invented in 1962 while he was at Rohm and Haas.[1] The process is alleged to be able to handle TDS in the range of 500–5000 ppm. [1–4] The process is shown below in Figure 11.1.

In operation, the following steps occur:

1. Inlet brackish water comes into the first vessel on the left, a weak base anion (WBA) already in the bicarbonate form.
2. All non-bicarbonate anions in the inlet are exchanged for bicarbonate. Outlet of the first WBA is mixture of Na, Ca, and Mg bicarbonates.
3. Second vessel is weak acid cation (WAC) exchanger in the hydrogen ion form.
4. WAC reacts with mixed bicarbonates, exchanging all cations for hydrogen. Outlet is strong solution of carbonic acid, essentially water and CO_2.
5. Third vessel is another WBA, but this one is in the free base form.
6. Inlet carbonic acid solution reacts quantitatively with WBA, converting it to bicarbonate cycle. Outlet is low TDS water. This completes one cycle.

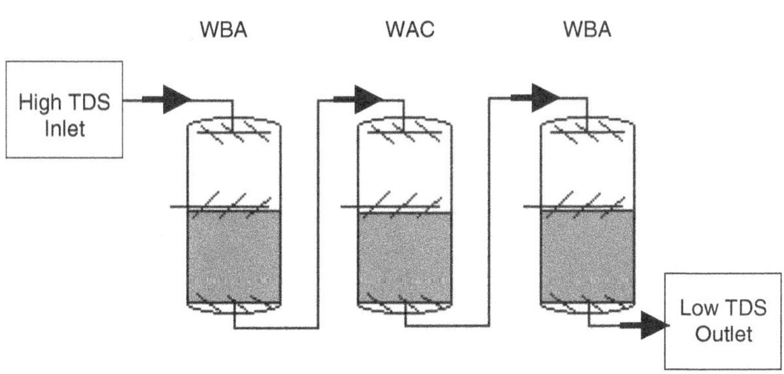

Figure 11.1 The Desal Process.

[1] Desal is a registered trademark of the Rohm and Haas Company, Philadelphia, PA now a subsidiary of Dow Chemical, Midland, MI.

7. Upon completion, the first WBA is regenerated with ammonia, caustic, or lime. The middle WAC unit is regenerated with strong acid, HCl or H_2SO_4.
8. The second cycle starts with a reversal of the flow. Inlet goes to what was third WBA, which was converted to bicarbonate form in step (6).
9. Pick up step (2), etc.

The overall service effect: brackish water is demineralized to low TDS water. The overall regeneration requirements: regenerate one WBA with ammonia, caustic, or lime; regenerate one WAC with HCl or H_2SO_4.

The claim is that the process would work up to 5,000 ppm TDS, encompassing brackish but not sea water. It is an elegant use of weak base and weak acid chemistries, and claims these additional advantages:

- uses low-cost regenerants
- rinse requirements are substantially lower (than a strong base, e.g.)
- regenerations are done close to minimum stoichiometry (maximum chemical efficiency)

There was a modification of the above system to handle acid mine drainage with a further twist involving aeration to break down the mixed metal bicarbonates from step (2) above, recovery of ammonia used for regeneration in step (7), and ultimate production of $CaSO_4$, gypsum, as the only process byproduct. [2]

Although the Desal Process chemistry was very clever, the capital requirement for the resin vessels was a major obstacle, especially at municipal-level flowrates. To the writer's knowledge (which excludes China), there are no operating Desal Process units of any significant size in the world. Again, resin technology was trumped by membrane technologies, once the latter became commercialized.

A few variations on the Desal Process were proposed; one, using continuous, counter-current injection of resin, was patented. [5] See below for the fate of many continuous resin inventions. A recent patent combines ion exchange and nanofiltration for desalinating brackish waters. [6] Its commercialization is unknown.

11.3 Life after RO

If, then, ion exchange is eclipsed by membrane technologies for desalination, is there a place for resins in this application? The answer is yes, ion exchange does have a place, but its role encompasses two very distinct levels of technology. To provide a context for the answer, consider the following range of ion exchange applications:

- At the simplest or humblest end is ion exchange softening. Only one resin, a strong acid cation exchanger, is used. Off spec resin works well. Regeneration is with ambient brine. No temperature control is needed. The concentration and dosage of brine can vary by a surprisingly wide margin and the technology still works.
- Next up the technology ladder is the combination of a cation and anion exchanger to make a demineralizer. Here the concentration of the regenerants and flowrates are more critical, and the caustic temperature becomes important.
- More sophisticated still is the use of weak and strong resin combinations, giving higher efficiencies but at the price of more details to watch, more complicated vessels and piping.
- Then we have the mixed bed polisher, a truly complicated and subtle machine with an unwieldy regeneration.
- Next up the ladder would be a mixed bed used as a condensate polisher, where diffusional and exchange kinetics enter the picture, in addition to the intrinsic complexities of the mixed bed geometry.
- More sophisticated still would be non-traditional water-treatment applications such as sugar syrup decolorization, non-aqueous catalysis, and antibiotic isolation in pharmaceutical applications.
- At the top of the list in terms of technical sophistication are metal-selective resins using exotic active groups and unusual regeneration protocols.

This review is relevant in that the role of ion exchange in desalination involves both the least and the most sophisticated ends of the resin technology spectrum outlined above. Ion exchange softening is used to pretreat RO feeds to lessen scaling of the main membrane-based desalination step. Element-selective ion exchange resins are currently used post-RO to remove an unexpected and pesky contaminant from sea water permeates, notably boron.

11.4 Ion Exchange Softening as Pre-Treatment

Ion exchange softening is a standalone water purification option for many industrial applications. It predates demineralization (which needed the strong base anion) by a decade and a half. And before the invention of the first cation exchange resins, just before World War II, green sand zeolites were being used to remove hardness from inlet waters. Softening is limited, however, to low pressure boilers. The process is shown in Figure 11.2. [7]

As shown in Figure 11.2, the chemistry is direct: a strong acid cation resin in the sodium form exchanges sodium for hardness ions (Ca, Mg, Ba, and Sr). The resin bed has a reasonably good but clearly finite capacity and, when exhausted, requires a quick regeneration with a brine solution. This reinforces the opening comments to this chapter, identifying the key limitations to resin systems: finite capacity and batch operation. But, the removal of hardness from low pressure boiler feedwater by way of resin softening allowed tens of thousands of such boilers to supply steam for early refineries, chemical plants, paper mills, and a host of general manufacturing applications ranging from food to automobile tires. In addition, the regenerant salt was extremely cheap and the systems

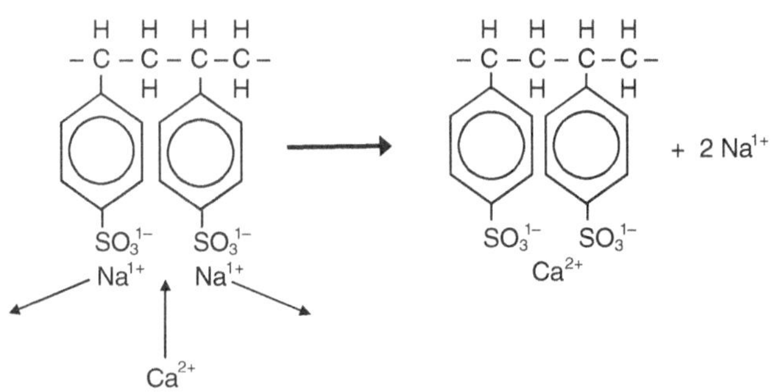

Figure 11.2 Basic ion exchange softening.

were robust. Many built in the 1950s are still operating today. The spent salt solutions were also more easily disposed of than today.

Softening for boiler feedwater makeup was eclipsed by full demineralization when the first strong base anion resins appeared in the mid 1950s. This also allowed the use of higher boiler pressures, generating steam for more efficient turbines, all the way up to super critical units.

On the non-commercial front, softening is wide spread in private homes to reduce pipe scaling and scum formation problems with detergents. Likewise, RO kits are offered for "under the sink" installation. It is ironic that a battle, which resins clearly lost in the commercial desalination arena, is still being fought under sinks and in home basements in the domestic market.

The main role of ion exchange in RO desalination is pre-softening all or part of the RO feed. Removal or reduction of inlet hardness directly translates to a lower potential for scaling of the membranes, typically the back-end elements. Partial softening can be effective with or without scale control additives. A fascinating new twist to softening chemistry here is the use of SWRO (Sea Water RO) rejects as potential resin regenerants. It literally is "free salt," productively using a stream which, in some geographical areas, is difficult to dispose of. [8]

11.5 Softening by Ion Exchange

There is an unusual combination of chemistries that supports highly efficient softening by a slightly different choice of resins than described in the previous section. The resin in Section 11.4 was a strong acid cation (SAC) exchanger with benzene sulfonic acid groups, shown on the left in Figure 11.3.

The SAC resin in the sodium form will react with inlet hardness, substituting two sodium ions for any divalent hardness ion, as described earlier. But it cannot be used to soften brackish solutions because the high salt concentrations favor the regeneration reactions, rather than the service/softening reactions. A weak acid resin incorporating carboxylic acid functionality, is shown on the right in Figure 11.3 On a practical basis, only the weak acid resin can remove hardness ions from brine or brackish waters. It is operated in the sodium form and the high density of carboxylic acid functionalities allows the active groups to act in concert, a pair

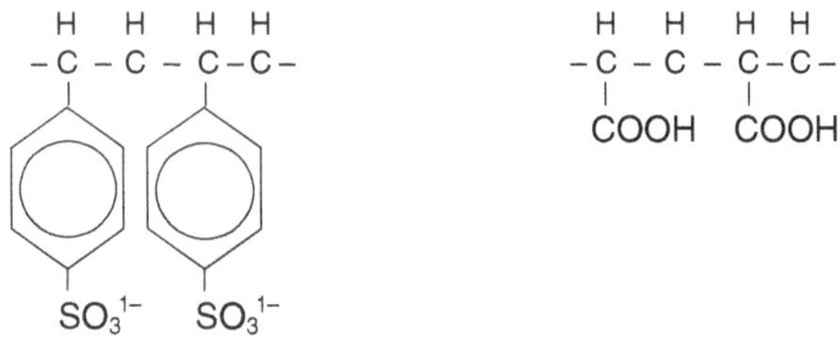

Strong acid cation exchange resin

- Polystyrene crosslinked with divinylbenzene
- Active group: sulphonic acid
- Reacts with all cations when in H^{+1} cycle
- Regeneration efficiency: 33%

Weak acid cation exchange resin

- Polyacrylic acid
- Active group: carbyxylic acid
- Selective reactions
- Regeneration efficiency: 100%

Figure 11.3 Strong and weak acid cation exchange resin.

of groups jointly holding a divalent hardness ion. This concerted co-operation is actually a chelation reaction. Chelants also have multiple carboxylic acid groups and when operated in the sodium form will strongly bond with divalent ions. The only problem here is the regeneration. The WAC resin saturated with hardness ions will not regenerate with brine. The regeneration has to be performed in a two-step process:

1. regeneration with 4% HCl to remove all hardness ions under a "safe" regeneration chemistry with does not imperil the production of calcium sulfate. This regeneration is stoichiometric. The HCl injection is followed by a rinse.
2. The second step is the treatment of the resin with 4% NaOH to put all the active groups in the Na cycle. This step is also stoichiometric.

Failure to do the NaOH step leaves the resin in the hydrogen or acid cycle. The resin will react with hardness ions in this cycle, but only to the extent that alkalinity is present. That is, it will not remove hardness above the level of alkalinity. Since sea water or brackish waters have only a modest alkalinity, this option is not viable.

The technical literature also recommends a merry go round system using three vessels alternating between lead, lag, and

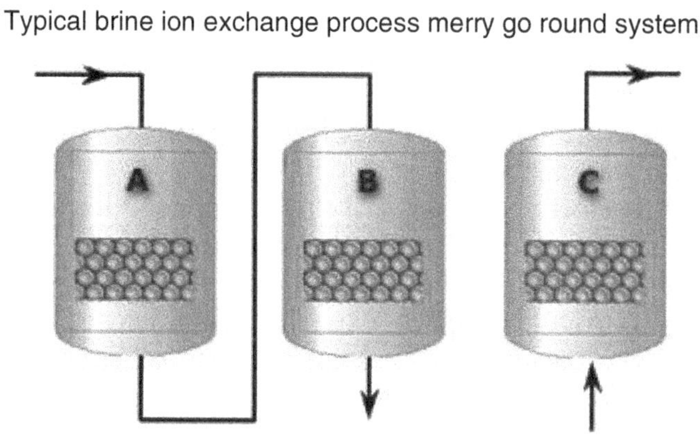

	Column Position		
Step	Loading	Polishing	Regenerating
1	A	B	C
2	B	C	A
3	C	A	B

Figure 11.4 Merry-Go-Round System.

regeneration positions. Merry go round operation is shown in Figure 11.4.

There is a rich literature on the use of ion exchange processes for pre-softening brackish waters prior to RO purification. [8–10]

11.6 Boron-Selective Ion Exchange Resins as Post-Treatment

The second area in which ion exchange resins can be used in desalination concerns selective removal of a problematic, usually metallic, contaminant. A broad resin technology exists to selectively remove metal X from mixture Y, usually with a stipulation of pH Z. Due to the toxicity of some heavy metals, their removal from a waste stream becomes environmentally mandatory. For example, the availability of a resin to selectively take out Cu or Cd from acid mine tailings, and not take out Na, Ca, and Mg, has strong regulatory and commercial appeal. [12] Other heavy metals are quite valuable and their removal and subsequent recovery is economically justified. Resins have long been used to isolate gold and uranium chloride complexes from ore extraction liquors. Thus, there are numerous niche markets for resins able to isolate chosen metals from otherwise useless, benign, or even dangerous solutions. One key resin application here is the selective removal of ambient As from potable water. [13]

Most metal-selective resins use a chelating or sequestering active group in place of conventional acid or amine functionalities. As mentioned earlier, a chelant is characterized as having a very high affinity for a particular species, usually multivalent cations, while not reacting with background chemicals. The most common chelant is ethylene-diamine-tetra(acetic acid), EDTA. The tetra sodium salt is normally used so the four acetates ionize in solution. The geometry of the molecule positions the ionized acetates so they can interact with cations, and in particular multivalent metal cations. Thus, EDTA has a much stronger affinity with Ca^{+2} than with Na^{+1} or K^{+1}, and stronger still with Fe^{+3} and other trivalent species. There are commercial resins in which a half-EDTA, imidodiacetic acid, is the active group attached to the plastic polymeric matrix. Other resins use phosphonic acid in lieu of sulfonic acid as the active group in a cation exchanger. The former can selectively pull Fe out of a solution containing Cu, and *v.v.*, representing yet another niche market.

One of the most complicated non-traditional active groups is that of a boron-selective resin. The active group is a substituted glucosan, *N*-methyl-glucamine, shown in Figure 11.5. [14]

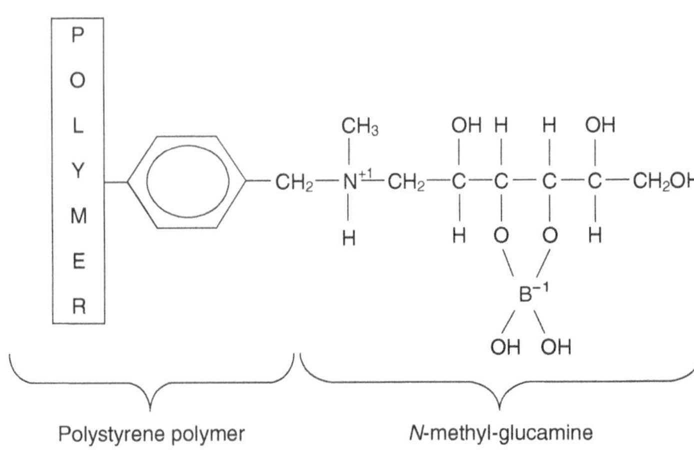

Figure 11.5 Active group in boron-selective resins.

The need for this resin in desalination arises from the very poor efficiency of boron removal in conventional membrane applications. Boron, like bicarbonate, can exist is several forms depending on pH. [15] The pK_a is 9.2. Below that pH, boric acid predominates; above that pH, monovalent anionic borate is the main species. Due to the modest alkalinities of brackish waters and the intrinsically low concentration of boron, boric acid is the main species and, like any unionized species, it shows a poor rejection. [16] Conversion to the ionized borate, which would show much higher rejection, requires a higher pH. The addition of caustic, however, also greatly increases the potential for scale formation, which must be countered by the feed of an antiscalant. Thus system operating costs keep rising.

The B problem arises from the need to meet governmental standards in water for potable and agricultural applications. The original, potable standard for B was set by the World Health Organization (WHO) at 0.5 mg/l, which then was raised to 2.4 mg/l in 2011. [17] But damage to certain crops, notably citrus and artichokes, can occur in irrigation water with B levels greater than 0.3 mg/l. Thus the need for post-RO polishing and the need for a boron-specific ion exchange resin.

A good case study for this application is from the Eilat desalination center in Israel. [18] The inlet B from the Mediterranean was 4.7 mg/l. RO permeate at the station was 3.5 mg/l B, above even the revised WHO standard, and greatly exceeding the toxicity level for irrigation. Membrane rejection of B was only 65%.

When a boron-selective resin was installed post-RO, B levels dropped to 0.2 mg/l. The key to this commercial success was the unusual selectivity of the resin *just for* B, and it's usefully high capacity. In addition, after exhaustion, the resin can be regenerated quantitatively with 5% sulfuric acid. The Eilat reference indicated the operating cost of adding the B-selective resin was only 4–6 US¢ per m^3. (The capital costs were not discussed.)

This installation used fixed bed resin vessels, directly paralleling conventional water treatment vessels. As with the Desal Process discussed earlier, the vessels for reasonable flowrates can be quite large and expensive. Trying to address this disadvantage, the technology has recently evolved in two new quite different directions. In one branch, the resin is added to the permeate as a very fine powder. Regular resins run between 300 and 1300 μ. This application uses pulverized resin, <20 μ in size. This particle size boosts the exchange kinetics quite nicely: the bead half-time removal reaction is between 30 and 45 minutes, whereas the pulverized resin reacts in under 10 minutes. [19] The pulverized resin is removed downstream by way of polypropylene hollow fiber ultrafilters. Due to the high speed kinetics, excellent absorption of B is achieved with very little resin and the material does not need to be reused. The application was a small scale test at Izmir, Turkey, but the claim was that in terms of B removal, it was superior to fixed bed ion exchange.

The second evolutionary branch uses a continuous contacting of resin beads and permeate. Continuous ion exchange systems have been pursued since the inception of ion exchange. If successful, they would overcome one of the limitations of resin-based systems identified earlier: batch/cycling operation. There are many ingenious geometries available to achieve continuous ion exchange. Very early in the development of ion exchange two designs predominated: the Asahi Process and the Higgins Loop. There are no known Asahi Process units left in North America. The Higgins Loop technology was acquired by Tetra Process Technologies in 1991, which was then purchased by Severn Trent. [20] Current applications include ammonium nitrate recovery from fertilizer plant waste waters, and the purification of phosphoric acid and silicic acid. There are no applications touching on desalination.

Another, older continuous designs use a carousel type arrangement of several dozen smaller columns of resin which rotated between metal plates containing service and regeneration zones. This was pioneered by ISEP, currently owned by Calgon. [21] Leaks and high maintenance were noted in these systems. The newest variant uses a *stationary* circular geometry of several dozen smaller columns all hard piped to a central drum which contains a rotating plate or valve assembly. One version of this is shown in Figure 11.6. The piping from the central drum goes to each of the stationary resin columns. [22]

Instead of having the columns rotate, the valving in the central drum rotates, sending the proper service and regeneration flows to the columns. There are far less leaks and an alleged reduced maintenance for the central drum geometry. We note the small size of the application in the picture.

This brings up an important observation regarding novel geometries for ion exchange applications, especially continuous processes. There are numerous lab- and pilot plant-scale studies for

Courtesy: www.puritech.be

Figure 11.6 Central drum with internal valve rotation.

continuous resin processes in the literature, some dating back decades. Many are patented. But for all the claims, citations, and patents, there is a noticeable lack of commercially operating, continuous ion-exchange systems. The few systems in operation are small, at least by desalination standards. On the other hand, the commercially successful, larger scale resin operations, such as B-removal, use conventionally designed, batch-sequenced vessels.

11.7 New Vessel Designs

There has been a sea change in conventional ion exchange vessel design, matching, in fact, a major change in resin bead sizing. Both factors can strongly improve the economics of using resins in desalination systems, not as the "main event," of course, but in the pre-softening and post-RO polishing positions. The change in vessel design was needed to operate beds in counter-current fashion more easily.

Older technology vessels had resin in the bottom half with the top half empty. This was a holdover from pressure filter designs, which required expansion space to purge the filter medium of particulates. Likewise, since resins are good filters, the first step in the regeneration was to backwash the bed to release any particulates filtered out by the beads. The backwash also served to unpack the bed which was likely pressed down during service by the high service flows over many hours. And, thirdly, backwash also helped to classify the resin bed by bead size: the larger beads would migrate to the bottom, smaller beads would migrate to the top, and good "open space" or void volume would be preserved up and down the bed.

Within the resin chemistry industry, it was well known that a slight gain in regeneration efficiency and a potentially much lower leakage from the columns can be obtained by introducing service and regenerant in a counter-current fashion. There have always been two options for counter-current ion exchange: Service Up, Regen Down, and Service Down, Regen Up.

A problem arises in counter-current operation with service down and regeneration up in conventional vessels. The upflowing acid or caustic tends to backwash the bed. If the bed levitates in the slightest, resin beads can move vertically, destroying the ionic layering effect which drives the benefits of counter-current regeneration. Clever methods have been developed over the years to "hold the bed down" during upflow regeneration. The most successful has been a flow of water from above the bed through the normal inlets. This is often referred to as a blocking flow, and it works quite well, although it adds significantly to the total volume of spent regeneration wastes. To avoid the latter problem, air is sometimes used to hold the bed in place, but the use of a water blocking flow is far more prevalent.

The radical change in vessel design was the introduction of the packed bed, which eliminates the half empty space in the older technology vessel design. See Figure 11.7.

There are clear advantages to a packed bed design:

- bed is held between two nozzle plates; resin cannot move during service or regeneration, preserving ionic layering, the key to successful counter-current operation
- vessel is smaller and less expensive for same volume of resin
- only nozzle plates are needed; no complicated laterals

And there is one glaring disadvantage: a packed bed vessel cannot be backwashed. As the lead vessel, we often see packed cation beds fouled with silt, debris, and other particulates from upstream

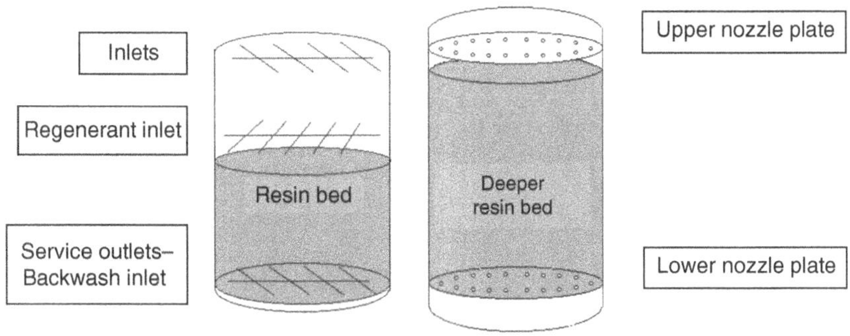

Figure 11.7 Conventional vessel and packed bed vessel.

equipment, especially in systems using surface water supplies. Some packed beds come with an external backwash or maintenance tank, but the resin has to be sluiced from the packed bed to the tank and back again, an operation which many plant personnel do not like to do. Some packed bed systems, victims of accountants, do not even have a maintenance tank! Be warned: The Achilles Heel of packed beds is particulates.

11.8 New Resin Bead Design

Almost contemporaneous with the introduction of packed beds, Dow pioneered the manufacture of Uniform Particle Size (UPS) resins. The older technology resins are made in a suspension polymerization process which first suspends the monomers (styrene and divinyl-benzene) in water to make a dispersion of monomer droplets. This is done in a batch reactor equipped with a paddle stirrer. Polymerization is initiated when the droplet size is correct, and each droplet polymerizes to a bead. The bead is then functionalized into a cation or anion exchanger. The key here is that the initial dispersion of droplets is inherently Gaussian or bell-curve shaped. The resulting resin batch will have the same particle size dispersion: mostly medium sized resin, with some smaller and some larger beads. The nominal commercial product extends from 0.3 mm to about 1.3 mm, as shown in the left-hand side of Figure 11.8. They are also termed hetero-disperse.

The newer manufacturing method uses a continuous process flow, rather than batch, and produces identically sized monomer droplets. When these are polymerized, the result is identically

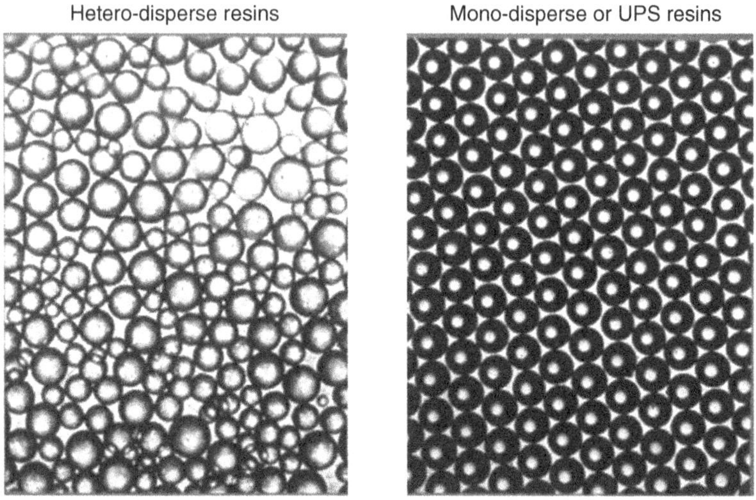

Figure 11.8 Hetero-disperse and mono-disperse or uniform particle size (UPS) resins.

sized resin beads, UPS, or mono-disperse materials, as seen in the right-hand picture in Figure 11.7. There are several advantages to UPS resins:

- the absence of large beads speeds up the final rinse step
- the absence of small beads reduces the flow-induced pressure drop through the bed
- the lower pressure drop allows much deeper beds than with hetero-sized resins
- there is an approximately 10% increase in volume capacity due to optimal packing of the uniformly sized beads

The main disadvantage used to be the higher price of UPS materials, but the prices have come down over the past fifteen years and are now virtually on a par with hetero-sized resins. In fact, some manufacturers are changing their supply strategies to offer only UPS materials in the conventional water treatment markets in certain geographies.

The combination of packed bed vessel design and UPS resins represents the state of the art in ion exchange technology for municipal-scale flowrate applications.

11.9 Conclusion

The use of ion exchange in desalination has had a fascinating history. Like the proto-star Jupiter in our solar system, the early potential star in the resin-based Desal Process never ignited. Had it burst into stardom (like Jupiter in Peter Hyams' *2010*), it still would have been rapidly eclipsed by the rise of commercially viable RO systems.

Resins can never replace membranes as the "main event" in desalination, but they do have a role in pre- and post-RO treatment. Ion exchange softening of RO feeds significantly reduces hardness and subsequent membrane scaling. After membrane desalination, ion exchange has a solid role in specific contaminant reduction, notably the current, selective removal of boron from sea water RO permeates. Any major, future development using ion exchange in desalination systems will likely focus on the removal of other permeate contaminants by way of the unlimited range of active groups which can be functionalized onto polymeric beads.

References

1. Rohm and Haas Amberlite/Duolite, *Ion Exchange Resins Engineering Manual*, Introduction, pp. 24–25, no declared publishing date.
2. Abandoned Mine Reclamation Clearing House, Treatment Options, at http://www.amrclearinghouse.org, pp. 18–21.
3. R. Kunin, *Amber-Hi-Lites*, Spring, 1963
4. Information is also available at the Dow Answer Center, dowmail1@mailwc.custhelp.com.
5. S.N. Zaharna, Desalting and Purifying Water by Continuous Ion Exchange, US Patent 3607739, assigned to Gulf Degremont, Inc., September 21, (1971).
6. A.K. SenGupta and S. Karkar, Brackish and Sea Water Desalination Using a Hybrid Ion Exchange-Nanofiltration Process, US Patent 7901577, assigned to Sengupta and Karkar, March 8, (2011).
7. W.E. Bornak, *Ion Exchange Deionization for Industrial Users*, Littleton, CO, Tall Oaks Publishing, (2003).
8. Among many, A.K. SenGupta and S.Sarkar. See [5].
9. G.K. Kunz, *Desalination* **3**, 363–372 (1967).
10. T. Vermeulen, B.W. Tleimat and G. Klein, Ion-exchange pretreatment for scale prevention in desalting systems, *Desalination* **47**, 149–159 (1983).

11. J.D. Pless, *Industrial and Engineering Chemistry Research* **45,** 4752–4756 (2006).
12. Among many, http://www.eichrom.com.
13. There are hundreds of references on this application. A good place to start is cfpub.epa.gov/safewater/arsenic/
14. C. Jacob, *Desalination* **205**, 47–52 (2007).A good place to start on this topic is N. Hilal, et al., *Desalination,* **273**, 23–35 (2011).
15. M. Busch, W.E. Mickols, S. Jons, J. Redondo and J. de Witte, *Boron Removal in Seawater Desalination,* IDA World Congress, BAH03-039, Bahrain, (2003).
16. Kabay N, E. Guler and M. Bryjak, Boron in seawater and its separation—a review *Desalination,* **261**, 212–217 (2010).
17. WHO *Guidelines for Drinking-Water Quality,* 4th edition, 2011.
18. N. Nadav, Boron Removal From Seawater Reverse Osmosis. Permeate Utilizing Selective Ion Exchange Resin. *Desalination* **124**, 131–135 (1999).
19. E. Güber, et al., *J. Membr. Sci* **375**, 249–257 (2011).
20. Tetra Process Technologies at SevernTrentservices.com
21. http://www.calgoncarbon.com
22. http://www.puritech.be

12

Electrosorption of Heavy Metals with Capacitive Deionization: Water Reuse, Desalination and Resources Recovery

Pei Xu[1*], Brian Elson[2] and Jörg E Drewes[3]

*Corresponding author: [1]*New Mexico State University, Las Cruces, New Mexico, USA*
[2]*The Water Company, Pueblo, Colorado, USA*
[3]*Urban Water Systems Engineering, Technische Universität München, Germany"*

Abstract
Capacitive deionization (CDI) is an emerging electrosorption process that simultaneously produces deionized water for reuse and recovers valuable heavy metals from waste streams. Bench-scale experiments were conducted to investigate the electrosorption selectivity of metals and cyanide by activated carbon electrodes. Over the course of 158 treatment cycles, CDI remained high efficiency in removing heavy metals and meeting drinking water standards. Within 25 to 30 minutes retention time, the percent removal of conductivity achieved 97.9–96.4% and the effluent conductivity increased slightly from 13 µS/cm in the beginning of experiments to 22 µS/cm at the end of the testing period. Activated carbon electrodes adsorbed trivalent and tetravalent ions preferentially compared to divalent ions. Uranium-238 (IV), iron (III) and chromium (III) were reduced by 98.9±0.6%, 99.9±0.06% 99.6±0.3% while nickel (II), copper (II), lead (II) and zinc were retained 89.0±7.3%, 94.8±2.7%, 94.0±3.3%, 85.2±9.0%, respectively. Nearly all the metal concentrations in the deionized water met the USEPA drinking water standards, except lead (II). CDI reduced the lead concentration from 19 mg/L in feed to 0.4–2 mg/L in treated water that exceeds the drinking water standards of zero lead. Cyanide removal was affected by varying pH in the electrolyte and the electrosorption effectiveness declined as cyanide ion converted to hydrogen cyanide at low pH.

Jane Kucera (ed.) Desalination, (521–548) 2014 © Scrivener Publishing LLC

Keywords: Capacitive deionization, wastewater treatment, removal of heavy metals, cyanide removal, water reuse, electrosorption, desalination

12.1 Introduction

12.1.1 Removal of Heavy Metals from Aqueous Solutions

The release of heavy metals into public sewer system, aquatic and terrestrial environment has posed major risks to human health and adverse effects on ecosystem. Because of the high mobility of metals in environment, heavy metals tend to be absorbed by living organisms and accumulated in plants, aquatic life and human bodies. Mining process is a major source for the contamination of heavy metals in environment. The pathways include runoff of mere soil or rock debris, discharge of large amounts of water produced from mine drainage, mine cooling, aqueous extraction and other mining processes. Other sources contributing to heavy metal water contamination include metal plating, smelting, battery manufacture, tanneries, petroleum refining, printing, and photographic industries.

A number of technologies have been employed to remove heavy metals in water and wastewater. Chemical precipitation and coagulation-flocculation operating at high pH are often considered simple and low cost methods for treating wastewater having high metal concentrations [1, 2]. Large amount of chemicals is required to reduce the contaminants to marginal level, and excessive production of sludge requires proper disposal to reduce long-term environmental impact. In general, precipitated effluent is unable to meet stringent discharge limits; subsequent treatment using other physico-chemical processes is obligated in order to comply with discharge regulations.

Ion exchange is an industrial standard widely used to remove heavy metals from wastewater due to its high efficiency, fast kinetics, less waste volume produced and compliance with strict discharge requirements [3, 4]. During ion exchange process, the metal ions in waste solution are exchanged by the ions contained in the fixed matrix of ion exchange resin. Depending upon the characteristics of the ion exchange resins, operating pH is often acidic ranging from pH 2 to 6. Once the ion exchange resin reaches capacity, it is regenerated with a concentrated electrolyte such as a strong acid.

The heavy metals of value in concentrated eluate may be recovered with suitable reagents [5]. The wastewater usually needs extensive pretreatment to prevent fouling and scaling on ion exchange resins.

Adsorption has also been proved an effective and economical method for heavy metal wastewater treatment. The adsorption process offers high flexibility in design and operation, and in many cases production of high-quality treated effluent [6]. A broad spectrum of sorbents have been investigated, including the universal adsorbent - activated carbon [7], carbon nanotubes [8, 9], derivatives of carbon products (such as additives and activated carbon composite) [10–13]. In recent years, inexpensive alternatives to activated carbon as sorbent to heavy metals have been studied, such as natural zeolite [14, 15], bentonite [16], and various chemically modified plant wastes [17–21]. The limitations of adsorption method include the low adsorption capacity, low selectivity, and slow pore diffusion of ions. In most cases adsorption is irreversible and adsorbents laden with heavy metals have to be disposed of.

In recent years, membrane processes have been increasingly used for removal of heavy metals present in the form of suspended solids and dissolved ionic species. Microfiltration (MF) and ultrafiltration (UF) utilizes permeable membranes to separate particulate and colloidal heavy metals. To improve the removal efficiency, MF or UF membranes can be modified to improve metal-binding and complexation [22–26], or combined with other processes such as adsorption, flotation and coagulation [16, 27–30]. Nanofiltration (NF) and reverse osmosis (RO) membranes are used to remove dissolved metal ions. The rejection efficiency depends on feed pH, pressure, permeate flux, metal concentrations, and membrane properties such as membrane pore size and surface charge. For loose NF membrane, the rejection increased when the feed pH was below the isoelectric point due to increased electrical repulsion between the positively charged metal ions and membrane surface [31]. However, the rejection of heavy metals as negatively charged species such as As(III) exhibited poor rejection by loose NF membrane, which was also attributed to larger molecular weight cut-off of the membrane. Using ultra-low-pressure RO membrane (ULPRO), over 95% of removal was achieved for separating divalent (Cu^{2+}, Ni^{2+}) and hexavalent (Cr^{6+}) heavy metals from wastewater [32]. Despite the high removal efficiency, membrane scaling is a considerable challenge and rigorous pretreatment is required for treatment of heavy metal contaminated wastewater.

Electrodialysis (ED) is another membrane process that uses charged ion exchange membranes to separate charged heavy metals from aqueous solution [33–35]. A pilot study showed that ED effectively reduced Cr(VI) concentration from 10 mg/L to 0.1 mg/L [36]. In general, treatment efficiency is a function of initial metal concentration, flow rate, operating pH, temperature and electrical current density [33–36]. However, ED is not effective in providing highly purified water with metal concentrations down to low microgram per liter level. Recently Mohmoud and Hoadley evaluated the feasibility of a bench-scale hybrid system combining ion exchange (IX) with ED to remove heavy metals from simulated dilute industrial wastewater [37]. The IXED process was found to have the combined advantages of the two individual techniques and the energy consumption was less than 32–44% of the energy required by ED process.

Although many techniques have been employed and studied for the treatment of heavy metal contaminated water, the ideal treatment should be not only suitable, appropriate and applicable to the local conditions, but also be able to meet discharge or reuse standards as well as recover heavy metals as valuable resources. This study investigated an innovative capacitive deionization method to purify wastewater to meet drinking water standards and recover heavy metals for potential resources recovery.

12.1.2 Capacitive Deionization

Capacitive deionization (CDI) is an electrosorption process in which ions are adsorbed onto the surface of porous electrodes under an electrical field. When aqueous solution is flowing between the porous electrodes, the positively charged ions such as calcium, magnesium, sodium, copper, lead adsorb onto the electric double layer (EDL) of negatively charged electrodes (cathode) while negatively charged ions such as sulfate, chloride, nitrate, cyanide, arsenate adsorb onto the EDL of positively charged electrode (anode). The major mechanisms related to the removal of charged constituents during water treatment are physisorption, chemisorption, electrodeposition, and/or electrophoresis. Once the adsorption capacity of the electrodes is achieved, adsorbed ions can desorb from the surface of the electrodes by eliminating (or reversing) the electric field, resulting in the regeneration of the electrodes.

During the last few decades, substantial research has been conducted to develop various electrode materials and configurations to enhance the CDI performance. A critical component of a CDI system is electrodes. The electrode materials should have high electrical conductance, high specific surface area, high chemical and electrochemical stability, and easily shaped based on design requirements [38]. The electrode materials that have been investigated include carbon aerogels [39–43], activated carbon [44, 45], graphene [46–48], nanostructured carbon cloth [49–51], nanoporous silicon dioxide or alumina surface-doped with magnesium [52], and carbon nanotubes and nanofibers [53–58].

Several configurations have been designed for CDI units. The first pseudo-continuous operation was achieved by Oren and Soffer [38] with an electrochemical parametric pumping column for the separation of salt from water. The CDI system was consisted of two identical high-surface-area carbon black electrodes separated by an electrically inert porous barrier and filled with influent. Each electrochemical parametric pumping cycle comprised four consecutive operations – (1) adsorption to separate salts from the solution to the electrode phase, (2) forward axial pumping of the solution, (3) desorption, back transfer of the salts to the solution, and (4) backward axial pumping of the solution, now containing an excess of salt.

Frame and plate structure is the most commonly used configuration for the CDI systems because of its simplicity, easy to design and build [40, 59]. A CDI module is composed of electrode pairs separated by conductive spacers and non-conductive screens to prevent short-circuiting. The electrode stack is held together by a metal or plastic frame, through which the treated solution flows continuously. A complete treatment cycle may include four stages: pre-charge the electrodes by an external power supply; adsorption to separate ions from water; regeneration by removing the electrical potential and the adsorbed salts diffusing back to the solution; and final rinse to flush out the concentrate solution [40].

A spiral wound CDI configuration was patented by Shiue et al. [60]. In the flow-through capacitors (FTC), two electrodes and two porous inert dividers are wound together into a hollow-center roll. A liquid-feeding pipe is inserted to the central opening for delivering fluids to the FTC. Nanoparticles of hydrated iron compound with Fe_3O_4 as the main component or its composite powders are used as the active materials for the electrodes.

Membrane capacitive deionization (MCDI) is a modification of CDI by combining ion-exchange (IX) membranes with the carbon electrodes [45, 53, 61–65]. Specifically, an anion exchange membrane is placed in front of the anode while a cation exchange membrane is positioned in front of the cathode. MCDI is reported to provide large improvements in desalination efficiency because IX membranes block co-ions from leaving the electrodes, thus increasing the salt removal efficiency of the process. In addition, counter-ions can be more fully flushed from the electrode region during regeneration process where a reversed voltage is used for releasing ions, thereby increasing the driving force for ion removal in the next cycle [64]. A bench-scale testing by Kim and Choi showed that the salt removal of the MCDI cell was enhanced by 33–56% compared to the CDI cell, and current efficiencies increased from 36–43% for CDI cell to 84–91% for MCDI cell [62].

CDI and MCDI have been employed to investigate the treatability of different types of water, including brackish water [40, 66], hard water [67], oil contaminated water [45], thermal power plant wastewater [63], RO concentrate from a wastewater reclamation plant [68], and produced water generated during oil and gas production [69].

CDI processes demonstrate several advantages such as a simple, modular design, no need for a high-pressure pump or heater, as well as operation at ambient conditions and low voltages. Previous studies have shown that CDI technology is cost competitive to RO at low TDS range (<3,000 mg/L) [69–71]. Field testing of CDI treating produced water showed that low fouling is a considerable advantage of CDI compared to membrane processes [40]. Therefore, the pretreatment for CDI is minimum (e.g. cartridge filter to remove particles), and the chemicals required for scaling control and cleaning are optional and site dependent.

Although substantial research has contributed to improve the efficiencies of CDI and related processes, few studies have focused on treating heavy metal contaminated wastewater. CDI is attractive for achieving multiple purposes simultaneously – adsorption of valuable heavy metals onto carbon electrodes for potential recovery; and production of highly purified water for water reuse. After extraction of heavy metals, the regenerated electrode material will be continuously used for the next cycles of treatment; therefore, no waste or residual is generated. Because of low fouling and scaling propensity of carbon electrodes, CDI is a promising alternative to membrane and ion exchanges processes for desalination of difficult to treat wastewaters.

The objectives of the study are to investigate the selectivity of a CDI process for the treatment and recovery of wastewater contaminated by heavy metals, radionuclides (uranium-238), and cyanide. Bench-scale experiments were conducted using a CDI cell equipped with porous activated carbon electrodes. A synthetic solution was prepared to represent acid mine drainage with major contaminants of chromium (Cr^{3+}), copper (Cu^{2+}), iron (Fe^{3+}), nickel (Ni^{2+}), lead (Pb^{2+}), zinc (Zn^{2+}) and uranium-238 (U^{4+}). Additional experiments were conducted to evaluate the electrosorption capacity of CDI to remove cyanide in wastewater.

12.2 Experimental Methods

12.2.1 CDI Treatment System

The experimental apparatus consisted of a 7-gallon feed water reservoir, a CDI cell, a DC power supply (Sorensen LH110-3, Eagar, Malaysia), and a customized data log and control system (Figure 12.1). The CDI testing cell was divided into three chambers and separated by two fine meshes. About 750 grams and 540 grams of activated carbon powders were packed tightly in the electrode chambers and compressed against the anode and cathode current collectors, respectively. The electrolytic chamber held 300 mL wastewater during treatment cycle. The system was automatically controlled, and all data (e.g., treatment time, voltage, current, resistance) were recorded at 1-minute interval and downloaded as data files. Once the resistance of the treated water reached the set-point of the control system, the water was drained for sampling and stored in an effluent container. The CDI cell was refilled with raw water from the feed water reservoir by gravity and a new treatment cycle started. The positively charged ions passed through the mesh and adsorbed onto the carbon powders in the cathode chamber using stainless steel as current collector. The negatively charged ions passed through the mesh and adsorbed onto the carbon powders in the anode chamber using carbon plate as current collector. To avoid gas and corrosive solutions accumulating in the adjacent areas of the anode and the cathode, a small amount of water was drained continuously from each side of the electrode chamber.

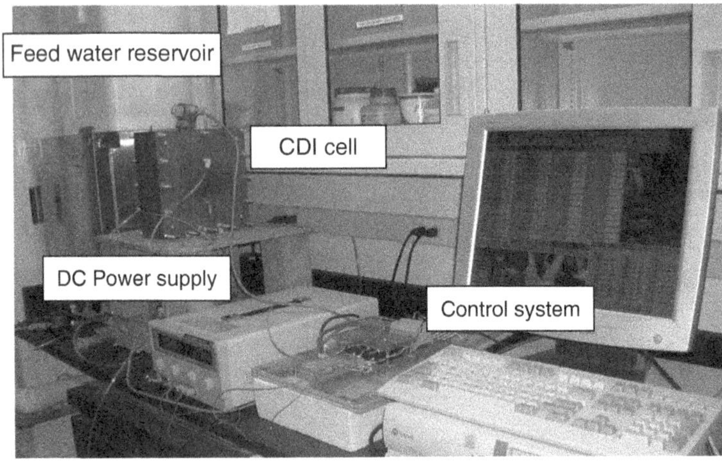

Figure 12.1 Photograph of the CDI treatment system.

The removal process was conducted by applying a constant current of 0.5 Amps across the cell until the cell potential reached 25 Volts. The cell potential was optimized through previous experiments to achieve high salt removal rate and reduce treatment time. The cell potential remained constant at 25 Volts until the retention time reached a desired test time (e.g., 5, 10, 15, 20, and 30 minutes), or a desired effluent water quality determined by the cell electrical resistance (e.g., voltage/current = 500 ohms).

12.2.2 Feed Water Quality and Sample Analysis

To assess the removal efficiency of metals by the CDI technology, a synthetic solution was prepared in the laboratory with deionized water and salts of the metals for evaluation. Ions selected for experiments included chromium (Cr^{3+}), copper (Cu^{2+}), iron (Fe^{3+}), nickel (Ni^{2+}), lead (Pb^{2+}), zinc (Zn^{2+}), and uranium-238 (U^{4+}). Table 12.1 summarizes the water matrix and drinking water standards for each of the ions. Reagents for the experiments were all certified ACS grade or above, including cupric nitrate, nickelous nitrate hexahydrate, zinc nitrate, chromium nitrate, ferric nitrate 9-hydrate, lead nitrate, and high purity uranium 238 in nitric acid solution. The feed water solution had a pH value of 2.73 and conductivity 615 μS/cm, representative of acid mine wastewater. It should be noted that all the metal salts in the study were soluble in the testing conditions. A series of 158 experiments was conducted to evaluate the effects of retention time and the properties of metals on the electrosorption capacity of the carbon electrodes. In total, 42 liters of metal contaminated solution were processed.

Water samples were taken at varying time intervals within different treatment cycles, and throughout the 158 testing runs to compare the treatment efficiency. Samples of the drains from anode and cathode chambers were collected between cycles #32 to #44. For quality assurance and quality control, deionized water used to prepare feed water solutions was taken as blank sample for analysis. No metals were detected in the blank sample.

The metals were analyzed by an inductively coupled plasma optical emission spectroscopy (ICP-OES) (Optima 5300, Perkin-Elmer, Fremont, CA). Water samples were filtered using a Pall Supor® 0.45 μm hydrophilic polyethersulfone membrane prior to analysis. Samples with uranium concentration below the detection limit of ICP-OES were further analyzed by inductively coupled plasma mass spectrometry (ICP-MS, Perkin Elmer Nexion 300Q, Waltham, MA). Conductivity and pH were measured by a conductivity meter (Model 431–61, Cole-Parmer, Vernon Hills, IL), and a pH meter with temperature gauge (Oakton 300 Series, Eutech Instruments, Singapore).

The relative constant effluent conductivity measurement indicated the CDI stack capacity remained fairly stable throughout the experiments. There was no poisoning or degradation of activated carbon electrode under these operating conditions. However, metal precipitation was observed on the mesh separators, resulting in slower refill of the raw solution to the electrolyte chamber. The clogging on the mesh also affected slightly the water transport from the middle electrolyte chamber to cathode and anode chambers.

Additional experiments were conducted to assess the removal efficiency of cyanide by the CDI technology (Table 12.1). The synthetic solution was prepared with deionized water, ACS grade sodium chloride and potassium cyanide. The pH of the feed water solution was adjusted to 11.0 to keep cyanide solution stable using 5N sodium hydroxide solution. The conductivity of the feed solution was 425 μS/cm. Eight runs were carried out, and the pH and conductivity of the effluents were monitored. The samples collected for total cyanide analysis were adjusted to pH above 12 for preservation. The samples were then stored at 4°C and analyzed within 16 hours of holding time. Hach instrument DR2800 was used to analyze total cyanide with the method 8027 (Pyridine-Pyrazolona

Table 12.1 Experimental water matrices and drinking water regulations.

Ions	Measured concentration in feed solution (mg/L)	Chemicals used	Analytical detection limit (mg/L)	Primary MCL* (mg/L)	Secondary standards (mg/L)
Metal Experiments					
Cr^{3+}	11.15±0.10	$Cr(NO_3)_3$	0.0057	0.1	
Cu^{2+}	15.56±0.07	$Cu(NO_3)_2$	0.0042	1.3	1.0
Fe^{3+}	16.09±0.80	$Fe(NO_3)_3\ 9H_2O$	0.0056		0.3
Ni^{2+}	18.74±0.06	$Ni(NO_3)_2\ 6H_2O$	0.0039		
Pb^{2+}	19.03±0.13	$Pb(NO_3)_2$	0.0123	0–0.015	
Zn^{2+}	12.09±0.01	$Zn(NO_3)_2$	0.0034		5
U^{4+}-238	1.85±0.08	UO_2-238 in nitric acid solution	0.050 (ICP) 0.0006 (ICP-MS)	0.03	
Cyanide Experiments					
CN^-	2.03±0.04	KCN	0.001	0.2	

Note: * USEPA National Primary/Secondary Drinking Water Regulations, Maximum Contaminant Levels (MCL).

Method). As gaseous hydrogen cyanide may form during the testing, a vacuumed funnel was installed on the top of the testing cell to collect the gas escaped from the three chambers (electrolytic chamber and two carbon electrode chambers). The collected gas was absorbed in 100 mL 5N NaOH solution. After the testing, the total cyanide concentration in the solution was analyzed as well.

12.3 Results and Discussions

12.3.1 CDI Voltage and Current Profiles

Figure 12.2 presents the voltage and current profiles of the testing runs of #3, #36, and #155, during treatment of metal contaminated water. The potential between the carbon and stainless steel current collectors increased gradually after applying the constant current. When the cell voltage achieved 25 Volts, the monitored current rapidly decreased. Both the increasing potential and decreasing current values resulted in an increase in resistance, indicating a desired decrease in the electrolyte conductivity, which was caused by driving the anions and cations in feed water into charged carbon adsorbents under the applied electrical current. With the increasing runs of treatment cycles, the time to achieve the desired voltage of 25 Volts decreased, indicating the overall electrical conductance of the carbon materials increased as a result of metal adsorption. To reach the desired resistance of 500 ohms, corresponding to 12–20 µS/cm

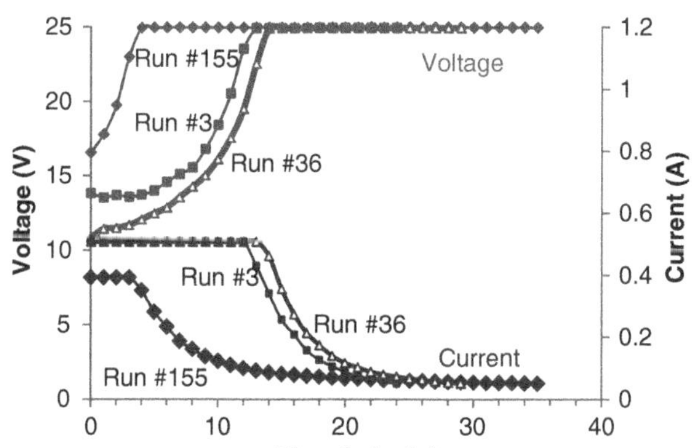

Figure 12.2 Voltage and current profiles during the treatment cycles of metals experiments.

in treated water, the energy demand in cycle #155 was approximately 30% lower than in the cycles of #3 and #36. It implies that the electrode efficiency was improved due to electrosorption of heavy metals.

12.3.2 Removal of Heavy Metals from Electrolytes

The kinetics of metal removal was examined by taking water samples at 0, 5, 10 15, 20 and 30 minutes of treatment in the beginning (Runs #5 to #9) and at the middle of the testing (Runs #72 to #76). The concentrations of the metal ions and conductivity at different treatment time are summarized in Table 12.2. The variation of conductivity as a function of retention time is shown in Figure 12.3. The percent removal of the heavy metals is illustrated in Figure 12.4.

The conductivity profiles in Runs #5–9 and Runs #72–76 show a similar removal trend (Figure 12.3). After 5 minutes retention, the conductivity of the treated water increased from 615 µS/cm in the solution to 700 µS/cm and 638 µS/cm for Run #5 and Run #72, respectively. This increase of conductivity in the electrolyte was likely attributed to desorption of pre-adsorbed ions (during previous runs) in the carbon electrodes back to bulk solution at low electrical potential. After 10 minutes treatment, the conductivity of the treated water decreased to 569 µS/cm and 535 µS/cm for Run #6 and Run #73 respectively. After 15 minutes, Run #7 showed a better salt removal than Run #74. In 30 minutes, the conductivity reached 13.5 µS/cm and 28.3 µS/cm, achieving 97.8% and 95.4% of removal for Run #9 and Run #76, respectively. The results indicate that, with the loading of heavy metals in the porous carbon electrodes, the electrosorption rate of ions reduced slightly.

The percent removal of all the metals exhibited the same trend, increasing with longer treatment time (Figure 12.4). However, the metal removal rates were higher in the beginning of testing for the Runs #5 to #9 than during later Runs #72 to #76. During the Runs #5 to 9, all metals achieved over 62% removal in 5 minutes although the conductivity of the electrolyte increased slightly due to the desorption of the pre-adsorbed ions from the carbon electrodes to the bulk electrolyte solution. The percent removal of all the metals increased to greater than 81%, 94%, and 98% in 10, 15 and 20 minutes contact time, respectively. During the Runs #72 to #76, the removal decreased to between 9% to 31% for the 5 minutes contact time, with copper having the lowest removal rate (8.7%), followed

Table 12.2 Summary of heavy metals concentrations at different treatment time.

Run No.	Run time (min)	Conductivity (μS/cm)	Cr (mg/L)	Cu (mg/L)	Fe (mg/L)	Ni (mg/L)	Pb (mg/L)	Zn (mg/L)	U-238 (mg/L)
Feed	0	615	11	15.55	15.52	18.7	18.94	12.09	1.79
#5	5	703	3.89	5.66	4.44	6.79	5.88	4.60	0.66
#6	10	569	1.67	2.43	2.50	2.96	2.28	1.90	0.34
#7	15	346	0.61	0.88	0.88	1.10	0.80	0.69	0.12
#8	20	103	0.18	0.31	0.18	0.34	0.38	0.22	0.043
#9	30	13	0.059	0.29	0.017	0.32	0.36	0.22	0.005
Feed	0	620	11.22	15.60	16.65	18.78	19.12	12.08	1.91
#72	5	638	9.23	14.24	11.54	16.11	14.74	10.55	1.52
#73	10	535	4.66	7.22	5.26	8.38	6.99	5.48	0.86
#74	15	410	2.00	3.41	1.12	4.09	3.62	2.68	0.32
#75	20	183	1.29	2.61	0.55	3.17	3.19	2.06	0.15
#76	30	28.3	0.032	1.47	0.010	2.63	2.17	2.25	0.010

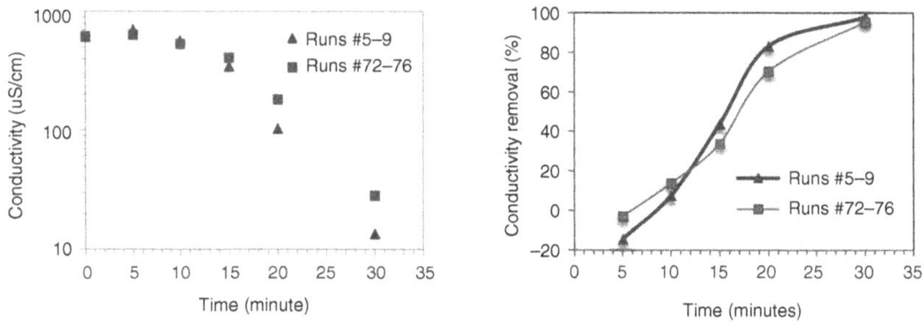

Figure 12.3 Variation of conductivity of the electrolyte during experimental Runs #5–9 and #72–76.

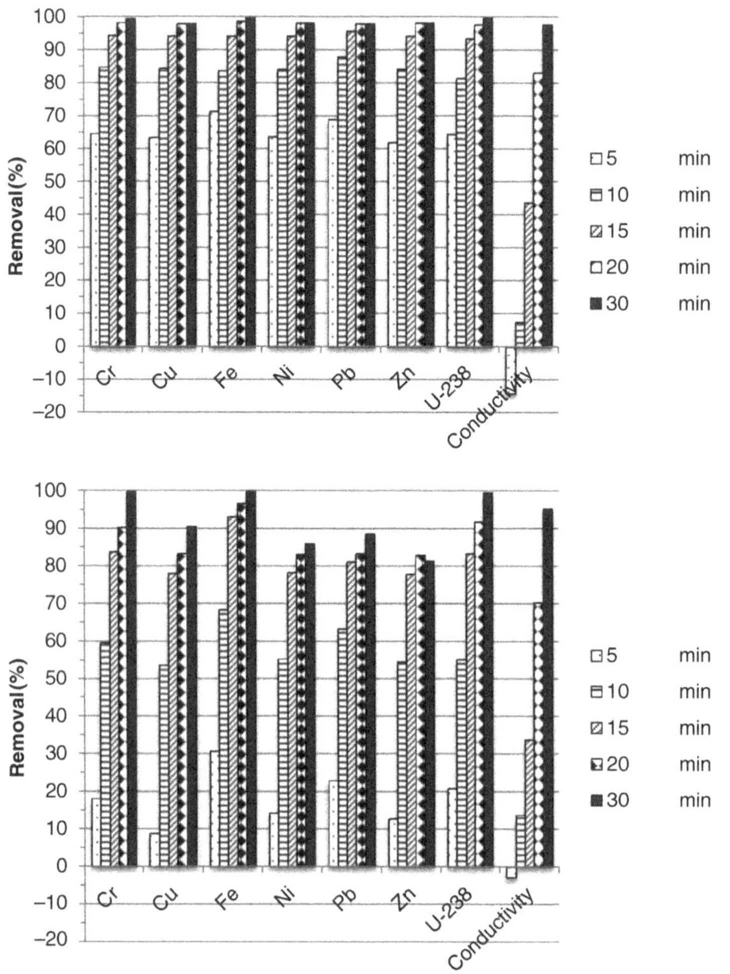

Figure 12.4 Removal percentage of the heavy metals from electrolytes during experimental Runs #5 to #9 (top) and Runs #72 to 76 (bottom).

by zinc and nickel. The metal removal rate increased to between 53% and 68% at 10 minutes contact time, and between 55% and 93% at 15 minutes contact time. At 30 minutes, the metal removal was greater than 98% for the beginning Run #9 while the removal was in the range of 81%–99.5% for later Run #76.

Figure 12.4 shows that the electrosorption capacities of iron, chromium and uranium-238 were consistently higher than those of zinc, nickel, lead and copper. At 30 minutes contact time, the removal of Fe^{3+}, Cr^{3+} and U^{4+}-238 reached 99.9%, 99.5% and 99.7% in the Run #9; and 99.9%, 99.7%, and 99.5% during the Run #76. The removal efficiency of divalent ions Cu^{2+}, Ni^{2+}, Pb^{2+} and Zn^{2+} was slightly lower than trivalent and tetravalent cations during the beginning of the testing, in the range of 98.1% to 98.3%. The differences between divalent ions and trivalent and tetravalent ions are more remarkable at the later testing period, with the percent removal of Cu^{2+}, Ni^{2+}, Pb^{2+} and Zn^{2+} 90.6%, 86.0%, 88.7% and 81.4%. This infers that the higher removal efficiency in the beginning of the experiments was attributed to the combined adsorption capacity of the virgin activated carbon material and the effect of electrosorption of EDL in the supercapacitor. With the sorption capacity of the activated carbon exhausted, the total removal efficiency decreased and was mainly attributed to electrosorption.

As seen in Figure 12.4, the treatment efficiency of CDI is determined by the adsorptive species, i.e, the chemical properties of the solutes relevant to the adsorption process. More specifically, differences in the extent of electrosorption are associated with the chemical state of the metals in the electrolyte solutions. Electrosorption rates of the ions Fe^{3+}, Cr^{3+} and U^{4+}-238 are the highest among all the metal ions tested, which can be attributed to their higher valence of the positively charged ions and small effective radii of 0.62 Å, 0.66 Å, 1.08 Å and 1.08 Å for hydrated ions $Cr(H_2O)_6{}^{3+}$, $Fe(H_2O)_6{}^{3+}$, $U(H_2O)_9{}^{4+}$ and $UO_2(H_2O)_5{}^{2+}$, respectively (Table 12.3).

In addition to the valence, hydrated radius of the ions with same valence affects electrosorption. For instance, the electrosorption selectivity of divalent ions on molar basis decreases with increasing hydrated radius: Ni^{2+} 0.319 mmol (0.72 Å) > Cu^{2+} 0.245 mmol (0.73 Å) > Zn^{2+} 0.185 mmol (0.74 Å) > Pb^{2+} 0.092 mmol (1.20 Å). The percent removal of divalent ions in terms of hydrated radius is however, not conclusive: Cu^{2+} 90.6 % (0.62 Å) > Pb^{2+} 88.7% (1.20 Å)> Ni^{2+} 86.1% (0.72 Å) > Zn^{2+} 81.4% (0.74 Å). This observation is in agreement with the experimental results of Gabelich et al. who reported

Table 12.3 Physico-chemical properties of the studied ions.

Ions	Charge	Atomic weight	Feed concentration (mmol)	Hydrated ions and radius (Å)		
Metal experiments						
Cr^{3+}	+3	52.0	0.214	$Cr(H_2O)_6^{3+}$	0.62	Ref [72]
Cu^{2+}	+2	63.5	0.245	$Cu(H_2O)_6^{2+}$	0.73 (Shannon)	Ref [72]
Fe^{3+}	+3	55.8	0.288	$Fe(H_2O)_6^{3+}$	0.66	Ref [72]
Ni^{2+}	+2	58.7	0.319	$Ni(H_2O)_6^{2+}$	0.715	Ref [72]
Pb^{2+}	+2	207.2	0.092	$Pb(H_2O)_6^{2+}$	1.20	Ref [72]
Zn^{2+}	+2	65.4	0.185	$Zn(H_2O)_6^{2+}$	0.74	Ref [72]
U^{4+}-238	+4	238.0	0.0078	$U(H_2O)_9^{4+}$ $UO_2(H_2O)_5^{2+}$	1.08 1.08	Ref [72]
Cyanide experiments						
CN^-	-1	26.0	0.078		~3.32	Ref [73]
OH^-	-1	17.0	1.000		3.00	Ref [74]
Cl^-	-1	35.5	3.419		3.32	Ref [74]

that the ion selectivity of carbon aerogel electrodes was based on ionic hydrated radius [39].

Figure 12.5 shows the variation of conductivity and the metal concentrations in the treated water throughout the testing period. The electrical resistance setpoint of these experiments was 500 ohms in the experiments, the treatment time increased gradually from 25 minutes in the beginning of the testing to 30 minutes at the end of the experiments. The conductivity of the effluent increased slightly from 13 µS/cm to 22 µS/cm at the end of the testing.

There are no drinking water standards set by the USEPA for nickel in water. Over the course of 158 runs, the nickel concentration in the treated water increased from 0.074 mg/L in the first few runs to between 2–5 mg/L for the Runs #50 to #158. CDI achieved 89.0±7.3% removal of nickel with the feed concentration of 18.74 mg/L.

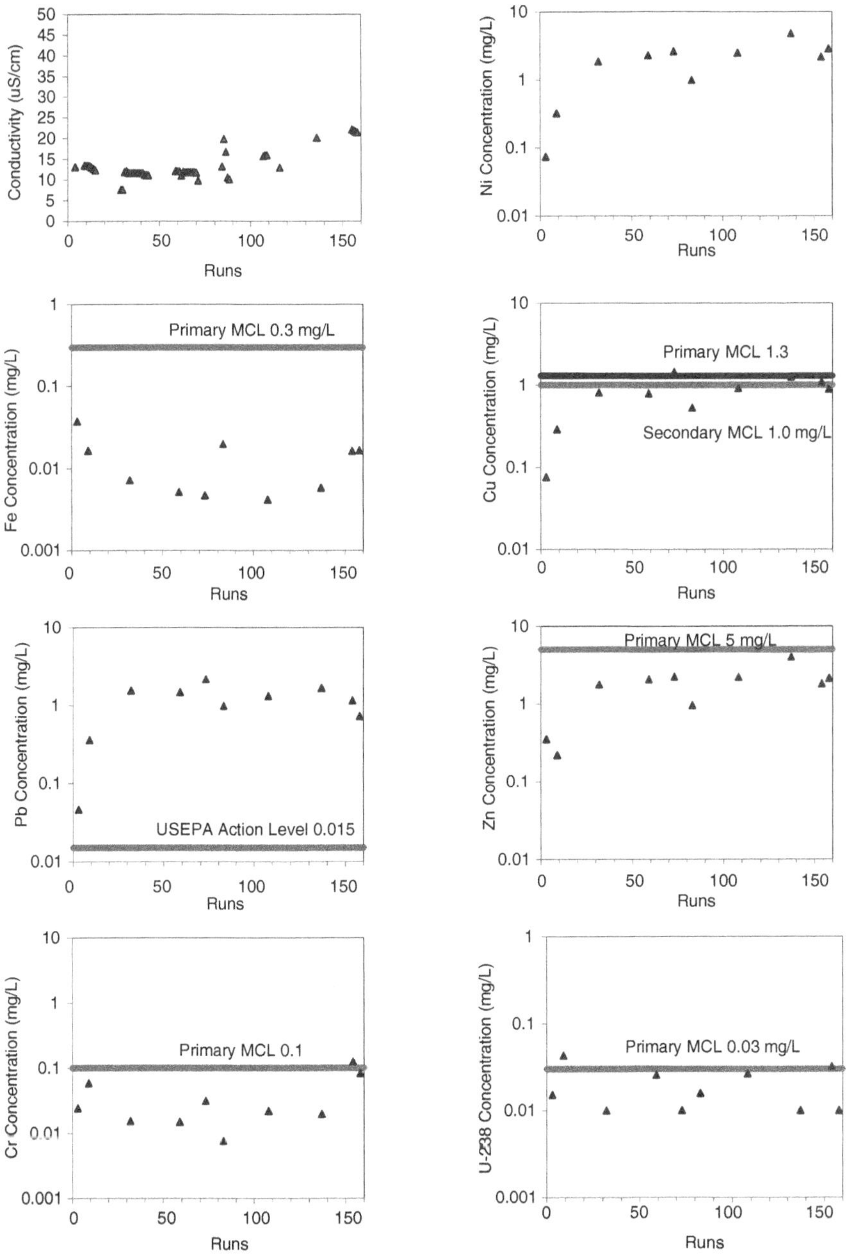

Figure 12.5 Variation of conductivity and metal concentrations in the treated electrolyte solutions throughout the 158 runs (setpoint: 500 ohms, corresponding to 25–30 minutes contact time).

Within 150 runs of the testing, the chromium concentrations decreased from 11.15 mg/L in the feed solution to below the MCL level of 0.1 mg/L regulated in the USEPA Primary Drinking Water Standards. At the end of the testing, the chromium concentration reached close to the MCL. CDI achieved 99.6±0.3% of chromium removal throughout the experiments.

During the first 60 cycles, the copper concentrations were reduced from 15.56 mg/L in the feed solution to below the USEPA Drinking Water Secondary Standards of 1.0 mg/L in the treated water. After 75 runs, the copper concentrations in the effluent exceeded the Primary MCL of 1.3 mg/L. The average removal rate of copper was 94.8±2.7% throughout the 158 runs. In order to meet the desired removal of contaminant from the water, increase of the retention time is required. An alternative option is to have a number of CDI cells arranged in series using the end cells to meet the drinking water standards and the first few cells for recovery of metals from water.

Throughout the testing, the iron concentrations in the effluent were below 0.04 mg/L, much lower than the Primary MCL of 0.3 mg/L. With the feed concentration of 16.09 mg/L, the percent removal of iron by CDI was 99.9±0.06% over the158 runs.

USEPA set a lead concentration of zero in drinking water standards and an action level of 0.015 mg/L. The lead concentration in the treated water was between 0.4 and 2 mg/L exceeding the drinking water standards. With a feed concentration of 19.03 mg/L, CDI achieved 94.0±3.3% removal of lead during the testing.

The zinc concentrations in the effluent increased with the treatment runs from the beginning of 0.2–0.3 mg/L to between 1–4 mg/L in the later runs. The percent removal of zinc by CDI was 85.2±9.0% throughout the testing and the zinc concentration was reduced from 12.09 mg/L to below the Primary MCL of 5 mg/L.

During the experiments the uranium concentrations were reduced from 1.85 mg/L in the feed solution to the range of 0.005 to 0.043 mg/L in the treated water, achieving 98.9±0.6% removal. However, approximately 20% of the effluent samples slightly exceeded the Primary uranium MCL of 0.03 mg/L.

Table 12.4 lists the metal concentrations in the samples collected from the anode and cathode drains. Because metals are positively charged ions and attracted by negative electrode, the metals moving towards the cathode adsorbed on the carbon materials and were removed from water solution. Thus the metals concentrations

Table 12.4 Metal concentrations measured in the electrode drain samples.

Ions	Concentration in drains from anode chamber (mg/L)	Concentration in drains from cathode chamber (mg/L)
Cr	0.71	0.002
Cu	1.13	0.004
Fe	1.19	0.005
Ni	1.75	5.160
Pb	0.32	<detection limit 0.0123
Zn	1.23	1.285
U238	0.19	0.028

in the drains from the cathode chamber were low. However, nickel and zinc were found at concentrations of 5.16 mg/L and 1.29 mg/L, respectively, corresponding to their relatively low electrosorption capacities. Small amount of metals was detected in the drains from the anode chamber at concentrations between 0.19 and 1.75 mg/L.

12.3.3 Removal of Cyanide

The removal of cyanide in sodium chloride solution was examined by collecting samples at 5, 10, 15, 20 and 30 minutes of treatment during the testing. The variation of conductivity and cyanide versus treatment time is shown in Figure 12.6. The removal of salt (in terms of conductivity) and cyanide increased with increasing treatment time. After 15 minutes of treatment, conductivity decreased from 425 µS/cm to 30 µS/cm, and cyanide concentration decreased from 2.06 mg/L to approximately 1.2 mg/L. After 30 minutes of treatment, cyanide concentration remained as high as 0.85 mg/L in the effluent, exceeding the Primary MCL of 0.2 mg/L regulated by the USEPA National Drinking Water Standards.

During the electrosorption process, chloride and hydroxide ions may compete with the sorption of cyanide. The hydrated cyanide ion is larger (~3.32 Å) than the hydroxide ion (3.00 Å) and slightly larger than chloride ions (3.32 Å). The molar concentrations of

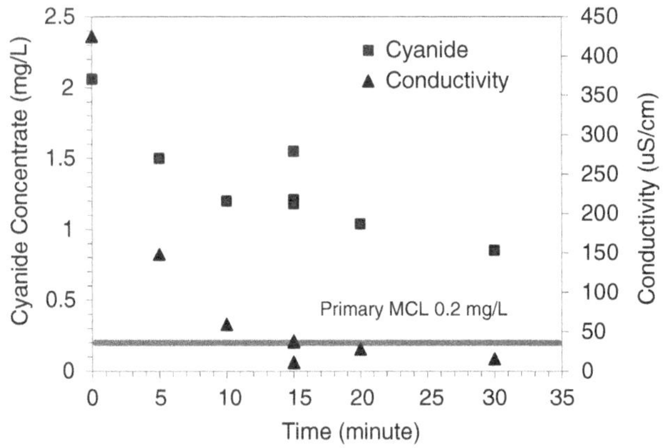

Figure 12.6 Variation of conductivity and cyanide in feed and treated water solutions at different retention time.

chloride and hydroxide were 44 and 13 times greater than the cyanide concentration in the electrolyte. Previous studies have found that in a multi-solute system, the adsorption capacity of an individual ion decreased although the total salt adsorption capacity of the adsorbent increased [75]. In addition, cyanide ion does not form strong hydrogen bonds. Its enthalpy of hydration is consequently considerably less negative than that of hydroxide ion, which reduces the electrosorption capacity of cyanide to carbon electrode.

The low removal rate of cyanide by the purification technology may also be attributed to the complex cyanide chemistry in water solution at different pH. In the beginning of the experiments, the feed water pH was adjusted to 11.0 to keep the cyanide ion (CN^-) the predominant stable form of free cyanide. During the treatment process, the pH of the water solution gradually declined to 5.7. As the pH drops, CN^- converts to hydrogen cyanide (HCN). The percentage of HCN continues to increase as the pH drops further, until at a pH of 7.0, where about 99.5 percent of the cyanide exists as HCN. At a pH below 7.0, essentially all dissolved cyanide is present as HCN. HCN is in a neutral form and not readily removed by the electrochemical sorption process. A portion of the formed HCN transforms to gas and was collected by the vacuum absorption system. The analytical results and mass balance estimated 3.8% of the cyanide entered in the CDI system during the eight treatment runs escaped from the CDI cell as gaseous hydrogen cyanide and captured by the vacuum sorption system.

Besides the above observations, several other reactions may occur in the electrochemical process. Under the initial alkaline conditions, cyanide may be oxidized to cyanogen chloride (ClCN), an intermediate toxic compound, which is then converted to cyanate (NCO^-). Under later acid conditions, free cyanide may also be oxidized to cyanogen (NC-CN), which may be difficult to remove from water by the electrochemical process.

There are a number of studies using electrochemical processes to remove cyanide from aqueous solutions. Szpyrkowicz et al. studied simultaneous electroxidation of cyanide and recovery of copper as a metallic deposition on the cathode from weak concentration rinse wastewaters using plate stainless steel electrodes [76]. Removal of copper cyanide from wastewater can be carried out effectively only at pH 13. Öğütveren and Koparal investigated the feasibility of treating synthetic solution with high strength cyanide using a bipolar trickle tower electrochemical reactor consisting of graphite Raschig rings as the electrodes [77]. The removal rate of cyanide decreased with decreasing concentration in the effluent, and more energy was required to reach lower cyanide concentration. This bench scale study found anodic oxidation of cyanide was effective for the removal of cyanide from waste stream. The pH of the solution decreased from 11 but above 8–9 in the treated water [77].

These studies indicate operating pH is critical for effective removal of ionic species with electrochemical and electrosorption processes. This is consistent with our experimental results that the cyanide removal rate decreased with the decreasing pH because of the conversion of negative cyanide ions to neutral hydrogen cyanide.

12.4 Conclusions

Capacitive deionization provides a comprehensive approach for contaminants removal and providence of deionized water for reuse without consumption of chemicals and generating sludge or residuals. The adsorbed metals can be recovered from activated carbon electrodes using appropriate chemicals for potential extraction therefore minimizing long-term environmental impact. Based on the experimental results, the major conclusions are as follows:

- The performance of the activated carbon electrodes was consistent throughout the bench-scale experiments.

Over the course of the 158 cycles, deterioration or scaling of the carbon materials, stainless steel and carbon current collectors due to the precipitation of the metals was not observed. Metal precipitation however was observed on the two meshes separating feed water cell and the carbon electrodes.

- The purification technology exhibited very high removal of iron, chromium, and uranium-238, with average removal efficiency of 99.9%, 99.6%, and 99%, respectively, over the 158 runs. The average removal of divalent ions were lower than trivalent and tetravalent ions achieving 94.8%, 85.3%, 89%, and 94%, for copper, zinc, nickel, and lead, respectively.
- Among the tested metals, only lead concentration in the treated water could not meet the MCL of the USEPA Primary Drinking Water Regulations.
- The electrosorption capacity of copper, zinc, lead, and nickel was observed to be exhausted faster than that of iron, chromium, and uranium-238.
- The removal of the metals increased with increasing retention time. With the increase in run cycles, the treatment time may need to be prolonged to meet the target water quality.
- The removal rate of cyanide declined during the electrosorption process and the treated water could not meet the cyanide MCL of Primary Drinking Water Regulations. This is mainly caused by the conversion of ionic cyanide to neutral hydrogen cyanide as a result of decreasing pH in the electrolyte.

The efficiency and electrosorption capacity of the CDI process merit further optimization and improvement before it becomes an economically feasible commercial technology. The following presents future research and development needed for the evaluation and optimization of CDI in the applications of heavy metal wastewater treatment:

- Test a variety of carbon materials, and develop cost-effective and high-efficient carbon electrodes.
- Conduct systematic experiments to better understand the removal mechanisms and kinetics of different heavy metals and the impact of pH.

- Optimize operational parameters (such as targeted resistance, retention time, applied current and voltage), and cell design (such as cell configuration, dimension of carbon electrodes).
- Develop control strategies to prevent potential scaling and corrosion of electrodes and related materials.
- Investigate and improve energy efficiency of the treatment process.
- Investigate the extraction and regeneration methods of the carbon electrodes for the recovery of heavy metals.

References

1. H.A. Aziz, M.N. Adlan and K.S. Ariffin, Heavy metals (Cd, Pb, Zn, Ni, Cu and Cr(III)) removal from water in Malaysia: Post treatment by high quality limestone, *Biores. Technol* **99,** 1578–1583 (2008).
2. T.A. Kurniawan, G.Y.S. Chan, W.-H. Lo and S. Babel, Physico-chemical treatment techniques for wastewater laden with heavy metals, *Chem. Eng. J* **118,** 83–98 (2006).
3. S. Rengaraj, K.-H. Yeon and S.-H. Moon, Removal of chromium from water and wastewater by ion exchange resins, *J. Hazard. Mater* **87,** 273–287 (2001).
4. B. Alyuz and S. Veli, Kinetics and equilibrium studies for the removal of nickel and zinc from aqueous solutions by ion exchange resins, *J. Hazard. Mater* **167,** 482–488 (2009).
5. A. Dabrowski, Z. Hubicki, P. Podkoscielny and E. Robens, Selective removal of the heavy metal ions from waters and industrial wastewaters by ion-exchange method, *Chemosphere* **56,** 91–106 (2004).
6. F. Fu and Q. Wang, Removal of heavy metal ions from wastewaters: A review, *J. Environ. Manage* **92** 407–418 (2011).
7. A. Jusoh, L. Su Shiung, N.Ä. Ali and M.J.M.M. Noor, A simulation study of the removal efficiency of granular activated carbon on cadmium and lead, *Desalination* **206,** 9–16 (2007).
8. H. Wang, A. Zhou, F. Peng, H. Yu and J. Yang, Mechanism study on adsorption of acidified multiwalled carbon nanotubes to Pb(II), *J. Colloid Int. Sci* **316** 277–283 (2007).
9. K. Pillay, E.M. Cukrowska and N.J. Coville, Multi-walled carbon nanotubes as adsorbents for the removal of parts per billion levels of hexavalent chromium from aqueous solution, *J. Hazard. Mater* **166** 1067–1075 (2009).
10. H.G. Park, T.W. Kim, M.Y. Chae and I.-K. Yoo, Activated carbon-containing alginate adsorbent for the simultaneous removal of heavy metals and toxic organics, *Process Biochem* **42,** 1371–1377 (2007).

11. A. Ucer, A. Uyanik and S.F. Aygun, Adsorption of Cu(II), Cd(II), Zn(II), Mn(II) and Fe(III) ions by tannic acid immobilised activated carbon, *Sep. Purifi. Technol* **47,** 113–118 (2006).
12. H. Yanagisawa, Y. Matsumoto and M. Machida, Adsorption of Zn(II) and Cd(II) ions onto magnesium and activated carbon composite in aqueous solution, *Appl.Surf. Sci* **256,** 1619–1623 (2010).
13. C.K. Ahn, D. Park, S.H. Woo and J.M. Park, Removal of cationic heavy metal from aqueous solution by activated carbon impregnated with anionic surfactants, *J. Hazard. Mater* **164** 1130–1136 (2009).
14. U. Wingenfelder, C. Hansen, G. Furrer and R. Schulin, Removal of Heavy Metals from Mine Waters by Natural Zeolites, *Environ. Sci. Technol* **39** 4606–4613 (2005).
15. R. Egashira, S. Tanabe and H. Habaki, Adsorption of heavy metals in mine wastewater by Mongolian natural zeolite, *Procedia Eng* **42,** 49–57 (2010).
16. E. Katsou, S. Malamis and K.J. Haralambous, Industrial wastewater pre-treatment for heavy metal reduction by employing a sorbent-assisted ultrafiltration system, *Chemosphere* **82** 557–564 (2010).
17. O.S. Amuda, A.A. Giwa and I.A. Bello, Removal of heavy metal from industrial wastewater using modified activated coconut shell carbon, *Biochem. Eng. J* **36,** 174–181 (2007).
18. M. Sciban, B. Radetic, Z. Kevresan, M. Klasnja, Adsorption of heavy metals from electroplating wastewater by wood sawdust, *Bioresour. Technol* **98,** 402–409 (2007).
19. W.S. Wan Ngah and M.A.K.M. Hanafiah, Removal of heavy metal ions from wastewater by chemically modified plant wastes as adsorbents: A review, *Bioresour Technol* **99,** 3935–3948 (2010).
20. K.K. Wong, C.K. Lee, K.S. Low and M.J. Haron, Removal of Cu and Pb by tartaric acid modified rice husk from aqueous solutions, *Chemosphere* **50,** 23–28 (2003).
21. V.K. Gupta, C.K. Jain, I. Ali, M. Sharma and V.K. Saini, Removal of cadmium and nickel from wastewater using bagasse fly ash-a sugar industry waste, *Water Res* **37,** 4038–4044 (2003).
22. J. Landaburu-Aguirre, V. García, E. Pongrácz and R.L. Keiski, The removal of zinc from synthetic wastewaters by micellar-enhanced ultrafiltration: statistical design of experiments, *Desalination* **240,** 262–269 (2009).
23. X. Li, G.M. Zeng, J.H. Huang, C. Zhang, Y.Y. Fang, Y.H. Qu, F. Luo, D. Lin and H.L. Liu, Recovery and reuse of surfactant SDS from a MEUF retentate containing Cd2+ or Zn2+ by ultrafiltration, *J. Membr. Sci* **337,** 92–97 (2009).
24. M.K. Aroua, F.M. Zuki, N.M. Sulaiman, Removal of chromium ions from aqueous solutions by polymer-enhanced ultrafiltration, Journal of Hazardous Materials, 147 (2007) 752–758.

25. M.A. Barakat and E. Schmidt, Polymer-enhanced ultrafiltration process for heavy metals removal from industrial wastewater, *Desalination* **256,** 90–93 (2010).
26. M.A.M. El Zeftawy and C.N. Mulligan, Use of rhamnolipid to remove heavy metals from wastewater by micellar-enhanced ultrafiltration (MEUF), *Sep. Purif. Technol* **77,** 120–127 (2011).
27. M. Al-Abri, A. Dakheel, C. Tizaoui and N. Hilal, Combined humic substance and heavy metals coagulation, and membrane filtration under saline conditions, *Desalination* **253,** 46–50 (2011).
28. C. Blocher, J. Dorda, V. Mavrov, H. Chmiel, N.K. Lazaridis and K.A. Matis, Hybrid flotation-membrane filtration process for the removal of heavy metal ions from wastewater, *Water Res* **37,** 4018–4026 (2003).
29. V. Mavrov, T. Erwe, C. Bl√∂cher and H. Chmiel, Study of new integrated processes combining adsorption, membrane separation and flotation for heavy metal removal from wastewater, *Desalination* **157,** 97–104 (2003).
30. V. Mavrov, S. Stamenov, E. Todorova, H. Chmiel and T. Erwe, New hybrid electrocoagulation membrane process for removing selenium from industrial wastewater, *Desalination* **201,** 290–296 (2006).
31. B.A.M. Al-Rashdi, D.J. Johnson and N. Hilal, Removal of heavy metal ions by nanofiltration, *Desalination*, In press (2012).
32. H. Ozaki, K. Sharma and W. Saktaywin, Performance of an ultra-low-pressure reverse osmosis membrane (ULPROM) for separating heavy metal: effects of interference parameters, *Desalination* **144** 287–294 (2002).
33. L. Cifuentes, I. Garcia, P. Arriagada and J.M. Casas, The use of electrodialysis for metal separation and water recovery from CuSO4,ÄìH2SO4,ÄìFe solutions, *Sep. Purif. Technol* **68** 105–108 (2009).
34. J. Lambert, M. Avila-Rodriguez, G. Durand and M. Rakib, Separation of sodium ions from trivalent chromium by electrodialysis using monovalent cation selective membranes, *J. Membr. Sci* **280** 219–225 (2006).
35. T. Mohammadi, A. Moheb, M. Sadrzadeh and A. Razmi, Modeling of metal ion removal from wastewater by electrodialysis, *Sep. Purif. Technol* **41,** 73–82 (2005).
36. S.K. Nataraj, K.M. Hosamani and T.M. Aminabhavi, Potential application of an electrodialysis pilot plant containing ion-exchange membranes in chromium removal, *Desalination* **217,** 181–190 (2007).
37. A. Mahmoud and A.F.A. Hoadley, An evaluation of a hybrid ion exchange electrodialysis process in the recovery of heavy metals from simulated dilute industrial wastewater, *Water Res* **46,** 3364–3376 (2012).
38. Y. Oren, Capacitive deionization (CDI) for desalination and water treatment -- past, present and future (a review), *Desalination* **228,** 10–29 (2008).
39. C.J. Gabelich, T.D. Tran and I.H.M. Suffet, Electrosorption of inorganic salts from aqueous solution using carbon aerogels, *Env. Sci. Technol* **36,** 3010–3019 (2002).

40. P. Xu, J.E. Drewes, D. Heil and G. Wang, Treatment of brackish produced water using carbon aerogel-based capacitive deionization technology, *Water Res* **42** 2605–2617 (2008).
41. R.W. Pekala, J.C. Farmer, C.T. Alviso, T.D. Tran, S.T. Mayer, J.M. Miller and B. Dunn, Carbon aerogels for electrochemical applications, *J. Non-Crystalline Solids* **225,** 74–80 (1998).
42. J.C. Farmer, D.V. Fix, G.V. Mack, R.W. Pekala and J.F. Poco, Capacitive deionization of NH4ClO4 solutions with carbon aerogel electrodes, *J. Appl. Electrochem* **26,** 1007–1018 (1996).
43. C.-M. Yang, W.-H. Choi, B.-K. Na, B.W. Cho and W.I. Cho, Capacitive deionization of NaCl solution with carbon aerogel-silicagel composite electrodes, *Desalination* **174,** 125–133 (2005).
44. J.-H. Choi, Fabrication of a carbon electrode using activated carbon powder and application to the capacitive deionization process, *Sep. Purif. Technol* **70,** 362–366 (2010).
45. Y.-J. Kim, J. Hur, W. Bae and J.-H. Choi, Desalination of brackish water containing oil compound by capacitive deionization process, *Desalination* **253,** 119–123 (2010).
46. H. Li, T. Lu, L. Pan, Y. Zhang and Z. Sun, Electrosorption behavior of graphene in NaCl solutions, *J. Mater. Chem* **19,** 6773–6779 (2009).
47. H. Li, L. Pan, C. Nie, Y. Liu and Z. Sun, Reduced graphene oxide and activated carbon composites for capacitive deionization, *J. Mater. Chem* **22** 15556–15561 (2012).
48. H. Wang, D. Zhang, T. Yan, X. Wen, L. Shi and J. Zhang, Graphene prepared via a novel pyridine-thermal strategy for capacitive deionization, *J. Mater. Chem* **22,** 23745–23748 (2012).
49. H.-J. Ahn, J.-H. Lee, Y. Jeong, J.-H. Lee, C.-S. Chi and H.-J. Oh, Nanostructured carbon cloth electrode for desalination from aqueous solutions, *Mater. Sci. Eng A* **449–451**, 841–845 (2007).
50. H.-J. Oh, J.-H. Lee, H.-J. Ahn, Y. Jeong, Y.-J. Kim and C.-S. Chi, Nanoporous activated carbon cloth for capacitive deionization of aqueous solution, *Thin Solid Films* **515,** 220–225 (2006).
51. M.-W. Ryoo, J.-H. Kim and G. Seo, Role of titania incorporated on activated carbon cloth for capacitive deionization of NaCl solution, *J. Colloid Int. Sci* **264,** 414–419 (2003)..
52. K.C. Leonard, J.R. Genthe, J.L. Sanfilippo, W.A. Zeltner and M.A. Anderson, Synthesis and characterization of asymmetric electrochemical capacitive deionization materials using nanoporous silicon dioxide and magnesium doped aluminum oxide, *Electrochim Acta* **54,** 5286–5291 (2009).
53. H. Li, Y. Gao, L. Pan, Y. Zhang, Y. Chen and Z. Sun, Electrosorptive desalination by carbon nanotubes and nanofibres electrodes and ion-exchange membranes, *Water Res* **42,** 4923–4928 (2008).

54. L. Pan, X. Wang, Y. Gao, Y. Zhang, Y. Chen and Z. Sun, Electrosorption of anions with carbon nanotube and nanofibre composite film electrodes, *Desalination* **244,** 139–143 (2009).
55. Z. Peng, D. Zhang, L. Shi and T. Yan, High performance ordered mesoporous carbon/carbon nanotube composite electrodes for capacitive deionization, *J. Mater. Chem* **22,** 6603–6612 (2012).
56. G. Wang, Q. Dong, Z. Ling, C. Pan, C. Yu and J. Qiu, Hierarchical activated carbon nanofiber webs with tuned structure fabricated by electrospinning for capacitive deionization, *J. Mater. Chem* **22** 21819–21823 (2012).
57. L. Wang, M. Wang, Z.-H. Huang, T. Cui, X. Gui, F. Kang, K. Wang and D. Wu, Capacitive deionization of NaCl solutions using carbon nanotube sponge electrodes, *J. Mater. Chem* **21** 18295–18299 (2011).
58. D. Zhang, T. Yan, L. Shi, Z. Peng, X. Wen and J. Zhang, Enhanced capacitive deionization performance of graphene/carbon nanotube composites, *J. Mater. Chem* **22,** 14696–14704 (2012).
59. J.C. Farmer, S.M. Bahowick, J.E. Harrar, D.V. Fix, R.M. Martinelli, A.K. Vu and K.L. Carroll, Electro- sorption of chromium ions on carbon aerogel electrodes as a mean of remediating ground water, *Energy Fuels* **11,** 337–347 (1997).
60. L.R. Shiue, A. Sun, C.C. Shiue, S. Wang, F.C.Y. Hsieh, W.T. Lo and Y.H. Hsieh, Replacable flow- through capacitor for removing charged species from liquids. EP 1291323 A1. Available at: <http://www.google.com/patents/EP1291323A1?cl=en%3E>, (2001).
61. P. Dlugolecki, P. Ogonowski, S.J. Metz, M. Saakes, K. Nijmeijer and M. Wessling, On the resistances of membrane, diffusion boundary layer and double layer in ion exchange membrane transport, *J. Membr. Sci* **349** 369–379 (2010).
62. Y.-J. Kim and J.-H. Choi, Enhanced desalination efficiency in capacitive deionization with an ion-selective membrane, *Sep. Purif. Technol* **71** 70–75 (2010).
63. J.-B. Lee, K.-K. Park, H.-M. Eum and C.-W. Lee, Desalination of a thermal power plant wastewater by membrane capacitive deionization, *Desalination* **196,** 125–134 (2006).
64. P.M. Biesheuvel and A. van der Wal, Membrane capacitive deionization, *J. Membr. Sci* **346,** 256–262 (2010).
65. R. Zhao, P.M. Biesheuvel and A. van der Wal, Energy consumption and constant current operation in membrane capacitive deionization, *Energy Environ. Sci* **5** 9520–9527 (2012).
66. R.T. Mayes, C. Tsouris, J.O. Kiggans Jr, S.M. Mahurin, D.W. DePaoli and S. Dai, Hierarchical ordered mesoporous carbon from phloroglucinol-glyoxal and its application in capacitive deionization of brackish water, *J. Mater. Chem* **20,** 8674–8678 (2010).

67. S.-J. Seo, H. Jeon, J.K. Lee, G.-Y. Kim, D. Park, H. Nojima, J. Lee and S.-H. Moon, Investigation on removal of hardness ions by capacitive deionization (CDI) for water softening applications, *Water Res* **44** 2267–2275 (2010).
68. L.Y. Lee, H.Y. Ng, S.L. Ong, G. Tao, K. Kekre, B. Viswanath, W. Lay, and H. Seah, Integrated pretreatment with capacitive deionization for reverse osmosis reject recovery from water reclamation plant, *Water Res* **43,** 4769–4777 (2009).
69. R. Atlas, Purification of Brackish Water using Hybrid CDI-EDI Technology, IPEC Conference, (2007).
70. T.J. Welgemoed, C.F. Schutte, Capacitive Deionization Technology™: An alternative desalination solution, *Desalination* **183,** 327–340 (2005).
71. P. Xu, J.E. Drewes, D. Heil and G. Wang, Treatment of brackish produced water using carbon aerogel-based capacitive deionization technology, *Water Res* **40** 2605–2617 (2008).
72. I. Persson, Hydrated metal ions in aqueous solution: How regular are their structures?, *Pure Appl. Chem* **82,** 1901–1917 (2010).
73. M. Meot-Ner, S.M. Cybulski, S. Scheiner and J.F. Liebman, Is cyanide significantly anisotropic? Comparison of cyanide vs chloride: clustering with hydrogen cyanide and condensed-phase thermochemistry, *J. Phys Chem* **92,** 2738–2745 (1988).
74. E.R. Nightingale, Phenomenological Theory of Ion Solvation. Effective Radii of Hydrated Ions, *J. Phys Chem* **63** 1381–1387 (1959).
75. T. Depci, A.R. Kul and Y. Onal, Competitive adsorption of lead and zinc from aqueous solution on activated carbon prepared from Van apple pulp: Study in single- and multi-solute systems, *Chem. Eng. J* **200–202,** 224–236 (2012).
76. L. Szpyrkowicz, F. Zilio-Grandi, S.N. Kaul and S. Rigoni-Stern, Electrochemical treatment of copper cyanide wastewaters using stainless steel electrodes, *Water Sci. Technol* **38** 261–268 (1998).
77. Ü.B. Öğütveren, S.K. E Törü, Removal of cyanide by anodic oxidation for wastewater treatment, *Water Res* **33,** 1851–1856 (1999).

SECTION V

RENEWABLE ENERGY SOURCES TO POWER DESALINATION

13

Solar Desalination

Mohammad Abutayeh, Chennan Li, D. Yogi Goswami and Elias K. Stefanakos

Khalifa University, Abu Dhabi, UAE
Aspen Technology, Houston, Texas Clean Energy Research Center, University of South Florida, Tampa, Florida

Abstract

Extracting fresh water from seawater requires a great deal of energy, both thermal and mechanical. Renewable energy driven desalination is becoming more viable despite its expensive infrastructure because it employs free natural energy sources and releases no harmful effluents to the environment. Solar radiation is usually chosen over other renewable energy sources because its thermal energy can be directly applied to drive desalination systems without the inevitable energy loss associated with energy conversion according to the second law of thermodynamics.

Solar desalination systems are classified into direct and indirect processes depending on the energy path to fresh water. Direct solar desalination systems combine solar energy collection and desalination in one process producing fresh water distillate by directly applying collected solar energy to seawater. Solar distillation using a solar still is an example of direct solar desalination. Indirect solar desalination systems comprise two sub–systems: a solar collection system and a desalination system. The solar collection sub–system is used either to collect heat using solar collectors and supply it via a heat exchanger to a thermal desalination process or to convert electromagnetic solar radiation to electricity using photovoltaic cells to power a physical desalination process. The desalination sub–system can be any conventional desalination system.

Keywords: Solar desalination, solar distillation, solar collectors, photovoltaic

Jane Kucera (ed.) Desalination , (551–582) 2014 © Scrivener Publishing LLC

List of Symbols

Nomenclature

ABHP	absorption heat pump
ADHP	adsorption heat pump
AGMD	air gap membrane distillation
CAOW	closed–air open–water cycle
Cp	heat capacity
CPC	compound parabolic collector
CSP	concentrated solar power
CWOA	closed–water open–air cycle
DCMD	direct contact membrane distillation
DNI	direct normal irradiance
ED	electro–dialysis
ETC	evacuated tube collector
FPC	flat plate collector
HDH	humidification–dehumidification
HP	heat pump
HTF	heat transfer fluid
k	thermal conductivity
LCZ	lower convecting zone
MD	membrane distillation
MED	multiple–effect distillation
MSF	multi–stage flash
MVC	mechanical vapor compressor
NCZ	non convecting zone
ORC	organic Rankine cycle
PTC	parabolic trough collector
PV	photovoltaic
RO	reverse osmosis
SGMD	sweeping gas membrane distillation
T_m	melting point
TVC	thermal vapor compressor
UCZ	upper convecting zone
VMD	vacuum membrane distillation
ΔH^L	latent heat
ρ	density

13.1 Introduction

Fresh water demand is persistently increasing as populations around the world keep growing and as existing fresh water reserves keep declining due to consumption and pollution. Figure 13.1 shows the estimated water consumption of US counties for the year 2000 [1]. Marine waters represent an infinite water source since about 98 % of all global water is present in oceans; therefore, seawater desalination is the logical approach to meet the rising fresh water demand.

Energy demand is also continually increasing due to relentless global industrialization. Fossil fuels remain the primary sources of energy for most of the world; however, their reserves are dwindling, production is peaking, and consumption is harming the environment. Renewable energy sources are continually replenished by cosmic forces and can be used to produce sustainable and useful forms of energy with minimum environmental impact.

Developing an economically–viable and environmentally–friendly desalination system involves lowering its energy demand and employing renewable energy to drive its operation; moreover, selecting the suitable desalination process requires several design considerations and knowledge of its design limitations. Serious economic and social disruptions are unfolding over the finite water and energy resources; hence, securing fresh water supply and

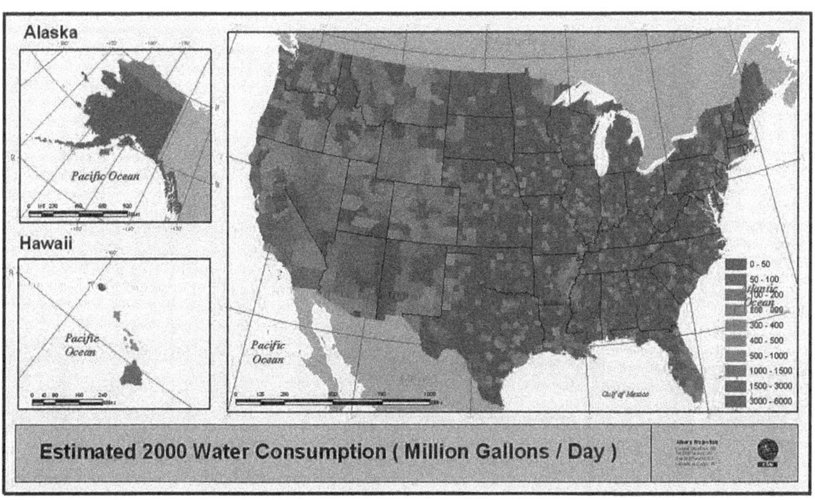

Figure 13.1 Estimated water consumption of US counties for 2000 [1]

employing renewable energy sources will help avoid catastrophic conflicts, continue modern lifestyles, and circumvent global warming and pollution [2].

Desalination can be accomplished by separation techniques developed over the years to produce potable water. The most widespread desalination methods are well documented and explained in literature. Momentous amounts of energy are required in all desalination processes; therefore, reducing energy demand, as well as employing renewable energy, is imperative to developing viable desalination processes. Various desalination systems driven by renewable energy have been developed over the last few years; nonetheless, most have not yet been commercially implemented due to the high capital cost associated with utilizing renewable energy.

Solar radiation is a very appealing source of energy because it is available at no cost; furthermore, exploiting it has no notable adverse effect on the environment. Plenty of research and development have been undertaken to utilize this free form of energy to develop more efficient sustainable processes such as water desalination and power generation. Figure 13.2 illustrates the US share of solar radiation [3]. Solar energy is intermittent and would probably require storage; however, maximizing its use alongside developing energy efficient processes can greatly diversify energy resources, save the environment, and reduce the imposed social cost [4].

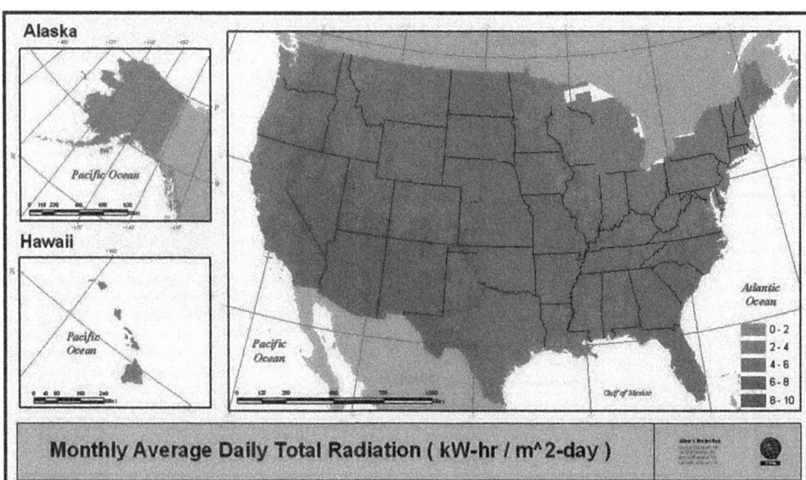

Figure 13.2 Monthly average daily solar radiation in the US [3]

Solar desalination is essentially a small–scale duplicate of the natural hydrological cycle that produces rain, which is the primary source of fresh water worldwide. Scholarly research and industrial experience have shown that thermal desalination schemes are more suitable than mechanical desalination schemes for large–scale applications. Solar energy is well suited to drive thermal desalination processes because it can be directly applied without entailing energy conversion losses; furthermore, existing thermal desalination installations can be easily retrofitted to accept solar heat.

13.2 Direct Solar Desalination

13.2.1 Solar Pond

Solar ponds are pools of water with a darkened bottom to maximize light absorption. They are designed to have increasing salinity with depth creating a density gradient that inhibits natural convection currents. Water absorbs solar radiation going through it causing its temperature to rise. The shorter the wave length of sunlight, the deeper it can penetrate the water column as shown in Table 13.1 [5]. The amount of absorbed energy increases with depth producing a vertical temperature incline causing a density gradient decreasing with depth. Conversely, salinity increases with depth producing a vertical salinity incline causing a density gradient increasing with depth. The final outcome of these disagreeing events is a stratified pond with increasing temperature and salinity with depth, as shown in Figure 13.3 [6]. Solar ponds function as both solar collectors and thermal energy storage media.

Table 13.1 Spectral absorption of solar radiation in water [5]

Wavelength (μm)	Layer Depth				
	0	1 cm	10 cm	1 m	10 m
0.2–0.6	23.7	23.7	23.6	22.9	17.2
0.6–0.9	36.0	35.3	36.0	12.9	0.9
0.9–1.2	17.9	12.3	0.8	0.0	0.0
> 1.2	22.4	1.7	0.0	0.0	0.0
Total	100.0	73.0	54.9	35.8	18.1

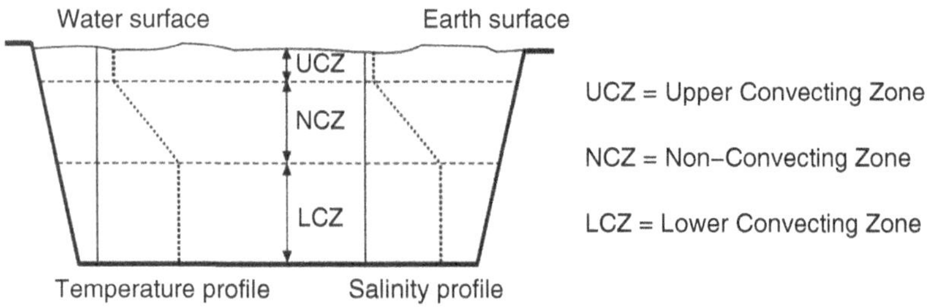

Figure 13.3 Vertical cross section of a solar pond

13.2.2 Solar Still

In a solar still, the heat collection and distillation processes occur within the same structure where solar energy is used directly for distillation by means of the greenhouse effect. Seawater is placed in a blackened basin inside an air tight transparent structure where it evaporates due to absorption of solar radiation then condenses on the sloping structure by losing its latent heat of condensation to the surroundings. Condensed droplets run down the cover to accumulating troughs to be collected as fresh water.

Solar stills are a small scale hydrological cycle, and their efficiency is dependent on meteorological limitations such as solar radiation, sky clearness, ambient temperature, wind velocity, and others. Furthermore, the output of solar stills is affected by many operating factors such as brine depth, vapor leakage, thermal insulation, cover slope, shape material, and others [7, 8]. The latent heat of condensation is normally wasted on the cover; therefore, system efficiency is relatively low with a daily production of about 3–4 l/m^2 [9]. Solar stills require large collection areas to maximize radiation absorption and are usually combined with other desalination methods to increase their efficiency.

Solar stills have been studied extensively resulting in the development of many different solar still configurations as illustrated in Figure 13.4. Combining solar stills with economizers and condensers as well as employing flat and concentrating solar collectors, integrating passive vacuum and heat recovery into solar stills, and augmenting conventional thermal desalination systems with solar stills are all different techniques to increase the feasibility of solar stills. Solar stills are relatively inexpensive to build and maintain making them a viable option for underdeveloped communities

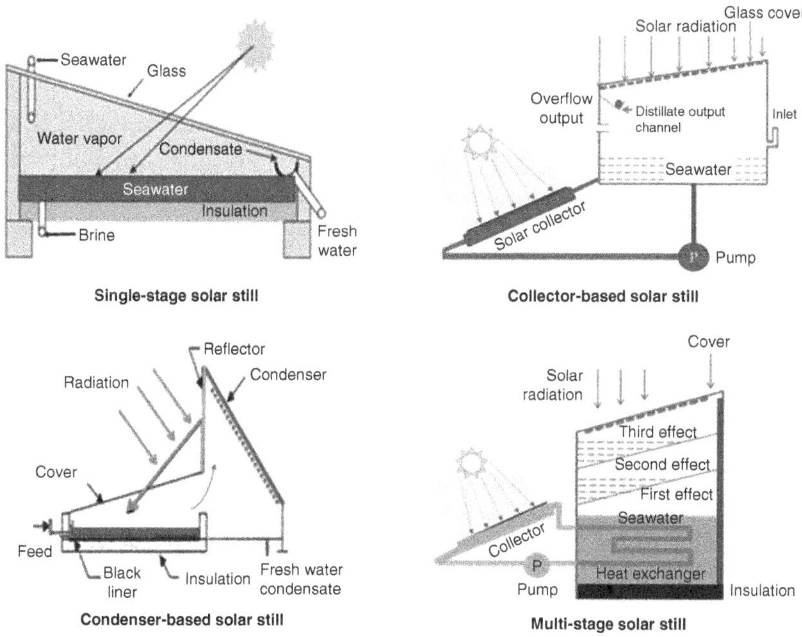

Figure 13.4 Different configurations of solar stills

with good solar resource and for emergency situations resulting in power outages.

13.3 Indirect Solar Desalination

13.3.1 Conventional Desalination

Desalination is very energy intensive and requires costly infrastructure; therefore; several desalination processes have been developed over the years to produce fresh water from seawater economically. These can be classified according to the applied separation scheme into thermal, physical, and chemical processes.

Thermal desalination processes produce a fractional phase change of liquid seawater to either vapor or solid. The new phase is then separated from the bulk brine water producing fresh water, while the latent heat of phase change is reclaimed. Multiple effect evaporation, multi–stage flash, vapor compression, and indirect contact freezing are examples of thermal desalination processes.

Physical desalination processes extract fresh water from seawater by applying pressure or electric potential across a membrane. Either fresh water or solute ions travel through the semi–permeable membrane due to the mechanically induced gradient yielding the desired separation. Reverse osmosis, electro–dialysis, and nano–filtration are examples of physical desalination processes.

Chemical desalination processes extract fresh water from seawater by precipitating its salts due to chemical reactions. These processes are less common because they are usually too expensive to produce fresh water. Ion exchange, gas hydrate, and liquid–liquid extraction are examples of chemical desalination processes.

13.3.2 Renewable Energy Driven Desalination

Fresh water and energy reserves are increasingly exhausted as was mentioned earlier; hence, seawater desalination using renewable energy sources is a very appealing research area. In addition, desalination is an enormously energy exhaustive process making fossil fuel based conventional desalination methods extremely unpopular especially in light of the growing impact of environmental pollution and global warming.

The worldwide capacity of desalination using renewable energy is less than 1 % of that of conventional desalination due to high capital and maintenance costs associated with using renewable energy [10]. Several renewable energy driven desalination plants were designed and constructed; however, most were geographically customized and built on pilot scale. A detailed record of these plants was put together by Tzen and Morris [11].

Wind energy can be utilized to generate electricity via turbines to run physical and chemical desalination plants, while geothermal energy can be utilized to generate heat via underground heat exchangers to run thermal desalination plants. Solar energy is the most promising renewable energy source due to its ability to drive the more popular thermal desalination systems directly through solar collectors and to drive physical and chemical desalination systems indirectly through photovoltaic cells.

Indirect solar desalination systems comprise two sub–systems: a solar collection system and a desalination system. The solar collection sub–system is used either to collect heat using solar collectors and supply it via a heat exchanger to a thermal desalination process or convert heat to electricity using photovoltaic cells to power a physical desalination process. The desalination sub–system can be any of the previously mentioned conventional desalination systems.

13.3.3 Thermal Driven Processes

13.3.3.1 Solar Assisted Multi–Stage Flash

MSF processes use both thermal and electrical energy which can be drawn from solar radiation in various schemes, as shown in Figure 13.5. Szacsvay et. al. [12] described a self–regulating Auto–flash multi–stage desalination system coupled with a solar pond. They expected that additional cost reductions, through upscaling and serial manufacture, could lead to the production of solar desalinated water at competitive prices. Posnansky [13] presented computer simulations and experimental results of a small solar pond for the performance data of the coupled MSF unit. The economic assessment shows that solar desalination has already become competitive for medium sized installations at remote locations. Tahri [14] tested a multiple distillation and flashing plant coupled with both a solar pond and an existing thermal plant in order to recover

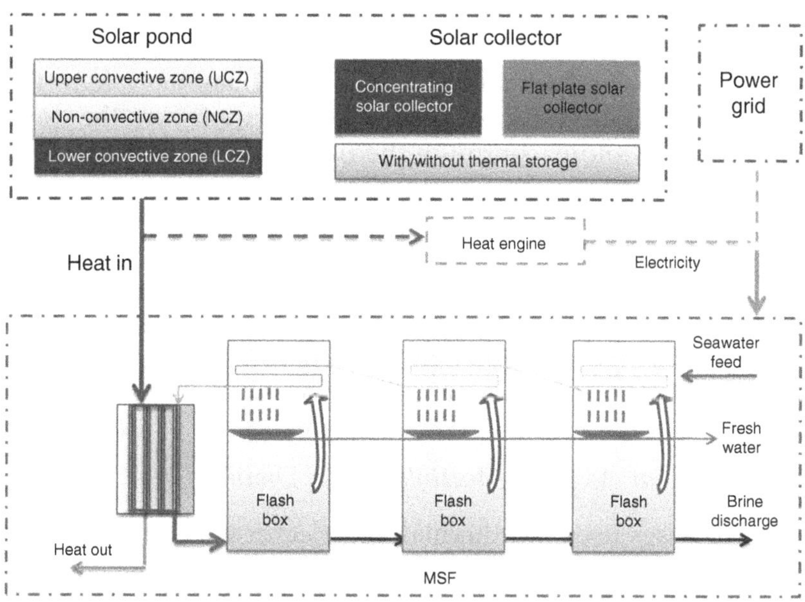

Figure 13.5 Solar assisted multi–stage flash desalination

heat from the exhaust gas of a thermal power plant. The total cost was \$2.10/m^3 of fresh water making the solar pond desalination technology viable for medium-sized desalination plants.

Safi et al. [15] simulated a solar pond assisted MSF system and calculated a cost of \$1.30/m^3 of fresh water. Abu–Jabal et al. [16] reported a pilot scale zero liquid discharge 4–stage MSF unit driven by solar thermal flat collectors with a 0.2 m^3/day maximum daily fresh water production. Agha [17] combined a solar pond with a MSF process and confirmed that solar desalination is a capital incentive enterprise. A wide variation in output between summer and winter exists due to the intermittent nature of solar energy; therefore, the solar pond had to be oversized.

Farwati [18] concluded from experiment that 1 m^2 of flat plate collector operating at 80 °C could produce 8.2 m^3 of distilled water annually, while the same size compound parabolic collector working at 122 °C could produce 13.2 m^3 of distilled water. Joseph et al. [19] experimentally studied a single–stage flash desalination system working with flat plate solar collectors obtaining a maximum distillate yield of 8.5 liter/day with a collector area of 2 m^2. Shaobo et al. [20] used pinch analysis to optimize a solar MSF desalination process and concluded that a wide working temperature range is needed, no distilled water should be pumped out at middle stages, and the same temperature difference should be kept stable in order to enhance performance. Jiang et al. [21] combined MSF with MED and used a direct solar thermal heating desalination system. They concluded that low flash evaporation pressure favors the desalination rate; therefore, control of the flash pressure is very important to the overall system. The desalination rate increased almost 5 times by reducing the flash evaporation pressure from 0.014 MPa to 0.010 MPa.

MSF requires precise pressure control to keep a steady differential between stages which conflicts with the varying nature of solar radiation; thus, it is imperative to employ an effective thermal energy storage system to compliment solar assisted MSF processes.

13.3.3.2 Solar Assisted Multiple–Effect Distillation

MED processes use both thermal and electrical energy which can be drawn from solar radiation in various schemes, as shown in Figure 13.6. A solar pond assisted MED is similar to a solar pond assisted MSF but more feasible due to the lower temperature

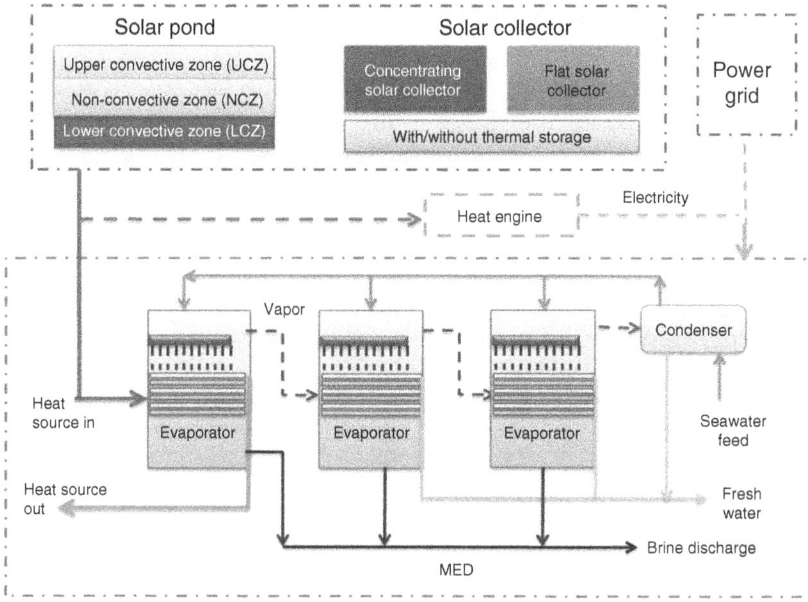

Figure 13.6 Solar assisted multi–stage flash desalination

requirements. Hawaj and Darwish [22] found that intermediate steam temperatures supplied between 80–90 °C are best suited to operate solar assisted MED systems. They also found that a large ratio of solar pond surface area to MED heat transfer area leads to a continuous increase in pond temperature. Garman and Muntasser [23] found optimum thickness of upper convective, non–convective, and lower convective zones to be 0.3 m, 1.1 m, and 4 m, respectively, for a solar pond assisting a low temperature MED system.

Solar collector assisted MED seawater desalination processes have been studied extensively. Technical feasibility and reliability solar collector assisted MED have been proven based on long term investigations. The Abu Dhabi Solar Desalination Plant at Abu Dhabi, United Arab Emirates and the Solar Thermal Desalination Project at Platforma Solar de Almeria, Spain are examples of solar collector assisted MED process.

The Abu Dhabi Solar Desalination Plant was operated from 1984 to 2002 where it employed evacuated tube solar collectors to collect the thermal energy needed to run an MED system. Researchers [24] developed a simulation program, called SOLDES, to predict the part load performance of the plant and to optimize the operating parameters in an effort to maximize fresh water production for a given solar resource every month of the year. Some plant maintenance operations were needed, most notably was maintaining clean collectors. Dust deposition was found to cause water production to drop by 40% [25]. Feasibility studies showed that it is not worth operating the desalination system solely on solar energy due to the high percentage of inactive time [26].

The Solar Thermal Desalination Project entailed 2 phases. Phase I used parabolic trough collectors to provide thermal energy to a 14 stage forward feed MED system with a capacity of 3 m^3/hour [27]. Phase II, also known as AQUASOL, used an advanced prototype of LiBr–H_2O double effect absorption heat pump, a 500 m^2 stationary compound parabolic concentrating collector field, a 24 m^3 water–based thermal energy storage system, and a smoke tube gas boiler to guarantee uninterrupted desalination operation. Researchers [28] found that the connection between the heat pump and the MED unit should not be direct by means of a closed water circuit, heated by the heat pump and cooled by the MED unit, but rather indirect by means of two auxiliary tanks.

13.3.3.3 Solar Assisted Heat Pumps

HP processes are generally used for small to medium scale applications and they are normally combined with other thermal processes to boost their efficiencies by recovering low grade steam and heating it to higher temperatures. Moreover, this low grade steam recovery reduces the overall cooling water demand resulting in less electricity use. There are four basic types of HP desalination processes shown in Figure 13.7: thermal vapor compressor (TVC), mechanical vapor compressor (MVC), absorption heat pump (ABHP), and adsorption heat pump (ADHP).

TVC could be integrated with different size MED and MSF desalination plants where steam compression is carried out by an ejector and vapor from the last effect or stage is carried by a motive stream back to the first effect [29]. MVC is widely studied and used because of its simplicity and relative low energy demand. MVC can be attached to MED and MSF desalination plants to compress the vapor of the last effect or stage to recover heat in the rejected brine and distillate product streams [30]. ABHP absorbs the vapor of the last effect or stage through LiBr–H_2O and discharges steam for use by the first effect or stage [31]; while ADHP adsorbs the vapor of the last effect or stage through zeolite–H_2O or other pairs

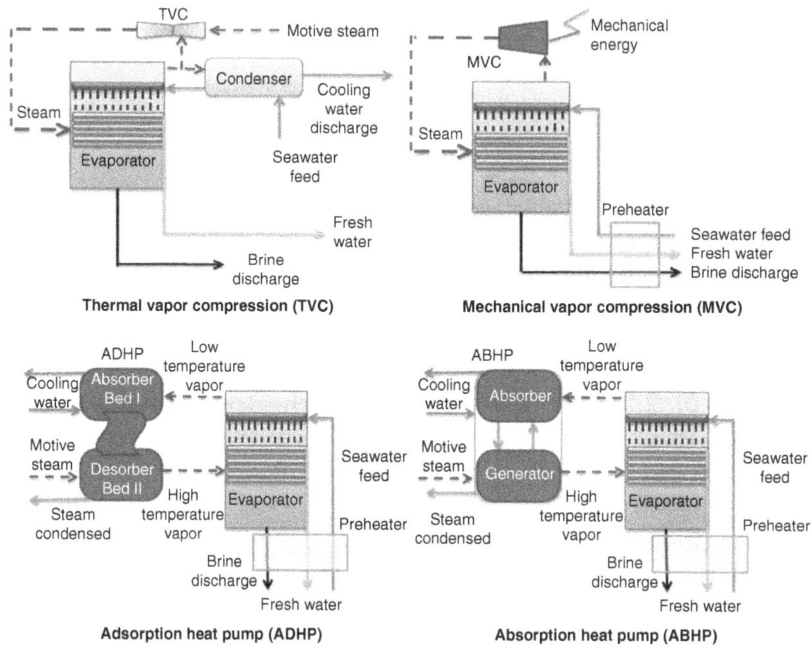

Figure 13.7 Different configurations of heat pumps

and generates high temperature steam for use by the first effect or stage through a desorber bed [32]. ABHP and ADHP have higher potential for applications in desalination than TVC and MVC; however, there are no commercial applications at the present time [33].

Solar assisted heat pumps can be combined with other desalination processes in many configurations as shown in Figure 13.8. Photovoltaics are used for MVC–based systems since they are driven by mechanical energy. In contrast, solar thermal processes are used for TVC, ABHP, and ADHP systems since they are driven by thermal energy.

13.3.4 Mechanical and Electrical Power Driven Processes

13.3.4.1 Solar Driven Reverse Osmosis

Osmosis is a natural phenomenon in which water passes through a membrane from a lower to a higher concentration solution. The flow of water can be reversed if a pressure larger than the osmotic pressure is applied on the lower concentration side. In RO desalination systems, seawater pressure is raised above the natural osmotic

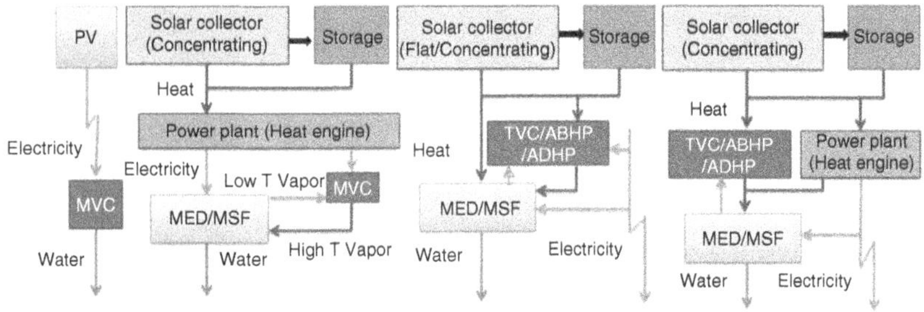

Figure 13.8 Possible configurations of solar assisted heat pumps systems

pressure of 2.5 MPa but kept below the membrane tolerance pressure, usually 6 to 8 MPa, forcing pure water through membrane pores to the fresh water side.

RO is the most common desalination process in terms of installed capacity due to its superior energy efficiency compared to thermal desalination systems despite requiring extensive water pretreatment; furthermore, part of the consumed mechanical energy can be reclaimed from the rejected concentrated brine with a suitable energy recovery device. Electricity generated by PV arrays or mechanical energy produced by solar ponds and collector–attached heat engines, such as Sterling engines or Rankine engines, can be employed to drive RO desalination as shown in Figure 13.9.

PV powered RO desalination is common in demonstration plants due to the modularity and scalability of both PV and RO systems. A lot of research has been conducted on combining RO with energy recovery device, battery, hybrid wind–diesel power source, or other desalination methods [34]. A detailed economic analysis [35] and a thorough optimization strategy [36] of PV powered RO desalination were carried out with favorable results reported. Economic and reliability considerations are the main challenges to improving PV powered RO desalination systems; however, the sharply dropping PV panel costs are making this desalination option ever more feasible.

Contrary to PV powered RO desalination which is commercially available in small scale, solar thermal assisted RO desalination is still far from commercialization. Combining solar augmented organic Rankine cycle (ORC) with RO desalination have been theoretically and experimentally studied [37]. The advantage of coupling ORC with RO is that seawater provides a heat sink for the ORC condenser while it is preheated to increase the RO membrane

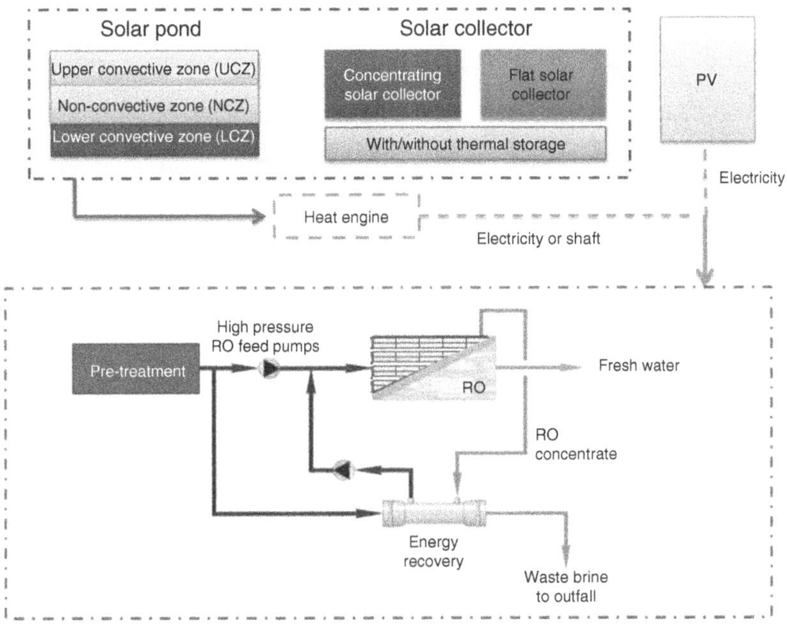

Figure 13.9 Solar driven reverse osmosis

permeability leading to reduced power consumption. Flat plate, evacuated tube, or concentrating solar collectors are used to heat the working fluid of the ORC to a design temperature and a fraction of the mechanical energy generated by the ORC power cycle can be used to drive the attached RO desalination system.

13.3.4.2 Solar Driven Electro–Dialysis

ED is an electrically driven desalination process suitable only for brackish water due to its high operating cost and expensive ion exchange membranes. Saline water is passed through an electro–dialysis stack consisting of alternating layers of cationic and anionic ion exchange membranes in an electrical field. Cations and anions then migrate in opposite directions through ion selective membranes and away from the saline feed in response to applied voltage across the electro–dialysis stack, producing fresh water in the intermediary channels. The electro–dialysis stack can be arranged in series to increase purification and in parallel to increase output. An ED schematic is shown in Figure 13.10.

PV arrays have been used by researchers to supply ED processes with their electrical load. In one instance, seawater was circulated a

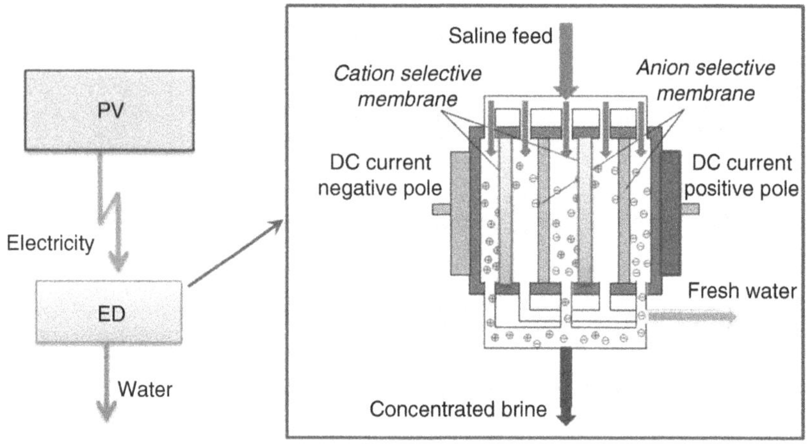

Figure 13.10 PV assisted electro–dialysis

few times to obtain the desired water quality in a small scale 10 m^3/day PV driven ED plant experiment [38].

13.3.4.3 *Solar Driven Membrane Distillation*

MD desalination requires both thermal and mechanical energies; therefore, driving it with solar energy is similar to solar assisted MSF or MED, as shown in Figure 13.11, which could use low grade heat from solar collectors or solar ponds and electricity from a PV system or a power grid. MD is a separation process that uses hydrophobic micro–porous membranes to prevent seawater from passing through membrane pores and only allow generated vapor to transfer to the other side. MD is a thermally driven process where vapor pressure gradient is the driving force for mass transfer across the membrane. There are four basic MD configurations: direct contact (DCMD), air gap (AGMD), sweeping gas (SGMD), and vacuum (VMD).

There are some disagreements among researchers about MD energy consumption and cost. Some believe that MD is unfavorable when compared with MED and MSF because of the additional resistance to mass transport and the reduced thermal efficiency due to heat conductivity losses resulting in increased energy demand [39], while others claim that the MD energy consumption is comparable to that of MSF plants with even less pumping power requirements [40]. In addition, MD uses membranes that are robust and

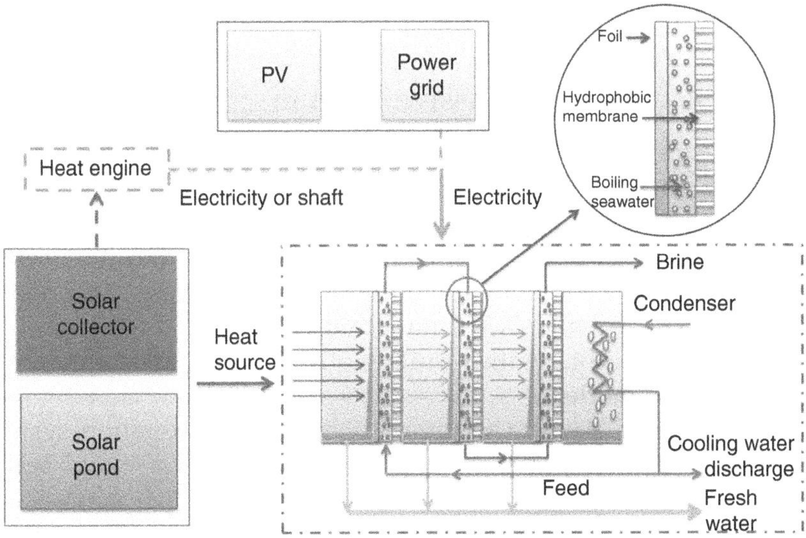

Figure 13.11 Solar driven membrane distillation

cheap and it operates at temperatures below 80 °C; therefore, MD saves on chemical usage, seawater pretreatment, and fuel costs.

Solar pond driven MD has been shown to be feasible; furthermore, modeling results have shown that augmenting MD with solar collectors would enhance membrane permeate flux [41]. There are only a few reports on solar driven MD desalination costs despite numerous published conventional MD desalination cost estimations. Banat et. al. estimated the water cost in the SMADES project to be $\$15/m^3$ for a 100 L/day system using a 10 m^2 membrane and 5.73 m^2 flat panel collectors, $\$18/m^3$ for a 500 L/day system using 40 m^2 membranes and 72 m^2 PV–flat panel collectors, and showed that cost could be further reduced by increasing reliability and plant lifetime [42].

Solar driven MD is still under development with reports on novel processes, experimentally confirmed modeling, and pilot plant evaluations continue to appear in the literature. MD has the disadvantage of additional resistance to mass transfer due to its membranes compared to MED and MSF; however, this disadvantage can be overcome by using more area for heat and mass transfer given the low cost of MD materials. In addition, MD could be used for high recovery or highly concentrated seawater that RO systems cannot handle and that normally require high energy consumption.

13.4 Non–Conventional Solar Desalination

13.4.1 Solar Assisted Passive Vacuum

The most common thermal desalination technique, multi–stage flash, was modified to have its system vacuum created passively and to have its thermal energy requirements drawn from solar insolation [43]. The modifications furthered feasibility and broadened applicability of the conventional desalination process. Theoretical [44] and experimental [45] simulations were run under various analogous conditions with results matching well, thus validating the developed model.

The proposed passive vacuum solar flash desalination system consists of a saline water tank, a concentrated brine tank, and a fresh water tank placed on ground level plus an evaporator and a condenser placed at least 10 meters above ground. The ground level tanks are open to the atmosphere, while the evaporator-condenser assembly, or flash chamber, is insulated and sealed to retain both heat and vacuum as shown in Figure 13.12.

The operation consists of pumping seawater through the condenser, preheating it before flowing it through the channels of a solar heater to reach a desired temperature by varying its flow rate in relation to available solar insolation. Consequently, the flash temperature would be kept constant via a variable speed feed pump that has

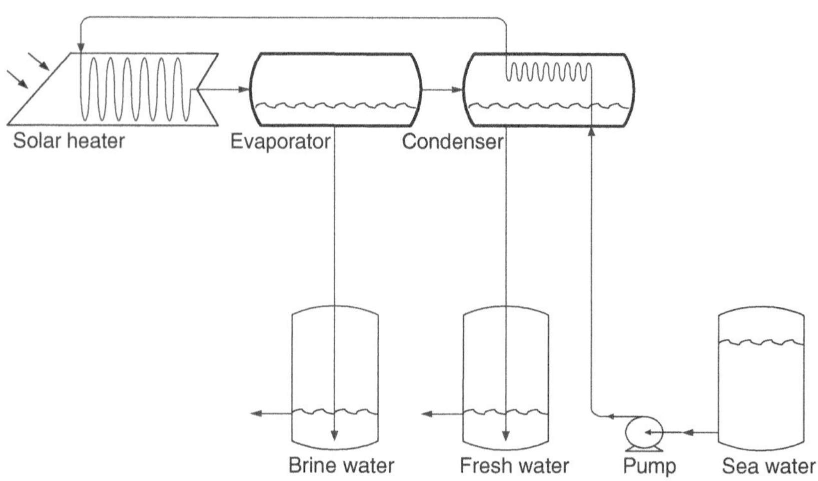

Figure 13.12 Single–stage solar flash desalination system

a control loop including temperature feedback and solar radiation feedforward components. Hot seawater then flashes into an insulated vacuumed evaporator through an expansion orifice or a pressure–reducing valve, producing water vapor and concentrated brine.

Produced water vapor flows to the condenser due to a vapor pressure gradient and condenses by losing its latent heat of condensation to feed seawater passing through the condenser while concentrated brine remains in the evaporator. Fresh water condensate as well as concentrated brine flow down to fresh water and brine water ground tanks, respectively, due to gravity through linking pipes that stretch down till just above the bottom of the tanks. Fresh water and brine water ground tanks have discharge pipes positioned slightly higher than the lip of the linking pipes, keeping their levels constant to maintain a vacuum in the flash chamber by a hydrostatic balance with their levels.

Multi–stage process can be implemented by flashing saline water in sequentially lower pressure flash chambers, as shown in Figure 13.13. Applying the multi–stage strategy to solar assisted passive vacuum will result in more evaporation and better recovery of the heat of condensation resulting in more water output at a better thermal efficiency.

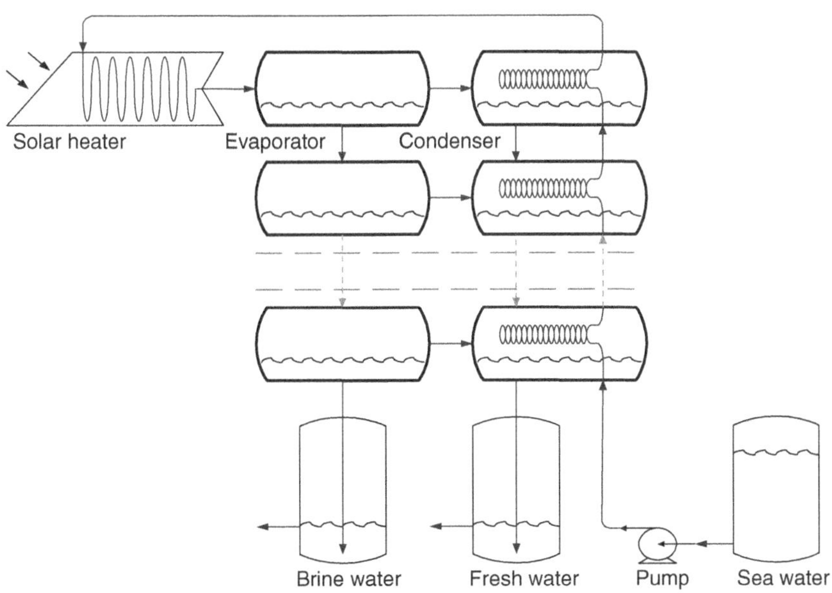

Figure 13.13 Multi–stage solar flash desalination system

The process was more feasible if operated at higher temperatures and moderate flow rates. Higher flash temperatures will result in more evaporation and subsequent condensation, resulting in more fresh water production. In addition, the increased amount of heat reclaimed from the condensing vapor reduced the overall heater load and thus required less solar collection area. The collective outcome of increased fresh water output and decreased heater load is a significant decrease in prime energy consumption of the desalination unit, making it more economically viable. In addition, most fresh water production occurs at the beginning of the operation, where vacuum pressure is lowest. The collective outcome of applying passive vacuum generation and solar heating schemes is a significant decrease in prime energy consumption due to lower energy input and higher product output, furthering the feasibility of the proposed desalination unit.

13.4.2 Solar Driven Humidification–Dehumidification

The HDH process, which uses low grade heat that could be supplied by solar collectors, is based on the fact that saturation humidity roughly doubles for every 10 °C increase in temperature. For example, air at 900 °C can hold five times more water than air at 70 °C. Air extracts some water vapor when it comes in contact with seawater at the expense of its sensible heat causing seawater to cool. On the other hand, the distilled water is recovered by maintaining humid air in contact with the cooling surface, thus recovering the latent heat of condensation from the vapor.

The main configurations of HDH are closed–air open–water cycle (CAOW) and closed–water open–air cycle (CWOA). The multi–effect CAOW water heated system is regarded as the most energy efficient among all the different HDH configurations. The schematic of one solar assisted multi–effect CAOW can be seen in Figure 13.13. . The basic cycle has a solar collector as the heat source, a humidifier, and a dehumidifier. Seawater passes through the collector where temperature rises and then through the humidifier where water vapor and heat are given up to the counter–current air stream cooling down the brine. Finally, air passes over the dehumidifier cooled by fresh or seawater [46].

The seawater greenhouse is another solar assisted HDH application shown in Figure 13.15. Seawater greenhouse produces fresh water plus cools and humidifies the crop growing environment.

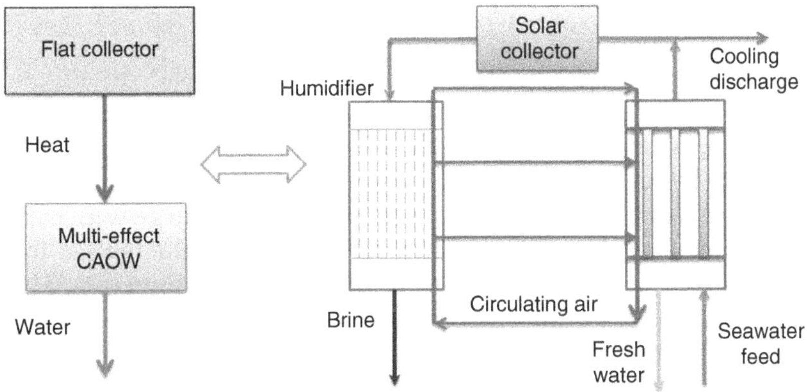

Figure 13.14 Solar assisted multi–effect CAOW system

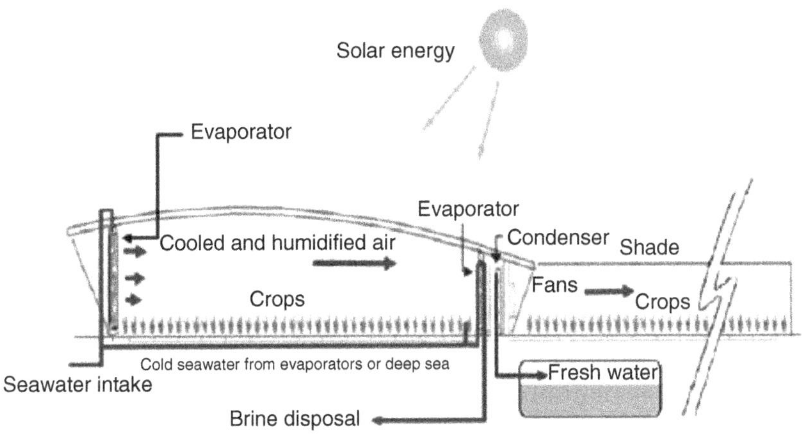

Figure 13.15 Solar assisted seawater greenhouse

It is suitable for arid regions because the plastic cover entraps long wave radiation and reduces transpiration; so, fresh water is produced and the environment is humidified. The seawater greenhouse is especially suitable for remote arid areas because it provides additional water supplies without relying on scarce groundwater; in essence, it makes agriculture immune to climatic variations.

Some of the optimization steps taken to improve HDH processes include: developing better simulation models, employing corrosion–free collectors, incorporating a cooling tower into the schematic, using pinch analysis to improve performance, adjusting the seawater/air flow ratio, and many others [47].

13.4.3 Power–Water Cogeneration

A lot of research has been conducted on using solar desalination in remote arid areas that normally use small scale desalination systems. Hybrid thermal desalination systems generally have a relatively lower cost compared to similar capacity solar only desalination systems; however, most of them tend to be large with capacities larger than 100 m^3/day. Hybrid systems could reduce fossil fuel energy consumption, save on fuel transportation costs, avoid the drawback of an intermittent solar source, and provide crucial water that is not weather and season dependent. Nonetheless, very few studies have been reported on small scale solar assisted hybrid desalination systems leaving a big opportunity for research and development of such systems.

It is worth mentioning that thermal energy obtained from solar collectors is not really equivalent to thermal energy from waste heat. Waste heat sources are considered once–through heat sources as opposed to the solar resource being a recirculating source. This difference could impact the choice of heat engines for the desalination system; therefore, the two types of heat sources should be analyzed separately.

CSP plants produce electricity via generators joined to steam turbines that are supplied with solar generated steam. Numerous sun tracking mirrors focus sunlight onto an aperture producing the necessary heat to generate steam to drive the steam turbines of conventional Rankine cycle power plants. Part of this solar heat can be diverted to desalinate seawater using a thermal process such as MSF or MED. The steam produced from seawater can then be used in the power cycle to preheat feed water going to the evaporator before it is condensed to produce fresh water. Solar generated steam can also be expanded in steam turbines producing power then condensed producing fresh water.

13.5 Process Evaluation

13.5.1 System Integration

There are many kinds of desalination processes and solar technologies. Selecting a suitable desalination process requires several design criteria, such as seawater quality, desired fresh water quality, energy efficiency, environmental impact, capital cost, operation and maintenance cost, plus other site–specific factors [48]. Selecting a suitable solar system requires many considerations, such as available solar resource, location, energy storage method, operating temperature range, plant configuration, solar collector type, working fluids, and others. Most of the reported solar desalination systems are not developed as a single system but are integrations of components developed independently though some systems requires some minor changes for better integration.

Solar assisted desalination means that either solar generated electricity is used to power mechanical processes or that solar heat is employed in thermal processes. Solar chimneys, solar dish, and photovoltaic cells directly convert solar radiation to electricity; therefore, they are suitable to be combined with mechanical processes like MVC and RO. FPC, ETC, CPC, PTC, and solar dish collectors plus Fresnel mirrors, solar tower, and solar ponds can be used to simultaneously generate heat and electricity; therefore, they could be combined with any kind of desalination technology based on the design. Probable processes of different solar technologies combined with seawater desalination technologies are shown in Figure 13.16.

13.5.2 Solar System Considerations

Solar system cost could range from 17–77% of total system costs based on configuration. All solar assisted desalination processes

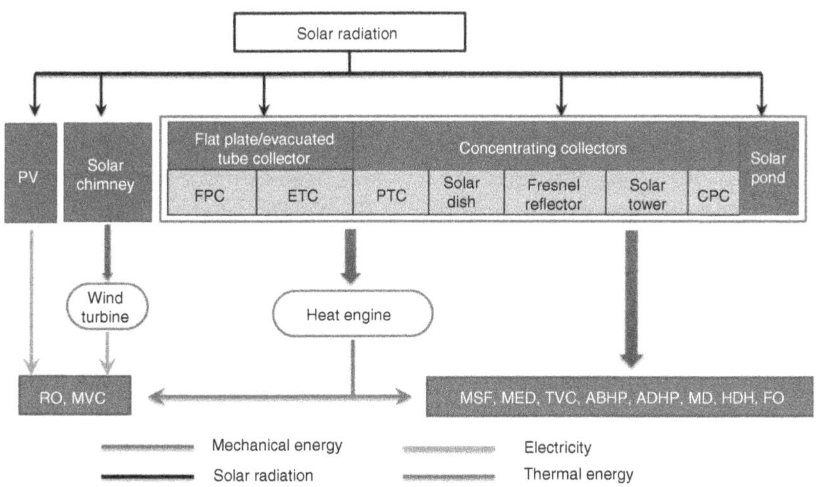

Figure 13.16 Possible solar desalination arrangements

have more than 25% additional cost spent for solar collectors with the exception of the solar pond driven desalination that does not require solar collectors. In general, the solar system represents a very large part of the overall system cost and needs to be carefully selected. Location is one of the most important factors when selecting a solar system or a desalination scheme. A certain solar desalination system will provide higher water production rates at locations with higher solar radiation; therefore, lowering the overall water cost.

Solar ponds possess large surface areas while pond evaporation rates sometimes exceed water production rates making solar ponds inappropriate. For organic Rankine cycle assisted RO systems, exergy destruction in the power plant subsystem was shown to be almost ten times greater than that in the RO subsystem indicating that the overall efficiency depends more on the solar plant and less on the desalination system [49]. A general rule of thumb for simple solar stills is to aim for at least 3–5 liter of water per day per m^2 of solar still area. For example, a small family that consumes water at a rate of 0.6 m^3/day will need 120–200 m^2 land area of a simple solar still [50].

Compared to PV, CSP has the advantage of using a thermal energy storage system to extend the hours of operation beyond sunset and the ability to use a backup fuel for unexpected conditions. CSP is a more appropriate solar power technology for large scale power production; nevertheless, CSP plants would have little output on cloudy or foggy days due to very low direct normal radiation. Thermal energy storage is very important and new materials are always under research and development for CSP applications.

DNI between 1900–2100 kWh/m^2–year is considered the viability threshold for CSP systems whereas PV is considered suitable at much lower radiation levels because it exploits both direct and diffuse irradiance [51]. Coastal areas usually have higher humidity thus lowering the DNI; furthermore, coastal areas normally have higher land value due to tourism and population density. CSP requires large flat land which could increase the cost of CSP assisted desalination plants if built in coastal areas. A CSP plant combined with an RO system is regarded as one of the best choices for solar desalination.

Photovoltaic cells are made from common semiconductor compounds and can directly convert solar radiation into useful electricity. Cells are arranged to form modules that are combined to form panels. Photovoltaic systems include an array of joined panels to produce the required electrical output. Photovoltaics can be employed independently or jointly with other sources to generate electricity needed to power desalination systems.

PV is usually used to power mechanical desalination processes such as MVC or RO. PV module cost is the determining factor for feasibility of PV assisted desalination system. PV module prices have been dropping dramatically in recent years due to mass production making PV powered desalination more of an attractive option. In addition, PV systems require very little maintenance compared to CSP thus increasing their popularity. The modular nature of PV systems simplifies their scale up so as to allow a project to be built in phases. PV systems do not require water at all; therefore, solar and desalination subsystems are developed independently. RO desalination entails stringent pretreatment that normally require skillful labor; therefore, the application of PV driven desalination needs to be carefully evaluated in remote areas where it is hard to find skillful operators and where both energy and water are not very abundant.

13.5.3 Solar Collectors

The solar collection sub–system of an indirect solar desalination system is essentially a solar collector that absorbs incident solar radiation and transfers heat to a fluid flowing through it. The working fluid of the collector can either be a medium to transfer heat to the process or to a thermal energy storage reservoir, or it can be the seawater itself before going through a thermal desalination system. Solar collectors can be either stationary or tracking. Tracking solar collectors can be designed to go after the rays of sunlight by moving around either a single axis or double axes.

Solar collectors can also be classified as concentrating and non–concentrating types. The concentration ratio of a solar collector is the relative amount of the solar flux on the receiver to flux on the aperture. Concentrating collectors have a highly reflective surface to reflect and concentrate solar radiation onto a receiver or an absorber, while non–concentrating collectors have a highly absorptive surface with low emittance to maximize heat transfer to the working fluid. Solar collectors are chosen according to the desired process temperature. Table 13.2 includes an extensive list of solar collectors and their operational temperature ranges [6].

13.5.4 Photovoltaics

Photovoltaic (PV) cells can be made from common semiconductors like silicon or germanium or semiconductor compounds such as GaAs, CdTe, CuIn, GaSe …etc. PV cells in their simplest form are

Table 13.2 Solar collectors [6]

Tracking	Collector Type	Absorber	Concentration Ratio	Operational Range
Stationary	Flat plate	Flat	1	30–80 °C
	Evacuated tube	Flat	1	50–200 °C
	Compound parabolic	Tubular	1–5	60–240 °C
Single–axis	Compound parabolic	Tubular	5–15	60–300 °C
	Linear Fresnel	Tubular	10–40	60–250 °C
	Parabolic trough	Tubular	15–45	60–300 °C
	Cylindrical trough	Tubular	10–50	60–300 °C
Double–axis	Parabolic dish	Point	100–1000	100–500 °C
	Heliostat field	Point	100–1500	150–2000 °C

large area electronic semiconductor diodes allowing current to flow in the reverse direction in the presence of light. PV cells can directly convert solar radiation into useful electricity, as shown in Figure 13.17 [5]. Cells are connected in series and/or parallel configurations to form a PV module or panel. Photovoltaic panels can be designed for specific voltage and current output when the sun rays strike the module vertical to its surface and it has an intensity of 1,000 watts per square meter. Under these conditions the PV module power output is expressed in peak watts or peak kilowatts. Photovoltaic systems include an array of joined panels to produce the required electrical output. Figure 13.18 [5] shows a schematic of a PV system that includes storage (batteries) for a stand-alone operation. Since the PV systems generates a DC power output, an inverter is used to convert DC to AC. Photovoltaics can be employed independently or jointly with other sources to generate the electricity needed to power desalination systems.

13.5.5 Environmental Impact

The brine discharged from desalination plants has higher temperature and higher salinity than the seawater surrounding the

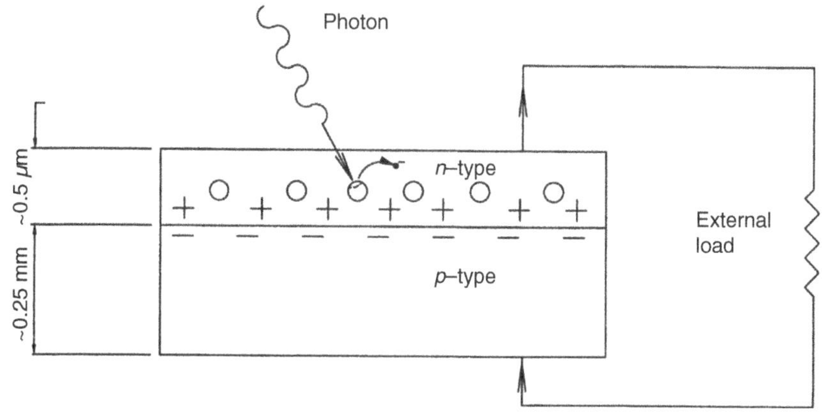

Figure 13.17 Photovoltaic cell schematics

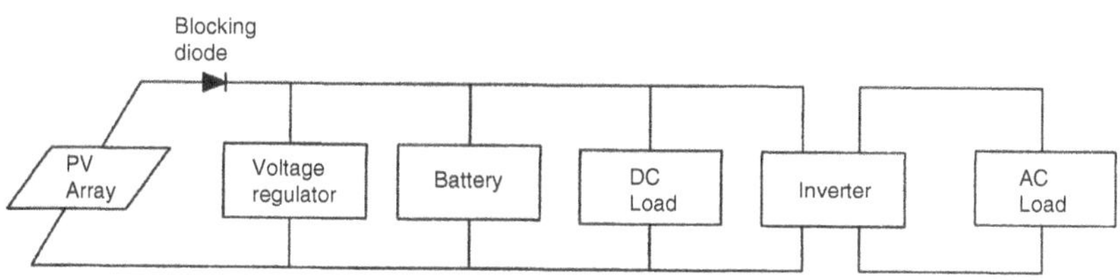

Figure 13.18 Schematic of a stand–alone photovoltaic system

plant; therefore, many researchers have expressed concerns about the environmental impact caused by desalination plants especially environmental and ecological impacts that have occurred around older MSF plants discharging to the sea with little flushing [52]. Some ocean animals could not tolerate high salinity environments such as oceanic Posidonia, which could only tolerate a maximum salinity of 39 g/L NaCl while most discharge brine salinities are higher than 60 g/L NaCl; therefore, high recovery or near zero liquid discharge technologies need to be further developed [52]. Brauns studied the energy collected from salinity gradient power by reverse electro–dialysis combined with a seawater desalination unit [53]. Similar ideas could be expanded to combining a solar driven MED with a MVC system during the day and operating just the MVC system at night. Temperature change is also a big concern for ocean animals since most desalination plants discharge brine at a temperature higher than 40 °C. Many desalination plants mix untreated seawater with rejected brine in order to lower the discharge temperature; however, this will involve more plant; therefore, many researchers have expressed concerns about the environmental impact caused by desalination plants especially environmental and ecological impacts that have occurred around older MSF plants discharging to the sea with little flushing [52]. Some ocean animals could not tolerate high salinity environments such as oceanic Posidonia, which could only tolerate a maximum salinity of 39 g/L NaCl while most discharge brine salinities are higher than 60 g/L NaCl; therefore, high recovery or near zero liquid discharge technologies need to be further developed [52]. Brauns studied the energy collected from salinity gradient power by reverse electro–dialysis combined with a seawater desalination unit [53]. Similar ideas could be expanded to combining a solar driven MED with a MVC system during the day and operating just the MVC system at night. Temperature change is also a big concern for ocean animals since most desalination plants discharge brine at a temperature higher than 40 °C. Many desalination plants mix untreated seawater with rejected brine in order to lower the discharge temperature; however, this will involve more pumping power further increasing the desalination cost. Alarcón et. al. incorporated an absorption heat pump (LiBr–H_2O) to partially recover the heat rejected from the MED unit so that the heat discharged to the environment could be significantly reduced [54]. As for the potential environmental hazards caused by solar power systems, battery banks and heat transfer fluids are the main concerns. Implementing thermal energy storage systems can replace the need for batteries while including a properly sized HTF containment structure in the plant will help eliminate the hazard of an accidental HTF release.

References

1. National Atlas of the United States. *Water Resources of the United States*; U. S. Geological Survey: Reston, VA (2005).
2. P.E. Minton, *Handbook of Evaporation Technology*; Noyes Publications: Westwood, NJ (1986).
3. E. Maxwell, R. George and S. Wilcox *Climatological Solar Radiation Model*; the National Climatic Data Center: Asheville, NC (1998).
4. A. Culp, *Principles of Energy Conversion*; McGraw Hill: New York, NY (1991).
5. D.Y. Goswami, F. Kreith and J.F. Kreider, *Principles of Solar Engineering*, second edition; Taylor & Francis: Philadelphia, PA (2000).

6. S.A. Kalogirou, Seawater desalination using renewable energy sources, *Prog. Energy Combustion Sci* **31(3)**, 242–281 (2005).
7. R. Tripathi, G.N. Tiwari, Effect of water depth on internal heat and mass transfer for active solar distillation. *Desalination* **173**, 187–200 (2005).
8. H.N. Singh, G.N. Tiwari, Monthly performance of passive and active solar stills for different Indian climatic conditions. *Desalination* **168**, 145–150 (2004).
9. S. Kalogirou. Survey of solar desalination systems and system selection. *Energy* **22**, 69–81 (1997).
10. E. Delyannis, Historic Background of Desalination and Renewable Energies, *Solar Energy* **75(5)**, 357–366 (2003).
11. E. Tzen, R. Morris, Renewable Energy Sources for Desalination, *Solar Energy* **75(5)**, 375–379 (2003).
12. T. Szacsvay, P. Hofer-Noser and M. Posnansky, Technical and economic aspects of small-scale solar-pond-powered seawater desalination systems. *Desalination* **122**, 185–193 (1999).
13. M. Posnansky, Technical and economical aspects of solar desalination with particular emphasis on solar pond powered distillation plants. *Desalination* **67**, 81–95 (1987).
14. K. Tahri, The prospects of fresh water supply for Tan Tan City from non-conventional water resources. *Desalination* **135**, 43–50 (2001).
15. M.J. Safi, A. Korchani, Cogeneration applied to water desalination: Simulation of different technologies. *Desalination* **125**, 223–229 (1999).
16. M.S. Abu-Jabal, I. Kamiya and Y. Narasaki, Proving test for a solar-powered desalination system in Gaza-Palestine. *Desalination* **137**, 1–6 (2001)..
17. K.R. Agha, The thermal characteristics and economic analysis of a solar pond coupled low temperature multi stage desalination plant. *Solar Energy* **83**, 501–510 (2009).
18. ,M.A. Farwati, Theoretical study of multi-stage flash distillation using solar energy. *Energy* **22**, 1–5 (1997).
19. J. Joseph, R. Saravanan and S. Renganarayanan, Studies on a single-stage solar desalination system for domestic applications. *Desalination* **173**, 77–82 (2005).
20. H. Shaobo, Z. Zhang, Z. Huang and A. Xie, Performance optimization of solar multi-stage flash desalination process using Pinch technology. *Desalination* **220**, 524–530 (2008).
21. J. Jiang, H. Tian, M. Cui and L. Liu, Proof-of-concept study of an integrated solar desalination system. *Renewable Energy* **34**, 2798–2802 (2009).
22. O. Al Hawaj, M.A. Darwish, A solar pond assisted multi-effect desalting system. *Desalination* **99**, 119–135 (1994).
23. M.A. Garman, M.A. Muntasser, Sizing and thermal study of salinity gradient solar ponds connecting with the MED desalination unit. *Desalination* **222**,689–695 (2008).

24. A.M. El-Nashar, Validating the performance simulation program "SOLDES" using data from an operating solar desalination plant. *Desalination* **130**, 235–253 (2000).
25. A.M. El-Nashar, Seasonal effect of dust deposition on a field of evacuated tube collectors on the performance of a solar desalination plant. *Desalination* **239**, 66–81 (2009).
26. S. Kalogirou, Economic analysis of a solar assisted desalination system. *Renewable Energy* **12**, 351–367 (1997).
27. E. Zarza, J. Ajona, J. León, A. Gregorzewski and K. Genthner, Solar thermal desalination project at the Plataforma Solar de Almeria. *Solar Energy Mater* **24**, 608–622 (1991).
28. D.C. Alarcón-Padilla, L. García-Rodríguez and J. Blanco-Gálvez, Design recommendations for a multi-effect distillation plant connected to a double-effect absorption heat pump: A solar desalination case study. *Desalination* **262**, 11–14 (2010).
29. R. Kouhikamali, M. M, Process investigation of different locations of thermo-compressor suction in MED-TVC plants. *Desalination* n.d.; In Press, Corrected Proof.
30. H. Ettouney, Design of single-effect mechanical vapor compression. *Desalination* **190**, 1–15 (2006).
31. V.V. Slesarenko, Desalination plant with absorption heat pump for power station. *Desalination* **126**, 281–285 (1999).
32. A. Al-Ansari, H. Ettouney and H. El-Dessouky, Water-zeolite adsorption heat pump combined with single effect evaporation desalination process. *Renewable Energy* **24**, 91–111 (2001).
33. D. Zejli, R. Benchrifa, A. Bennouna and O.K. Bouhelal, A solar adsorption desalination device: first simulation results. *Desalination* **168**, 127–135 (2004).
34. B. Parida, S. Iniyan and R. Goic, A review of solar photovoltaic technologies. *Renewable Sustainable Energy Rev* **15**, 1625–1636 (2011).
35. G. Fiorenza, V.K. Sharma and G. Braccio, Techno-economic evaluation of a solar powered water desalination plant. *Energy Conversion Manage* **44**, 2217–2240 (2003).
36. H.M. Laborde, K.B. França, H. Neff and A.M.N. Lima, Optimization strategy for a small-scale reverse osmosis water desalination system based on solar energy. *Desalination* **133**,1–12 (2001).
37. D. Manolakos, G. Papadakis, E.S. Mohamed, S. Kyritsis and K. Bouzianas, Design of an autonomous low-temperature solar Rankine cycle system for reverse osmosis desalination. *Desalination* **183**, 73–80 (2005).
38. O. Kuroda, S. Takahashi, S. Kubota, K. Kikuchi, Y. Eguchi, Y. Ikenaga, N. Sohma, K. Nishinoiri, S. Wakamatsu and S. Itoh. An electrodialysis sea water desalination system powered by photovoltaic cells. *Desalination* **67**, 33–41 (1987).

39. S. Al-Obaidani, E. Curcio, F. Macedonio, G. Di Profio, H. Al-Hinai and E. Drioli,. Potential of membrane distillation in seawater desalination: Thermal efficiency, sensitivity study and cost estimation. *J. Membr. Sci* **323**, 85–98 (2008).
40. A.M. Alklaibi and N. Lior Membrane-distillation desalination: Status and potential. *Desalination* **171**, 111–131 (2005).
41. J.-P. Mericq, S. Laborie and C. Cabassud, Evaluation of systems coupling vacuum membrane distillation and solar energy for seawater desalination. *Chem. Eng. J* **166**, 596–606 (2011).
42. F. Banat, N. Jwaied, M. Rommel, J. Koschikowski and M. Wieghaus, Performance evaluation of the "large SMADES" autonomous desalination solar-driven membrane distillation plant in Aqaba, *Jordan. Desalination* **217**, 17–28 (2007).
43. M. Abutayeh, "Theoretical and Experimental Simulation of Passive Vacuum Solar Flash Desalination," University of South Florida (2010).
44. M. Abutayeh and D.Y. Goswami, Passive vacuum solar flash desalination. *AIChE J* **56**, 1196–1203 (2009).
45. M. Abutayeh, D.Y. Goswami, Experimental Simulation of Solar Flash Desalination. *J. Sol. Energy Eng* **132**, 041015–041017 (2010).
46. H. Müller-Holst, M. Engelhardt, M. Herve and W. Schölkopf, Solar thermal seawater desalination systems for decentralised use. *Renewable Energy* **14**, 311–318 (1998).
47. J. Orfi, M. Laplante, H. Marmouch, N. Galanis, B. Benhamou, S.B. Nasrallah and C.T. Nguyen, Experimental and theoretical study of a humidification-dehumidification water desalination system using solar energy. *Desalination* **168**, 151–159 (2004).
48. Minton, E. Paul, *Handbook of Evaporation Technology* (1986).
49. N. Bouzayani, N. Galanis and J. Orfi. Thermodynamic analysis of combined electric power generation and water desalination plants. *Appl. Thermal Eng* **29**, 624–633 (2009).
50. R. Semiat, Energy Issues in Desalination Processes. *Environ. Sci. Technol* **42**, 8193–8201 (2008).
51. Technology Roadmap - Concentrating Solar Power 2010.
52. T. Mezher, H. Fath, Z. Abbas and A. Khaled, Techno-economic assessment and environmental impacts of desalination technologies, *Desalination* **266(1–3)**, 263–273 (2011).
53. E. Brauns, Towards a worldwide sustainable and simultaneous large-scale production of renewable energy and potable water through salinity gradient power by combining reversed electrodialysis and solar power. *Desalination* **219(1)**, 312-323 (2008).
54. D. Alarcón-Padilla and L. García-Rodríguez, Application of absorption heat pumps to multi-effect distillation: a case study of solar desalination, *Desalination* **212(1)**, 294-302 (2007).

SECTION VI

FUTURE EXPECTATIONS

14

Future Expectations

Patrick V. Brady and Michael M. Hightower

Sandia National Laboratories Albuquerque, New Mexico, USA

Abstract

Because of increased water demands, advanced water treatment will increasingly target impaired waters - municipal wastewaters, agricultural return flows, oilfield produced waters, inland brines, and seawater - for drinking water, industrial processes, and energy extraction and production. Better membranes and new technologies for concentrate management will be critical to future desalination and reuse.

Keywords: Desalination, enhanced oil recovery, produced water, nanotechnology, advanced water treatment, hydrofracking, impaired water reuse, water availability, silica removal.

14.1 Introduction

In the coming years demand for water for drinking, agriculture, and industry can be expected to grow. Because primary water supplies are increasingly allocated, or over-allocated, there will be greater reliance on advanced water treatment that targets non-traditional, impaired waters - municipal wastewaters, oilfield produced waters, inland brines, and seawater. Membrane filtration will play a central role as will its associated technologies of pre-treatment, scale and biofouling inhibition, and concentrate disposal. The first half of this chapter documents increased future water requirements and explores several industrial water uses likely to loom large in the coming decades, including hydrofracking and chemical

Jane Kucera (ed.) Desalination , (585–618) 2014 © Scrivener Publishing LLC

waterflooding for gas and oil, and use in cooling towers for electricity production. The second half of this chapter identifies the most pressing water treatment research needs, and then examines water production technologies presently being developed in the lab that might provide 'new water' in the future.

14.2 Historical Trends in Fresh Water Supply Development

Over the past 50 years, the demand for fresh water for drinking water and domestic needs, for irrigated agriculture and livestock production, energy production and generation, and commercial and industrial applications has grown significantly as economic development and population have also continued to grow. The demand for fresh water resources is often met through one of four typical water development strategies:

1. Developing new local fresh surface or ground water resources to increase water supplies,
2. Importing fresh water from other areas to supplement local supplies,
3. Increasing water use efficiency and reducing water demand in water use intensive sectors such as agriculture, domestic uses, and industrial and energy applications, or
4. The treatment or substitution of marginal quality or impaired waters for fresh water to meet specific needs – such as for lower quality industrial water needs, cooling water, etc.

Since the early 1900's the major water development strategies in most regions of the world focused on developing local and regional fresh water resources and reallocating those fresh water resources within a region or watershed. While these two common approaches significantly improved water supplies around the globe in the 20th century, the approach has led to increasingly allocated, and often the over-allocation, of fresh surface and ground water resources. This is highlighted in Figures 14.1 and 14.2, which show the trends in development and utilization of fresh surface water and fresh ground waters supplies in the U.S. since

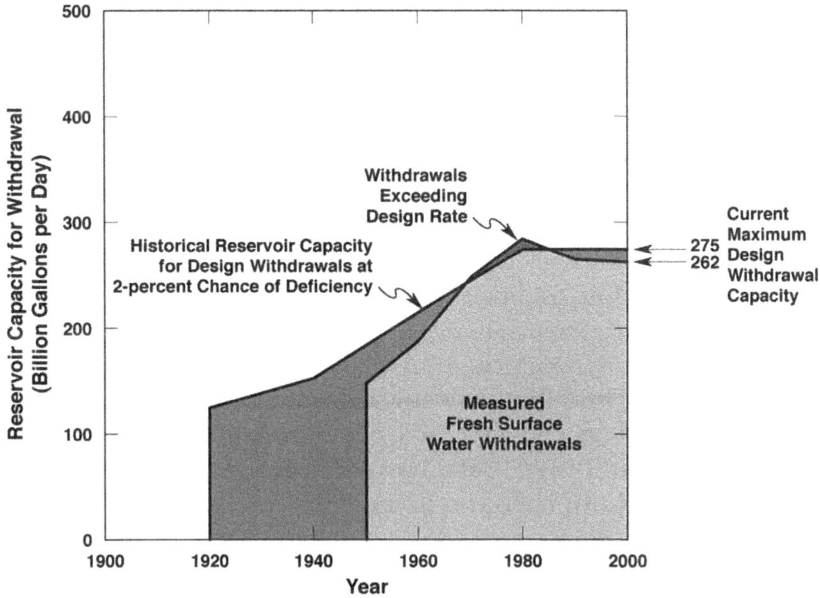

Figure 14.1 U.S. fresh surface water withdrawal capacity 1920-2000 (from Sandia National Laboratories 2009).

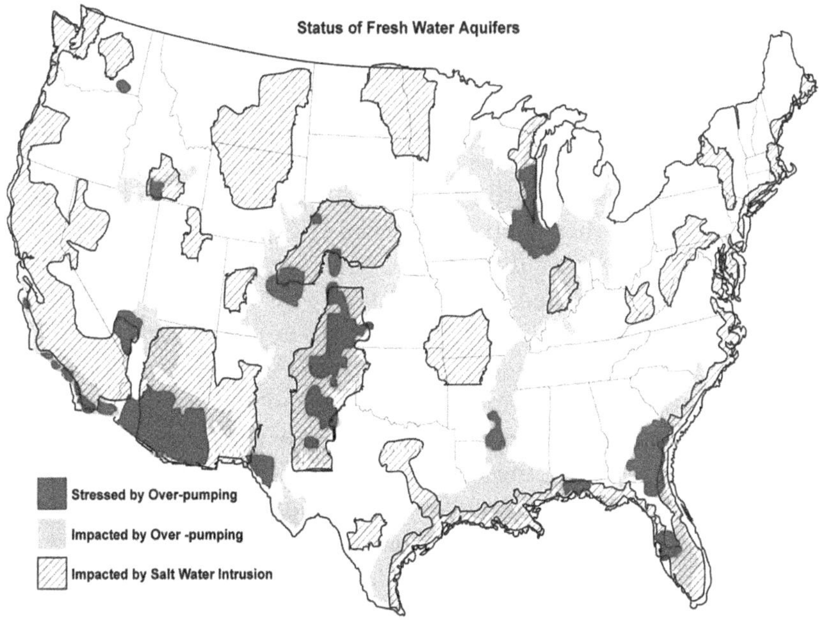

Figure 14.2 Major ground water aquifers impacted by overpumping (from Sandia National Laboratories 2009).

1920, when many of the large major dams in the U.S. started to become operational.

Several thousand dams were built in the U.S. in the 20th century – adding several million acre-feet of fresh surface water storage, and tripling fresh surface water withdrawal capacity in the U.S. The stored surface water was used for multiple needs and uses including agriculture, hydroelectric power, and domestic, commercial, and industrial uses. These dams were also typically built to help reduce flooding and better manage water runoff, also provide irrigation and/or recreation, or (in part early on) as public works projects. As shown in Figure 14.1, by 1980 very few new large dams were being built, and the fresh surface water withdrawal capacity in the U.S., based on a 50–year or 2-percent chance of drought deficiency design withdrawal rate, has remained constant since then. Currently, few additional dams to increase fresh surface water storage capacity are being pursued, but rather dam removal efforts are being considered, suggesting that in the next several decades, fresh surface water withdrawal capacity will actually be reduced through future dam removals, siltation of current reservoirs, and the projected reduction in fresh surface water inflows through the impacts of climate change. Now dams are filling up with sediment at a rate that is substantially faster than new ones are being built (at least in the U.S.). The storage capacity – in essence, the "slack" in the surface hydrologic system - built over the 20th century is rather quickly being lost due to siltation and the public resistance to further dam building. This loss of system flexibility comes at a time when it will be tested hardest as runoff is expected to be increasingly more erratic due to climate change.

To meet the growing demand for fresh water that could not be met by available surface water supplies, by 1950 and 1960 many locations and municipalities turned to the development and use of fresh ground water resources across the U.S. As shown in Figure 14.2, this development went somewhat unchecked in many regions, and has led to significant overpumping of many major aquifers across the U.S. This overpumping has led to a number of issues including salinization of the ground water in the aquifers, salt water intrusion into the fresh water aquifers along the coasts, significant water table draw down in many regions, and land subsidence of ten to more than thirty feet in many regions with dropping water tables.

Figures 14.1 and 14.2 show that the U.S. has developed much of the easily or readily available renewable fresh surface water resources and has overpumped much of the readily available fresh ground water aquifers. This has put many states in the position of pushing the availability and sustainability limits of many local and regional water supply availability in many areas, with little excess capability to meet water needs in drought conditions.

Figure 14.3 identifies states that expect to experience local and regional water shortages by 2013 under average conditions, as identified by state water managers in a 2003 GAO study [1]. As shown in Figure 14.3, most of the 47 water managers that participated in the study expected local and regional water shortages by 2013. In the same study, almost half of the respondents expected regional or statewide water shortages 2013 under drought conditions.

Therefore, the fresh water supply infrastructure, as well as fresh water resources, in many regions are less resilient now because of continued overuse and over demands. This has forced many regional and state water management agencies and professionals to consider other options such as increasing fresh water use efficiency, importing fresh water from greater distances, and greater reliance on advanced water treatment that increasingly targets impaired waters - municipal wastewaters, municipal storm water, industrial waste water, oilfield produced waters, inland brackish ground water, minewater, seawater; or other nontraditional water resources – to create additional or new fresh water supplies.

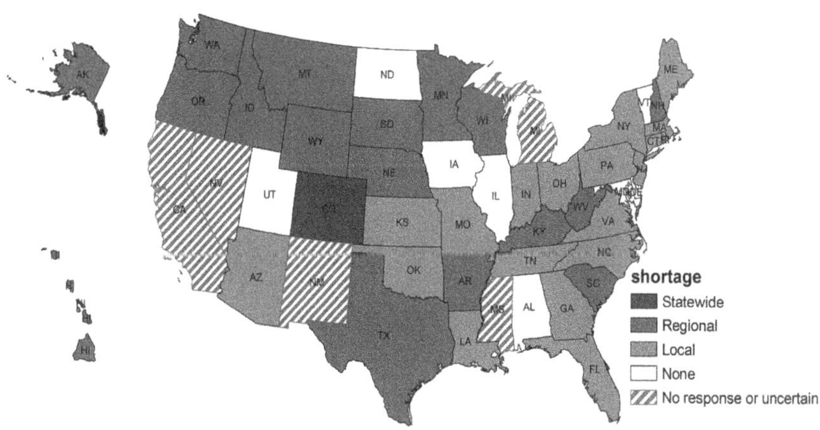

Figure 14.3 Expected water supply shortages by 2013 under average conditions (from Sandia National Laboratories 2009).

14.3 Emerging Trends and Directions in Alternative Water Supply Development

An evaluation conducted by Sandia National Laboratories [2] looked at the general scale of the energy requirements of some different water supply development options. Since energy use is a major and recurring expense in many alternative water supply development approaches and is one of the largest cost elements for these approaches, energy use provides a reasonable comparison of the relative costs of various alternative water supply options.

Water conservation and improved water use efficiency require almost no energy to implement and are therefore the most cost effective fresh water supply augmentation strategies. Relative to several other options, as highlighted in Table 14.1, waste water reuse and storage, and brackish water desalination and treatment are less energy intensive than fresh water importation from long distances. Even seawater desalination is competitive in energy demand to importing fresh water supplies from hundreds of miles away. These results show that energy demands for these advanced treatment oriented options will vary by location and can overlap, and therefore the most cost-effective approach will depend on

Table 14.1 Energy Demand for Several Alternative Water Supply Enhancement Options [2].

Water Supply Options	Energy Demand (kWhr/kgal)
Fresh Water Importation (100–300 miles)	10–18
Seawater Desalination w/Reverse Osmosis	12–20
Brackish Groundwater Desalination	
Reverse Osmosis Treatment	7–9
Pumping/concentrate management	1–3
Total	8–12
Waste Water Aquifer Storage and Recovery	3–4
Pre-treatment (as needed)	3–4
Post-treatment (as needed)	2–3
Pumping	5–11
Total	

site-specific conditions, but the results highlight that desalination and impaired water reuse technologies are not only cost-effective relative to other approaches, but in many cases can be the most cost effective alternative for supplementing fresh water resources in many regions.

From the results in Table 14.1, one might expect that water reuse and desalination would be an emerging trend and direction in new water supply development. Figure 14.4 shows the current and projected growth in water treatment of impaired supplies, though waste water reuse and desalination in the U.S. over the past two decades. Not surprisingly, as suggested by the information in Table 14.1, the lowest energy use and low water treatment cost waste water reuse is growing at approximately 15 percent per year, while the slightly higher energy cost desalination is growing at approximately 10 percent per year, which is still a very fast pace. Therefore, at these rates of growth, desalination and water reuse, both technologies which currently rely heavily on the same type of advanced membrane treatment technologies, could emerge as the major "new" water supply resources in the future.

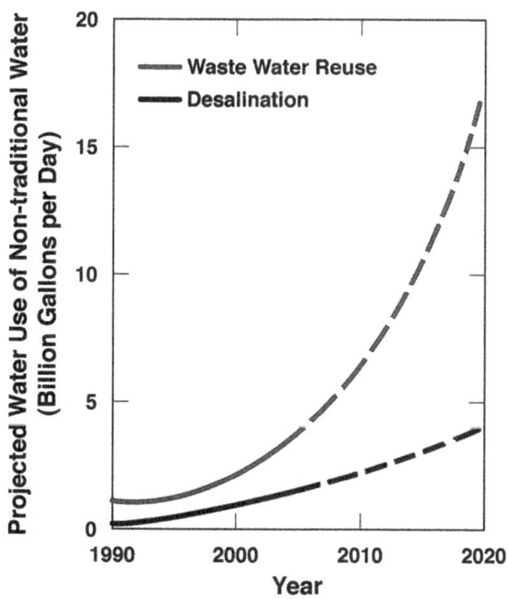

Figure 14.4 Growth in waste water reuse and desalination in the U.S. since 1990 (from Sandia National Laboratories 2009).

Also not surprisingly, these lower energy cost water supply development technologies, are seeing significant growth in providing alternative sources of fresh water supplies to cities, and regions around the globe. Therefore, membrane filtration and desalination will play a central role as will its associated technologies of pretreatment, scale and biofouling inhibition, and concentrate disposal in the next several decades in addressing the needs of many states, and municipalities, in developing new and long-term sustainable water supplies.

14.3.1 Desalination of Impaired Waters

Over-allocated freshwater supplies will increasingly highlight the reclamation of "Impaired waters" - including inland brines, municipal wastewaters, agricultural return water, and waters produced from oil and gas operations. Table 14.2 gives representative compositions of each, along with freshwater for comparison. Seawater is included as it is also a non-traditional source of drinking water.

Table 14.2 Impaired water compositions.

	Produced water	Ag Return water	Waste water	Seawater	Fresh water
Na	982	2182	76	10752	41
Ca	40		76	416	18
Mg	13	554	43	1295	.76
HCO_3	1100	270	396	145	92
Cl	920	239	102	19345	22
SO_4	110	4730	68	2701	31
SiO_2	120	37	17	10	16
NO_3		48			
Total PO_4		2	6	0.1	
TDS	3879	9723	869	~35,000	297
BOD	30	3	8		

Notes: Produced, Ag return, Reclaimed municipal effluent (secondary treated water) and Fresh water are from [3].

Figures 14.5 and 14.6 show the opportunities and challenges with treating and utilizing inland brackish ground water to supplement drinking water supplies. Figure 14.5 shows the locations in states where brackish water has been identified, while Figure 14.6 shows

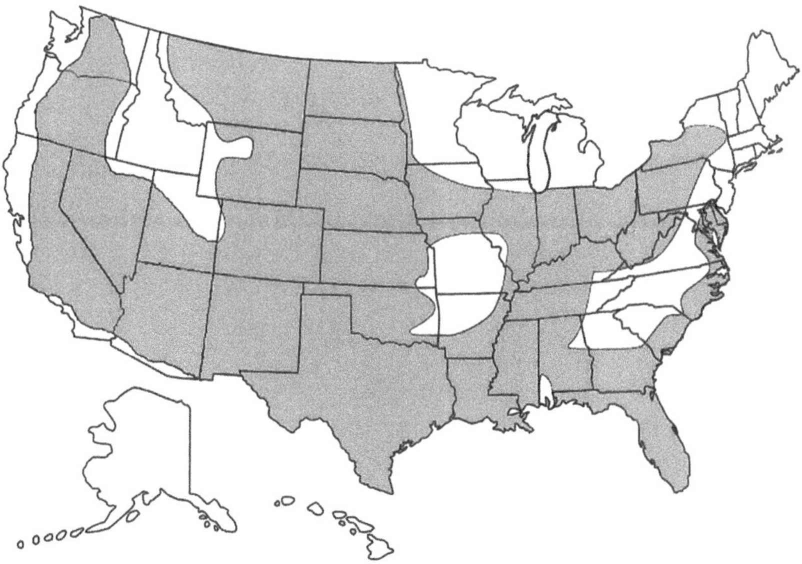

Figure 14.5 Locations of identified brackish ground water resources.

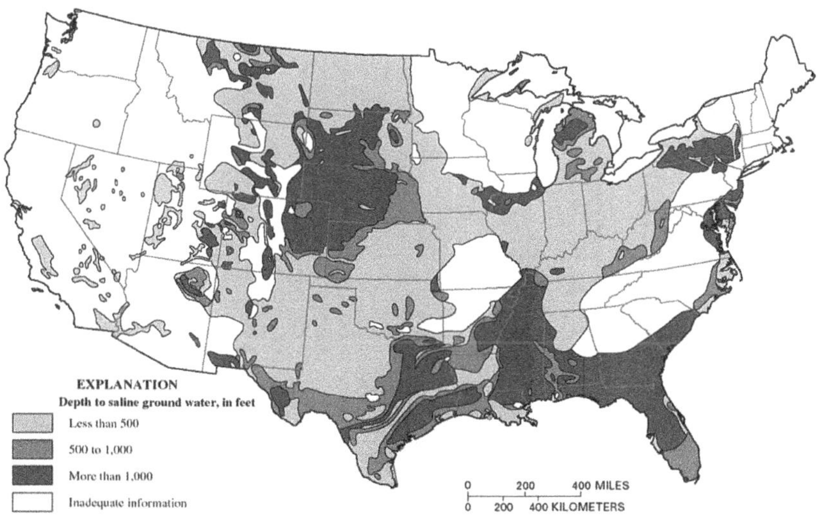

Figure 14.6 Depth to saline groundwater in the U. S. (from U. S. Geological Survey, 2002).

the areas where the brackish water has been even moderately characterized.

The volume of saline water in some states exceeds the volume of drinkable groundwater making it a target for drinking water desalination. Desalination of impaired waters is relatively expensive (see the chapter by Pankratz) but can provide system resilience because it is independent of the weather and because it adds to water system reserves at a time when water storage is declining.

14.3.1.1 El Paso's Kay Bailey Hutchison Desalting Plant

The largest inland desalination effort in the U.S. is El Paso Water Utilities' (EPWU) Kay Bailey Hutchison Desalting Plant which is operated in partnership with Ft. Bliss. EPWU uses RO to treat 18 million gallons per day (MGD) of brackish water that, after blending, supplies 27.5 MGD of water to El Paso and Ft. Bliss. Desalting was prompted by long-term pumping and lowering of local water tables which increasingly drew in high TDS groundwater. Surface water is limited and imported water is expensive (Figure 14.7).

Recovery at the plant is 70–80%; salt rejection reaches 93%. The concentrate is piped 22 miles north to near the Texas-New Mexico border (Figure 14.8) where it is injected ~ 4000′ into the Fusselman Dolomite. The connate water in the dolomite is 6,000 – 9,000 mg/L TDS which is saltier than the KBH concentrate which is 2500 – 5100 mg/L TDS.

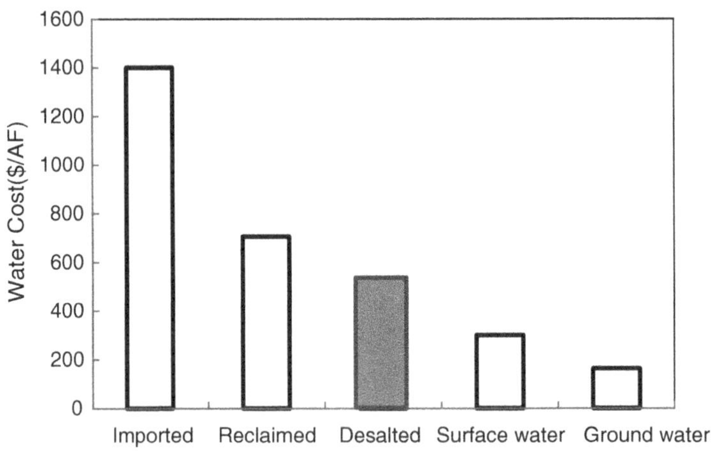

Figure 14.7 Estimated water costs for El Paso (from Hutchinson 2009).

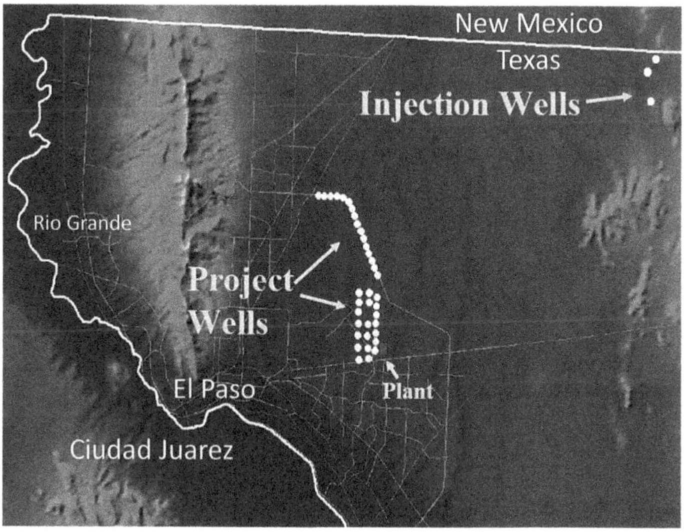

Figure 14.8 El Paso desalting operation (from Hutchinson 2009).

The El Paso inland desalination effort had a total capital cost of 91M$: 40M$ plant; 32M$ for production wells and collectors, and 19M$ for concentrate disposal. Disposal is a smaller fraction of the operating costs. The amortized capital and O&M cost of water is 534$/AF, which is broken down into: $189 for wells and collectors; $42 to Ft. Bliss for water and land, $232 for the desalination plant; $49 for disposal, and $22 for the finished pipeline [4].

Although disposal costs are a relatively minor fraction of the overall operation, reducing the amount of concentrate disposed would lengthen the lifetime of the disposal wells while resulting in fuller water recovery. Further concentration of the concentrate is limited by the potential formation of silica and calcium sulfate scale. Silica scale prevention has been the particular focus of EPWU efforts to lower disposal volumes. Because silica scale formation limits desalination of many inland brines, the general controls over silica removal deserve closer examination. While anti-scalants effectively inhibit formation of crystalline calcium salts of carbonate, phosphate, and sulfate, they are less effective against formation of amorphous silica scales. Many natural and industrial process waters contain dissolved silica levels that are a large fraction of the solubility limit of silica (~120–150 ppm at 25°C), hence little further concentration is required for silica scale to grow. Once formed, silica scale is hard and resistant to most forms of chemical attack (e.g. acid washing) and far harder to remove than, for example, calcite scale. The low thermal conductivity of silica scale, 0.2–0.5 kcal/m^{2}h per °C for silica (compared to 0.5–1 for calcite, and 1–2 for gypsum) disproportionately decreases boiler output if silica scale is present. Approaches to prevent silica scaling will be repeatedly considered below.

14.3.2 Impaired Water Usage in Energy Production

Thermoelectric power generation accounts for 49% of US water withdrawals and 3% (3 billion gallons per day) of consumptive use (USGS, 2005). Growing energy demand, combined with an already allocated water supply will likely force greater reliance on impaired waters to cool power plants. For example, more than 22,000 MW of new and replacement power were planned or under construction in California in 2001, which would require the equivalent annual water use of a city of a half a million people in a state where new water supplies are limited. The primary technical challenges will be: how to prevent scale formation in the face of the high salt concentrations that go with multiple loop water use; and how to dispose of the concentrate brine blowdown that remains. Municipal wastewater, agricultural return flows, inland brines, and produced waters are the most likely candidates for cooling tower use (e.g. EPRI 2003).

The produced water in Table 14.2 has high salinity, silica, and hardness; the agricultural return flow is similar, with the addition of high phosphate; the municipal wastewater has low to moderate salinity, silica, and hardness, but high concentrations of phosphate and ammonia. The produced water composition is similar to that of an inland brine. The primary scale-forming minerals will be calcium carbonate, calcium phosphate, and calcium sulfate, and silica (along with magnesium silicate). Preventing their formation at present would involve a combination of side-stream softening through e.g. nanofiltration [5, 6], ion exchange, or lime addition, and/or use of anti-scalants. Lime softening tends to remove calcium carbonate, calcium phosphate, and silica (see below). Nanofiltration removes Ca and Mg while lowering TDS. As noted previously, anti-scalants are effective against calcium sulfate and calcium carbonate scale, less so against silica scale.

While impaired water use in power plants is technically feasible, the overall costs of water treatment and sludge disposal are large.

EPRI (2003) calculated treatment+disposal costs for the waters in Table 14.2 for a hypothetical 500 MW combined cycle gas power plant. In each case, the blowdown was evaporated and disposed of as a sludge. For the produced water, overall costs came to \$3.39/1000 gallons; agricultural return water cost \$4.56/1000 gallons; and wastewater cost \$1.85/1000 gallons. The numbers include different scenarios for blowdown evaporation – e.g. inland desert versus Central Valley and differences in the cost of the feedwater, freshwater being the most expensive. The central conclusion of the EPRI study was that water costs associated with degraded water are at least 1.5 to 2.5 times the costs associated with fresh water at inland plants and 1.1 to 1.2 times that of fresh water at coastal plants. New methods are needed to reduce this difference.

One of the keys to lowering costs is to decrease the volume of blowdown generated by increasing cycles of concentration. Methods that are able to concentrate the blowdown further lower disposal costs by decreasing the amount of land needed for evaporation ponds and/or capital outlays for salt crystallizers. Preventing the formation of the limiting scale-forming mineral is one way to concentrate blowdown. Anti-scalants for calcium minerals are able to raise by several multiples the allowable Ca levels in blowdown, which would tend to make silica a more frequent limiting mineral. One non-anti-scalant silica control method is lowering the fluid pH [7]. Low pH kinetically inhibits the formation of silica nuclei for several hours. This is shown in Figure 14.9 where silica levels 10–20 times the silica saturation value can be seen to prevent silica formation in a mildly acid solution (pH 3.6). Silica formation at higher pHs is quite rapid. Lowering pH through acid additions [8], or by simply recirculating the CO_2-rich flue gas might allow

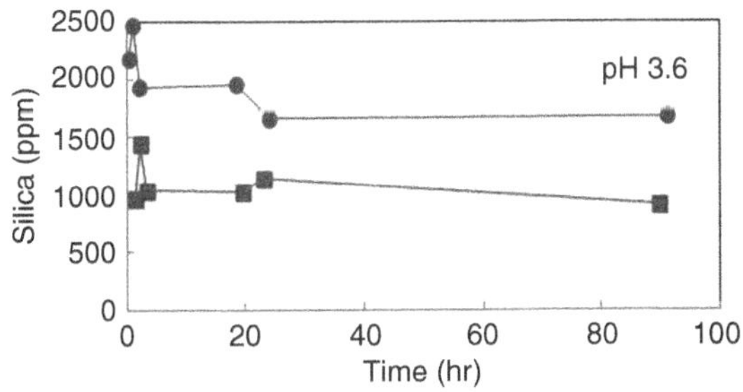

Figure 14.9 Inhibition of silica precipitation at low pH.

substantial concentration of blowdown and proportional lowering of blowdown disposal costs. Lower pH would also prevent formation of calcium carbonate and calcium phosphate scale, but not calcium sulfate. If low pH inhibition of silica scale formation becomes widespread, future research must focus on inhibiting the low pH formation of calcium sulfate. This task will be made difficult by the fact that at low pHs many of the phosphate and carboxylate functional groups on anti-scalants that might otherwise bind to calcium minerals, and thereby inhibit scale formation, will be protonated and less inclined to link to calcium.

In general, the controlled and selective precipitation of salts would be a useful tool for management of many desalination concentrates, particularly if it allowed the mining of brines for valuable elements. One example of the latter is the selective precipitation of silica from geothermal brines to extract lithium used in batteries [9].

14.3.2.1 Palo Verde Nuclear Power Plant, Arizona

An example of how impaired water might be used for energy generation in the future is that of the Palo Verde Nuclear Power Plant west of Phoenix, Arizona. Palo Verde Plant is the largest nuclear power plant in the U.S. and Palo Verde's zero liquid discharge facility is the largest facility of its type in the U.S. as well [10]. For cooling water, Palo Verde uses wastewater effluent from the 91st Avenue Wastewater Treatment Plant, City of Phoenix.

Average water input has been ~ 70,000 acre feet (~ 25 billion gallons) per year. Trickling filters reduce ammonia through biological denitrification, followed by lime softening to lower hardness. CO_2 injection is followed by soda ash softening. The two softening steps generate 100 tonnes of sludge per day. The remaining fluid goes through ~ 25-fold concentration in the course of cooling the reactors and the blowdown (~ a billion gallons per year; or ~ 4% the volume of the treated water) is ultimately evaporated in ponds. The evaporation rate in the area is 60–72 inches/yr.

The industrial scale production of homogenous lime softening residuals and blowdown begs for a reuse application. None has been found to date. Residuals reclamation will continue to be the focus of future research as new power plants and municipalities follow the example of Palo Verde.

Figure 14.10 Palo Verde Water Treatment Plant (from Lott 2011).

14.3.3 Salinization

While desalination will maximize the beneficial use of water, unless there is a salt export mechanism, salinization will follow. Municipal water use results in a 'salt pickup' – a salinity increase of 300–500 mg/L seen in wastewater. Some of the salt comes from evaporative concentration of irrigation water, some from food salt and cleaners. Some of the pickup comes from the use of water softeners which are ion exchangers that add dissolved sodium or potassium to waters while removing calcium and magnesium.

The US Bureau of Reclamation has estimated that 14 million acres in the US (25% of the total) have been lost or limited by salinity and that the worldwide losses are approximately 190 million acres – with the number growing 2.5 to 5 million acres per year. Salt accumulation costs residential users in several ways. The lifetimes of water-using appliances (e.g. water heaters, faucets, dish washers and evaporative collers) are shortened. Purchase of bottled water as an alternative to more saline tap water and increased water softening adds to the cost.

Increased salinity decreases the efficiency of cooling and therefore adds to the water costs of commercial and industrial users. Water treatment facilities themselves are affected by the TDS of input waters; there appears to be a decrease in the expected life of a facility that is linearly related to the TDS [11]. For example, overall costs of salinization in Southern California and the Colorado River Basin have been estimated to amount to respectively 100 million and 300 million dollars per year. Of the latter number, $176 million came from crop damages; $81 million came from damages to households; $38 million came from damages to commercial and industrial facilities; and $35 million came from damages to utilities [12]. Southern California TDS increases of 100 mg/L were calculated to result in damages of roughly 100 million dollars per year. 100 mg/L TDS increases of Salt River Project and Central Arizona Project water were calculated to cause roughly $15–30 million worth of damages in Central Arizona [11] – which includes Phoenix and Tucson.

Roughly 1.3 million m^3/yr of surface water is brought into Central Arizona each year by the Salt (~ 800,000 m^3/yr) and Verde Rivers (~ 500,000 m^3/yr). The Gila River makes a minor contribution as well. TDS levels for the Salt River are 750–900 mg/L; for the Verde River TDS is 300–400 mg/L. The Central Arizona Project brings ~ 700,000 m^3/yr of water into the basin from the Colorado River which has TDS = 580–630 mg/L. Phoenix is calculated to receive roughly 1.5 million tons of salt per year, 50% from the CAP and 30% from the Salt River. Since only 400,000 tons leaves, the remainder - ~ 1.1 million tons is calculated to remain [12]. This amounts to roughly 900 lbs per person per year of salt that ends up in appliances, in soils, or in groundwater.The absence of salt export mechanisms may ultimately limit the economic potential of Central Arizona.

14.4 Desalination for Oil and Gas

Future oil and gas production will increasingly rely on advanced treatment to: reclaim/dispose produced water, recycle water during oil sands extraction, and design waters for enhanced oil recovery. Roughly 150 million barrels of drilling waste accumulated in 1995 [13] plus 18 billion barrels of produced water and 20 million barrels of associated waste (1 barrel = 42 U.S. gallons = 159 liters). Drilling waste is produced proportionally to drilling activity and therefore shows up early on in development of an oil field. Produced water accumulates on the back end and is greatest in older fields. Drilling waste accumulates at roughly 1.2 barrels per foot drilled. Produced water is volumetrically the most imposing waste product and averages roughly 6 barrels per barrel of oil produced. Over time for a given field this number increases to upwards of 100, though most fields become uneconomical when the water to oil ratio exceeds 10 to 20. Produced water disposal costs range from 0.1 to 4$ per barrel and an estimated 15 billion barrels are generated per year in the United States. Producing wells typically last 15–30 years. Over the last decade two new oil and gas sources; oil sands and shale gas, have prompted new water treatment challenges that can be expected to grow in the future.

14.4.1 Water Treatment and the Oil Sands

The oil sands of Alberta have proven oil reserves of 1.75 trillion barrels, reserves only exceeded by those of Saudia Arabia. The oil is in the form of solid bitumen that is heated with steam to make easier its collection when it is at depth, or mined if it is near the surface. Water is used in both cases – for steaming at depth and for washing sand and clays from surface-mined bitumen. One means for steaming, Steam Assisted Gravity Drainage (SAGD), is depicted in Figure 14.11. Limited water availability has prompted reuse. But steam injected into bitumen-bearing formations dissolves silica which

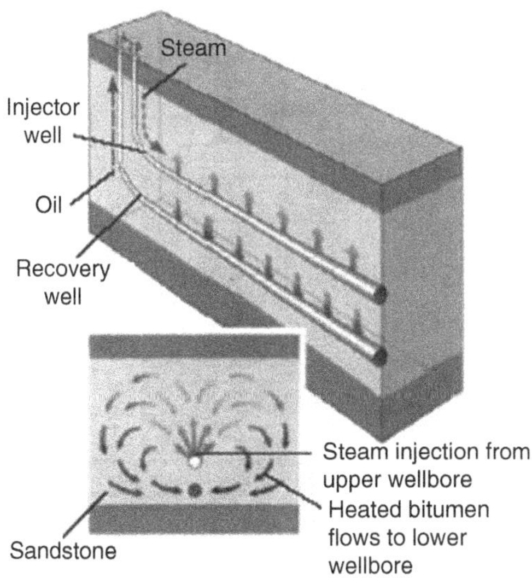

Figure 14.11 Schematic of SAGD process (source: http://pubs.usgs.gov/fs/fs070-03/fig3.jpg).

must be removed before water reuse, otherwise silica (and calcite) scale forms and reduces heat transfer in once-through steam generators limiting their effectiveness and increasing costs.

Figure 14.12 shows how warm lime softening is practiced in the SAGD process to remove scale-forming elements. A portion of the boiler blowdown is recycled and combined with silica-rich steam condensate from the free water knockout and low silica makeup water, followed by warm lime softening in a clarifier reactor. The non-recycled portion of the blowdown is typically disposed of by deep well injection. After warm lime softening, filtration, then treatment by weak acid cation exchange resins produces a low hardness, low silica water for further steam generation. Warm lime softening involves adding slaked lime and soda ash to raise the wastewater pH and precipitate hardness as calcite, and addition of magnesium oxide to aid precipitation of silica. A flocculant is used to stabilize and enhance the floc that forms; and a portion of the floc is recirculated to nucleate further precipitation. The process is

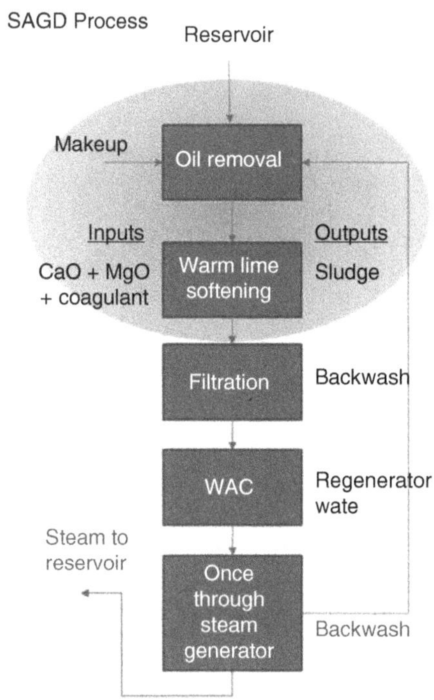

Figure 14.12 Warm lime softening treatment for SAGD.

run at temperatures of 70–90°C to take advantage of faster reaction rates and the lower solubilities of the metal hydroxide and carbonate minerals that form. A relatively high volume of warm lime softening sludge is produced and ultimately landfilled. Better silica removal approaches would allow greater water reuse and less waste disposal.

At the mechanistic level, warm lime softening removes silica as follows: slaked lime (and MgO and soda ash) addition raises the pH; high pH deprotonates silicic acid to silicate ion which combines with a metal cation (Ca^{+2} or Mg^{+2} from dissolution of the lime and MgO), or the positively charged surface of MgO, to form a poorly defined amorphous, hydrated metal-silicate mixture. A number of other metals are also known to precipitate/coagulate silica.

A number of alternatives to warm lime softening exist, such as;

1. Ion exchange to remove all multivalent cations [14, 15] would make unnecessary a silica removal step because silica scale formation requires multivalent cations to accelerate polymerization, coagulate silica polymers, and form metal silicates. Ion exchange should therefore allow much higher silica waters to be used in steam generators.
2. Mechanical Vapor Compression (MVC) produces an essentially pure feedwater and a blowdown that is ~ 33-fold more concentrated than the makeup water (e.g. [14, 16]) and which must be treated before deepwell injection, or evaporated using a crystallizer to a cake for zero liquid discharge landfilling (e.g. [17]). Silica removal is a central requirement of MVC concentrate management (e.g. [14]).
3. Silica scale inhibitors might allow higher silica levels to be present during steam generation [18]. Lowering solution pH might also apply.

The primary drawbacks to warm lime softening are extensive chemical inputs (lime, soda ash, MgO, coagulants, regenerants for the ion exchange step that follows warm lime softening) and management of the high water sludges that are produced. An important advantage of warm lime softening is the long industrial experience with the process. Also, managing warm lime softening sludges above ground might be more easily done than assuring the sustainable injection of e.g. MVC concentrate at depth.

Oil sands that are processed ex situ pose an even larger water treatment challenge. Oil sands are washed in a dilute base to loosen the bitumen from the sand. The resulting wastewater includes toxic naphthenic acids and mineral fines – clay particles that settle very slowly and prevent the rapid reuse and/or disposal of the water. The wastewaters are piped to areally extensive settling ponds that pose a health threat to wildlife. Settling/coagulating the clay fines rapidly would allow much smaller ponds. Presently, a slurry of gypsum provides some settling of fines. Better methods are needed though. Note that any treatment alternative must be relatively inexpensive because of the very large water volumes involved.

14.4.2 Treatment of HydrofrackingFlowback

Hydrofacturing ("hydrofracking") of shales has caused a remarkable expansion of gas production in the US – from 0.1 Trillion cubic feet (TCF) in 2000 to 3 TCF, or 14% of US gas production, in 2009. Because of hydrofracking, US recoverable gas reserves have grown by 70% since 1990 [19]. Shale gas hydrofracking was originally developed by US engineers but is now being explored by a number of countries (e.g. China, Poland, Argentina, Great Britain) as a means for boosting their own natural gas production and reserves. Climate change concerns should drive an even larger future market for natural gas since methane produces less greenhouse gas per BTU produced than oil or coal. For example, the US could reduce its emission of greenhouse gas by 8% simply by switching to natural gas combined cycle power plants [19]. Increased adoption of compressed natural gas as a transportation fuel could further decrease greenhouse gas emissions.

Figure 14.13 shows the locations of the major U.S. shale gas plays and basins. Figure 14.14 shows the basics of the hydrofrack process: 1–3 million gallons of water are mixed with an assortment of HF chemicals (Table 14.3) in stages to: clean the well bore, impose semi-vertical fractures in the shale, and then emplace proppants to hold the fractures open and allow methane to escape. Slickwater is the fluid used to impose the fractures and deliver proppant. Flowback is the water that is pumped back out of the well and is a mixture of shale water (the fluid that was originally in contact with the shale), hydrofrack additives, and residual slickwater. Flowback is held in

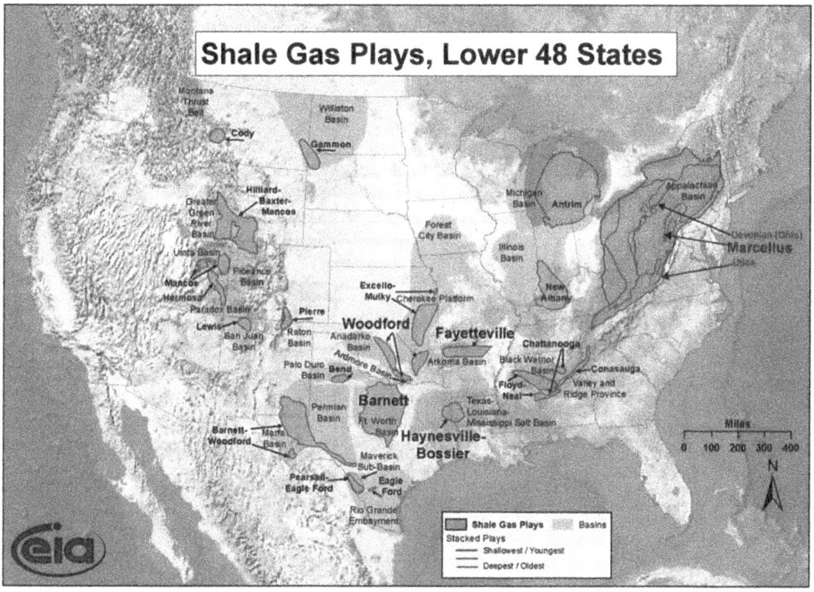

Figure 14.13 U.S. Gas plays and basins (from U.S. Environmental Protection Agency 2011).

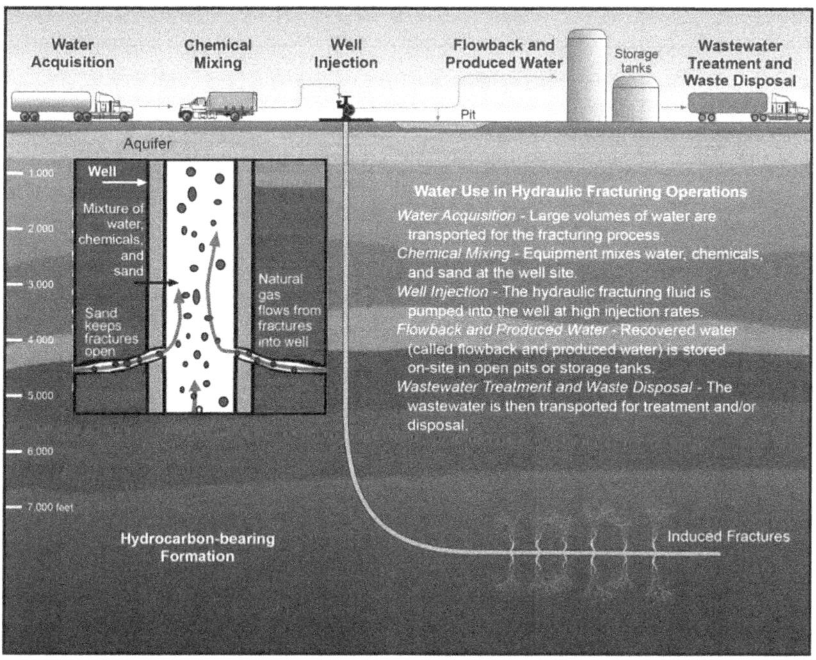

Figure 14.14 Schematic of hydrofrack process (from U.S. Environmental Protection Agency 2011).

Table 14.3 Frack fluid additives [20].

Additive	Chemical	Action
Diluted acid	HCl	Helps dissolve minerals and initiate cracks in the rock
Biocide	Glutaraldehyde	
Breaker	Ammonium persulfate	Allows delayed breakdown of gel polymer chains
Corrosion inhibitor	N,n-dimethyl formamide	
Crosslinker	Borate salts	Maintains fluid viscosity
Friction reducer	Polyacrylamide, mineral oil	Minimizes friction between fluid and pipe
Gel	Guar gum or hydroxy-ethyl cellulose	Thickens fluid and suspends proppant
Iron control	Citric acid	
KCl		Prevents clay swelling
Oxygen scavenger	Ammonium bisulfite	Corrosion protection
pH adjusting agent	Na_2CO_3	
Proppants	Quartz sand, ceramic	Hold fractures open
Scale inhibitor	Ethylene glycol	
Surfactant	Isopropanol	Increases fluid viscosity

ponds and then either recycled, injected into Class II UIC wells, or treated and discharged to surface waters.

The HF stages include (Additives in parentheses):

1. Acid stage – cleans near wellbore (Acid)
2. Slickwater pad stage – opens the formation (Friction reducer, KCl)
3. Proppant stage – inserts proppant (Proppant, Gelling agent, pH adjusting agent, Crosslinker (early), Breaker (late))
4. Flushing stage – flushes excess proppant from wellbore.

Biocides, Scale inhibitors, and iron control agents are also introduced to prevent the formation of sulfate and carbonate minerals, and iron (hydr)oxide. Fracking is typically done 2,000 – 6,000 feet below overlying drinking water aquifers which is thought to be sufficiently deep to prevent drinking water contamination.

Hydrofracking has prompted an enormous outcry from the environmental community which claims that hydrofracking contaminates drinking water aquifers, wells, and surface waters with HF chemicals and salts, and wells with natural gas. States such as New York imposed temporary bans on hydrofracking. National legislation has been introduced to restrict the practice, for example, the Senate Bill 1215: Fracturing Responsibility and Awareness of Chemicals (FRAC) Act of 2009. Organizations that have issued studies and/or taken a position on the environmental impacts of hydrofracking include MIT, The State of New York, DOE (twice), New York City, the State of Pennsylvania, the Texas Railroad Commission, the New York Times, the National Resources Defense Council, the Groundwater Protection Council, and the National Association of Regulatory Utility Commissioners.

Any future hydrofracking regulations will probably require treatment of flowback salinity to drinking water standards (~ 500 mg/L) if it is to be discharged to surface waters or wastewater treatment plants. The high salinities of the flowback makes discharging it untreated to traditional wastewater treatment plants lethal to their existing biodegradation processes. Deep well injection, when available, is the most effective means for disposing of flowback.

Because of its expense, treatment will probably only be pursued where injection wells are absent. The key will be to modify traditional treatment steps (e.g. biodegradation, coagulation) to work in the high TDS flowback. Treating flowback water from hydrofracking poses an enormous technical challenge primarily because of the extreme salt loads (200,000 mg/L and higher) that approach halite saturation. Naturally occurring radioactive material - radium, thorium, uranium, and radon - are also present along with mercury, lead, and arsenic. High sulfate levels complicate treatment further by prompting the formation of sulfate scales containing Ca, Ba, and Ra. Reduced iron is present at the tens of mg/L level in formation waters and can be expected to form ferric hydroxide flocs once exposed to oxygen during treatment. Oil/grease and the frack additives would, from a less saline fluid, be most easily removed by some combination of biodegradation, flocculation, and oxidation (+ filtration). Developing stand-alone treatment steps to break down/remove organics in the face of high TDS is an important research goal.

Ultimately, the salts themselves must be removed. New methods are needed to achieve zero-liquid discharge, the turning of the brine into a sludge (Sludges typically face fewer regulatory hurdles when landfilled than waste brines do when discharged to surface waters). Figure 14.15 emphasizes the fact that traditional water treatment approaches such as EDR, RO, and ion exchange cannot be used on flowback. The very high salt loads instead require evaporation and distillation approaches. Figure 14.16 schematically outlines the relative costs of concentrate management. Keep in mind that the low-cost solutions: surface water and sewer discharge are neither likely to be allowed for flowback disposal. Consider that half a billion gallons of water would be required to dilute a million gallons of 250,000 mg/L flowback down to 500 mg/L.

Desalination is likely to also be applied to provide the water for the frack jobs in areas where fresh makeup water is unavailable, too expensive to transport, and where saline groundwaters are available locally. Transportation costs in the Bakken, for example, can be three to four times the cost of procuring or disposing of the water [21] point out the possibility of desalinating non-potable water from the underlying Dakota Sandstone to provide hydrofrac fluids while avoiding transportation costs.

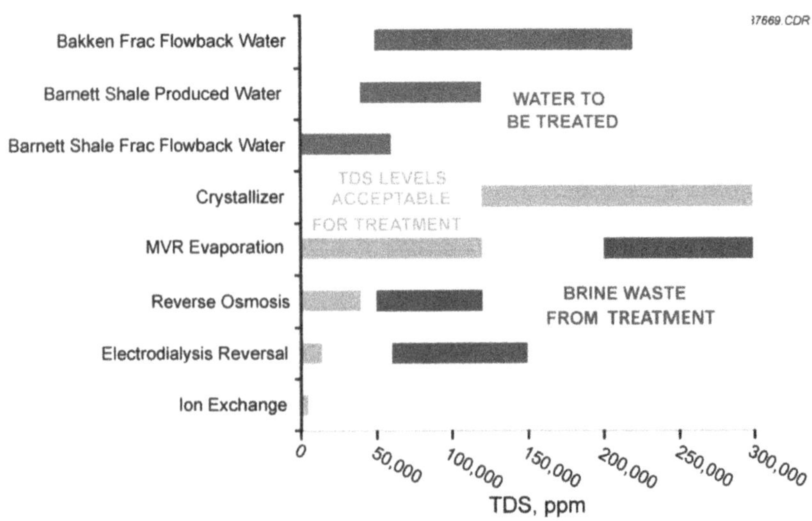

Figure 14.15 Salt levels in Bakken Play fluids (from Energy & Environmental Research Center 2010).

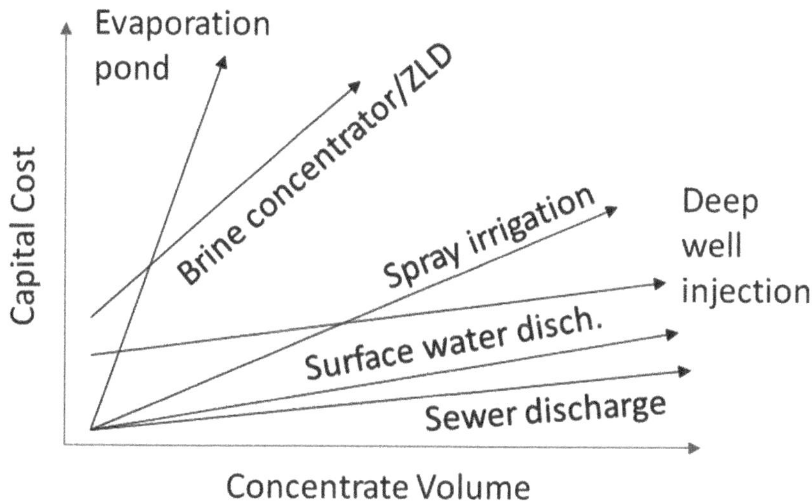

Figure 14.16 Concentrate management costs (after Mickley 2006).

14.4.3 Chemical Waterflooding for Enhanced Oil Recovery

An important future goal of water treatment will be to produce precisely designed fluids that chemically enhance the mobilization of oil from sandstone or limestone reservoirs. Depleted oil reservoirs typically still retain 2/3 of their original oil that is difficult to extract because the oil is bound to tightly to the reservoir. A 1% increase in oil recovery would result in 80 billion barrels of additional oil – roughly 8 trillion dollars.

Part of the challenge of extracting the residual oil is physical, e.g. reaching oil trapped in inaccessible pores. Much of oil adhesion though is caused by chemical bonding between the oil and the reservoir surfaces. While waterfloods have traditionally been used only to pressurize reservoirs, evidence is increasing that oil recovery can be enhanced by waterfloods whose bulk composition (pH, hardness, sulfate, and so on) is chemically tailored to mobilize the oil [22]. Specific compositions of waterfloods that eliminate the electrostatic bonds that otherwise bind oil to sandstone [23–25] and limestone [26, 27] reservoir surfaces might allow enhanced recovery that is less expensive than traditional enhanced oil recovery methods involving injection of steam, CO_2, and/or surfactants. Roughly 60% of the world's oil is in limestone reservoirs; the remaining 40% is in sandstones.

Traditional waterflooding requires roughly 150% of total pore space of the field and should include the volume of any adjacent overlying gas interval or basal water zone (http://www.ogs.ou.edu/PTTC/pwm/pw_s4.htm); a 1 km^2 field with a producing zone 100 m thick with 20% porosity would require 8 million gallons of water. A useful rule of thumb is that 8 gallons of waterflood is required for each additional gallon of oil recovered. Usually more than half of the water that is used is produced water from elsewhere in the field. Recall that far more water is produced at oilfields than oil. Presently, water treatment comes into the picture only when the potential for the injectate to cause clays to swell, and damage the formation, trigger biologic activity, and/or precipitate minerals is considered. Present-day treatment is done to remove particulates, dispersed oil, and oxygen. Particulates are removed by settling tanks or through filtration. Dispersed oil reduces formation water permeability, adheres to scale deposits, and provides a food source for microorganisms. Dispersed oil is removed by demulsification chemicals. Oxygen, CO_2, and H_2S all lead to corrosion. Oxygen scavengers (e.g. bisulfite salts) are used to remove oxygen; oxygen, SO_2, or hypochlorite are used to oxidize H_2S; air stripping with an inert gas such as N_2 can remove CO_2. Microorganisms range in size from 0.2 to 10 microns and can either be filtered out or subjected to a biocide. Removal of scale-forming dissolved solids containing e.g. Ca^{+2}, silica, or sulfate can be done through a combination of ion exchange, lime softening, and selective precipitation/coagulation. The nature and extent of treatment is obviously driven by economics.

Water treatment in the future will focus as well on altering salinity, hardness, sulfate, pH, and alkalinity levels in chemical waterfloods to chemically loosen oil from reservoir surfaces. In water-wet reservoirs, a thin layer of water separates oil from the reservoir. Increased electrostatic attraction between the two thins the water layer and makes oil adhesion more likely, and oil recovery harder [28–30]. The identity of the relevant electrostatic bonds are becoming clearer [26, 22, 25, 27] suggesting that soon it will be possible to precisely design water flood chemistries for oil recovery. The simplest way to make up chemical waterfloods will be to mix the waters on hand – for example, seawater, produced water, desalinated brine - in proportions designed to eliminate the electrostatic bonds identified above. Ca^{+2} levels might also be reduced for sandstone waterfloods through ion exchange and/or lime softening. Sulfate levels can be raised for limestone waterfloods by using larger proportions of sulfate-rich seawater and/or produced water. Obviously, reservoir mineralogy will also play an important role. For example, equilibrium with calcite in limestone reservoirs will influence the final pH and Ca^{+2} levels in a limestone waterflood. Similarly, equilibrium with gypsum will set upper limits on the amount of sulfate in a waterflood.

14.5 The Future of Desalination Technologies

Seawater RO presently requires an energy expenditure in the range of 9–60 kJ/kg at a cost of $2 to $4 per 1000 gallons of water produced. The theoretical minimum energy expenditure is 3–7 kJ/kg [31]. The fact that energy consumption is nearly half the overall water production costs (see Figure 14.17) means that order of magnitude reductions in desalination costs probably won't occur in the future but that low energy desalination will be an important goal for the future.

The need for improvements in multiple areas is emphasized in the research targets identified in the National Desalination Roadmap Implementation Report [32]. The research targets can be grouped into Membranes, Concentrate Management, and Alternative Technologies.

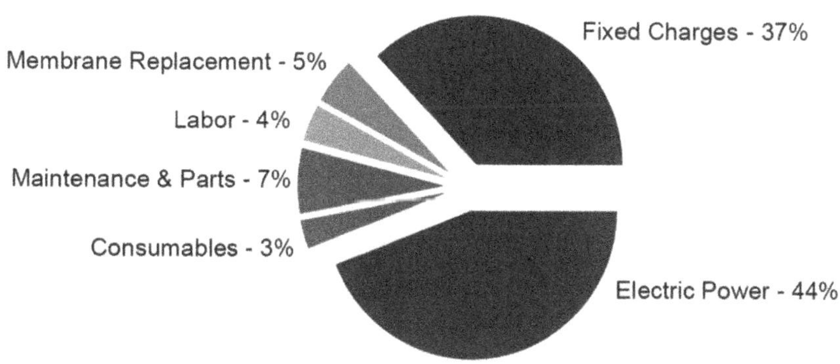

Figure 14.17 Cost breakdown for seawater reverse osmosis (from Miller 2003).

Membranes

1. Seawater pretreatment conditioning optimization,
2. Integrated treatment approaches for desalination-specific problem contaminants (e.g. boron, NDMA).
3. Alternatives to spiral membranes/spiral membrane configuration optimization,
4. Mechanistic understanding of high-pressure membranes,
5. Pretreatment to prevent membrane biofouling,
6. Boron removal by membranes,
7. Improved feedwater recovery for membrane desalination,
8. Fouling characterization, sensing, and prediction,
9. Post-treatment stabilization of desalinated water,
10. Variable operating rate capability desalination plants,
11. Biomimetic desalination,

Alternative Technologies

1. Hybrid reverse osmosis/electrodialysis desalination,
2. Membrane distillation,
3. Forward osmosis,
4. Nanostructured high flux membranes,

Concentrate Management

1. Hybrid techniques for zero liquid discharge,
2. Self-sealing evaporation ponds,
3. Evaporation enhancement (for concentrate management),
4. Selective contaminant removal from concentrate for resale,
5. Silica removal,
6. Selective precipitation of concentrate salts,

Other

1. Minimization of anti-scalants and biocides,
2. Energy recovery from desalination, and
3. Co-siting of desalination with wastewater treatment, power generation, oil and gas, and other industrial facilities.

A number of the above have been touched upon in this chapter, including: silica removal, anti-scalants and biocides, selective precipitation of concentrate salts, and co-siting of desalination facilities. Energy recovery is discussed in the chapter by Pankratz. Forward osmosis is examined in the chapter by McCutcheon. Membrane distillation is covered in the chapter by Cath. Electrodialysis and electrodeionization are explored in the chapters by Moon and Lee and Wood and Gifford. Many of the research targets involve biofouling are discussed in the several chapters. And the remainder of non-traditional desalination approaches (e.g. clathrates, wind and wave-driven desalination, osmotic pumps, offshore desal, fog collection, solar stills, centrifugal desalination) have been described in detail by [31].

Improvements in RO technology focus on high flux, low energy membranes in particular biomimetic desalination and nanostructured membranes, and supporting efforts in computational materials chemistry. We anticipate that national and international scientific initiatives in nanotechnology and biotechnology will result in improved desalination technologies in the future. Nanotechnology—the use of materials and processes that operate over a length scale of roughly 1 to 100 molecular diameters—is the target of over $10 billion in R&D investment each year. Global biotech R&D investments amounted to roughly $30 billion in 2010. While a relatively small fraction of these efforts focus on water treatment, the broad scale technical advances that are occurring in both fields are expected to benefit desalination [33].

A key goal of desalination research is to build a molecular level understanding of water behavior at membrane interfaces and use it to precisely design high-flux membranes [34]. Membranes might be designed that prevent attachment of biofilms, or that don't degrade in biofilm-destroying oxidants. The expectation is that parallel advances in materials development will make the transition from the lab to the market rapid for each technology. Note thought that existing processes, such as coagulation, lime softening, micropollutant filtration, and ion exchange for example, already rely on nanoscale processes. By the same token the design of existing RO membranes involves an ample contribution of nanoscale coordination chemistry. Recent emphasis on better theoretical understanding and resolution of these processes at the molecular level is what is expected to drive the respective technologies to higher levels of performance and lower costs.

14.5.1 Biomimetic and Nanotech Membranes

Examples of biological desalination include mangroves, which separate H_2O from seawater and secrete salt crystals in their roots, and seabirds far from shore that separate water in a gland above their beak and sneeze out the concentrate. Peter Agre received the Nobel Prize in Chemistry in 2003 for his co-discovery of aquaporins "the plumbing system for cells". Cell walls of most organisms contain specialized protein channels, 'aquaporins', that quickly transport water or ions selectively across the cell membrane [35]. Water is transported in a hydrophobic channel a few Ångstroms in diameter, while hydrogen bonding interactions with functional groups on the channel walls preferentially orient the molecules in single-file fashion inside the pore. Ions are effectively excluded because of the large increase in free energy associated with penetrating a channel due to loss of its hydration sphere and the associated decrease in entropy.

The effectiveness of aquaporins in shuttling water through cell membranes has motivated the search for aquaporin-assisted membranes, and for synthetic analogues. While natural aquaporin proteins extracted from living organisms can be incorporated into a lipid bilayer membrane or a synthetic polymer matrix, porous inorganic membranes modified to provide aquaporin-like function may provide a more robust alternative. These include carbon nanotubes, double-walled carbon nanotubes, and metal oxide frameworks. Making carbon nanotubes is presently an involved technical process. Typically a substrate containing metal seeds of the same diameter as the nanotubes are heated to 600 to 900°C, and then a carbon-containing gas such as methane or alcohol is added. Nanotubes then grow from the metal seeds. The metal from the seeds are problematic in that the metal can later occlude nanotubes. Large-scale manufacture of nanotube-based membranes will require vastly improved manufacturing processes to be economical.Self-assembly and template directed synthesis of aquaporin-simulating materials is one potential way out.

14.6 Summary

As presented in this chapter, supplies of fresh water have become limited in many regions of the work and the U.S. due to economic growth and development. To meet the growing water needs of industrialized and developing nations, increased efficiency in fresh water use and increased utilization of impaired waters through advanced water treatment processes will be required. While the use of impaired or nontraditional waters, such as waste water reuse and desalination started to grow significantly in the U.S. in the 1990's, that trend started in many other regions of the world in the 1970's.

All indications are that this trend in the treatment and use of impaired waters to supplement fresh water resources and to better utilize non fresh water resources where applicable and fresh water is not needed, such as in many industrial applications, will continue to grow as demands on the worlds fresh water resources continue to grow. Therefore, development of approaches to more efficiently and effectively treat and utilize impaired waters such as seawater, brackish ground water, oil and gas produced water, and domestic and industrial waste water, will continue to grow and expand.

Many of the impaired and nontraditional waters mentioned have increased levels to even very high levels of salts that will need to be treated, meaning that desalination technologies and improvements in desalination approaches and processes will be vital to enable the U.S. and most other countries in meeting future global sustainable water resource and supply needs.

References

1. Government Accountability Office. Fresh Water Supply: States' Views of How Federal Agencies Could Help Them Meet the Challenges of Expected Shortages, USGA, ed., Report # GAO-03-514. Washington, DC, p. 110, Jul (2003).
2. Sandia National Laboratories. Energy-Water Challenges and Research and Development Issues. Albuquerque, NM, Sandia National Laboratories (2009).
3. EPRI, *Use of Degraded Water Sources as Cooling Water in Power Plants*, California Energy Commission (2003).
4. W. Hutchinson,Desalination of brackish groundwater and deep well injection of concentrate in El Paso, Texas http://www.gwpc.org/meetings/uic/2009/proceedings/Hutchison,%20Bill.pdf. Groundwater Protection Council (2009).
5. J.V. Mattson and T.G.I. Harris.Zero Discharge of Cooling Water by Sidestream Softening. Water Pollution Control Federation**51**, 2602–2614 (1979).
6. S.J. Altman, R. P. Jensen, M.A. Cappelle A.L. Sanchez, R.L. Everett, H.L. Anderson and L.K. McGrath, Nanofiltration Treatment of Side-Stream

Cooling Tower Water for Reduction of Water Usage. *Desalination* **285**, 177–183 (2011).
7. J.L. Featherstone, Geothermal plant silica control system. US Patent 4429535 (1984).
8. L.M. Costa and P.J. McCabe. US Patent 7514001 B2 High recovery reverse osmosis process and apparatus (2009).
9. W.L. Bourcier, M. Lin, G. Nix, Recovery of minerals and metals from geothermal fluids https://e-reports-ext.llnl.gov/pdf/324646.pdf. Livermore, CA, Lawrence Livermore National Laboratory (2005).
10. R. Lott, Water and energy in Arizona http://www.gwpc.org/meetings/forum/2011/Proceedings/pdf_presentations/9f_Lott_Bob.pdf. Groundwater Protection Council (2011).
11. T. Poulson, Future Trends Analysis (Draft Phase I Document). U.S. Department of Interior, Bureau of Reclamation, Central Arizona Salinity Study (2003).
12. M. Gritzuk, Planned desalination programs by the City of Phoenix. Bi-National Desalination Conference (2004).
13. U.S. Environmental Protection Agency. Draft plan to study the potential impacts of hydraulic fracturing on drinking water resources (2011).
14. Portelance, S. N. (2000). Waste Water Treatment Options for a SAGD oil production facility; IWC-07-26. International Water Conference, Pittsburgh, PA.
15. Bridle, M. (2005). "Treatment of SAGD-produced waters without lime softening." SPE/PS-CIM/CHOA 97686 PS2005-317: 1-5.
16. Heins, W. F. (2010). "Is a paradigm shift in produced water treatment technology occurring at SAGD facilities?" Journal of Canadian Petroleum Technology 49: 10-15.
17. Heins, W. F., R. McNeill, et al. (2005). World's first SAGD facility using evaporators, drum boilers, and zero discharge crystallizers to treat produced water. Paper 2005-115. Canadian International Petroleum Conference.
18. Pedenaud, P., C. Goulay, et al. (2006). "Silica-scale inhibition for steam generation in OTSG boilers." SPE Production and Operations February 2006: 26-32.
19. MIT,The Future of Natural Gas. http://web.mit.edu/mitei/research/studies/documents/natural-gas-2011/NaturalGas_Report.pdf (2011)
20. U.S. Department of Energy - All Consulting. Modern Shale Gas Development in the United States: A Primer (2009).
21. D.J. Stepan, R.E. Shockley, Bakken water opportunities assessment - Phase 1, 2010-EERC-04-03; http://www.nd.gov/ndic/ogrp/info/g-018-036-fi.pdf (2010).
22. T.A. Austad,A. RezaiDoust and T.Puntervold.Chemical mechanism of low salinity water flooding in sandstone (SPE 129767).SPE Improved Oil Recovery Symposium: 1–17 (2010).

23. A. Lager, K.J. Webb, C.J.J. Black,M. Singleton and K.S. Sorbie,Low salinity oil recovery - an experimental investigation; SCA2006-36 (2006).
24. I.R. Collins, G.R. Jerauld, A. Lager and P.L. McGuire, INTERNATIONAL PATENT WO 2008/029131 A1/Hydrocarbon Recovery (2008).
25. P.V. Brady and J.L. Krumhansl,A Surface Complexation Model of Oil-Brine-Sandstone Interfaces at 100oC: Low Salinity Waterflooding. *J. Petroleum Sci. Eng.* **18**, 171–176 (2012).
26. A. Rezaei Doust, T. Puntervold, S. Strand and T. Austad, Smart water as wettability modifier in carbonate and sandstone: A discussion of similarities/differences in the chemical mechanisms. *Energy Fuels* **23**, 4479–4485 (2009).
27. P.V. Brady, J.L. Krumhansl, and P.E. Mariner, Surface complexation modeling for improved oil recovery (SPE153744) SPE Improved Oil Recovery Symposium: 1-10, (2012)
28. J.S.K. Buckley, K. Takamura and N.R. Morrow,Influence of electrical surface charges on the wetting properties of crude oil.*SPE Reservoir Eng***4**, 332–340 (1989).
29. S.T. Dubey and P.H. Doe,Base number and wetting properties of crude oils *SPE Reservoir Eng* **8(3),**195–199 (1993).
30. J.S. Buckley, Y. Liu and S. Monsterleet, Mechanisms of wetting alteration by crude oil.*SPE J***3(1)**, 54–61 (1998).
31. J.E. Miller, Review of Water Resources and Desalination Technologies http://www.sandia.gov/water/docs/MillerSAND2003_0800.pdf. Albuquerque, NM, Sandia National Laboratories (2003).
32. American Water Works Association Research Foundation, WateReuse Foundation, et al. (2010). National Desalination Roadmap Implementation Report http://www.sandia.gov/water/docs/Desal Implement Road map 1-26-2010_c_web.pdf.
33. M.A. Shannon, P.W. Bohn,E. Elimelech,J.G. Georgiadis, B.J. Mariñas and A.M. Mayes, Science and technology for water purification in the coming decades. *Nature* **452**, 301–310 (2008).
34. Cygan, R. T., C. J. Brinker, M.D. Nyman, K. Leung and S.B. Rempe,A molecular basis for advanced materials in water treatment, *MRS Bulletin* **33**, 42–47(2008).
35. M. Borgnia, S. Nielsen, A. Engel and P. Agre, Cellular and molecular biology of the aquaporin water channels. *Annu. Rev. Biochem* **68**, 425–458 (1999).
36. M.E. Mickley, State of the Science Report to the Joint Water Reuse and Desalination Task Force. Concentrate Management (2006).
37. Energy & Environmental Research Center. Bakken Water Opportunities Assessment - Phase 1; Summary Report http://www.nd.gov/ndic/ogrp/info/g-018-036-fi.pdf. Grand Forks, ND, Energy & Environmental Research Center (2010).

Index

Also of Interest

Check out these other titles from Scrivener Publishing

By the Same Author:

Reverse Osmosis, by Jane Kucera, ISBN 9780470618431. The only comprehensive in-depth coverage of reverse osmosis and the benefits of its use in industrial water management. *NOW AVAILABLE!*

Also Available from Scrivener Publishing:

Electrochemical Water Processing, by Ralph Zito, ISBN 9781118098714. Two of the most important issues facing society today and in the future will be the global water supply and energy production. This book addresses both of these important issues with the idea that non-usable water can be purified through the use of electrical energy, instead of chemical or mechanical methods. *NOW AVAILABLE!*

Biofuels Production, Edited by Vikash Babu, Ashish Thapliyal, and Girijesh Kumar Patel, ISBN 9781118634509. The most comprehensive and up-to-date treatment of all the possible aspects for biofuels production from biomass or waste material available. *NOW AVAILABLE!*

Biogas Production, Edited by Ackmez Mudhoo, ISBN 9781118062852. This volume covers the most cutting-edge pretreatment processes being used and studied today for the production of biogas during anaerobic digestion processes using different feedstocks, in the most efficient and economical methods possible. *NOW AVAILABLE!*

Bioremediation and Sustainability: Research and Applications, Edited by Romeela Mohee and Ackmez Mudhoo, ISBN 9781118062845.

Bioremediation and Sustainability is an up-to-date and comprehensive treatment of research and applications for some of the most important low-cost, "green," emerging technologies in chemical and environmental engineering. *NOW AVAILABLE!*

Sustainable Energy Pricing, by Gary Zatzman, ISBN 9780470901632. In this controversial new volume, the author explores a new science of energy pricing and how it can be done in a way that is sustainable for the world's economy and environment. *NOW AVAILABLE!*

Green Chemistry and Environmental Remediation, Edited by Rashmi Sanghi and Vandana Singh, ISBN 9780470943083. Presents high quality research papers as well as in depth review articles on the new emerging green face of multidimensional environmental chemistry. *NOW AVAILABLE!*

Energy Storage: A New Approach, by Ralph Zito, ISBN 9780470625910. Exploring the potential of reversible concentrations cells, the author of this groundbreaking volume reveals new technologies to solve the global crisis of energy storage. *NOW AVAILABLE!*

Bioremediation of Petroleum and Petroleum Products, by James Speight and Karuna Arjoon, ISBN 9780470938492. With petroleum-related spills, explosions, and health issues in the headlines almost every day, the issue of remediation of petroleum and petroleum products is taking on increasing importance, for the survival of our environment, our planet, and our future. This book is the first of its kind to explore this difficult issue from an engineering and scientific point of view and offer solutions and reasonable courses of action. *NOW AVAILABLE!*